ALGEBRA
A Graduate Course

I. Martin Isaacs
University of Wisconsin, Madison

Brooks/Cole Publishing Company
Pacific Grove, California

I(T)P™ The trademark ITP is used under license.

Brooks/Cole Publishing Company
A Division of Wadsworth, Inc.

Printed in the United States of America
10   9   8   7   6   5   4   3   2   1

Library of Congress Cataloging-in-Publication Data
Isaacs, I. Martin, [date]–
    Algebra, a graduate course / I. Martin Isaacs.
        p.   cm.
    Includes index.
    ISBN 0-534-19002-2
    1. Algebra.   I. Title.
QA154.2.I83   1993
512'.02— dc20

512.02
I 13

93-21157
CIP

Publisher: *Gary W. Ostedt*
Marketing Representative: *Jay Honeck*
Editorial Assistants: *Beth Wilbur* and *Carol Ann Benedict*
Production Coordinator: *Marlene Thom*
Production Assistant: *Tessa A. McGlasson*
Manuscript Editor: *Carol Reitz*
Interior and Cover Design: *Vernon T. Boes*
Interior Illustration: *Techsetters, Inc.*
Typesetting: *Techsetters, Inc.*
Printing and Binding: *Arcata Graphics/Fairfield*

To Deborah

Preface

When I started graduate school at Harvard in 1960, I knew essentially no abstract algebra. I had seen the definition of a group when I was an undergraduate, but I doubt that I had ever seen a factor group, and I am sure that I had never even heard of modules. Despite my ignorance of algebra (or perhaps because of it), I decided to register for the first half of the graduate algebra sequence, which was being taught that year by Professor Lynn Loomis. I found the course exciting and beautiful, and by the end of the semester I had decided that I wanted to be an algebraist. This decision was reinforced by an equally spectacular second semester.

I have now been teaching mathematics for more than a quarter-century, and I have taught the two-semester first-year graduate algebra course many times. (This has been mostly at the University of Wisconsin, Madison, but I also taught parts of the corresponding courses at Chicago and at Berkeley.) I have never forgotten Professor Loomis's course at Harvard, and in many ways, I try to imitate it. Loomis, for example, used the first semester mostly for noncommutative algebra, and he discussed commutative algebra in the second half of the course. I too divide the year this way, which is reflected in the organization of this book: Part 1 covers group theory and noncommutative rings, and Part 2 deals with field theory and commutative rings.

The course that I took at Harvard "sold" me on algebra, and when I teach it, I likewise try to "sell" the subject. This affects my choice of topics, since I seldom teach a definition, for example, unless it leads to some exciting (or at least interesting) theorem. This philosophy carries over from my teaching into this book, in which I have tried to capture as well as I can the "feel" of my lectures. I would like to make my students and my readers as excited about algebra as I became during my first year of graduate school.

Students in my class are expected to have had an undergraduate algebra course in which they have seen the most basic ideas of group theory, ring theory, and field theory, and they are also assumed to know elementary linear algebra and matrix theory. Most important, they should be comfortable with mathematical proofs; they should know how to read them, invent them, and write them. I do not require, however, that my students actually remember the theorems or even the definitions from their undergraduate algebra courses. Given my own lack of preparation as a first-year graduate student, I am well aware that a few in my audience may be completely innocent of algebra, and I want to conduct the course so that a student such as I was can enjoy it. But this is a graduate course, and it would not be fair to the majority to go on endlessly with "review" material. I resolve this contradiction by making my presentation complete: giving all definitions and basic results, but I do this quickly, and I intersperse the review material with ideas that very few of my audience have seen before. I have attempted to do the same in this book.

By the end of a year-long graduate algebra course, a good student is ready to go more deeply into one or more of the many branches of algebra. She or he might enroll in a course in finite groups, algebraic number theory, ring theory, algebraic geometry, or any of a number of other specialized topics. While I do not pretend that this book would be suitable as a text for any of these second-year courses, I have attempted to include some of the important material from many of them. I hope that this provides a convenient way for interested readers to sample a number of these topics without having to cope with the somewhat inconsistent notations and different assumptions about readers' backgrounds that are found in the various specialized books. No attempt has been made, however, to designate in the text which chapters and sections are first-year material and which are second; this is simply not well defined. Lecturers who teach from this book undoubtedly do not agree on what, precisely, should be the content of a first-year course. In addition to providing opportunities for students to sample advanced topics, the additional material here should provide some flexibility for instructors to construct a course compatible with their own tastes. Also, those with very well-prepared and capable students might elect to leave out much of the "easy stuff" and build a course consisting largely of what I think of as "second-year" topics.

Since it is impossible, in my opinion, to cover all of the material in this book in a two-semester course, some topics must be skipped, and others might be assigned to the students for independent reading. Perhaps it would be useful for me to describe the content of the course as I teach it at Madison.

I cover just about all of the first four chapters on basic group theory, and I do most of Chapter 5 on the Sylow theorems and p-groups, although I omit Theorem 5.27 and Section 5D on Brodkey's theorem. I do Chapter 6 on symmetric and alternating groups except for Section 6D. In Chapter 7, I cover direct products, but I omit Theorem 7.16 and Section 7C on semidirect products. In Chapter 8 on solvable and nilpotent groups, I omit most of Sections 8C and 8D and all of 8F. I do present the Frattini argument (8.10) and the most basic definitions and facts about nilpotent groups. Chapter 9 on transfer theory I skip entirely.

Chapters 10 and 11 on operator groups are a transition between group theory and module theory. I cover Sections 10A and 10B on the Jordan-Hölder theorem for operator groups, but I omit Section 10C on the Krull-Schmidt theorem. I cover

Sections 11A and 11B on chain conditions, but I touch 11C only lightly. I discuss Zorn's lemma (11.17), but I do not present a proof.

Chapter 12 begins the discussion of ring theory, and some readers may feel that there is a downward "jump discontinuity" in the level of sophistication at this point. As in Chapters 1 and 2, where the definitions and most basic properties of groups are presented (reviewed), it seems that here too it is important to give clear definitions and discussions of elementary properties for the sake of those few readers who may not be comfortable with this material. Section 12A could be assigned as independent reading, but I usually go over it quickly in class. I cover all of Chapter 12 in my course. I do Sections 13A and 13B on the Jacobson radical completely, but sometimes I skip Section 13C on the Jacobson density theorem. (I often find that I am running short of time when I get here.) I do Sections 13D and 14A and as much of 14B on the Wedderburn-Artin theorems as I have time for, and I omit the rest of Chapter 14 and all of Chapter 15.

In the second semester, I start with Chapter 16, and I cover almost all of that, except that I go lightly over Section 16D. I construct fraction fields for domains, but I do not discuss localization more generally. Chapters 17 and 18 discuss basic field extension theory and Galois theory; I cover them in their entirety. I discuss Section 19A on separability, but usually I do only a small amount of 19B on purely inseparable extensions, and I skip 19C. I cover Sections 20A and 20B on cyclotomic extensions, but I skip 20C and go very quickly over 20D on compass and straightedge constructions. In Chapter 21 on finite fields, I cover only Sections 21A and 21D: the basic material and the Wedderburn theorem on finite division rings. In Chapter 22, I omit Sections 22C and 22E, but, of course, I cover 22A and 22B on the solvability of polynomials thoroughly, and I present the fundamental theorem of algebra in 22D. In Chapter 23, I do only Sections 23A, 23B, and 23C, discussing norms and traces, Hilbert's Theorem 90, and a very rudimentary introduction to cohomology. I generally skip Chapter 24 on transcendental extensions completely, and I almost completely skip Chapter 25. (I may mention the Artin-Schreier theorem, but I never discuss formally real fields.)

Chapter 26 begins the discussion of the ideal theory of commutative rings. I cover the first two sections, but I skip Section 26C on localization. In Chapter 27 on noetherian rings, I usually cover only the first two sections and seldom get as far as 27C on the uniqueness of primes in the Lasker-Noether theorem. (I wish that I could discuss Krull's results on the heights of prime ideals in Section 27E, but it seems impossible to find the time to do that.) In Chapter 28 on integrality, I cover only the first three sections. I try to cover at least Section 29A, giving the basic properties of Dedekind domains, but often I find that I must skip Chapter 29 entirely because of time pressure. I always leave enough time, however, to prove Hilbert's Nullstellensatz in Section 30A, and that completes the course.

The user of this book will choose what to read (or teach) and what to skip, but I, as the author, was forced to make other choices. For most of these, there were arguments in both directions, and I am certain that very few will agree with all of my decisions, and perhaps I cannot even hope for a majority agreement on each of them separately. I elected not to include tensor products, for example, because

there just didn't seem to be much interesting that one could say about them without going deeply either into the theory of simple algebras or into homological algebra. Somewhat similarly, I decided not to discuss injective modules. It would have taken considerable effort just to prove that they exist in most cases, and there did not seem much that one could do with them without going into areas of ring theory beyond what I wished to discuss. Also, I did not discuss fully the characterization of finitely generated modules over PIDs, but I did include what seem to be the two most important special cases: the fundamental theorem for finite abelian groups and the fact that torsion-free, finitely generated modules over PIDs are free.

Some of my inclusions (for example, the Berlekamp factorization algorithm and character theory) may seem too specialized for a book of this sort. I discussed the factorization algorithm for polynomials over finite fields, for instance, mostly because I think it is a really slick idea, but also because computer algebra software has become widely available recently, and it seemed that students of "theoretical" algebra ought to receive at least a glimpse of the sort of algorithms that underlie these programs. I included some character theory (Chapter 15 and part of Chapter 28) partly because it provides nice applications of some of the theory, but also, I must admit, because that is my own primary research interest.

I also had to make decisions about notation. I suspect that the most controversial are those concerning functions and function composition. To me (and I think to most group theorists) it seems more natural to write "fg" rather than "gf" to denote the result of doing first function f and then function g. It did not seem wise to use left-to-right composition in the group theory chapters and then to switch and use right-to-left notation in the rest of the book, and so I decided to make consistency a high priority. Since I am a group theorist (and group theory is the first topic in the book), I elected to use left-to-right composition everywhere.

Customarily, when functions are composed left-to-right, the name of the function is written to the right of the argument. A critic of my left-to-right composition challenged my claim to consistency by betting that I did not write $(x)\sin$ and $(y)(d/dx)$ to denote the sine of x or the derivative of y with respect to x. He was right, of course. Like most mathematicians, I write most functions on the left. Nevertheless, I always (in this book at least) compose left-to-right. By my notational scheme, therefore, the following silly looking equation is technically correct: $f(g(x)) = (gf)(x)$. In order to avoid such strange looking formulas, I take the view that the function name may be written on *either* side of its argument, whichever is most convenient at the moment. In contexts where function composition is important, I nearly always choose to put the function name on the right, but I am perfectly comfortable in writing $\sin(x)$, since we are not generally interested in compositions of trigonometric functions. No information is lost by allowing the same function to appear sometimes on the left and sometimes on the right because the composition rule is always left-to-right and is independent of how the function is written.

Another of my decisions that will not meet with universal approval is that by my definition, rings have unity elements. There are a few places where this is not a good idea: one cannot conveniently state the theorem, for example, that a right artinian "ring" with no nonzero nilpotent ideals must have a unity. Most of the time, however, assuming the existence of a unity is a convenience, and so we have built it into the definition. We have also required in the definition of a module that the unity of the ring act as an identity on the module.

Whatever notational scheme one adopts, it is important that students learn of the existence of common alternatives; how else could they read the literature or attend courses from other lecturers? For this reason, I have attempted to mention competing notations and definitions whenever appropriate in the text.

At the end of each chapter, there is a fairly extensive list of problems. Few of these are routine exercises, and some of them I consider to be quite difficult. The purpose of these problems is not just to give practice with the definitions and with understanding the theorems. My hope is that by working these problems, students will get a feeling of what it is actually like to *do* algebra, and not just to learn it. (I should mention that when I teach my algebra course, I assign five problems per week.)

This is not a "scholarly" book; I have not attempted to trace back to their sources the various definitions, lemmas, theorems, and ideas presented here. I have credited items to individuals in cases such as the "Sylow theorems," the "Jacobson radical," and the "Hilbert basis theorem" where such attribution is standard and well known and in other situations where it seemed appropriate. Even in these cases, however, I have not given bibliographic references to the original sources.

I am grateful to the following reviewers for their helpful comments: Michael Aschbacher, California Institute of Technology; Carl Widland, Indiana University; Edward Green, Virginia Polytechnic Institute and State University; Seth Warner, Duke University; E. Graham Evans, Jr., University of Illinois, Urbana-Champaign; Robert L. Griess, Jr., University of Michigan, Ann Arbor; Ancel Mewborn, University of North Carolina at Chapel Hill; Peter Norman, University of Massachusetts, Amherst; and Gerald Janusz, University of Illinois, Urbana-Champaign.

Let me close this preface by expressing my hope that this book will engender, in some readers at least, the same excitement and love for algebra that I received from Professor Loomis in my first year of graduate school.

I. Martin Isaacs
Madison, WI
1992

Contents

PART TWO
Commutative Algebra   231

PART ONE

Noncommutative Algebra

CHAPTER ONE

Definitions and Examples of Groups

1A

From the abstract, axiomatic point of view that prevails today, one can argue that group theory is, in some sense, more primitive than most other parts of algebra and, indeed, the group axioms constitute a subset of the axiom systems that define the other algebraic objects considered in this book. Things we learn about groups, therefore, will often be relevant to our study of modules, rings, and fields. In addition, group theory has considerable indirect connection to these other areas. (The most striking example of this is probably the use of Galois groups to study fields.) It is largely for these reasons that we begin this book on algebra with an extensive study of group theory. (If the whole truth were told, the fact that the author's primary research interest and activity are in group theory would be seen as relevant, too.)

The subject we call "algebra" was not born abstract. In its youth, algebra was the study of concrete objects such as polynomials, rather than of things defined by axiom systems. In particular, early group theory was concerned with groups of mappings, known as "transformation groups." (In the early literature, for instance, the elements of a group were referred to as its "operations.")

For at least two reasons, we begin our study of group theory by (temporarily) adopting this nineteenth-century point of view. First, mappings of one kind or another are ubiquitous throughout algebra (and most of the rest of mathematics, too) and so it makes sense to begin with them. Furthermore, some of the most interesting examples of groups are best constructed and visualized as transformation groups.

We begin our study of mappings with some notation and definitions. (It is this author's belief that mathematics at its best consists of theorems and examples. Definitions are often dull, although they are a necessary evil, especially near the beginning of an expository work. We pledge that the balance of theorems and examples versus definitions will become more favorable as the reader progresses through the book.)

The notation $f : A \to B$ means that f is a mapping (that is, a function) from the set A to the set B. (If either A or B is empty, there are no mappings, and so the existence of f implies that A and B are both nonempty.) The set A is the *domain* of f and B is the *target*. The *image* or *range* of f is denoted $f(A)$. It is the subset

$$\{f(a) \mid a \in A\} \subseteq B.$$

The map f is *onto* or *surjective* if its image is all of the target B. It is *one-to-one* or *injective* if distinct elements of A map to distinct elements of B. If f is both injective and surjective, we say that it is a *bijection*.

Note that we have not specified whether our functions "act on" the right or the left, although in writing "$f(a)$" above, we seem to be implying an action on the left. Our point of view on this question is perhaps slightly unconventional, but it will, we hope, be quite comfortable for the student.

We maintain that functions do not "act" on any particular side, and so it is permissible to write $f(a)$ or $(a)f$, whichever is more intelligible in a given context. Both notations mean precisely the same thing: namely, the result of applying f to a. What is the cost of this freedom of notation? Confusion could enter when two mappings are composed. For instance, if

$$f : A \to B \quad \text{and} \quad g : B \to A,$$

does fg mean "f then g" or does it mean "g then f"? Proponents of "action on the right" would say the former and "leftists" would choose the latter. Our convention throughout this book is that fg always means "f then g." This does not, however, constrain us to write the mappings on the right, but in a setting in which function composition is important, it will usually enhance clarity to do so, and so we shall. According to our notation, therefore, we have

$$(a)(fg) = ((a)f)g,$$

but it would be equally correct (though more confusing) to write

$$(fg)(a) = g(f(a)).$$

For any nonempty set A, we write i_A (or sometimes just i) to denote the identity map. Thus $i_A(a) = a$ for all $a \in A$, and it is clear that

$$i_A f = f \quad \text{and} \quad g i_A = g$$

for arbitrary maps $f : A \to B$ and $g : B \to A$. Note that the associative law for maps is a triviality. If $f : A \to B$, $g : B \to C$, and $h : C \to D$, then $f(gh)$ and $(fg)h$ are equal, since both are the map obtained by doing first f, then g, and then h.

(1.1) LEMMA.   *Let $f : A \to B$.*

 a. *f is injective iff there exists $h : B \to A$ such that $fh = i_A$.*
 b. *f is surjective iff there exists $g : B \to A$ such that $gf = i_B$.*

c. *If f is a bijection, then the maps g and h above are uniquely determined and equal.*

Proof.   Suppose f is injective. Fix an element $a \in A$ and define $h : B \to A$ by

$$(b)h = \begin{cases} a & \text{if } b \notin (A)f \\ x & \text{if } (x)f = b \text{ for some } x \in A. \end{cases}$$

Note that by the injectivity of f, there is at most one element $x \in A$ such that $(x)f = b$. Also, the mapping h is unambiguously defined, since the two cases are mutually exclusive and exhaust the possibilities.

Conversely, if $h : B \to A$ and $fh = i_A$, we wish to show that f is injective. Suppose $(x)f = (y)f$. Then $x = (x)fh = (y)fh = y$, as required.

Now suppose f is surjective. For each $b \in B$, choose $a \in A$ with $(a)f = b$, and once this choice is made, define $g : B \to A$ by $(b)g = a$. Clearly $gf = i_B$. Conversely, suppose $gf = i_B$. Then

$$B = (B)i_B = (B)gf \subseteq (A)f \, ,$$

since $(B)g \subseteq A$. It follows that $(A)f = B$, as required.

Finally, assume f is a bijection so that maps h and g as in parts (a) and (b) exist. Then

$$g = gi_A = g(fh) = (gf)h = i_B h = h \, .$$

In particular, g is uniquely determined, since it must equal any valid choice for h. Similarly, h is uniquely determined.   ∎

In our proof that g exists when f is assumed to be surjective, we needed to make some choices; the definition of g was not forced. In fact, if B is an infinite set, we would need to make infinitely many choices. Some mathematicians feel that a definition that requires infinitely many choices is somewhat suspect, and they have created an additional axiom of set theory, called the "axiom of choice," to deal with this situation. (It is precisely this axiom to which we implicitly appealed in the preceding proof.) It has been proved that the axiom of choice is not a consequence of the rest of set theory, but that it can be assumed without introducing any contradictions into mathematics. Most mathematicians (except those working in set theory itself) freely assume and use the axiom of choice whenever it is convenient to do so, and we will follow that policy here. We shall have a little more to say about the axiom of choice in Chapter 11, when we prove Zorn's lemma.

In Lemma 1.1(a) we say that h is a *right inverse* of f and in (b), that g is a *left inverse*. In the case that f is a bijection, the unique left and right inverse of f is simply called the *inverse* of f and it is denoted f^{-1}. It is interesting to observe the striking symmetry in the statement (though not in the proof) of Lemma 1.1. The conditions that f has right and left inverses are essentially mirror images, although there is no such relationship apparent between the equivalent conditions that f be injective or surjective, respectively. We shall see more of this "duality" later.

1B

Let X be an arbitrary nonempty set. We denote by $\text{Sym}(X)$ the set of all bijections from X to itself. (These bijections on X are also called *permutations*, and if X is finite, this is, in fact, the more common term.) The object $\text{Sym}(X)$ is called the *symmetric group* on X. (Note that if X is finite, containing n elements, say, then $\text{Sym}(X)$ consists of precisely $n!$ permutations.)

(1.2) COROLLARY. *Let $G = \text{Sym}(X)$.*

    a. $i_X \in G$.
    b. g^{-1} *exists and lies in G for each* $g \in G$.
    c. $gh \in G$ *for each* $g, h \in G$.

Proof. Part (a) is immediate, and (b) follows from Lemma 1.1. It is not hard to prove (c) directly, but we prefer the following argument. Given $g, h \in G$, we see that

$$(gh)(h^{-1}g^{-1}) = i = (h^{-1}g^{-1})gh\,,$$

and thus gh has a left and right inverse. It follows by Lemma 1.1 that $gh \in \text{Sym}(X)$. ∎

(1.3) DEFINITION. Let X be any set. A *permutation group* on X is any nonempty subset $G \subseteq \text{Sym}(X)$ such that

    i. $g^{-1} \in G$ for each $g \in G$ and
    ii. G is closed under function composition.

Note that for $g \in G$, there is no question that the mapping g^{-1} exists and lies in $\text{Sym}(X)$. The point of condition (i) is that g^{-1} actually lies in the subset G. Conditions (i) and (ii), together with the assumption that $G \neq \varnothing$, imply that $i_X \in G$, and so this need not be assumed.

Given a nonempty set G of mappings on some set X, perhaps the best strategy for showing that G is a permutation group is first to verify that each element $g \in G$ has both a left and a right inverse in G. From this it follows that $G \subseteq \text{Sym}(X)$ and this condition need not be verified separately. All that remains, then, is to check the closure condition.

Some obvious examples of permutation groups are $\text{Sym}(X)$ and the singleton set $\{i_X\}$ for arbitrary nonempty X. We devote the next few pages to descriptions of several more interesting examples.

Consider a square $ABCD$ and let $X = \{A, B, C, D\}$ be its vertex set (see Figure 1.1). Within $\text{Sym}(X)$, let G denote the set of permutations of X that can be realized by a physical motion of the square through 3-space. For instance, imagine the square being rotated 90° counterclockwise about its center. This brings A to the position formerly occupied by D, and so on, and the associated element of G is the map $g : A \mapsto D \mapsto C \mapsto B \mapsto A$. Similarly, a "flip" about the vertical axis v yields the map $h : A \mapsto B \mapsto A; C \mapsto D \mapsto C$. If we first do the rotation and then the flip, the result is the same as a flip about the diagonal axis d, and the corresponding element of G is precisely the composition $gh : A \mapsto C \mapsto A; B \mapsto B; D \mapsto D$.

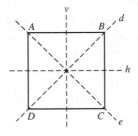

Figure 1.1

The reader should be warned of a possible source of confusion here. One might be tempted to say that the effect of the 90° counterclockwise rotation is that position A is now occupied by vertex B and so the associated map ought to take A to B. With this scheme, this rotation would yield the permutation $g': A \mapsto B \mapsto C \mapsto D \mapsto A$. The flip about axis v still yields h, but the combined operation, the flip about d, does *not* yield the composition $g'h$. (In fact, it yields hg'.) We conclude that if we want function composition, which is the group operation, to correspond to "composition of rotations," we should use the convention given earlier: The mapping g associated with a physical motion satisfies $(x)g = y$ if x goes to the position where y was.

A few moments of reflection should convince the reader that G is a group and that it contains exactly eight elements. One way to obtain the count is to focus on a particular edge, say AB. After a rotation, AB will coincide with the original position of one of the four sides, and in that position it can be in either of two orientations. This yields a total of eight alternatives for how to place AB, and each of these uniquely determines the locations of all four corners.

The eight elements of G are the permutations induced by the four "flips" (about axes h, v, e, and d)) and the four "planar rotations" of 0°, 90°, 180°, and 270°. The standard name for our group G is the *dihedral group* of order 8 and it is denoted D_8. The word "dihedral," meaning two-sided, refers to the front and back sides of the square, which are interchanged by half the elements of the group.

In general, the *order* of a group G is its cardinality (number of elements), and we write $|G|$ to denote this number. (We also write $|X|$ to denote the cardinality of any set X, although the word "order" is generally reserved for groups.)

If, instead of a square, we had started with a regular n-gon ($n \geq 3$), the resulting dihedral group would be D_{2n} of order $2n$. As with D_8, half the elements of D_{2n} correspond to flips and half (counting the identity) correspond to plane rotations. We should mention that many users of group theory write D_n to refer to the dihedral group of order $2n$, whereas most group theorists use the notation we have presented here.

As a further source of interesting examples of groups, let us move up to three dimensions. Consider the vertex set X of a regular polyhedron. The permutations of X induced by physical rotations of the object form a group called the *rotation group* of the object. A usually larger group is the *full group of symmetries*, which consists of all permutations of X realizable by geometric congruences of the polyhedron.

Consider the case of a cube. The full group of the symmetries includes the "antipodal map" τ, which reflects each vertex through the center of the cube. (Thus $(A)\tau = F$ and $(C)\tau = H$ in Figure 1.2, for instance.) The reader should check that τ does not correspond to any rotation. Note that there is no antipodal map for the regular tetrahedron, although it is true for that figure too that there are symmetries that are not rotations. In fact, in this case, the full group of symmetries is the full symmetric group on the vertex set, of order $4! = 24$.

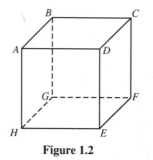

Figure 1.2

Let us compute the order of the rotation group R of a cube. After a rotation, face $ABCD$ can coincide with any of the six faces of the original cube, and in each location, it can have any of the four rotational orientations. It follows that $|R| = 6 \cdot 4 = 24$. The full group of symmetries S, on the other hand, has order 48. (We leave this as an exercise.) What are the 24 symmetries that are not rotations? Among these are the reflections in the nine planes of symmetry of the cube. These planes of symmetry are of two types: six that contain four vertices (for instance, the planes determined by B, D, G, E or by A, E, B, F) and three that are parallel to faces of the cube. A tenth nonrotational symmetry is the antipodal map τ. The remaining 14 nonrotational symmetries are rather hard to visualize and we shall not discuss them further now. The product (composition) of each of the nine reflections with τ yields a rotation of order 2. (The *order* of an element g of a group, denoted $o(g)$, is the least positive integer n, if it exists, such that g^n is the identity. If there is no such n, we say that g has *infinite order* and write $o(g) = \infty$. Elements of order 2 are usually called *involutions*.) A good exercise is to count how many elements of each order there are in the rotation group of a cube.

We shall briefly mention three more examples before proceeding with our study of groups in general. The first example is the "general linear" group $GL(V)$, where V is a vector space. This is the group of all nonsingular (invertible) linear transformations of V. It should be obvious that $GL(V) \subseteq \text{Sym}(V)$ is, in fact, a group.

Next we consider the "affine group" of the line. This is the set of all mappings on the real numbers $\mathbb{R}$ that are of the form $x \mapsto ax + b$, where $a, b \in \mathbb{R}$ and $a \neq 0$. The reader should check that this really is a group.

Our final example is the group associated with the Rubik cube puzzle. (We assume that the reader has some familiarity with this object.) Of the 54 colored squares on the surface of the cube, six may be viewed as never moving from their initial positions (although they do rotate). In other words, if we start with the red face on top and the green face in front, then all interesting cube moves can be made while keeping the red center square on top and the green center square in front. (Of course, this prohibits rotations of the entire cube, but such rotations are not strictly necessary for solving the puzzle.) Now let G be the group of those permutations on the $54 - 6 = 48$ colored squares that can be realized by some sequence of cube twists. How does one compute $|G|$?

As a first approximation, consider disassembling the cube. When this is done, one obtains eight small "corner" cubes having three colored faces each, and 12 small "edge" cubes having two colored faces each. The six face-centers remain attached to one another, and we view them as being fixed in space. To reassemble the cube, we can permute the corner cubes in 8! ways and the edge cubes in 12!. In addition, each corner cube can occur in three different orientations and each edge cube in two different orientations. This yields a total of $8! \cdot 12! \cdot 3^8 \cdot 2^{12}$ ways to reassemble the cube. It turns out (although it is not trivial to prove) that only one-twelfth of these are attainable via legal moves without doing violence to the puzzle. The order of the Rubik cube group, then, is given by

$$|G| = \left(\tfrac{1}{12}\right) 8! \cdot 12! \cdot 3^8 \cdot 2^{12} = 43{,}252{,}003{,}274{,}489{,}856{,}000.$$

1C

Throughout most of the nineteenth century, the word "group" meant "permutation group." We are now ready to give the modern definition, attributed to the English mathematician Arthur Cayley. Recall that a *binary operation* on a set G is a rule that assigns to each ordered pair of elements $x, y \in G$ another element of G. If $\circ$ is a binary operation, we write $x \circ y$ to denote the result of applying this rule to x and y, and we say that $\circ$ is *associative* if $x \circ (y \circ z) = (x \circ y) \circ z$ for all elements $x, y, z \in G$.

(1.4) DEFINITION.   A *group* is a set G together with an associative binary operation $\circ$ defined on G such that there exists $e \in G$ with the following properties:

i. For each $x \in G$, $x \circ e = x = e \circ x$.
ii. For each $x \in G$, there exists $y \in G$ such that $x \circ y = e = y \circ x$.

Note that the "closure" condition that $x \circ y \in G$ whenever $x, y \in G$ need not be stated explicitly, since it is subsumed in the assumption that $\circ$ is a binary operation on G. Observe also that any permutation group is a group with respect to the operation of function composition. In addition to permutation groups, Definition 1.4 allows such objects as the additive group of the integers, the multiplicative group of the positive rationals, and the groups of $n \times n$ nonsingular matrices over fields (with respect to matrix multiplication).

We shall usually follow the custom of suppressing the symbol "$\circ$" and writing xy in place of $x \circ y$. The operation is usually called "multiplication," and xy is referred to as the "product" of x and y.

(1.5) LEMMA. *Let G be a group. Then, for a, b ∈ G, there exist unique elements x, y ∈ G such that*

$$ax = b \quad and \quad ya = b.$$

In particular, the element e is unique, and for each $x \in G$, the element y of Definition 1.4(ii) is unique.

Proof.   Choose z such that $az = e = za$. Now

$$a(zb) = eb = b,$$

and so we can take $x = zb$.

For uniqueness, if $ax = ax'$, we have

$$x = ex = zax = zax' = ex' = x',$$

as required. The existence and uniqueness of y are proved similarly.   ■

In a permutation group, the unique element satisfying condition (ii) of Definition 1.4 is, of course, the identity map i. By analogy, this special element in an abstract group is called the *identity element* of the group and it is customarily denoted 1. The reader should note that the identity of a permutation group is defined by what it *is* (a particular mapping), whereas the identity of an abstract group is defined by how it *behaves* with respect to the group operation. Similarly, the element y of a permutation group that satisfies condition (ii) with respect to x is the inverse map, x^{-1}, and by analogy, in an abstract group, y is said to be the *inverse element* of x and the notation x^{-1} is used in this case, too.

In fact, the conditions of Definition 1.4 are more stringent than they really need to be.

(1.6) THEOREM.   *Let G be a set with an associative multiplication and suppose there exists $e \in G$ with the following properties:*

i.   *$xe = x$ for all $x \in G$ and*
ii. *for each $x \in G$, there exists $y \in G$ with $xy = e$.*

Then G is a group.

Proof.   Let $x \in G$ and choose y according to property (ii). It suffices to show that $ex = x$ and $yx = e$.

Use property (ii) to find $z \in G$ with $yz = e$. We have

$$x = xe = x(yz) = (xy)z = ez,$$

and so

$$yx = y(ez) = (ye)z = yz = e,$$

as required. Now

$$ex = (xy)x = x(yx) = xe = x,$$

and the proof is complete.   ■

We should mention that the "elementwise" calculations in the preceding proof are not typical of most of algebra. The proof of Theorem 1.6, in fact, could almost serve as a model of what algebra is not, or at least should not be, in the opinion of the author.

One way to describe the operation (multiplication) in an abstract group G is via a *multiplication table*. This is a square array, with rows and columns labeled by the elements of G and where the position in row x and column y is occupied by the element xy. Generally, it is neither useful nor practical to actually write down a multiplication table for G, but we can think of G as being defined by such a table.

One of the advantages of thinking about groups abstractly, as in Definition 1.4, is that it allows us to see that certain groups, perhaps defined very differently, are essentially "the same." Suppose, for example, that we rename all the elements of some group G, and that we use these new names to relabel the rows and columns of the multiplication table of G and also to replace the entries in the table. The result will be the multiplication table of a group that is not, in any essential respect, different from G. We can make this notion of "essential sameness" more precise, as follows.

(1.7) DEFINITION.   Let G and H be two groups and suppose $\theta : G \rightarrow H$ is a bijection. We say that θ is an *isomorphism* if

$$\theta(xy) = \theta(x)\theta(y)$$

for all $x, y \in G$. We say that G and H are *isomorphic*, and we write $G \cong H$ if an isomorphism between them exists.

If θ is an isomorphism from G to H, then θ induces a match-up of the elements of G with the elements of H that causes their multiplication tables to coincide. To the extent that we view groups as being defined by their multiplication tables, we see that isomorphic groups are essentially "the same." All "group theoretic" questions will have the same answers in G and H. For example, each of G and H will have equal numbers of elements of any given order, and G will be abelian iff H is abelian. (A group is said to be *abelian* if all of its elements commute, if $xy = yx$ for all elements x, y.)

As a concrete example, consider the group R of rotations of a cube and $S =$ Sym(4). (We write Sym(n) as a shorthand for Sym($\{1, 2, \ldots, n\}$).) We have seen that $|R| = 24$ and, of course, $|S| = 4! = 24$. In fact, we will see that $R \cong S$, and so these differently constructed objects are group theoretically identical. (Note that R permutes eight objects, the vertices of a cube, and S permutes $\{1, 2, 3, 4\}$. As permutation groups, therefore, R and S are quite different.)

In the cube of Figure 1.2, there are four "major diagonals," AF, BE, CH, and DG. Each element of R corresponds to a rotation of the cube and each such rotation induces a permutation of these four diagonals. If we fix an assignment of the numbers 1, 2, 3, and 4 to the four diagonals, then each element of R determines a particular element of $S =$ Sym(4). To see that the corresponding mapping $\theta : R \rightarrow S$ is an isomorphism, we need to establish that θ is a bijection. It is not very hard to see (although we will not write a formal proof) that θ is injective. In other words, two different rotations cannot induce the same permutation of the diagonals. Since $|R| = 24 = |S|$, it follows that θ maps onto S. Because the multiplications in both R and S come about by simply following one operation by another, it should now be fairly clear that θ is an isomorphism.

Note that if $\theta : G \to H$ is an isomorphism, then $\theta^{-1} : H \to G$ is an isomorphism also. Furthermore, if $\varphi : H \to K$ is another isomorphism, it is routine to check that $\theta\varphi : G \to K$ is an isomorphism. It follows from all this that isomorphism of groups is an equivalence relation.

Problems

1.1 A permutation group G on a set X is said to be *transitive* if for every two elements $x, y \in X$, there exists $g \in G$ with $(x)g = y$. Also, G is *regular* if it is transitive and there is a unique element that carries x to y for all $x, y \in X$. Show that a transitive abelian permutation group is necessarily regular.

1.2 Let G be any group. For $x \in G$, let r_x and l_x be the mappings $G \to G$ defined by

$$(g)r_x = gx \quad \text{and} \quad (g)l_x = xg,$$

or in other words, by right and left multiplication by x on G. Let $R = \{r_x \mid x \in G\}$ and $L = \{l_x \mid x \in G\}$. Show that R and L are permutation groups on G and that $R \cong G \cong L$.

NOTE: The fact that every group is isomorphic to a permutation group is known as Cayley's theorem.

1.3 Let G, R and L be as in Problem 1.2. Show that

$$L = \{f \in \text{Sym}(G) \mid fr = rf \text{ for all } r \in R\}.$$

1.4 Let G be a group of mappings on a set X with respect to function composition.
  a. Find an example where $G \not\subseteq \text{Sym}(X)$ and $|G| \geq 2$.
  b. Show that if G contains some injective function, then $G \subseteq \text{Sym}(X)$.

1.5 Let G be the dihedral group D_{2n}. Let $t \in G$ correspond to a "flip" and let $r \in G$ correspond to a "plane rotation." Show that $trt = r^{-1}$. Conclude that if n is odd, then only the identity of G commutes with all elements of G.

1.6 Decide whether or not D_{24} is isomorphic to the group of rotations of a cube. Prove your answer.

1.7 Let V be an n-dimensional vector space over a field F with (a finite number) q elements. One writes $GL(n, q)$ to denote $GL(V)$. Show that

$$|GL(n, q)| = (q^n - 1)(q^n - q)(q^n - q^2) \cdots (q^n - q^{n-1}).$$

1.8 Let G be a group in which every nonidentity element is an involution. Show that G is abelian.

NOTE: An abelian group in which every nonidentity element has the same prime order p is called an *elementary abelian p-group*.

1.9 Consider the eight objects ± 1, $\pm i$, $\pm j$ and $\pm k$ with multiplication rules:

$$ij = k \qquad\qquad jk = i \qquad\qquad ki = j$$
$$ji = -k \qquad\qquad kj = -i \qquad\qquad ik = -j$$
$$i^2 = j^2 = k^2 = -1\,,$$

where the minus signs behave as expected and 1 and -1 multiply as expected. (For example, $(-1)j = -j$ and $(-i)(-j) = ij = k$.)   Show that these objects form a group containing exactly one involution.

NOTE: This is called the *quaternion group* and is denoted Q_8.

CHAPTER TWO

Subgroups and Cosets

2A

What sorts of questions should we ask about a group G? What can we hope to answer? What do we need to know to claim that we "understand" G? In most cases, it would not be very practical (or interesting) to write down the multiplication table for the group, but we can get considerable insight into the "structure" of G by investigating its subgroups.

(2.1) DEFINITION. Let G be a group. A subset $H \subseteq G$ is a *subgroup* if H is closed under multiplication in G and forms a group with respect to this multiplication.

For instance, the permutation groups $G \subseteq \text{Sym}(X)$ are precisely the subgroups of the full symmetric group $\text{Sym}(X)$. For another example, view the integers $\mathbb{Z}$ as a group with respect to addition. Then, for each $n \in \mathbb{Z}$, the set $n\mathbb{Z}$ of all multiples of n is a subgroup of $\mathbb{Z}$. (In fact, these are all of the subgroups of $\mathbb{Z}$.) Of course, obvious examples of subgroups for any group G are G itself and the singleton subgroup $\{1\}$. We shall (in the hope that this will not cause confusion) write 1 in place of $\{1\}$ to denote this trivial subgroup of any group. Also, if G is a group and we write $H \subseteq G$, we generally intend this to mean that H is a subgroup of G unless we explicitly allow the possibility that H is merely a subset.

If $H \subseteq G$ is a subgroup, then H must contain some element e that acts as an identity element for H. In particular, $ee = e$. Since $e1 = e$ also (where 1 is the identity of G), we conclude that $1 = e$ by Lemma 1.5, and thus $1 \in H$. Now if $h \in H$, then there must exist $h' \in H$ with $hh' = 1$, and it follows that $h' = h^{-1}$ (where h^{-1} is the inverse of h in G). We have now shown that subgroups of a group G are closed under taking inverses (in G) as well as under multiplication.

Conversely, we have the following lemma.

(2.2) LEMMA.   *Let G be a group and let $H \subseteq G$ be a nonempty subset. Suppose $xy^{-1} \in H$ for all $x, y \in H$. Then H is a subgroup of G. In particular, any nonempty subset of G closed under multiplication and taking inverses in G is a subgroup.*

Proof.   Choose $h \in H$. Then $1 = hh^{-1} \in H$ by hypothesis. For $y \in H$, we have $y^{-1} = 1y^{-1} \in H$, and if also $x \in H$, then $xy = x(y^{-1})^{-1} \in H$. Therefore, the G-multiplication does define an operation on H and the associative property is inherited from G. Since $1 \in H$ and $y^{-1} \in H$ for all $y \in H$, we see that H has an identity and inverses and so is a group.   ∎

(2.3) COROLLARY.   *Suppose that $\mathcal{H}$ is a collection of subgroups of some group G and let*

$$D = \bigcap_{H \in \mathcal{H}} H \, .$$

Then D is a subgroup of G.

Proof.   Since each $H \in \mathcal{H}$ contains 1, we have $1 \in D$ and, in particular, $D \neq \varnothing$. Now if $x, y \in D$, then $x, y \in H$ for all $H \in \mathcal{H}$ and so $xy^{-1} \in H$ for all such H. Thus, $xy^{-1} \in D$ and D is a subgroup.   ∎

As a convenient notational shorthand, we will often write

$$\bigcap \mathcal{H} \quad \text{in place of} \quad \bigcap_{H \in \mathcal{H}} H \, .$$

How can we construct subgroups for a group? Much of group theory is concerned with variations on this question, but we will discuss a few such constructions now. Given any subset $X \subseteq G$, we can consider the family $\mathcal{H}$ of all subgroups $H \subseteq G$ such that $X \subseteq H$. (Note that $G \in \mathcal{H}$.) The subgroup $\bigcap \mathcal{H}$ is called the subgroup *generated* by X and is denoted $\langle X \rangle$. This subgroup is characterized by two properties:

1. $X \subseteq \langle X \rangle$.
2. If $X \subseteq H$ and H is a subgroup of G, then $\langle X \rangle \subseteq H$.

In other words, the group generated by X is the smallest subgroup of G that contains X (where the word "smallest" should be understood in the sense of containment). Note that if $X \subseteq G$ is itself a subgroup, then $\langle X \rangle = X$.

There is a more explicit (though somewhat less "clean") alternative construction of $\langle X \rangle$.

(2.4) LEMMA.   *Let G be a group and suppose that $X \subseteq G$ is an arbitrary subset. Then $\langle X \rangle$ is the set of all finite products*

$$u_1 u_2 u_3 \, \cdots \, u_n$$

of elements $u_i \in G$ such that either u_i or $u_i^{-1} \in X$. (The "empty product" with $n = 0$ is understood to equal 1.)

Proof.   Let S be the set of all finite products as in the statement of the lemma. Note that $1 \in S$ and so $S \neq \varnothing$ (even if $X = \varnothing$). Now S is clearly closed under multiplication, and since

$$(u_1 u_2 u_3 \cdots u_n)^{-1} = u_n^{-1} u_{n-1}^{-1} \cdots u_1^{-1} \in S,$$

it follows that S is a subgroup.

Since $X \subseteq S$, we have $\langle X \rangle \subseteq S$. On the other hand, since $X \subseteq \langle X \rangle$ and $\langle X \rangle$ is closed under multiplication and inverses, it follows from the definition of S that $S \subseteq \langle X \rangle$. The proof is complete.   ∎

If X is given as an explicitly listed set, for instance, $X = \{a, b, c\}$, then it is customary to omit the braces and write $\langle a, b, c \rangle$ instead of $\langle \{a, b, c, \} \rangle$. An important case of this is when $|X| = 1$. A group G is said to be *cyclic* if there exists some $g \in G$ with $\langle g \rangle = G$. In general, for any element g of any group, the subgroup $\langle g \rangle$ is cyclic. The following result is immediate from Lemma 2.4. (Note that for negative integers n, the power g^n is defined as $(g^{-1})^{-n}$.)

(2.5) COROLLARY.   *Let $g \in G$. Then $\langle g \rangle = \{g^n \mid n \in \mathbb{Z}\}$.*   ∎

Cyclic groups are ubiquitous, since they occur as subgroups in every group. We shall therefore take the time to study them in some detail.

(2.6) LEMMA.   *Let $G = \langle g \rangle$, so that G is cyclic. Let $H \subseteq G$ be a subgroup and suppose that $g^n \in H$, where n is the smallest positive integer that makes this true. Then*

a. *for $m \in \mathbb{Z}$, we have $g^m \in H$ iff n divides m and*
b. *$H = \langle g^n \rangle$.*

Note that if g has infinite order and $H = 1$, then there is no positive integer n such that $g^n \in H$. (Recall that $o(g) = \infty$ means that no positive power of g is 1.) In all other cases, if either $H > 1$ or $o(g) < \infty$, then there does exist a positive integer m with $g^m \in H$, and so the integer n of the lemma does exist. To see this, observe that if $o(g) < \infty$, we can take $m = o(g)$, and if $H > 1$, then if $1 \neq h \in H$, it follows that either h or h^{-1} will be of the form g^m for $m > 0$.

Proof of Lemma 2.6.   If $n \mid m$ (n divides m), we write $m = nq$, with $q \in \mathbb{Z}$. Then $g^m = (g^n)^q \in H$. Conversely, suppose $g^m \in H$. Write $m = qn + r$ with $0 \leq r < n$. Then

$$g^r = g^m (g^n)^{-q} \in H,$$

and by the minimality of n, it follows that $r = 0$ and n divides m, as required.

Statement (b) follows, since certainly $\langle g^n \rangle \subseteq H$ and if h is any element of H, then $h = g^m$ for some m, and so by part (a), $m = qn$ and $h = (g^n)^q \in \langle g^n \rangle$.   ∎

(2.7) COROLLARY. *Every subgroup of a cyclic group is cyclic.*   ∎

(2.8) LEMMA. *Let $g \in G$ with $o(g) = n < \infty$. Then*

   a. $g^m = 1$ *iff* $n \mid m$,
   b. $g^m = g^l$ *iff* $m \equiv l \bmod n$ *and*
   c. $|\langle g \rangle| = n$.

Proof. Apply Lemma 2.6(a) to the group $\langle g \rangle$ with $H = 1$. This yields part (a). Part (b) follows from (a) since $g^m = g^l$ iff $g^{m-l} = 1$. Finally, by part (b), the elements of $\langle g \rangle$ are in one-to-one correspondence with the residue classes of integers mod n, and there are exactly n of these.   ∎

Note that if $g \in G$ and $o(g) = \infty$, then all powers of g are distinct, since if $g^m = g^l$ with $m > l$, then $g^{m-l} = 1$ and g has finite order. We can thus write $|\langle g \rangle| = o(g)$ in all cases.

(2.9) THEOREM. *Let G be a finite cyclic group of order n. Then G has exactly one subgroup of order d for each divisor d of n, and G has no other subgroups.*

Proof. Write $G = \langle g \rangle$ so that $o(g) = n$ by Lemma 2.8(c). For each divisor d of n, we write $e = n/d$ and put $H_d = \langle g^e \rangle$. It is easy to see that $o(g^e) = d$ and thus $|H_d| = d$ by Lemma 2.8(c). What remains is to show that every subgroup $H \subseteq G$ is one of the H_d.

If $H \subseteq G$, then by Lemma 2.6, $H = \langle g^e \rangle$ for some integer e that divides every integer m such that $g^m \in H$. Since $g^n = 1 \in H$, we conclude that e divides n, and thus $H = H_d$, where $d = n/e$.   ∎

We mention that the additive groups of the integers and of the integers mod n are examples of cyclic groups. In fact, it is easy to prove (and we shall do so later) that every cyclic group is isomorphic to one of these.

To state our final results about cyclic groups in this chapter, we remind the reader that if a and b are integers that are not both zero, then their *greatest common divisor*, denoted $\gcd(a, b)$ is the largest integer that divides both a and b. Also, Euler's totient function $\varphi(n)$ is defined for positive integers n by $\varphi(n) = |U_n|$, where

$$U_n = \{ r \in \mathbb{Z} \mid 0 \leq r < n \text{ and } \gcd(r, n) = 1 \}.$$

(2.10) THEOREM. *Let G be cyclic of finite order n. Then G contains precisely $\varphi(n)$ elements of order n, and these are the elements g^r for $r \in U_n$, where g is any element of order n in G.*

Proof. By Lemma 2.8(c), the elements $x \in G$ of order n are just those elements for which $\langle x \rangle = G$. Let g be any such element, so that the powers g^r for $0 \leq r < n$ are the n distinct elements of G. We need to show that $o(g^r) = n$ iff $\gcd(r, n) = 1$.

Suppose first that $\gcd(r, n) > 1$. Then $d = n/\gcd(r, n) < n$ and n divides rd. It follows that $1 = (g^r)^d$ and so $o(g^r) \leq d < n$, as required. Now suppose $\gcd(r, n) = 1$ and let e be the least positive integer such that $g^e \in \langle g^r \rangle$. By

Lemma 2.6(a), we see that e divides r and also (since $g^n = 1 \in \langle g^r \rangle$) e divides n. Thus, $e = 1$ and $g \in \langle g^r \rangle$. Therefore, $\langle g^r \rangle = G$ and $o(g^r) = n$. $\blacksquare$

(2.11) THEOREM.   *Let B and C be cyclic of order $n < \infty$. Then $B \cong C$ and there are exactly $\varphi(n)$ different isomorphisms that map B to C.*

Proof.   Fix $b \in B$ such that $B = \langle b \rangle$. If $\theta : B \to C$ is any isomorphism, write $\theta(b) = c$. Then $\theta(b^m) = c^m$ for all $m \in \mathbb{Z}$, and thus θ is completely determined on all of B once we are given $c = \theta(b)$. Also, since θ is surjective, every element of C must have the form c^m for some $m \in \mathbb{Z}$, and thus c is a generating element of C.

We have now constructed an injective map from the set of all isomorphisms $\theta : B \to C$ into the set of all generating elements c of C; this map carries θ to the generator $c = \theta(b)$ of C. Since the total number of generating elements of C is $\varphi(n)$ by Theorem 2.10, it suffices to show that for every choice of generator c, there exists an isomorphism $\theta : B \to C$ such that $\theta(b) = c$.

The isomorphism we seek will necessarily map b^m to c^m, and so we will define θ by $\theta(b^m) = c^m$ for $m \in \mathbb{Z}$. The problem with this is that the element b^m of B might also be called b^l for some other integer l. We need to show that the value of θ at this element is unambiguously defined. We need, in other words, to show that $c^m = c^l$. Since $o(b) = |B| = n$ by Lemma 2.8(c), the equation $b^m = b^l$ yields that $m \equiv l \pmod{n}$ by Lemma 2.8(b). Thus, $c^m = c^l$ by Lemma 2.8(b) and (c). We now know that θ is well defined, and what remains is to show that θ is an isomorphism.

Since every element of C has the form $c^m = \theta(b^m)$, we see that θ is surjective. It is thus necessarily injective, since $|B| = |C| < \infty$. Finally,

$$\theta(b^m b^l) = \theta(b^{m+l}) = c^{m+l} = c^m c^l = \theta(b^m)\theta(c^m),$$

and so θ really is an isomorphism. $\blacksquare$

2B

Recall that a group G is abelian if $xy = yx$ for all $x, y \in G$. (Note that cyclic groups are automatically abelian.) If G is nonabelian, we might wish to consider for some $g \in G$, the set

$$\mathbf{C}_G(g) = \{x \in G \mid xg = gx\}$$

of all elements that commute with g. This set is the *centralizer* of g in G, and what makes it especially useful is that it is a subgroup of G.

(2.12) LEMMA.   *Let $g \in G$. Then $\mathbf{C}_G(g)$ is a subgroup of G.*

Proof.   Since $1 \in \mathbf{C}_G(g)$, the centralizer is nonempty and it is easy to see that it is closed under multiplication. If $x \in \mathbf{C}_G(g)$, then $xg = gx$, and multiplying by x^{-1} from both the left and right yields $x^{-1}(xg)x^{-1} = x^{-1}(gx)x^{-1}$. Thus, $gx^{-1} = x^{-1}g$ and $x^{-1} \in \mathbf{C}_G(g)$, as required. $\blacksquare$

We can define the centralizer of an arbitrary subset $X \subseteq G$ by

$$\mathbf{C}_G(X) = \{y \in G \mid xy = yx \text{ for all } x \in X\}.$$

Thus,

$$\mathbf{C}_G(X) = \bigcap_{x \in X} \mathbf{C}_G(x),$$

and so the centralizer of any subset of a group is a subgroup, by Corollary 2.3. In particular, taking $X = G$, we get the *center* of G, denoted $\mathbf{Z}(G)$. Thus

$$\mathbf{Z}(G) = \mathbf{C}_G(G) = \{y \in G \mid xy = yx \text{ for all } x \in G\}$$

is a subgroup.

Note that $\mathbf{Z}(G)$ is an abelian group and that G is abelian iff $G = \mathbf{Z}(G)$. Of course, it can happen (and often does) that the center of a group is trivial. For instance, for the dihedral groups,

$$|\mathbf{Z}(D_{2n})| = \begin{cases} 1 & \text{if } n \text{ is odd} \\ 2 & \text{if } n \text{ is even.} \end{cases}$$

The rotation groups of the five regular polyhedra all have trivial centers, but the full groups of symmetries of four of these objects have centers of order 2. (Which one is the exception, and why?)

The following is an example that shows how one can use the fact that centralizers are not merely sets of elements but are subgroups.

(2.13) LEMMA. *Let $X \subseteq G$ be a subset such that $xy = yx$ for all $x, y \in X$. Then $\langle X \rangle$ is abelian.*

Proof. This follows fairly easily from Lemma 2.4, but we prefer this argument. By hypothesis, $X \subseteq \mathbf{C}_G(X)$. Since $\mathbf{C}_G(X)$ is a subgroup, we conclude that $\langle X \rangle \subseteq \mathbf{C}_G(X)$ and so $X \subseteq \mathbf{C}_G(\langle X \rangle)$. As above, this yields $\langle X \rangle \subset \mathbf{C}_G(\langle X \rangle)$ and so $\langle X \rangle$ is abelian. ∎

If $\theta : G_1 \to G_2$ is an isomorphism, it should be clear that $\theta(\mathbf{Z}(G_1)) = \mathbf{Z}(G_2)$. Although this can be proved by a routine computation, we hope the reader will see that this has to be true because the center is a "group theoretic" object, and isomorphisms capture all group theoretic information.

An important special case is where $G_1 = G_2$. An isomorphism from a group G to itself is called an *automorphism* of G. (Note that the identity map on G is an automorphism, but most groups have many other automorphisms, too.) Since isomorphisms carry centers to centers, it follows that every automorphism of G maps $\mathbf{Z}(G)$ to itself. A subgroup $H \subseteq G$ with the property that $\theta(H) = H$ for every automorphism θ of G is said to be *characteristic* in G, and we write H char G.

Not only is the center of a group characteristic, but generally any subgroup uniquely defined by group theoretic properties and not dependent on arbitrary choices or on the names of elements is also characteristic. A good rule of thumb is that any subgroup described by the definite article "the" is characteristic. In Problem 2.7, for example, we shall define the "Frattini subgroup" of a group. Without referring to the definition, the reader should understand that the Frattini subgroup of any group is characteristic.

An important example of an automorphism of G is the *inner automorphism* θ_g induced by an element $g \in G$. This is the map

$$\theta_g(x) = g^{-1}xg.$$

(The reader should check that θ_g is really an automorphism.) A fairly standard notation that we shall adopt is

$$x^g = g^{-1}xg$$

for $x, g \in G$. The element x^g is said to be the *conjugate* of x with respect to g. In this language, the inner automorphism induced by g is the corresponding conjugation map. Observe that if x and g commute, then $x^g = x$, and thus in an abelian group, inner automorphisms are trivial. (As if to compensate for this, another type of automorphism exists only in abelian groups: this is the map $\theta(x) = x^{-1}$ for $x \in G$.)

The set $\text{Aut}(G)$ of all automorphisms of G is a subgroup of $\text{Sym}(G)$, and the set $\text{Inn}(G)$ of inner automorphisms is a subgroup of $\text{Aut}(G)$. (The reader should check these assertions.)

Let us go back to the situation of an isomorphism $\theta : G_1 \rightarrow G_2$. It should be clear that if $H \subseteq G_1$ is a subgroup, then $\theta(H)$ is a subgroup of G_2. In particular, automorphisms map subgroups to subgroups. The subgroup

$$H^g = \{h^g \mid h \in H\}$$

is a subgroup *conjugate* to H. It is, of course, the image of H under the inner automorphism induced by g.

Since characteristic subgroups are fixed by all automorphisms, they are surely fixed by inner automorphisms, and so if C char G, then $C = C^g$ for all $g \in G$. (Note that this is completely obvious in the case $C = \mathbf{Z}(G)$, since then $x^g = x$ for all $x \in C$. In general, the equation $C^g = C$ does not imply that $x^g = x$ for all $x \in C$.)

This leads us to the definition of what is certainly one of the most important concepts in group theory.

(2.14) DEFINITION.   A subgroup $N \subseteq G$ is *normal* if $N^g = N$ for all $g \in G$. We write $N \triangleleft G$ in this situation.

In other words, the normal subgroups of a group are precisely those subgroups fixed by all inner automorphisms. All characteristic subgroups are normal and all subgroups of abelian groups are normal. Of course, the subgroups 1 and G are always normal in any group G.

(2.15) LEMMA.   *Let $H \subseteq G$ be a subgroup. Then, $H \triangleleft G$ if $H^g \subseteq H$ for all $g \in G$.*

The reader should be warned that this lemma does not state that $H^g = H$ whenever $H^g \subseteq H$. Since the inner automorphism induced by the element g is a bijection, it is certainly true that $|H^g| = |H|$, and if H is finite, this equality of orders together with the containment $H^g \subseteq H$ certainly does imply that $H^g = H$. For infinite subgroups, however, this does not follow and is not generally true. (An example is given in the problems at the end of this chapter.)

Proof of Lemma 2.15.   We must show that $H^g = H$ for all $g \in G$. Since $H^g \subseteq H$ for all elements g, it follows that

$$H = (H^g)^{g^{-1}} \subseteq H^{g^{-1}}$$

for all $g \in G$. Applying this result with the element g^{-1} in place of g, we obtain

$$H \subseteq H^{(g^{-1})^{-1}} = H^g$$

and thus $H = H^g$.   ∎

For example, consider the case $G = D_{2n}$, the dihedral group, and let H be the set of plane rotations in G. Since H is closed under multiplication, we have $H^g \subseteq H$ if $g \in H$. On the other hand, if $g \notin H$, then g is a "flip" that interchanges the front and back of the n-gon. In this case $g^{-1} = g$, and for $h \in H$ we have $h^g = ghg$, which does not interchange front and back. Thus, $h^g \in H$ for all $h \in H$, and it follows that $H \triangleleft G$.

Now H is cyclic of order n and it follows that each subgroup C of H is characteristic in H. This is so since if $\theta \in \mathrm{Aut}(H)$, then $\theta(C)$ is a subgroup of H such that $|C| = |\theta(C)|$. It follows by Theorem 2.9 that $C = \theta(C)$, as required. Thus, C char H and $H \triangleleft G$. The next result shows that $C \triangleleft G$.

(2.16) LEMMA.   *Let $N \triangleleft G$ and suppose that C char N. Then $C \triangleleft G$.*

Proof.   Let $g \in G$. Since $N \triangleleft G$, the conjugation map (inner automorphism) induced by g maps N to itself and, in fact, defines an automorphism of N. (Caution: It may not be an inner automorphism of N.) Since C is characteristic in N, this automorphism of N maps C to itself, and so $C^g = C$, as required.   ∎

In contrast with Lemma 2.16, it does not follow that $C \triangleleft G$ if all that is known is that $C \triangleleft N$ and $N \triangleleft G$ (or even that N char G).

We give one more example of a normal subgroup now.

(2.17) THEOREM.   *Let G be any group. Then $\mathrm{Inn}(G) \triangleleft \mathrm{Aut}(G)$.*

Proof.   Let $\theta \in \mathrm{Inn}(G)$ and $\sigma \in \mathrm{Aut}(G)$. By Lemma 2.15, it suffices to show that $\theta^\sigma \in \mathrm{Inn}(G)$ for any choice of θ and σ.

We can write $\theta = \theta_g$ (the conjugation map induced by $g \in G$). To compute θ^σ, we apply it to $x \in G$.

$$(x)\theta^\sigma = (x)\sigma^{-1}\theta_g\sigma = (g^{-1}(x\sigma^{-1})g)\sigma = (g^{-1})\sigma \cdot x \cdot (g)\sigma ,$$

where the last equality follows since σ is an automorphism. We have

$$(x)\theta^\sigma = (g\sigma)^{-1} \cdot x \cdot (g\sigma),$$

and so $\theta^\sigma = \theta_{(g)\sigma}$, the inner automorphism induced by $(g)\sigma \in G$.   ∎

2C

Let $X, Y \subseteq G$ be any two subsets. We write

$$XY = \{xy \mid x \in X, y \in Y\}.$$

Even if X and Y are both subgroups, it does not follow that XY is a subgroup.

(2.18) LEMMA. *Let $H, K \subseteq G$ be subgroups. Then HK is a subgroup iff $HK = KH$.*

Proof.   Assume that HK is a subgroup. Since $1 \in H$, we have $K \subseteq HK$ and similarly $H \subseteq HK$. It follows that $KH \subseteq HK$ since HK is closed under multiplication. Also, if $x \in HK$, then $x^{-1} \in HK$, and we can write $x^{-1} = hk$ for some $h \in H$ and $k \in K$. It follows that

$$x = (hk)^{-1} = k^{-1}h^{-1} \in KH$$

and thus $HK \subseteq KH$. This proves that $HK = KH$.

Conversely, assume $HK = KH$. To prove that this set is a subgroup, let x and y be any two elements and write

$$x = h_1 k_1 \qquad \text{and} \qquad y = k_2 h_2$$

for $h_1, h_2 \in H$ and $k_1, k_2 \in K$. Then

$$xy^{-1} = h_1 k_1 h_2^{-1} k_2^{-1}.$$

However, $k_1 h_2^{-1} \in KH = HK$, and we can write $k_1 h_2^{-1} = h_3 k_3$ with $h_3 \in H$ and $k_3 \in K$. We now have

$$xy^{-1} = (h_1 h_3)(k_3 k_2^{-1}) \in HK$$

and thus HK is a subgroup.   ∎

In the case where $X = \{x\}$, we write xY or Yx instead of $\{x\}Y$ or $Y\{x\}$.

(2.19) DEFINITION.   Let $H \subseteq G$ be a subgroup. If $g \in G$, then the sets

$$Hg = \{hg \mid h \in H\}$$

and

$$gH = \{gh \mid h \in H\}$$

are, respectively, the *right coset* and the *left coset* of H determined by g.

Note that if $g \notin H$, then also $g^{-1} \notin H$, and it follows that $1 \notin Hg$ and $1 \notin gH$. In particular, the cosets Hg and gH are not subgroups in this case. If $g \in H$, on the other hand, then $Hg = H = gH$, and thus the subgroup H is one of its own right cosets and left cosets. Also note that for any element $g \in G$, we have $g \in gH$ and $g \in Hg$. This shows that G is the union of all the right cosets and also of all the left cosets of any subgroup.

(2.20) LEMMA.   *Let $H \subseteq G$ be a subgroup.*

  a. *If $Hx \cap Hy \neq \varnothing$, then $Hx = Hy$.*
  b. *If $xH \cap yH \neq \varnothing$, then $xH = yH$.*

Proof.   First note that $Hh = H$ for $h \in H$. (This is really part of Lemma 1.5 applied to H.) Thus

$$H(hx) = (Hh)x = Hx,$$

and so if $g \in Hx \cap Hy$, we have

$$Hg = Hx \quad \text{and} \quad Hg = Hy,$$

so that $Hx = Hy$, as desired. Part (b) is proved similarly.   ∎

(2.21) COROLLARY.   *Let $H \subseteq G$ be a subgroup. Then G is the disjoint union of the distinct right cosets of H. The analogous result also holds for left cosets.*   ∎

(2.22) LEMMA.   *Let $H \subseteq G$ be a subgroup. For every $g \in G$, we have*

$$|gH| = |H| = |Hg|.$$

Proof.   The map $\theta : H \to Hg$ defined by $(h)\theta = hg$ certainly maps onto Hg and it is injective by Lemma 1.5. It follows that $|H| = |Hg|$, and the other equality is proved similarly.   ∎

If $H \subseteq G$ is a subgroup, then the *index* of H in G, denoted $|G : H|$, is the number of distinct right cosets of H in G. As we shall see, the cardinality of the set of left cosets of H in G is equal to that of the right cosets, and so the index of a subgroup is, in fact, left-right symmetric.

In Theorem 2.9, we showed that if G is a finite cyclic group and $H \subseteq G$ is a subgroup, then $|H|$ divides $|G|$. We are now ready to prove this much more generally.

(2.23) THEOREM (Lagrange).   *Suppose $H \subseteq G$ is a subgroup. Then $|G| = |H||G : H|$. In particular, if G is finite, then $|H|$ divides $|G|$ and $|G|/|H| = |G : H|$.*

Proof.   The group G is the disjoint union of $|G : H|$ right cosets, each of cardinality equal to $|H|$.   ∎

Note that we could as well have worked with left cosets and concluded that if G is finite, then the "left index" equals $|G|/|H|$ and therefore the left and right indices are equal for subgroups of finite groups. The proof of this fact for arbitrary groups is left to the problems at the end of the chapter.

An important consequence of Lagrange's theorem is the following corollary.

(2.24) COROLLARY. *Let G be finite and let $g \in G$. Then $o(g)$ divides $|G|$ and $g^{|G|} = 1$.*

Proof.   We have $o(g) = |\langle g \rangle|$ by Lemma 2.8(c), and this divides $|G|$ by Theorem 2.23. The last assertion is immediate from Lemma 2.8(a).   ∎

As an application of Corollary 2.24 we mention the number theoretic result of Euler that $a^{\varphi(n)} \equiv 1 \bmod n$ for positive integers a and n such that $\gcd(a, n) = 1$. The trick here is to observe that

$$U_n = \{r \in \mathbb{Z} \mid 0 \le r < n \text{ and } \gcd(r, n) = 1\}$$

becomes a group under multiplication if we identify each element r with its residue class mod n. (A few things need to be checked, but we will not do so here.)

Observe that Euler's theorem is immediate by applying Corollary 2.24 to the group U_n.

2D

There is an important connection between the normality of a subgroup and the properties of its cosets.

(2.25) THEOREM.   *Let $H \subseteq G$ be a subgroup. Then the following are equivalent*:

i. $H \triangleleft G$
ii. $Hg = gH$ for all $g \in G$.
iii.   *Every left coset of H in G is a right coset.*
iv.   *The set of right cosets of H in G is closed under set multiplication.*

Proof.   First assume (i). Then $g^{-1}Hg = H$ for all $g \in G$, and multiplication by g on the left yields $Hg = gH$, proving (ii). That (ii) implies (iii) is obvious, so we assume (iii) and prove (iv).

If $x, y \in G$, we must show that $HxHy$ is a right coset. By (iii), however, $xH = Hg$ for some $g \in G$, and we have

$$HxHy = H(Hg)y = Hgy,$$

which is a right coset, as required.

Finally, assume (iv). Then $Hg^{-1}Hg$ is a right coset containing $g^{-1}g = 1$. Thus

$$g^{-1}Hg \subseteq Hg^{-1}Hg = H1 = H,$$

and H is normal by Lemma 2.15.   ∎

Note that item (ii) of Theorem 2.25 is left-right symmetric. It follows that we can get two additional conditions equivalent to H being normal by exchanging the words "left" and "right" in (iii) and (iv).

(2.26) COROLLARY. *Let $H \subseteq G$ be a subgroup. Then the following are equivalent*:

   i. $H \lhd G$.
   ii. *Every right coset of H in G is a left coset.*
   iii. *The set of left cosets of H in G is closed under set multiplication.*   ■

If $H \lhd G$, we use the notation G/H (read "G mod H") to denote $\{Hg \mid g \in G\}$. By Theorem 2.25, we know that G/H is closed under set multiplication.

(2.27) THEOREM. *If $H \lhd G$, then G/H is a group. The identity element of G/H is the coset H, and the inverse of the coset (Hx) in G/H is Hx^{-1}. Also*

$$(Hx)(Hy) = H(xy)$$

for all $x, y \in G$.

Proof. We have $H(Hx) = Hx$ and $(Hx)H = HxH = HHx = Hx$ since $xH = Hx$. Also, $xy \in (Hx)(Hy)$ and thus $(Hx)(Hy) = H(xy)$ by Lemma 2.20. In particular, $(Hx)(Hx^{-1}) = H = (Hx^{-1})(Hx)$.   ■

The group G/H is called the *quotient group* or *factor group* of G by H. For example, if $G = \mathbb{Z}$ (with respect to addition) and $H = n\mathbb{Z}$ (the multiples of n), then the (additive) coset $H + m$ is the residue class of m mod n and the factor group G/H is the additive group of residues mod n.

Note that if G is finite and $H \lhd G$, then $|G/H| = |G : H| = |G|/|H|$ by Lagrange's theorem.

The following is another consequence of Theorem 2.25.

(2.28) COROLLARY. *Let $N \lhd G$ and let $H \subseteq G$ be any subgroup. Then $HN = NH$ is a subgroup and it is normal if $H \lhd G$.*

Proof. We have

$$HN = \bigcup_{h \in H} hN = \bigcup_{h \in H} Nh = NH$$

by Theorem 2.25. It follows that HN is a subgroup by Lemma 2.18.

If $g \in G$, then since conjugation by g defines an automorphism of G, we have $(HN)^g = H^g N^g = H^g N$. If $H \lhd G$, then $H^g = H$ and the proof is complete.   ■

2E

Even if the subgroup $H \subseteq G$ is not normal, we may still be able to use some of our results about normality. The idea is to find some subgroup $K \subseteq G$ such that $H \lhd K$.

In fact, we shall see that for any subgroup $H \subseteq G$, there is a unique subgroup $K \supseteq H$ maximal with the property that $H \triangleleft K$.

It is convenient to work more generally and consider subsets that may not be subgroups. If $X \subseteq G$ is any subset, then we define the *normalizer* of X in G to be the set

$$\mathbf{N}_G(X) = \{g \in G \mid X^g = X\}.$$

(2.29) LEMMA. *The normalizer $\mathbf{N}_G(X)$ is a subgroup of G for every subset $X \subseteq G$. If X is a subgroup, then $X \subseteq \mathbf{N}_G(X)$.*

Proof. First, note that $X^1 = X$ and $(X^g)^h = X^{gh}$ for elements $g, h \in G$. It follows that $\mathbf{N}_G(X)$ is nonempty and that it is closed under multiplication. To see that it contains the inverse of each of its elements, suppose that $g \in \mathbf{N}_G(X)$. Then

$$X^{g^{-1}} = (X^g)^{g^{-1}} = X^{gg^{-1}} = X^1 = X$$

and thus $g^{-1} \in \mathbf{N}_G(X)$, as desired.

If X is a subgroup, then conjugation by any element $x \in X$ defines an automorphism of X and, in particular, the conjugation map is surjective. Thus $X^x = X$ for $x \in X$, and it follows that $X \subseteq \mathbf{N}_G(X)$, as required. ∎

(2.30) COROLLARY. *Suppose $H \subseteq G$ is a subgroup and write $N = \mathbf{N}_G(H)$. Then $H \triangleleft N$, and if $K \subseteq G$ is any subgroup containing H, then $H \triangleleft K$ iff $K \subseteq N$.* ∎

We saw in Corollary 2.28 that if $N \triangleleft G$, then $HN = NH$, and so NH is a subgroup of G. This can be generalized as follows.

(2.31) COROLLARY. *Let $H, K \subseteq G$ be subgroups. If $K \subseteq \mathbf{N}_G(H)$, then $HK = KH$ and HK is a subgroup of G.*

Proof. Since $H \triangleleft \mathbf{N}_G(H)$, we can apply Corollary 2.28 in the group $\mathbf{N}_G(H)$. ∎

The reader should note that although the condition $xH = Hx$ implies that $x \in \mathbf{N}_G(H)$, it does not follow from the equation $HK = KH$ that $K \subseteq \mathbf{N}_G(H)$.

Problems

2.1 Suppose $G = H \cup K$, where H and K are subgroups. Show that either $H = G$ or $K = G$.

2.2 Let G be a group with the property that there do not exist three elements $x, y, z \in G$, no two of which commute. Prove that G is abelian.

2.3 Suppose $\sigma \in \text{Aut}(G)$.
  a. If $x^\sigma = x^{-1}$ for all $x \in G$, show that G is abelian.
  b. If $\sigma^2 = 1$ and $x^\sigma \neq x$ for $1 \neq x \in G$, show that if G is finite, it must be abelian.

HINT: For part (b), show that the set $\{x^{-1}x^\sigma \mid x \in G\}$ is the whole group G. To do this, consider the map $x \mapsto x^{-1}x^\sigma$ for $x \in G$.

2.4 Suppose G has precisely two subgroups. Show that G has prime order.

2.5 A proper subgroup $M < G$ is *maximal* if whenever $M \subseteq H \subseteq G$, we have $H = M$ or $H = G$. Suppose that G is finite and has only one maximal subgroup. Show that $|G|$ is a power of a prime.

2.6 Let $H \subseteq G$ with $|G : H| = 2$. Show that $H \triangleleft G$.

2.7 The *Frattini subgroup* $\Phi(G)$ is the intersection of all maximal subgroups of G. (If there are none, then $\Phi(G) = G$.) We say that an element $g \in G$ is a *nongenerator* if whenever $\langle X \cup \{g\} \rangle = G$, we have $\langle X \rangle = G$ for subsets $X \subseteq G$. If G is finite, show that $\Phi(G)$ is the set of nongenerators of G.

2.8 If $H \subseteq G$, a *right transversal* for H in G is a subset $T \subseteq G$ such that each right coset of H in G contains exactly one element of T. Now let $H, K \subseteq G$ and let S be a right transversal for $H \cap K$ in K.
  a. Show that there exists a right transversal T for H in G with $T \supseteq S$.
  b. If T is as in part (a), show that $T = S$ iff $HK = G$.
  c. If $|G : H| < \infty$, show that $|K : H \cap K| \leq |G : H|$ with equality iff $HK = G$.
  d. If $|G| < \infty$ and $HK = G$, show that $|G| = |H||K|/|H \cap K|$.

2.9 (Dedekind's lemma) Let $H \subseteq K \subseteq G$ and $L \subseteq G$. Show that $K \cap HL = H(K \cap L)$.

2.10 Suppose G is finite and $G = H \cup K \cup L$ for proper subgroups H, K and L. Show that $|G : H| = |G : K| = |G : L| = 2$.

HINT: First get (say) $|G : H| = 2$ and then use Problem 2.8 to complete the proof.

2.11 Let G be finite and assume $H, K \subseteq G$ with $\gcd(|G : H|, |G : K|) = 1$. Show that $HK = G$.

HINT: If $U \subseteq V \subseteq G$, then $|G : U| = |G : V| \cdot |V : U|$. Compute $|G : H \cap K|$ and use Problem 2.8.

2.12 If $x, y \in G$, then the *commutator* of x and y, denoted $[x, y]$, is equal to $x^{-1}y^{-1}xy$. If also $z \in G$, then $[x, y, z]$ means $[[x, y], z]$. Note that $[x, y] = 1$ iff x and y commute. Prove the following commutator identities.
  a. $[x, y][y, x] = 1$
  b. $[xy, g] = [x, g]^y[y, g]$
  c. $[x, y^{-1}, z]^y[y, z^{-1}, x]^z[z, x^{-1}, y]^x = 1$

NOTE: Part (c) was discovered by P. Hall.

2.13 Let $H, K \subseteq G$ be subgroups. Write $[H, K]$ to denote the subgroup of G generated by all commutators $[h, k]$ with $h \in H$ and $k \in K$.

   a. Show that $H \subseteq \mathbf{C}_G(K)$ iff $[H, K] = 1$.
   b. Show that $H \subseteq \mathbf{N}_G(K)$ iff $[H, K] \subseteq K$.
   c. If $H, K \triangleleft G$ and $H \cap K = 1$, show that $H \subseteq \mathbf{C}_G(K)$.

2.14 Let $H, K \subseteq G$ be subgroups, and let $[H, K]$ be the subgroup defined in Problem 2.13.

   a. Show that $[H, K] = [K, H]$.
   b. Show that $H \subseteq \mathbf{N}_G([H, K])$.

HINT: For part (b), use Problem 2.12(b).

2.15 Let $H \subseteq G$ be a subgroup. Let $\mathcal{R}$ and $\mathcal{L}$ denote the sets of all right and left cosets of H in G, respectively.

   a. Show that there is a bijection $\theta : \mathcal{R} \to \mathcal{L}$ such that $\theta(Hx) = x^{-1}H$ for all $x \in G$.
   b. If there exists a bijection $\varphi : \mathcal{R} \to \mathcal{L}$ such that $\varphi(Hx) = xH$ for all $x \in G$, show that $H \triangleleft G$.

NOTE: Part (a) tells us that the "right index" and the "left index" of a subgroup are always equal.

2.16 Suppose $Z \subseteq \mathbf{Z}(G)$ and G/Z is cyclic. Show that G is abelian.

2.17 Let Q_8 be the group of Problem 1.9. Show that every subgroup of Q_8 is normal.

NOTE: It is a theorem that if $|G|$ is odd and every subgroup is normal, then G is abelian.

2.18 Let π be a set of prime numbers. A finite group is said to be a π-group if every prime that divides its order lies in π. If G is finite, show that G has a unique largest normal π-subgroup (which may be trivial and may be all of G).

NOTE: The largest normal π-subgroup of G is denoted $\mathbf{O}_\pi(G)$.

2.19 Let C be cyclic of order n. Show that $\mathrm{Aut}(C)$ is an abelian group of order $\varphi(n)$.

HINT: Take $B = C$ in Theorem 2.11.

NOTE: In fact, $\mathrm{Aut}(C) \cong U_n$. If n is an odd prime power this is cyclic but observe that U_8 is not cyclic.

2.20 Given a positive integer n, show that

$$n = \sum_{d|n} \varphi(d).$$

HINT: Let C be cyclic of order n. How many elements of order d are in C?

2.21 Suppose $A \triangleleft G$ is abelian and $AH = G$ for some subgroup H. Show that $A \cap H \triangleleft G$.

HINT: Show that $A \subseteq \mathbf{N}_G((A \cap H))$ and $H \subseteq \mathbf{N}_G((A \cap H))$.

NOTE: The computation of the normalizer of a subgroup is often a good way to prove normality.

2.22 Let G be the affine group of the line. (Recall that this is the set of all maps $\mathbb{R} \to \mathbb{R}$ of the form $x \mapsto ax + b$ with $a, b \in \mathbb{R}$ and $a \neq 0$. Show that G has a subgroup H such that H^g is a proper subgroup of H for some element $g \in G$.

HINT: Let H be the set of maps where $a = 1$ and $b \in \mathbb{Z}$.

CHAPTER THREE

Homomorphisms

3A

When considering mappings from one mathematical object to another of the same type, we usually give special attention to those maps that, in some sense, respect the structure of the objects. In particular, a map $\varphi : G \to H$ (where G and H are groups) is called a *homomorphism* if $\varphi(xy) = \varphi(x)\varphi(y)$ for all $x, y \in G$. It is these homomorphisms that are the "good" maps between groups.

Note that group isomorphisms are precisely those homomorphisms that happen to be bijections. At the other extreme, if G and H are any two groups, we can always define the trivial homomorphism $\varphi : G \to H$ by setting $\varphi(g) = 1$ for all $g \in G$. We'll mention a few more examples. Consider the group $GL(V)$ of nonsingular linear transformations on the finite-dimensional vector space V over a field F. (We can, of course, identify $GL(V)$ with the group of nonsingular $n \times n$ matrices over F, where $n = \dim_F V$.) For each $t \in G$, its determinant $\det(t) \in F - \{0\}$. The set, denoted $F^\times$, of nonzero elements of F forms a group with respect to multiplication, and the fact that $\det(st) = \det(s)\det(t)$ tells us that det $: GL(V) \to F^\times$ is a homomorphism.

We can map the additive group of the integers onto that of the integers mod n by sending each integer to its residue class. It should be clear that this defines a homomorphism. In fact, since residue classes are cosets, this is a special case of something more general, and we come to what is perhaps the most important example of a homomorphism.

Let $N \triangleleft G$ and consider the factor group G/N. Define $\pi : G \to G/N$ by $\pi(g) = Ng$. Since $NxNy = Nxy$ for all $x, y \in G$, we see that π is a homomorphism. This map is called the *canonical* homomorphism of G onto G/N, and we note that it does map *onto* G/N; it is surjective.

(3.1) LEMMA. *Let $\varphi : G \to H$ be a homomorphism. Then:*

a. $\varphi(1) = 1$ *and* $\varphi(x^{-1}) = \varphi(x)^{-1}$ *for all* $x \in G$.
b. $N = \{g \in G \mid \varphi(g) = 1\}$ *is a normal subgroup of* G.
c. $\varphi(x) = \varphi(y)$ *iff* $Nx = Ny$.
d. φ *is injective iff* $N = 1$.

Proof. We have $\varphi(1) = \varphi(1 \cdot 1) = \varphi(1)\varphi(1)$, and canceling $\varphi(1)$ yields $1 = \varphi(1)$. Now $1 = \varphi(1) = \varphi(x \cdot x^{-1}) = \varphi(x)\varphi(x^{-1})$. Thus $\varphi(x^{-1}) = \varphi(x)^{-1}$, and part (a) is proved.

Since $1 \in N$ by part (a), we see that $N \neq \varnothing$. If $x \in N$, then $\varphi(x^{-1}) = \varphi(x)^{-1} = 1^{-1} = 1$ and $x^{-1} \in N$. Also, if $x, y \in N$, then $\varphi(xy) = \varphi(x)\varphi(y) = 1 \cdot 1 = 1$ and $xy \in N$. This proves that N is a subgroup. For normality, compute that $\varphi(x^{-1}nx) = \varphi(x^{-1})\varphi(n)\varphi(x) = \varphi(x)^{-1} \cdot 1 \cdot \varphi(x) = 1$ for $n \in N$ and $x \in G$. Thus $N^x \subseteq N$ for all $x \in G$, and so $N \triangleleft G$.

For part (c), observe that if $\varphi(x) = \varphi(y)$, then $1 = \varphi(x)\varphi(y)^{-1} = \varphi(xy^{-1})$, and so $xy^{-1} \in N$ and $x \in Ny$. Therefore, $Nx = Ny$. Conversely, if $Nx = Ny$, then $x = ny$ for some $n \in N$, and we have $\varphi(x) = \varphi(ny) = \varphi(n)\varphi(y) = \varphi(y)$, since $\varphi(n) = 1$.

Finally, if φ is injective, then for $n \in N$ we have $\varphi(n) = 1 = \varphi(1)$ and so $n = 1$ and thus $N = 1$. Conversely, if $N = 1$ and $\varphi(x) = \varphi(y)$, we must have $x = y$ by part (c). $\blacksquare$

The normal subgroup $N = \{g \in G \mid \varphi(g) = 1\}$ is called the *kernel* of the homomorphism φ and is denoted $\ker(\varphi)$. For example, if we consider the determinant map $\det : GL(V) \to F^\times$, we see that $\ker(\det)$ is the group of all linear transformations of V with determinant 1. This subgroup of the general linear group is called the *special linear* group and is denoted $SL(V)$.

If $\pi : G \to G/N$ is the canonical homomorphism, where G is arbitrary and $N \triangleleft G$, then $\ker(\pi) = N$. To see this, observe that $\pi(g) = 1$ iff $Ng = N$, and this happens iff $g \in N$. (Recall that N is the "1" element in G/N.) This gives an interesting characterization of the set of all normal subgroups of a group.

(3.2) COROLLARY. *Let G be an arbitrary group. Then the normal subgroups of G are precisely the kernels of all homomorphisms defined on G.*

Proof. This is immediate from Lemma 3.1(b) and the fact that every $N \triangleleft G$ is the kernel of the associated canonical homomorphism $G \to G/N$. $\blacksquare$

In some sense, the canonical homomorphisms $\pi : G \to G/N$ for normal subgroups $N \triangleleft G$ are the only surjective homomorphisms. More precisely, we have the following Isomorphism theorem.

(3.3) THEOREM (Isomorphism). *Let $\varphi : G \to H$ be a surjective homomorphism and let $N = \ker(\varphi)$. Then $H \cong G/N$. In fact, there exists a unique isomorphism $\theta : G/N \to H$ such that $\pi\theta = \varphi$, where π is the canonical homomorphism $G \to G/N$.*

Before giving the proof, some discussion is appropriate. Figure 3.1 is a diagram showing the relevant groups and maps. The arrow representing θ is dashed rather than solid, since it is the existence of this map that is the assertion of the theorem. The equation $\pi\theta = \varphi$ states that starting with an element of G and following either of the two routes to H, either direct or via G/N, we reach the same element of H. In general, one says that a diagram of groups and homomorphisms is a *commutative diagram* if, whenever there are alternative routes between groups, the composite maps along the two routes are equal.

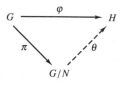

Figure 3.1

Note that what the commutativity of the diagram in Figure 3.1 really says is that if we were to identify G/N with H, using the map θ to make this identification, then φ and π would turn out to be the same map. In other words, every surjective homomorphism is "essentially" just a canonical homomorphism.

Proof of Theorem 3.3.   Since we require that $\pi\theta = \varphi$, we are forced to define $(Ng)\theta = g\varphi$, and hence θ is certainly unique. We need to check, however, that θ is well defined. Specifically, if $Nx = Ny$, we require that $x\varphi = y\varphi$. (If this were not the case, the definition of the image of this coset under θ would be ambiguous.)   In fact, we have the desired equality by Lemma 3.1(c). Since

$$(NxNy)\theta = (Nxy)\theta = (xy)\varphi = (x\varphi)(y\varphi) = (Nx)\theta(Ny)\theta\,,$$

we see that θ is a homomorphism. We certainly have $\pi\theta = \varphi$, and since φ is surjective, it follows that θ must be surjective too.

Finally, we check that θ is injective by showing that $\ker(\theta)$ is trivial (and using Lemma 3.1(d)). If $Ng \in \ker(\theta)$, then $1 = (Ng)\theta = g\varphi$, and so $g \in \ker(\varphi) = N$. Thus, $Ng = N$ (the identity of G/N) and $\ker(\theta)$ is trivial, as required.   ∎

The following application of the Isomorphism theorem (3.3) is a slight variation on Theorem 2.11.

(3.4) COROLLARY.   *Up to isomorphism, the only cyclic groups are the groups $\mathbb{Z}$ and $\mathbb{Z}/n\mathbb{Z}$ for positive integers n.*

Proof.   Let C be cyclic and write $C = \langle c \rangle$. Map $\varphi : \mathbb{Z} \to C$ by $\varphi(k) = c^k$. Observe that $\varphi(k + l) = c^{k+l} = c^k c^l$, and so φ is a homomorphism from the group $\mathbb{Z}$ (with respect to addition) to C. Clearly, φ maps onto C, and so by Theorem 3.3, we have $C \cong \mathbb{Z}/K$, where $K = \ker(\varphi)$.

The proof is completed by observing that the subgroup $K \subseteq \mathbb{Z}$ is cyclic (by Corollary 2.7), and thus $K = \langle n \rangle$ for some integer n. Since $\langle n \rangle = \langle -n \rangle$, we can take $n > 0$ if K is nontrivial. ∎

Next, we explore the connections between homomorphisms and the subgroup structures of the domain and target. Suppose $\varphi : G \to H$ is a homomorphism. If $U \subseteq G$, we write

$$\varphi(U) = \{\varphi(u) \mid u \in U\}.$$

We say that $\varphi(U)$ is the *image* of U under φ. Similarly, if $V \subseteq H$ we define the *inverse image* of V under φ by the formula

$$\varphi^{-1}(V) = \{u \in G \mid \varphi(u) \in V\}.$$

The inverse image is always defined, even though there may not be a map φ^{-1}. If φ is an isomorphism, so that φ^{-1} does exist, then the notation $\varphi^{-1}(V)$ may seem ambiguous since it might also mean the image of V under φ^{-1}. Because this is equal to the inverse image of V under φ in this case, no confusion should arise.

(3.5) LEMMA. *Let $\varphi : G \to H$ be a homomorphism.*

 a. *If $U \subseteq G$ is a subgroup, then $\varphi(U) \subseteq H$ is a subgroup.*
 b. *If $V \subseteq H$ is a subgroup, then $\varphi^{-1}(V) \subseteq G$ is a subgroup that contains* $\ker(\varphi)$.

Proof. This is completely routine using Lemma 3.1. ∎

An important example of the Lemma 3.5 is where $N \triangleleft G$ and $\pi : G \to G/N$ is the canonical homomorphism. If $H \subseteq G$, then NH is a group and $N \triangleleft NH$. The quotient group NH/N is a subgroup of G/N, and, in fact, NH/N is the image of H under π. To see this, note that

$$NH/N = \{Nnh \mid n \in N, h \in H\} = \{Nh \mid h \in H\} = \pi(H).$$

(3.6) THEOREM (Diamond). *Let $N \triangleleft G$ and $H \subseteq G$. Then $H \cap N \triangleleft H$ and* $H/(H \cap N) \cong NH/N$.

Proof. Let $\pi : G \to G/N$ be the canonical homomorphism and let φ be the restriction of π to the subgroup H. Thus, φ is a homomorphism of H onto $\pi(H) = NH/N$, and since φ is equal to π on the elements of H, we have

$$\ker(\varphi) = H \cap \ker(\pi) = H \cap N.$$

Thus $H \cap N \triangleleft H$ and

$$H/(H \cap N) = H/\ker(\varphi) \cong NH/N$$

by Theorem 3.3, the Isomorphism theorem. ∎

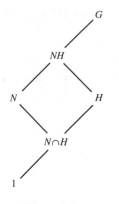

Figure 3.2

The diagram in Figure 3.2 is known as a "lattice diagram" and is entirely different from the mapping diagram in Figure 3.1. These lattice diagrams can be extremely useful, since with them it is possible to represent graphically some quite complicated interrelationships among subgroups. Indeed, it is occasionally possible to read an entire proof directly from a carefully drawn lattice diagram. We digress briefly to discuss these objects.

In a lattice diagram, the nodes represent a few of the subgroups of some group, and the line segments indicate containment: the lower node is a subgroup of the upper.

According to the standard conventions, Figure 3.3(a) indicates not only that $X \subseteq U$ and $X \subseteq V$ (allowing the possibility of equality) but also that $X = U \cap V$. We would draw diagram (b) if we did not know that X was the intersection, and the unlabeled node in (b) represents $U \cap V$. We do not intend diagram (b) to preclude the possibility that $U \cap V = X$, however.

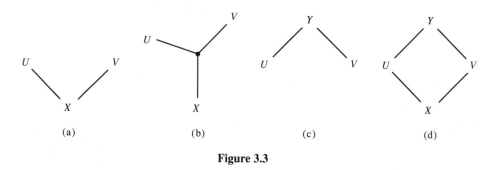

Figure 3.3

In Figure 3.3(c), the assertion is that Y is generated by U and V, and not merely that $U \subseteq Y$ and $V \subseteq Y$. An important special case is where $Y = UV$ is the product, and not just the group generated. In that case, we say that the four nodes $U, V, Y = UV$ and $X = U \cap V$ form a *diamond*, and it is customary to draw the figure as a parallelogram, as in Figure 3.3(d). Note that Figure 3.2, representing the situation in Theorem 3.6, contains a diamond. The conclusion of Theorem 3.6, the Diamond theorem, can be paraphrased as "opposite sides of a parallelogram are isomorphic." This requires some normality, of course.

In the case that we have a diamond as in Figure 3.3(d), there is further information available. By Problem 2.8(c), we have $|Y : V| = |U : X|$ and $|Y : U| = |V : X|$, and so if we imagine that the lengths of the line segments in a lattice diagram represent indices of subgroups, this is consistent with the fact that opposite sides of a parallelogram are equal.

We now end our digression and return to the study of homomorphisms. The point of the next theorem is that the subgroup structure of the image of a homomorphism from G is captured by that part of G containing the kernel.

(3.7) THEOREM (Correspondence). *Let $\varphi : G \to H$ be a surjective homomorphism and let $N = \ker(\varphi)$. Define the following sets of subgroups:*

$$\mathcal{S} = \{U \mid N \subseteq U \subseteq G\}$$

and

$$\mathcal{T} = \{V \mid V \subseteq H\}.$$

Then $\varphi(\ )$ and $\varphi^{-1}(\ )$ are inverse bijections between $\mathcal{S}$ and $\mathcal{T}$. Furthermore, these maps respect containment, indices, normality and factor groups.

The meaning of the last sentence is as follows: Suppose $U_1, U_2 \in \mathcal{S}$ and V_1, V_2 are the corresponding elements of $\mathcal{T}$. (Thus $V_i = \varphi(U_i)$ and $U_i = \varphi^{-1}(V_i)$.) Then, $U_1 \subseteq U_2$ iff $V_1 \subseteq V_2$, and in this case, $|U_2 : U_1| = |V_2 : V_1|$. Also, $U_1 \triangleleft U_2$ iff $V_1 \triangleleft V_2$, and in this case, $U_2/U_1 \cong V_2/V_1$.

Proof of Theorem 3.7.   By Lemma 3.5, $\varphi(\ )$ and $\varphi^{-1}(\ )$ do define maps between $\mathcal{S}$ and $\mathcal{T}$ and we need to show that they are inverses of each other. For $U \in \mathcal{S}$, it is immediate from the definitions that $U \subseteq \varphi^{-1}(\varphi(U))$. If $g \in \varphi^{-1}(\varphi(U))$, then $\varphi(g) \in \varphi(U)$ and thus $Ng = Nu$ for some $u \in U$ by Lemma 3.1(c). Thus, $g \in Nu \subseteq U$ since $N \subseteq U$, and we have $\varphi^{-1}(\varphi(U)) = U$.

Now let $V \in \mathcal{T}$. Clearly, $\varphi(\varphi^{-1}(V)) \subseteq V$. If $v \in V$, then since φ is surjective, there exists $g \in G$ with $\varphi(g) = v$. Thus, $g \in \varphi^{-1}(V)$ and $v = \varphi(g) \in \varphi(\varphi^{-1}(V))$. We have now established that $\varphi(\varphi^{-1}(V)) = V$, and so $\varphi(\ )$ and $\varphi^{-1}(\ )$ are indeed inverse bijections.

It is clear that these maps respect containment, and nothing further need be said about that. Now suppose $U_1 \triangleleft U_2$ with $U_i \in \mathcal{S}$. Then for all $x \in U_2$, we have $x^{-1}U_1x = U_1$, and applying φ, we get $\varphi(x)^{-1}\varphi(U_1)\varphi(x) = \varphi(U_1)$. This says that $\varphi(U_1) \triangleleft \varphi(U_2)$, as required.

Next, assume $U_1 \subseteq U_2$ with $U_i \in \mathcal{S}$, and suppose $\varphi(U_1) \triangleleft \varphi(U_2)$. We need to show that $U_1 \triangleleft U_2$ and so we let $x \in U_2$ and observe that

$$\varphi(x^{-1}U_1x) = \varphi(x)^{-1}\varphi(U_1)\varphi(x) = \varphi(U_1)$$

by the normality assumption. Since $x^{-1}U_1x \in \mathcal{S}$ and $\varphi(\ )$ is injective (because it has an inverse), we conclude that $x^{-1}U_1x = U_1$ and $U_1 \lhd U_2$, as required. The maps, therefore, do respect normality.

Next, we consider factor groups. Let $U_1 \lhd U_2$ with $U_i \in S$ and write $V_i = \varphi(U_i)$. To show that $U_2/U_1 \cong V_2/V_1$, we construct a homomorphism θ of U_2 onto V_2/V_1 and show that $\ker(\theta) = U_1$. What we want then will follow from the Isomorphism theorem.

Let $\pi : V_2 \to V_2/V_1$ be the canonical homomorphism, and let φ' be the restriction of φ to U_2. Write $\theta = \varphi'\pi$, so that $\theta : U_2 \to V_2/V_1$. Then θ is a homomorphism, and since $\theta(U_2) = \pi(\varphi(U_2)) = \pi(V_2) = V_2/V_1$, we see that θ is surjective. Now $\ker(\theta) = \varphi^{-1}(V_1) = U_1$ and it follows that $U_2/U_1 \cong V_2/V_1$ by the Isomorphism theorem.

Suppose $U_1 \subseteq U_2$ with $U_i \in S$ and $V_i = \varphi(U_i)$. We must show that $|U_2 : U_1| = |V_2 : V_1|$, and so we construct a bijection from $\{U_1x \mid x \in U_2\}$ onto $\{V_1z \mid z \in V_2\}$. Note that $\varphi(U_1x) = V_2\varphi(x)$ and so $\varphi(\ )$ maps cosets to cosets. To show that this map is surjective, let $z \in V_2$. Then $z = \varphi(x)$ for some $x \in U_2$, and we have $\varphi(U_1x) = V_1z$, as required. Finally, if $\varphi(U_1x) = \varphi(U_1y)$ for $x, y \in U_2$, we have $V_1\varphi(x) = V_1\varphi(y)$, and so $\varphi(x)\varphi(y)^{-1} \in V_1$. Thus $\varphi(xy^{-1}) \in V_1$ and $xy^{-1} \in \varphi^{-1}(V_1) = U_1$. Therefore, $x \in U_1y$ and $U_1x = U_1y$. This shows that our map is injective, and the proof is complete. ∎

The following corollary is very useful.

(3.8) COROLLARY. *Let $N \lhd G$. Then every subgroup of G/N has the form H/N for some (unique) subgroup $H \subseteq G$ with $H \supseteq N$.*

Proof. Apply Theorem 3.7 to the situation $\pi : G \to G/N$, where π is the canonical homomorphism. Note that if $H \subseteq G$ with $N \subseteq H$, then we have $\pi(H) = \{Nh \mid h \in H\} = H/N$. ∎

Another consequence of our theorem is this corollary.

(3.9) COROLLARY. *Let $N \subseteq M \lhd G$ with $N \lhd G$. Then $G/M \cong (G/N)/(M/N)$.*

Proof. Again, apply the Correspondence theorem to the canonical homomorphism $G \to G/N$. Since G corresponds to G/N and M corresponds to M/N, the desired isomorphism is just the statement that the maps in Theorem 3.7 respect factor groups. ∎

We comment briefly on the interpretation of the Correspondence theorem in terms of lattice diagrams, at least in the case where φ is the canonical homomorphism $G \to G/N$. If we have a lattice diagram for some of the subgroups of G, including N, then the part of the diagram above N is a valid lattice diagram for G/N. In fact, diamonds go over to diamonds, since if $N \subseteq U$ and $N \subseteq V$ and UV is a group, then in G/N we have $(U/N)(V/N) = (UV)/N$.

The Isomorphism theorem, the Diamond theorem, the Correspondence theorem and Corollary 3.9 are often referred to collectively as the "homomorphism theorems."

The three results we have named must surely be the most used theorems in algebra (if one includes their module and ring theoretic analogs). They are so basic, however, that they are seldom quoted explicitly in the professional literature. They are simply used without comment.

3B

In Problem 2.12 we defined the *commutator* $[x, y]$ of $x, y \in G$ by the formula $[x, y] = x^{-1}y^{-1}xy$, and we pointed out that x and y commute iff $[x, y] = 1$. The subgroup generated by all commutators in G is called the *commutator subgroup* or the *derived subgroup* of G and is denoted G'. (Note that, using the notation of Problem 2.13, we could write $G' = [G, G]$.) Since an isomorphism of one group to another clearly maps the commutator subgroup of the first group onto that of the second, it follows that automorphisms of G map G' to itself. In short, G' char G, and in particular $G' \triangleleft G$. Note that $G' = 1$ iff G is abelian.

(3.10) THEOREM.   *Let $N \triangleleft G$. Then G/N is abelian iff $G' \subseteq N$.*

Proof.   Let $\pi : G \to G/N$ be the canonical homomorphism. If $x, y \in G$, then $[\pi(x), \pi(y)] = \pi([x, y])$, and so $\pi(x)$ and $\pi(y)$ commute iff $\pi([x, y]) = 1$. Since this happens iff $[x, y] \in N$, it follows that G/N is abelian iff $[x, y] \in N$ for all $x, y \in G$.   ∎

(3.11) COROLLARY.   *Let $\varphi : G \to A$ be a homomorphism, where A is abelian. Then $G' \subseteq \ker(\varphi)$.*

Proof.   We have $G/\ker(\varphi) \cong \varphi(G) \subseteq A$ by the Isomorphism theorem (3.3). Therefore, $G/\ker(\varphi)$ is abelian and $G' \subseteq \ker(\varphi)$.   ∎

(3.12) COROLLARY.   *Let $G' \subseteq H \subseteq G$. Then $H \triangleleft G$.*

Proof.   Since G/G' is abelian, we have $H/G' \triangleleft G/G'$. By the Correspondence theorem (3.7) applied to the canonical homomorphism $\pi : G \to G/G'$, we conclude that $H \triangleleft G$.   ∎

We shall have more to say about derived subgroups later, when we consider "solvable" groups.

3C

This is perhaps a good place to introduce simple groups. We say that a nonidentity group is *simple* if it has just two normal subgroups, the identity and the group itself. For example, groups of prime order are automatically simple since their only subgroups, normal or otherwise, are the identity and the whole group. By Problem 2.4, these groups of prime order (which are necessarily cyclic) are the only abelian simple groups.

Since the normal subgroups of a group are precisely the kernels of homomorphisms, it follows that a group G is simple iff $G > 1$ and every nontrivial homomorphism defined on G is injective.

It turns out that nonabelian simple groups are really quite rare. For instance, there are just five (isomorphism classes of) nonabelian simple groups of order less than 1000; they have orders 60, 168, 360, 504, and 660. It is a good project for a student of group theory to try to prove this result. In other words, the task is to find some reason why each of the other numbers less than 1000 cannot be the order of a nonabelian simple group. In the next few chapters, we will prove a number of results that, when applied with enough perseverance and cleverness, should be sufficient to accomplish this goal. The reader is urged to attempt this project, but be warned that the number 720 is perhaps an order of magnitude more difficult to eliminate than are any of the others.

Various aspects of the problem of finding all finite nonabelian simple groups have been undergoing active research for the better part of a century, and recently the complete solution of this problem was announced. The so-called Classification theorem gives a list of a number of infinite families of simple groups and 26 exceptional "sporadic" groups, and it asserts that every nonabelian finite simple group lies on this list. The complete proof of this theorem runs to many thousands of pages written by dozens of authors. (We mention that the smallest of the sporadic groups has order 7920 and was discovered by E. Mathieu in 1861. The largest of these groups was discovered by B. Fischer and R. Greiss and was actually constructed by Greiss in 1982. Its order is 808, 017, 424, 794, 512, 875, 886, 459, 904, 961, 710, 757, 005, 754, 268, 000, 000, 000.)

Why have simple groups evoked so much interest and effort? Presumably, it is their comparative rarity among finite groups in general that has inspired most of this, but another reason is that in some sense, simple groups are the "bricks" from which all finite groups are constructed.

To make this precise, let G be any finite group and assume $G > 1$. Let N be a maximal normal subgroup of G. (In other words, $N \triangleleft G$ is a proper subgroup and there exists no subgroup $M \triangleleft G$ with $N < M < G$.) It follows by the Correspondence theorem (3.7) that G/N is simple. If $N > 1$, repeat this process by choosing some maximal normal subgroup of N (which, of course, may not be normal in G). Continuing in this way, we obtain a series of subgroups

$$\mathcal{S} : 1 = G_0 \triangleleft G_1 \triangleleft \cdots \triangleleft G_{n-1} \triangleleft G_n = G,$$

where each group is maximal normal in the next. Such a series is called a *composition series* for G. The factor groups G_i/G_{i-1} for $1 \leq i \leq n$ are all simple (possibly of prime order), and the group G may be viewed as being built from these simple *composition factors*. In Chapter 10, we prove the Jordan-Hölder theorem, which asserts that the composition factors of G are uniquely determined (up to isomorphism) and they are independent of the particular series $\mathcal{S}$. Different composition series for G will yield composition factors that are isomorphic to those from $\mathcal{S}$, although perhaps they occur in a different sequence.

We close this discussion of simple groups with the observation that there is a practical consequence of the fact that all finite groups are built from simple groups. Often, a theorem about general finite groups can be proved by an appeal to the simple group classification. As an example of this technique, let us consider groups of order relatively prime to 15. From the classification, the following is true:

(∗) If S is a finite simple group and $|S|$ is not divisible by 3 or 5, then $|S|$ is prime.

(3.13) THEOREM. *Assume fact (∗). Let $G > 1$ be a finite group with order relatively prime to 15. Then G has a nonidentity abelian normal subgroup.*

Proof. Let M be minimal normal in G. In other words, $1 < M \lhd G$, and there does not exist $K \lhd G$ with $1 < K < M$. (Note that the existence of M is assured by the finiteness of G.) Next, let N be maximal normal in M. By the Correspondence theorem (3.7), the group M/N is simple. Since $|M/N|$ divides $|G|$, it is relative prime to 15, and so by fact (∗), M/N has prime order. In particular, M/N is abelian and hence $M' \subseteq N$ by Theorem 3.10, and it follows that $M' < M$.

Now M' char $M \lhd G$ and thus $M' \lhd G$. It follows from the minimality of M that $M' = 1$, and thus M is abelian, as desired. ∎

Problems

3.1 Let G be any group. Show that $G/Z(G) \cong \mathrm{Inn}(G)$. Conclude that $\mathrm{Inn}(G)$ cannot be a nontrivial cyclic group.

3.2 Let Q_8 be the group described in Problem 1.9. Show that $G/Z(G)$ can never be isomorphic to Q_8.

HINT: If $G/Z(G) \cong Q_8$, show that G has two abelian subgroups of index 2.

3.3 Let φ be a homomorphism defined on a finite group G, and let $H \subseteq G$.
 a. Show that $|\varphi(G) : \varphi(H)|$ divides $|G : H|$.
 b. Show that $|\varphi(H)|$ divides $|H|$.

3.4 Let π be any set of prime numbers. A finite group H is a π-*group* if all primes that divide $|H|$ lie in π. If $|G| < \infty$, then a *Hall* π-subgroup of G is a π-subgroup H such that $|G : H|$ is divisible by no prime in π. Let φ be a homomorphism defined on G.
 a. If $H \subseteq G$ is a Hall π-subgroup, show that $\varphi(H)$ is a Hall π-subgroup of $\varphi(G)$.
 b. Show that $\varphi(G)$ is a π-group iff $HN = G$, where $N = \ker(\varphi)$.

NOTE: It is not true for every finite group G and every prime set π that G necessarily contains a Hall π-subgroup. The symmetric group $\mathrm{Sym}(5)$, for example, has no subgroup of order 15 and hence does not have a Hall $\{3, 5\}$-subgroup.

3.5 Let G be a finite group and suppose π is a set of primes.

    a. Show that G has a unique normal subgroup N, minimal with the property that G/N is a π-group.

    b. Let $\tau : G \to G/N$ be the canonical homomorphism. Show that every homomorphism φ of G to a π-group "factors through" τ. (This means that $\varphi = \tau\theta$ for some homomorphism θ.)

NOTE: The subgroup N of Problem 3.5 is usually denoted $\mathbf{O}^{\pi}(G)$.

3.6 In the lattice diagram in Figure 3.4(a), $UV = Y$ so that we have a diamond. Also, $U \subseteq A \subseteq Y$. Let $B = A \cap V$. Show that diagram (b) is a valid lattice diagram and that $\{A, Y, V, B\}$ and $\{U, A, B, X\}$ are both diamonds.

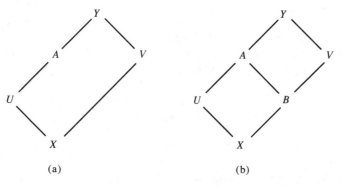

(a)   (b)

Figure 3.4

HINT: Use Dedekind's lemma (Problem 2.9).

3.7 In the lattice diagram in Figure 3.5(a), $\{U, Y, V, X\}$ is a diamond and $N \lhd Y$. Let $A = NU$ and show that diagram (b) is valid.

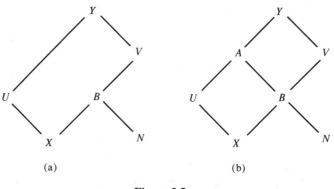

(a)   (b)

Figure 3.5

3.8 Let $N \lhd G$ and suppose $\varphi : G \to H$ is a surjective homomorphism such that $N \cap \ker(\varphi) = 1$. Let $x \in N$. Show that $\varphi(\mathbf{C}_G(x)) = \mathbf{C}_H(\varphi(x))$.

HINT: Consider commutators.

3.9 Let $H \subseteq G$. Show that $\mathbf{N}_G(H)/\mathbf{C}_G(H)$ is isomorphic to a subgroup of Aut(H).

NOTE: The result of Problem 3.9 is sometimes called the "N/C theorem." We shall have occasion to refer to it often.

3.10 A group G is called *metabelian* if there exists some abelian normal subgroup A of G such that G/A is also abelian.
  a. Show that G is metabelian iff $G'' = 1$. (We write G'' to denote $(G')'$, the commutator subgroup of the commutator subgroup of G.)
  b. Show that homomorphic images of metabelian groups are metabelian.
  c. Show that subgroups of metabelian groups are metabelian.

3.11 a. Suppose G'' is cyclic. Show that $G'' \subseteq \mathbf{Z}(G)'$.
  b. Suppose in addition that G'/G'' is cyclic. Show that $G'' = 1$.

  HINT: For part (a), consider $G/\mathbf{C}_G(G'')$ and Aut(G''). Problem 2.19 is relevant.

3.12 Let $\varphi : G \to H$ be a surjective homomorphism with G finite. Suppose $h \in H$ has an order that is a power of the prime p. Show that G has an element g of p-power order with $\varphi(g) = h$.

  HINT: Choose x with $\varphi(x) = h$ and take $g \in \langle x \rangle$.

3.13 Let G be finite and abelian and suppose p is a prime divisor of $|G|$. Show that G has an element of order p.

  HINT: Work by induction on $|G|$. If $1 < H < G$, apply the inductive hypothesis to H and to G/H.

3.14 Let G be finite and abelian and let m divide $|G|$. Show that G has a subgroup of order m.

  HINT: Use Problem 3.13 and induction.

CHAPTER FOUR

Group Actions

4A

Although there are many advantages in moving from the older definition of groups as permutation groups to the modern abstract axiomatic definition, there is also one significant potential disadvantage. With a permutation group one has the set being permuted, and this provides a tool for the study of the group. For instance, if the set has exactly n elements, then by Lagrange's theorem, we can conclude that the order of the group divides $|\mathrm{Sym}(n)| = n!$.

The idea behind the theory of group actions is to regain the advantages of working with permutation groups while continuing to deal with abstract groups. Actions provide techniques for producing subgroups and (especially important) normal subgroups. They can be used as an efficient counting tool to prove theorems about groups and to solve certain combinatorial problems involving symmetry.

(4.1) DEFINITION. Let G be a group and Ω any nonempty set. Assume that for each $g \in G$ and $\alpha \in \Omega$, there is defined a unique element $\alpha \cdot g \in \Omega$. Suppose the following conditions hold:

a. $\alpha \cdot 1 = \alpha$ for all $\alpha \in \Omega$ and
b. $(\alpha \cdot g) \cdot h = \alpha \cdot (gh)$ for all $\alpha \in \Omega$ and $g, h \in G$.

Then we say that G *acts* on Ω or that $\cdot$ is an *action* of G on Ω.

The prototype examples of a group action are permutation groups. Let $\Omega \neq \varnothing$ and suppose $G \subseteq \mathrm{Sym}(\Omega)$. Then, for $\alpha \in \Omega$ and $g \in G$, we can simply take $\alpha \cdot g = (\alpha)g$. (In other words, we simply evaluate the function g at α.) Condition (i) of Definition 4.1 is satisfied by the definition of the identity element of $\mathrm{Sym}(\Omega)$, and condition (ii) is satisfied by the definition of multiplication in this group.

If G is any group and $\Omega \neq \varnothing$ is arbitrary, we can define the *trivial action* of G on Ω by $\alpha \cdot g = \alpha$ for all $\alpha \in \Omega$ and $g \in G$. This rather uninteresting example demonstrates that an action need not be "faithful." (An action is *faithful* if the identity is the only element $g \in G$ such that $\alpha \cdot g = \alpha$ for all $\alpha \in \Omega$.) In general, the *kernel* of an action is the set of group elements that act like 1 and "fix" all $\alpha \in \Omega$. (We say *g fixes α* if $\alpha \cdot g = \alpha$.)

The most useful actions of finite groups are usually internal (in some sense) to the group. There are, for instance, two important ways in which a group G can act on itself. (In other words, we take $\Omega = G$.) The first of these is the *regular action* defined by $x \cdot g = xg$ for all $x \in G$ and $g \in G$. (Note that in this case, condition (ii) of Definition 4.1 is just the associative law.) The other important action of G on itself is the *conjugation action*, where we define $x \cdot g = x^g = g^{-1}xg$. (The reader should check that this really is an action, that conditions (i) and (ii) of Definition 4.1 are satisfied.) Observe that the regular action is faithful and that $\mathbf{Z}(G)$ is the kernel of the conjugation action of G.

If $X \subseteq G$ is any subset and $g \in G$, then as usual we define the product $Xg = \{xg \mid x \in X\}$. This can be used to define an action of G on the set of all subsets of G by setting $X \cdot g = Xg$. (Again, the reader should check that this really is an action. Usually, the verification that something is an action is utterly routine and will be omitted.) If $H \subseteq G$ is a subgroup, take Ω to be the set $\{Hx \mid x \in G\}$ of all right cosets of H in G. If $X \in \Omega$, then also $Xg \in \Omega$ since $(Hx)g = H(xg)$, and thus right multiplication defines an action of G on Ω. (Note that, in general, neither of the following will work to define an action of G on the set of left cosets of H in G:

1. $(xH) \cdot g = xHg$
2. $(xH) \cdot g = (xg)H$.

We leave it to the reader to discover what goes wrong in each case.)

(4.2) LEMMA. *Let G act on Ω. For each $g \in G$, define $\pi_g : \Omega \to \Omega$ by $(\alpha)\pi_g = \alpha \cdot g$. Then $\pi_g \in \mathrm{Sym}(\Omega)$ and the map $\theta : G \to \mathrm{Sym}(\Omega)$ defined by $\theta(g) = \pi_g$ is a homomorphism whose kernel is equal to the kernel of the action.*

Proof.   If $g, h \in G$ and $\alpha \in \Omega$, then

$$(\alpha)\pi_g\pi_h = (\alpha \cdot g)\pi_h = (\alpha \cdot g) \cdot h = \alpha \cdot (gh) = (\alpha)\pi_{gh},$$

and so $\pi_g\pi_h = \pi_{gh}$ for all $g, h \in G$. (Note that we have used condition (ii) from Definition 4.1.) Also, by condition (i),

$$(\alpha)\pi_1 = \alpha \cdot 1 = \alpha,$$

and so π_1 is the identity function i_Ω on Ω.

Now for $g \in G$, we have

$$\pi_g\pi_{g^{-1}} = \pi_1 = \pi_{g^{-1}}\pi_g,$$

and thus π_g is an element of $\mathrm{Sym}(\Omega)$ by Lemma 1.1(a) and (b).

We have

$$\theta(g)\theta(h) = \pi_g\pi_h = \pi_{gh} = \theta(gh)$$

and θ is a homomorphism. An element $g \in G$ lies in $\ker(\theta)$ iff $\pi_g = i_\Omega$, and this is equivalent to saying that $\alpha \cdot g = \alpha$ for all $\alpha \in \Omega$; that is, g is in the kernel of the action.   ∎

(4.3) COROLLARY.   *Let G act on Ω and let K be the kernel of the action. Then $K \lhd G$ and G/K is isomorphic to a subgroup of* $\mathrm{Sym}(\Omega)$.

Proof.   Let $\theta : G \to \mathrm{Sym}(\Omega)$ be as in Lemma 4.2. Then $K = \ker(\theta) \lhd G$ and $G/K \cong \theta(G)$ by the Isomorphism theorem (3.3).   ∎

In the search for simple groups, it is convenient to have nonsimplicity criteria in order to eliminate potential candidates and narrow the field. For this reason, we feel that any result that can be used to produce a nonidentity proper normal subgroup where one was not previously known to exist must be considered to be a good theorem. By this standard, the following is surely one of the most accessible good theorems.

(4.4) THEOREM.   *Let $H \subseteq G$ with $|G : H| = n < \infty$. Then there exists $N \lhd G$ such that*

a. $N \subseteq H$ *and*
b. $|G : N|$ *divides $n!$.*

In particular, if $n > 1$ and $|G|$ does not divide $n!$, then G is not simple.

Proof.   Let G act by right multiplication on the set $\Omega = \{Hx \mid x \in G\}$, and let N be the kernel of this action. By Corollary 4.3, $N \lhd G$ and G/N is isomorphically embedded in $\mathrm{Sym}(\Omega)$. Since $|\Omega| = n$, we have $|\mathrm{Sym}(\Omega)| = n!$, and Lagrange's theorem yields that $|G : N| = |G/N|$ divides $n!$.
   To see that $N \subseteq H$, observe that if $x \in N$, then since $H \in \Omega$, we have

$$x \in Hx = H \cdot x = H,$$

where the last equality holds by the definition of the kernel of an action.
   For the last statement, it suffices to show that $1 < N < G$. Since $n > 1$ we have $H < G$ and so $N < G$ by property (a). Finally, if $N = 1$, then $|G| = |G : N|$ divides $n!$, and this contradiction proves $N > 1$.   ∎

To use Theorem 4.4 as a nonsimplicity criterion, we need some way to find subgroups $H \subseteq G$ with relatively small index. So far, we have not seen much that can be used for this purpose, but in the next chapter we shall obtain the Sylow theorems; these provide some very powerful techniques for finding subgroups of finite groups and keeping control over their indices.
   The next result is an application of Theorem 4.4 that may be viewed as a generalization of Problem 2.6.

(4.5) COROLLARY.   *Let $H \subseteq G$, where G is finite and $|G : H| = p$ is the smallest prime divisor of $|G|$. Then $H \lhd G$.*

Proof.   Let $N \lhd G$ be as in Theorem 4.4. We have $N \subseteq H$ and we write $|H : N| = m$. Then $|G : N| = |G : H||H : N| = pm$, and by Theorem 4.4(b), this must divide $p! = p(p - 1)!$. It follows that m divides $(p - 1)!$, and so every prime divisor q of m satisfies $q \leq p - 1$. By Lagrange's theorem, however, q divides $|G|$ and this contradicts the assumption on p.

We conclude that q cannot exist and so $m = 1$. This yields $H = N \lhd G$.   ∎

Theorem 4.4 remains valid even if G is infinite, provided that $|G : H| < \infty$.

(4.6) COROLLARY.   *Let $H \subseteq G$ have finite index. Then there exists a normal subgroup N of finite index with $N \subseteq H$.*   ∎

Corollary 4.6 can sometimes be used to convert problems about infinite groups into the corresponding finite group problem. If G is finite, for instance, and $U \subseteq V \subseteq G$, then it is immediate by Lagrange's theorem that $|G : U| = |G : V||V : U|$. Suppose now that G is infinite but that $|G : U| < \infty$, with $U \subseteq V \subseteq G$ as before. It is still true that $|G : U| = |G : V||V : U|$. To see this, apply Corollary 4.6 to obtain $N \lhd G$ of finite index with $N \subseteq U$. Write $\overline{G} = G/N$ and let $\overline{U}$ and $\overline{V}$ denote the images of U and V under the canonical homomorphism $G \rightarrow \overline{G}$. Since $\overline{G}$ is finite, we have $|\overline{G} : \overline{U}| = |\overline{G} : \overline{V}||\overline{V} : \overline{U}|$. However, $|\overline{G} : \overline{U}| = |G : U|$ by the Correspondence theorem (3.7); similarly, $|\overline{G} : \overline{V}| = |G : V|$ and $|\overline{V} : \overline{U}| = |V : U|$. This proves the desired equality.

Before we leave Theorem 4.4 and its consequences, we present a result that gives some more precise information about the normal subgroup N.

(4.7) THEOREM.   *Let $H \subseteq G$ and let N be the kernel of the action of G on the right cosets of H (by right multiplication). Then*

a. $N = \bigcap_{x \in G} H^x$ *and*

b. *if $M \lhd G$ with $M \subseteq H$, then $M \subseteq N$.*

Proof.   Let $x, g \in G$. Then $(Hx) \cdot g = Hx$ iff $Hxg = Hx$, and this happens iff $xg \in Hx$. Thus, g fixes Hx iff $g \in x^{-1}Hx = H^x$. It follows that the kernel of the action is precisely the set of elements that lie in every one of the conjugates of H, and this proves property (a).

Now let M be as in (b). For $x \in G$, we have $M = M^x \subseteq H^x$ and thus

$$M \subseteq \bigcap_{x \in G} H^x = N,$$

as claimed.   ∎

In the situation of Theorem 4.7, the normal subgroup N is the largest normal subgroup of G contained in H. It is called the *core* of H in G, and we write $N = \text{core}_G(H)$.

Group actions can also be used to produce subgroups that are not necessarily normal. If G acts on Ω and $\alpha \in \Omega$, we write

$$G_\alpha = \{g \in G \mid \alpha \cdot g = \alpha\}\,.$$

This is called the *stabilizer* of α in G, and it is routine to check that G_α is always a subgroup of G. We consider some examples.

Let G act on itself via conjugation. If $x \in G$, then $G_x = \{g \in G \mid x^g = x\}$, and since $x^g = x$ iff $xg = gx$, we see that the stabilizer in G of $x \in G$ under conjugation is just $\mathbf{C}_G(x)$.

Next, let G act via conjugation on the set of all subsets of G. If X is one of these subsets, then $G_X = \mathbf{N}_G(X)$, the normalizer of X. (In this case, some authors call G_X the *setwise* stabilizer of X to distinguish it from $\mathbf{C}_G(X)$, which they call the *pointwise* stabilizer of X.)

For our final example, let G act by right multiplication on the set of right cosets of $H \subseteq G$ in G. Then, as we saw in the proof of Theorem 4.7, the stabilizer of the coset Hx is the conjugate H^x. In particular, H is its own stabilizer in this action.

4B

We return now to the general case of a group G acting on a set Ω. We say that the action is *transitive* if for every two elements α, $\beta \in \Omega$, there exists an element $g \in G$ with $\alpha \cdot g = \beta$. For instance, the regular action of G and the usual action on the right cosets of a subgroup are transitive. In general, the conjugation action of G on itself is not transitive, since if $x, y \in G$ have different orders, then there can exist no $g \in G$ with $x^g = y$. (Recall that conjugation by g defines an automorphism, and so it preserves the orders of elements.)

In general, if G acts on Ω, then the *orbits* of this action are the sets of the form $\{\alpha \cdot g \mid g \in G\} \subseteq \Omega$.

(4.8) LEMMA.   *Let G act on Ω. Then the orbits partition Ω. This means*

    a. Ω *is the union of the orbits and*
    b. *any two different orbits are disjoint.*

Proof.   Write $\mathcal{O}_\alpha = \{\alpha \cdot g \mid g \in G\}$. Since $\alpha \cdot 1 = \alpha$, we have $\alpha \in \mathcal{O}_\alpha$ and thus

$$\Omega = \bigcup_{\alpha \in \Omega} \mathcal{O}_\alpha \,,$$

proving part (a).

We show now that if $\gamma \in \mathcal{O}_\alpha$, then $\mathcal{O}_\gamma = \mathcal{O}_\alpha$. We have $\gamma = \alpha \cdot x$ for some $x \in G$, and thus

$$\gamma \cdot g = (\alpha \cdot x) \cdot g = \alpha \cdot xg \in \mathcal{O}_\alpha \,.$$

This yields $\mathcal{O}_\gamma \subseteq \mathcal{O}_\alpha$. Also, $\alpha = \gamma \cdot x^{-1}$, so that $\alpha \in \mathcal{O}_\gamma$ and hence the above argument yields $\mathcal{O}_\alpha \subseteq \mathcal{O}_\gamma$. We have shown that $\mathcal{O}_\alpha = \mathcal{O}_\gamma$, as claimed.

Finally, if $\mathcal{O}_\alpha \cap \mathcal{O}_\beta \neq \varnothing$, choose $\gamma \in \mathcal{O}_\alpha \cap \mathcal{O}_\beta$. Then $\mathcal{O}_\alpha = \mathcal{O}_\gamma = \mathcal{O}_\beta$, and part (b) is proved.   ∎

The partition of Ω by the orbits of an action is analogous to the partition of a group by the cosets of a subgroup. This is not entirely accidental, since if $H \subseteq G$, we can let H act on G by right multiplication. In this case, the orbit containing $g \in G$ is exactly the left coset gH. (The reader should try to find an action of H on G whose orbits are the right cosets of H.)

An important example of the partition of a set into orbits with respect to an action occurs in the conjugation action of a group on itself. In that case the orbits are called *conjugacy classes*. Note that the class (i.e., conjugacy class) of an element $x \in G$ consists of the element x alone iff $x \in \mathbf{Z}(G)$. As an example, the reader should check that the classes of D_8, the group of symmetries of a square, are the following. Each of the identity and the 180° rotation constitutes an entire class. The ±90° rotations form a class, the two diagonal flips form a class and the horizontal and vertical flips form a class. This gives a total of five classes for D_8.

The phrase "conjugacy class" is also sometimes applied to an orbit of the conjugacy action of G on its set of subgroups, and so one must examine the context to be certain of the meaning. Usually, if a class of subgroups is intended, this is stated explicitly.

4C

One of the major applications of actions is for counting. The key to this is the following theorem.

(4.9) THEOREM. *Let G act on Ω and let $\mathcal{O}$ be an orbit of this action. Let $\alpha \in \mathcal{O}$ and write $H = G_\alpha$, the stabilizer. Then there exists a bijection $\mathcal{O} \leftrightarrow \{Hx \mid x \in G\}$.*

Proof. We construct a map $f : \mathcal{O} \to \{Hx \mid x \in G\}$ as follows. If $\beta \in \mathcal{O}$, choose $x \in G$ with $\beta = \alpha \cdot x$, and set $f(\beta) = Hx$. We need to check that this is well defined. In other words, if also $\beta = \alpha \cdot y$, we must establish that $Hx = Hy$.

We have $\alpha \cdot x = \alpha \cdot y$, so that $\alpha \cdot xy^{-1} = (\alpha \cdot x) \cdot y^{-1} = (\alpha \cdot y) \cdot y^{-1} = \alpha \cdot 1 = \alpha$, and hence $xy^{-1} \in H$. Therefore, $x \in Hy$ and so $Hx = Hy$, as required.

It is clear that f maps onto $\{Hx \mid x \in G\}$, since for any x, we have $Hx = f(\alpha \cdot x)$. Finally, to show that f is injective, suppose that $f(\beta) = f(\gamma)$. Then $\beta = \alpha \cdot x$ and $\gamma = \alpha \cdot y$ with $Hx = Hy$. This yields $y = hx$ for some $h \in H$, and hence

$$\gamma = \alpha \cdot y = (\alpha \cdot h) \cdot x = \alpha \cdot x = \beta \,,$$

where the third equality holds since $h \in H = G_\alpha$.   ∎

The following is a restatement of Theorem 4.9 into a useful form that we have dubbed the "fundamental counting principle," or FCP.

(4.10) COROLLARY (FCP).   *Suppose G acts on* Ω, *and let* $\mathcal{O}$ *be an orbit. Then*

$$|\mathcal{O}| = |G : G_\alpha|,$$

where α *is any element of* $\mathcal{O}$. *If G is finite, we have*

$$|\mathcal{O}| = |G|/|G_\alpha|,$$

and in particular, $|\mathcal{O}|$ *divides* $|G|$.   ■

As our first application, we apply the FCP to get information about conjugacy classes.

(4.11) COROLLARY.   *Let* $g \in G$ *and let* cl(g) *denote the conjugacy class containing* g. *Then* $|\text{cl}(g)| = |G : \mathbf{C}_G(g)|$. *In particular, for a finite group, all class sizes divide the order of the group.*

Proof.   The class cl(g) is an orbit under the conjugation action of G on itself, and $\mathbf{C}_G(g)$ is the stabilizer of g. The result is thus just a special case of the FCP.   ■

(4.12) COROLLARY.   *Let G be finite and suppose that every two nonidentity elements of G are conjugate. Then* $|G| \leq 2$.

Proof.   Write $n = |G|$ and assume $n > 1$. Then G has a class of size $n - 1$, and hence $n - 1$ divides n. This yields $n \geq 2(n - 1)$ and so $n \leq 2$.   ■

In view of Corollary 4.12 it seems rather surprising that in fact there do exist infinite groups in which all nonidentity elements are conjugate. Our next result may be viewed as a generalization of Corollary 4.12 where we do not attempt to construct a precise bound.

(4.13) THEOREM (Landau).   *For each positive integer k there exists a bound B(k) such that a finite group G having exactly k conjugacy classes satisfies* $|G| \leq B(k)$.

We need a lemma.

(4.14) LEMMA.   *Given a positive integer k and a number A, there exist at most finitely many solutions in positive integers* x_i *for the equation*

$$\frac{1}{x_1} + \frac{1}{x_2} + \cdots + \frac{1}{x_k} = A.$$

Proof.   We may assume that $A > 0$ or else there are no solutions at all. Also, it is no loss to assume that x_k is the smallest of the x_i, so that

$$\frac{k}{x_k} \geq \frac{1}{x_1} + \cdots + \frac{1}{x_k} = A$$

and we have $1 \le x_k \le k/A$. There are, therefore, only finitely many possibilities for x_k.

If $k = 1$, the proof is complete, so we assume $k > 1$ and work by induction on k. For each possible value of x_k, the inductive hypothesis guarantees that there exist only finitely many solutions to

$$\frac{1}{x_1} + \frac{1}{x_2} + \cdots + \frac{1}{x_{k-1}} = A - \frac{1}{x_k},$$

and there are thus only finitely many solutions in all.   ∎

Proof of Theorem 4.13.   Suppose G has exactly k conjugacy classes and their sizes are $c_1, c_2, \ldots, c_k$. We then have the so-called class equation,

$$|G| = c_1 + c_2 + \cdots + c_k.$$

Since each class size divides $|G|$, we can write $|G| = c_i x_i$ for positive integers x_i. (In fact, $x_i = |\mathbf{C}_G(g_i)|$, where g_i is any element of the ith class.) This yields

$$1 = \frac{1}{x_1} + \cdots + \frac{1}{x_k},$$

and so by Lemma 4.14, there is a number $B(k)$, depending only on k, such that all $x_i \le B(k)$.

Since the identity is in a class by itself, we have (say) $c_1 = 1$ and so $x_1 = |G|$. Therefore $|G| \le B(k)$.   ∎

We can obtain further useful information by applying the FCP to the conjugation action of G on its set of subsets.

(4.15) COROLLARY.   *Let $X \subseteq G$ be a subset. Then the number of distinct G-conjugates of X (counting X itself) is $|G : \mathbf{N}_G(X)|$.*   ∎

As an application of Corollary 4.15, we give the following corollary.

(4.16) COROLLARY.   *Let $H \subseteq G$, where G is finite. If*

$$\bigcup_{g \in G} H^g = G,$$

then $H = G$.

Proof.   Let $N = \mathbf{N}_G(H)$. By Corollary 4.15, there are exactly $|G : N|$ distinct subgroups H^g, and so

$$|G| - 1 = |G - \{1\}| = \left| \bigcup_g (H^g - \{1\}) \right| \le |G : N|(|H| - 1)$$

since $|H^g| = |H|$ for each $g \in G$. Also, $N \supseteq H$ and so $|G : N| \le |G : H|$, and this yields $|G| - 1 \le |G : H|(|H| - 1) = |G| - |G : H|$. We conclude that $|G : H| \le 1$, as desired.   ∎

The following application of the FCP generalizes Problem 2.8(d).

(4.17) COROLLARY.   *Let $H, K \subseteq G$ be finite subgroups. Then*

$$|HK| = |H||K|/|H \cap K|.$$

Proof.   Let $\Omega = \{Hx \mid x \in G\}$ and let K act on Ω by right multiplication. Now

$$HK = \bigcup_{k \in K} Hk,$$

and since $Hk = H \cdot k$, we see that HK is the union of those cosets that are elements of the orbit $\mathcal{O}$ containing H. Therefore, since the cosets in $\mathcal{O}$ are disjoint and each contains exactly $|H|$ elements, we have $|HK| = |H||\mathcal{O}|$.

Finally, the stabilizer in G of H is the subgroup H itself, and hence the stabilizer in K of H is $H \cap K$. By the FCP, we have $|\mathcal{O}| = |K|/|H \cap K|$, and the result follows.   ∎

Note that we need not assume that HK is a group in Corollary 4.17.

4D

Now we wish to apply the FCP in a different way. Instead of computing the size of a single orbit, we will count the total number of orbits. To do this, we make a definition.

Let G act on a finite set Ω. The associated *permutation character* is the integer-valued function χ defined on G by the formula

$$\chi(g) = |\{\alpha \in \Omega \mid \alpha \cdot g = \alpha\}|.$$

For instance, the permutation character associated with the regular action of a finite group G is the function

$$\chi(g) = \begin{cases} 0 & \text{if } g \neq 1 \\ |G| & \text{if } g = 1, \end{cases}$$

and for the conjugation action of G on G we get

$$\chi(g) = |\mathbf{C}_G(g)|.$$

Our key result is the following theorem.

(4.18) THEOREM (Cauchy-Frobenius).   *Let G act on Ω, with both G and Ω finite. Then the total number n of orbits is given by*

$$n = \frac{1}{|G|} \sum_{x \in G} \chi(x).$$

Proof.   We define

$$S = \{(\alpha, g) \mid \alpha \in \Omega, \; g \in G, \; \alpha \cdot g = \alpha\}.$$

The cardinality of S can be computed in two different ways. First, we see that for each $\alpha \in \Omega$, there are exactly $|G_\alpha|$ elements g such that (α, g) lies in S, and this gives

$$|S| = \sum_{x \in G} |G_\alpha|.$$

Also, for each $g \in G$, the number of elements $\alpha \in \Omega$ paired with g is $\chi(g)$, and so we have

$$|S| = \sum_{g \in G} \chi(g).$$

We deduce that

$$\frac{1}{|G|} \sum_{g \in G} \chi(g) = \frac{1}{|G|} \sum_{\alpha \in \Omega} |G_\alpha| = \sum_{\alpha \in \Omega} \frac{1}{|\mathcal{O}_\alpha|},$$

where $\mathcal{O}_\alpha$ is the orbit containing α, and the second equality follows by the FCP. Since the sum of $1/|\mathcal{O}|$, taken as many times as the orbit $\mathcal{O}$ has elements, is equal to 1, we see that the sum on the right is equal to the number n of orbits.   ∎

A convenient paraphrase of the Cauchy-Frobenius formula is that the number of orbits is equal to the average value of the permutation character χ.

(4.19) COROLLARY.   *Suppose that the finite group G acts transitively on the set Ω and that $|\Omega| > 1$. Then there exists $g \in G$ fixing no point of Ω.*

Proof.   Since G is transitive, there is just one orbit and the average value of the permutation character is 1. Since $\chi(1) = |\Omega| > 1$, there must exist some $g \in G$ at which χ has a below-average value. For some $g \in G$, therefore, $\chi(g) < 1$ and so $\chi(g) = 0$.   ∎

Note that Corollary 4.19 provides an alternative proof of Corollary 4.16. If $H < G$, then setting $\Omega = \{Hx \mid x = G\}$, we have $|\Omega| > 1$ and G acts transitively on Ω. By Corollary 4.19, therefore, there exists $g \in G$ fixing no coset Hx. Since the stabilizer of Hx is H^x, it follows that g lies in no conjugate of H, and Corollary 4.16 follows.

In many books, the Cauchy-Frobenius theorem is incorrectly attributed to W. Burnside and is referred to as Burnside's orbit-counting formula. It is also sometimes associated with G. Pólya, who showed how to use it in certain problems of enumerative combinatorics. We close this chapter with an example of Pólya's method.

Consider the following problem. Suppose that we are in the business of man-
ufacturing colored squares out of bits of wire. We have a supply of 1-inch pieces
of wire in n different colors, and we make 1-inch squares by soldering four of these
together at a time. We wish to keep in stock a supply of every possible colored square
that a customer might request. We store our completed squares in bins, and we ask:
What is the smallest number of bins we need such that every request for a colored
square can be filled simply by reaching into the appropriate bin? At first glance, it
appears that the answer should be n^4, since there are n different colors possible for
each of the four sides. (We say that there are n^4 different *color configurations* for
the squares.)

We can manage with fewer than n^4 bins. The squares with a red top and the
three remaining sides blue, for example, need not be segregated from the squares
with a red bottom and the three remaining sides blue. (In fact, the bin containing
these two color configurations also contains two others.) Squares with one blue side
and three red sides, however, must be stored in a different bin. We need to count the
number of "fundamentally different" squares and not just the number of different
color configurations. A bit of experimentation shows that when $n = 2$, six bins are
required. Figure 4.1 shows the six fundamentally different squares in this case.

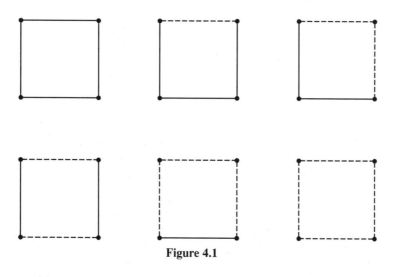

Figure 4.1

To put the above discussion into somewhat more precise language, we see that
the group D_8 of symmetries of a square acts on the set Ω of n^4 color configurations
of the square. What we are asked to count is the number of orbits of this action.

To solve the problem, we compute the value of the permutation character on
each of the eight elements of D_8. The identity fixes everything, of course, and so
$\chi(1) = |\Omega| = n^4$. A 180° rotation fixes a color configuration iff the top and bottom
have the same color and the two sides have the same color (which need not be
the same as that assigned to the top and bottom). It follows that the character value
for this element is n^2. The reader should have little difficulty in verifying the data
in Table 4.1.

TABLE 4.1

Type of Element	Number of This Type	Value of χ
Identity	1	n^4
180° rotation	1	n^2
90° rotation	2	n
Diagonal axis flip	2	n^2
Horizontal or vertical flip	2	n^3

By the Cauchy-Frobenius formula (4.18), therefore, the number of fundamentally different colored squares we can make with n colors is

$$\tfrac{1}{8}(n^4 + 2n^3 + 3n^2 + 2n).$$

Note that when $n = 2$, this is equal to 6, as we expected.

Problems

4.1 Let G act transitively on Ω and assume G is finite. Define an action of G on the set $\Omega \times \Omega$ by putting $(\alpha, \beta) \cdot g = (\alpha \cdot g, \beta \cdot g)$. Let $\alpha \in \Omega$. Show that G has the same number of orbits on $\Omega \times \Omega$ that G_α has on Ω.

4.2 An action of G on Ω is *doubly transitive* if G is transitive on the set of those ordered pairs (α, β) with $\alpha \neq \beta$ (with the action as in Problem 4.1). Show that G is doubly transitive on Ω if

$$\frac{1}{|G|} \sum_{g \in G} \chi(g)^2 = 2,$$

where χ is the associated permutation character. Do not assume G is transitive.

4.3 Let $H, K \subseteq G$ be finite subgroups. Show that $|HgK| = |H||K|/|K \cap H^g|$.
NOTE: A set of the form HgK is called a *double coset*.

4.4 Let $\varphi : G \to H$ be a surjective homomorphism with $|G|$ finite and let $g \in G$. Show that
$$|\mathbf{C}_G(g)| \geq |\mathbf{C}_H(\varphi(g))|.$$

HINT: Show that the conjugacy class of g in the inverse image in G of $\mathbf{C}_H(\varphi(g))$ has size $\leq |\ker(\varphi)|$.

4.5 Let G be finite. Show that the probability that two elements of G chosen at random commute with each other is equal to $k/|G|$, where k is the number of conjugacy classes of G. (The two elements are allowed to be equal; they are chosen randomly with replacement.)

4.6 Find a formula for the number of fundamentally different ways to color the six faces of a cube if there are n colors available. (Two colorings are considered to be the same if the corresponding colored cubes are indistinguishable after being shaken up in a box. Note that a colored cube may be distinguishable from its mirror image.)

4.7 How many different necklaces can be made by stringing r symmetrical beads on a cord and closing the cord into a loop? Assume that there are n different colors of beads available.

4.8 Let G act on Ω, where G and Ω are finite. For $g \in G$, write $a(g)$ to denote the number of orbits of $\langle g \rangle$ on Ω. Let $f_{G,\Omega}$ be the polynomial defined by the formula

$$f_{G,\Omega}(x) = \frac{1}{|G|} \sum_{g \in G} x^{a(g)}.$$

Show that $f_{G,\Omega}(x)$ is an integer whenever x is a positive integer.

NOTE: The assertion remains true if the word "positive" is dropped. Also, observe what happens if we take $\Omega = G$ to be a cyclic group of prime order p. In that case, $f_{G,\Omega}(x) = (1/p)((p-1)x + x^p)$, and the problem yields that $x^p - x$ is divisible by p for integers x.

4.9 (Cauchy). Let G be finite and let p be a prime divisor of $|G|$. Show that G has an element of order p.

HINT: Suppose the assertion is false and that G is a counterexample of minimum possible order. Show that p divides the size of each nontrivial conjugacy class of G. Use the class equation and Problem 3.13.

CHAPTER FIVE

The Sylow Theorems and p-groups

5A

So far, we have studied properties of subgroups, but we have not developed especially powerful techniques for finding or constructing subgroups. If G is finite, for instance, Lagrange's theorem tells us that for any subgroup $H \subseteq G$, we have that $|H|$ divides $|G|$. On the other hand, if we are given a divisor m of $|G|$, we have not discussed the question of whether there must exist a subgroup of order m. (In the comparatively easy case where G is abelian, this was done in Problem 3.14.) The main point of Sylow's theorem is the partial converse of Lagrange's theorem, which asserts that if m is a power of a prime number and is a divisor of $|G|$, then G does have a subgroup of order m. This result might well be called the "fundamental theorem of finite group theory." (We mention that the full converse of Lagrange's theorem is not, in general, true. There exists a group of order 12, for instance, that has no subgroup of order 6.)

We begin with a number theoretic lemma about binomial coefficients.

(5.1) LEMMA. *Let $n = p^a m$, where p is a prime. Then*

$$\binom{n}{p^a} \equiv m \mod p.$$

Proof. If $f(x)$ and $g(x)$ are two polynomials with integer coefficients, we use the notation $f(x) \equiv g(x) \mod p$ to mean that for every $j \geq 0$, the coefficients of x^j in $f(x)$ and $g(x)$ are congruent mod p. For instance, we have

$$(x + 1)^p \equiv x^p + 1 \mod p,$$

since $\binom{p}{j} \equiv 0 \mod p$ for $1 \leq j \leq p - 1$.

We now have

$$(x + 1)^{p^2} \equiv (x^p + 1)^p \equiv x^{p^2} + 1 \mod p,$$

and continuing like this, we obtain

$$(x + 1)^{p^a} \equiv x^{p^a} + 1 \mod p$$

and

$$(x + 1)^n = (x + 1)^{p^a m} \equiv (x^{p^a} + 1)^m \mod p.$$

Comparison of the coefficients of x^{p^a} on both sides yields

$$\binom{n}{p^a} \equiv \binom{m}{1} = m \mod p,$$

as required.   ∎

We are now ready to prove the Sylow existence theorem, or E-theorem.

(5.2) THEOREM (Sylow E).   *Let G be a finite group of order $p^a m$, where p is prime and $p \nmid m$. Then G has a subgroup H of order p^a.*

Proof (Wielandt).   Let Ω be the collection of all subsets $X \subseteq G$ with $|X| = p^a$. Then

$$|\Omega| = \binom{|G|}{p^a} \equiv m \not\equiv 0 \mod p,$$

where we have used Lemma 5.1 and the hypothesis that p does not divide m.

Let G act on Ω by right multiplication. (Since $|Xg| = |X|$, this does define an action.)   Because $p \nmid |\Omega|$, there must be some orbit $\mathcal{O}$ with $p \nmid |\mathcal{O}|$, and we choose $X \in \mathcal{O}$. By the FCP (Corollary 4.10), we have $|\mathcal{O}| = |G|/|H|$, where $H = G_X$ is the stabilizer of X in G. Since p^a divides $|G|$ but p does not divide $|\mathcal{O}|$, we must have $p^a \big| |H|$, and in particular, $p^a \leq |H|$.

Now fix $x \in X$. For $h \in H$, we have

$$xh \in Xh = X \cdot h = X,$$

since H is the stabilizer of X. It follows that the left coset $xH \subseteq X$ and thus

$$|H| = |xH| \leq |X| = p^a.$$

Combining this with our previous inequality yields $|H| = p^a$.   ∎

As we shall see later, the condition that $p \nmid m$ is not really necessary, and in fact G has a subgroup of order p^a whenever p is a prime and p^a divides $|G|$.

(5.3) DEFINITION.   Let G be a finite group and p a prime. A *Sylow p-subgroup* of G is a subgroup $P \subseteq G$ such that $|P| = p^a$ is the full power of p dividing $|G|$. The set of all Sylow p-subgroups of G is denoted $\mathrm{Syl}_p(G)$.

The content of Sylow's E-theorem (5.2) is that $\mathrm{Syl}_p(G)$ is nonempty for every finite group. (If $p \nmid |G|$, then the trivial subgroup is a Sylow p-subgroup, and so even in this case, $\mathrm{Syl}_p(G) \neq \varnothing$.)

In Problem 4.9, the reader was asked to prove Cauchy's theorem concerning the existence of elements of prime order in finite groups. Cauchy's result can also be obtained as an immediate corollary of Sylow's E-theorem.

(5.4) COROLLARY (Cauchy).   *Let G be finite with $p \mid |G|$, where p is a prime. Then G has an element of order p.*

Proof.   Let $P \in \mathrm{Syl}_p(G)$. Since $p \mid |G|$, we have $P > 1$ and we can choose $x \in P - \{1\}$. Then $o(x) = p^e$ for some $e > 0$, and thus $x^{p^{e-1}}$ has order p.   ∎

A (not necessarily finite) group P is said to be a *p-group* (where p is a prime) if every element has (finite) order a power of p.

(5.5) COROLLARY.   *A finite group is a p-group iff its order is a power of p.*

Proof.   This is immediate by Cauchy's theorem (5.4) and Corollary 2.24 to Lagrange's theorem.   ∎

We see that a Sylow p-subgroup of a finite group G is indeed a p-subgroup, and in fact by Lagrange's theorem, it has the largest order any p-subgroup of G could have. It does not follow (but it is true) that a subgroup of G maximal with the property that it is a p-group is necessarily Sylow. It is conceivable that one could have a p-subgroup not contained in any larger p-subgroup and yet with order too small to qualify as a Sylow p-subgroup. That this cannot actually happen is the essential content of the Sylow "development" or D-theorem.

(5.6) THEOREM (Sylow D).   *Let G be finite and let $P \subseteq G$ be a p-subgroup. Then there exists $S \in \mathrm{Syl}_p(G)$ with $P \subseteq S$.*

We shall soon prove Theorem 5.6, but first we wish to consider the relationship among the various Sylow p-subgroups of G for a fixed prime p.

The condition that a subgroup of G be a Sylow p-subgroup is just a condition on the order of the subgroup, and therefore any conjugate of a Sylow p-subgroup of G is again a Sylow p-subgroup. What is less trivial is that every two Sylow p-subgroups of G are conjugate. This is the essential content of the Sylow conjugacy or C-theorem.

(5.7) THEOREM (Sylow C).   *Let G be a finite group. Then the set $\mathrm{Syl}_p(G)$ is a single conjugacy class of subgroups of G.*

We can combine the Sylow D- and C-theorems into a single result.

(5.8) THEOREM.   *Let G be finite and suppose $P \subseteq G$ is a p-subgroup and $S \in \mathrm{Syl}_p(G)$. Then $P \subseteq S^x$ for some $x \in G$.*

Proof.   Let $\Omega = \{Sx \mid x \in G\}$ and let P act on Ω by right multiplication. We have $|\Omega| = |G : S|$, and this is not divisible by p since S is a Sylow p-subgroup.

It follows that some orbit of the action of P on Ω must have size not divisible by p.

Since P is a p-group, however, all of these orbits have p-power size by the FCP. Since the only p-power not divisible by p is the number 1, there must exist some orbit containing only a single element. It follows that P stabilizes Sx for some $x \in G$.

If $y \in P$, then

$$Sx = (Sx) \cdot y = Sxy,$$

and hence

$$S^x = x^{-1} Sx = x^{-1} Sxy = S^x y$$

and $y \in S^x$. Thus $P \subseteq S^x$, as required.   ∎

Note that since S^x is a Sylow p-subgroup of G containing P in Theorem 5.8, the Sylow D-theorem (5.6) is an immediate consequence. The Sylow C-theorem (5.7) also follows, since all we need to show is that if $P, S \in \mathrm{Syl}_p(G)$, then $P = S^x$ for some $x \in G$. By Theorem 5.8, we have $P \subseteq S^x$, and since $|P| = |S| = |S^x|$, equality follows. The Sylow D- and C-theorems (5.6 and 5.7) have now been proved.

Since $\mathrm{Syl}_p(G)$ is a single conjugacy class of subgroups, the following is immediate by Corollary 4.15.

(5.9) COROLLARY. *Let G be finite and let $P \in \mathrm{Syl}_p(G)$. Then*

$$|\mathrm{Syl}_p(G)| = |G : \mathbf{N}_G(P)|,$$

and in particular, this divides $|G : P|$.   ∎

(5.10) COROLLARY. *Let $S \in \mathrm{Syl}_p(G)$. The following are then equivalent:*

i. $S \triangleleft G$.
ii. *S is the unique Sylow p-subgroup of G.*
iii. *Every p-subgroup of G is contained in S.*
iv. *S is characteristic in G.*

Proof.   Assuming (i), we have $G = \mathbf{N}_G(S)$, and hence $|\mathrm{Syl}_p(G)| = |G : \mathbf{N}_G(S)| = 1$ by Corollary 5.9, proving (ii). That (ii) implies (iii) is immediate from the Sylow D-theorem. Now assume (iii). If $\sigma \in \mathrm{Aut}(G)$, then $\sigma(S)$ is a subgroup of G satisfying $|\sigma(S)| = |S|$, and in particular, $\sigma(S)$ is a p-subgroup. By (iii), therefore, we have $\sigma(S) \subseteq S$ and hence $\sigma(S) = S$, proving (iv). Finally, the fact that (iv) implies (i) is obvious.   ∎

We shall generally write $n_p(G)$ to denote the number $|\mathrm{Syl}_p(G)|$ of Sylow p-subgroups of G. By Corollary 5.9, if $|G| = p^a m$, then $n_p(G)$ must divide m. For instance, if $|G| = 360 = 2^3 \cdot 3^2 \cdot 5$, then $n_3(G)$, which is the number of subgroups of order 9 in G, is necessarily one of the numbers 1, 2, 4, 8, 5, 10, 20 or 40.

There is a further result that usually significantly shortens the list of possibilities for $n_p(G)$.

(5.11) THEOREM (Sylow counting).   *Let G be a finite group and write $n_p(G) =$ $|\mathrm{Syl}_p(G)|$. Then*

$$n_p(G) \equiv 1 \mod p.$$

In fact,

$$n_p(G) \equiv 1 \mod p^e$$

if $p^e \le |S : S \cap T|$ for all $S, T \in \mathrm{Syl}_p(G)$ with $S \ne T$.

For instance, if $|G| = 360$ as before, then the only possibilities for $n_3(G)$ are 1, 4, 10, and 40. Furthermore, since 4 and 40 are not congruent to 1 mod 9, the second assertion of the theorem tells us that if $n_3(G) \in \{4, 40\}$, then there exist distinct $S, T \in \mathrm{Syl}_3(G)$ such that $9 > |S : S \cap T|$. In particular, in this case $S \cap T > 1$ and the distinct Sylow 3-subgroups of G do not intersect trivially.

It is the first assertion of the theorem that is by far the most important, although we shall see later how the additional information about Sylow intersections contained in the second statement can be used. We have included this somewhat complicated refinement of the usual Sylow counting theorem in Theorem 5.11, since in fact, there is very little extra effort involved in its proof.

We begin with a lemma for which we offer two different proofs.

(5.12) LEMMA.   *Let G be finite and let $S \in \mathrm{Syl}_p(G)$. Suppose P is a p-subgroup of $\mathbf{N}_G(S)$. Then $P \subseteq S$.*

Proof.   Since $P \subseteq \mathbf{N}_G(S)$, we conclude that PS is a subgroup. (See Corollary 2.31.) Also, $|PS| = |P||S|/|P \cap S|$ by Corollary 4.17, and it follows that PS is a p-group. However, $PS \supseteq S$ and yet, since S is Sylow in G, it is maximal among p-subgroups. We conclude that $PS = S$, and it follows that $P \subseteq S$.   ∎

Alternative proof.   Since S is Sylow in G and $S \subseteq \mathbf{N}_G(S)$, we have that $S \in \mathrm{Syl}_p(\mathbf{N}_G(S))$ (by Lagrange's theorem). However, $S \triangleleft \mathbf{N}_G(S)$, and so by Corollary 5.10 applied to the group $\mathbf{N}_G(S)$, we conclude that $P \subseteq S$.   ∎

Proof of Theorem 5.11.   Let $P \in \mathrm{Syl}_p(G)$ and let P act by conjugation on the set $\mathrm{Syl}_p(G)$. (Note that conjugates of Sylow subgroups are certainly Sylow subgroups; this follows from the definition and does not require the Sylow C-theorem.)

Since P stabilizes itself under conjugation, it follows that $\{P\}$ is one of the orbits of this action. To prove that $n_p(G) \equiv 1 \mod p^e$, it suffices to show that every other orbit has size divisible by p^e. (If there are no other orbits, then $n_p(G) = 1$, and there is nothing to prove.)

Suppose, then, that $S \in \mathrm{Syl}_p(G)$ with $S \ne P$, and let $\mathcal{O}$ be the orbit containing S. Thus, $|\mathcal{O}| = |P : \mathbf{N}_P(S)|$ by the FCP. Also, $\mathbf{N}_P(S) \subseteq S$ by Lemma 5.12, and it follows that $\mathbf{N}_P(S) = P \cap S$. Thus $|\mathcal{O}| = |P : P \cap S|$.

If p^e is as in the statement of the theorem, then $p^e \le |P : P \cap S|$, and since this index is a p-power, we conclude that p^e divides $|P : P \cap S|$, as required. In particular, since $P \ne S$, we have $P \not\subseteq S$ and so $|P : P \cap S| > 1$ and hence $|P : P \cap S| \ge p$, and we can take $e = 1$, proving the first statement.   ∎

5B

To demonstrate the power of the Sylow theorems, we will give a number of applications that show how they can be used to analyze groups of specific orders.

(5.13) THEOREM. *Let $|G| = pq$, where $p > q$ are primes. Then G has a normal Sylow p-subgroup. Also, if G is nonabelian, then $q|(p-1)$ and G has exactly p Sylow q-subgroups.*

Proof. By Corollary 5.9, we have $n_p(G) = 1$ or q, and by the Sylow counting theorem, $n_p(G) \equiv 1 \bmod p$. Since $1 < q < p$, we have $q \not\equiv 1 \bmod p$, and we conclude that $n_p(G) = 1$ and G has a normal Sylow p-subgroup, P.

Since G/P is a group of prime order q, it is abelian and we have $G' \subseteq P$ by Theorem 3.10. Similarly, if G had a normal Sylow q-subgroup Q, then $G' \subseteq Q$, and we would have $G' \subseteq P \cap Q = 1$ and G would be abelian. It follows that if G is not abelian, then $n_q(G) > 1$ and thus $n_q(G) = p$. This implies that $p \equiv 1 \bmod q$, and the proof is complete. ∎

We remark that for any choice of the primes $p > q$, there always exists a cyclic group of order pq, and in fact (up to isomorphism), this is the only abelian group of this order. To see this, suppose G is abelian and $|G| = pq$. Then G has normal (hence unique) Sylow subgroups P and Q of orders p and q, respectively. If $x \in G$ with $o(x) = p$, then $|\langle x \rangle| = p$ and so $\langle x \rangle = P$ and $x \in P$. Similarly, if $o(x) = q$, then $x \in Q$. Since $P \cup Q < G$, we can find $x \in G$ with $x \notin P$ and $x \notin Q$, and the only possibility is that $o(x) = pq$, and so $G = \langle x \rangle$.

Theorem 5.13 allows the possibility (but does not prove the existence) of a nonabelian group of order pq whenever $q|(p-1)$. In fact, for every choice of such primes p and q, it is possible to construct a nonabelian group of order pq, and in fact, there is a unique such group up to isomorphism. We address this situation in Problem 5.13.

(5.14) THEOREM. *Let $|G| = p^2q$, where p and q are primes. Then G has a normal Sylow p-subgroup or a normal Sylow q-subgroup.*

Proof. If $n_q(G) = 1$, we are done by Corollary 5.10, and so we may assume $n_q(G) > 1$ and hence $p \neq q$ and $n_q(G) \in \{p, p^2\}$ by Corollary 5.9. We consider the two possibilities.

First, suppose $n_q(G) = p$. Then $p \equiv 1 \bmod q$, and in particular $p > q$. Therefore $q \not\equiv 1 \bmod p$ and we cannot have $n_p(G) = q$. We conclude that $n_p(G) = 1$, and we are done in this case.

In the case where $n_q(G) = p^2$, we count elements. If $Q_1, Q_2 \in \mathrm{Syl}_q(G)$ with $Q_1 \neq Q_2$, then since each of these subgroups has prime order q, we have $Q_1 \cap Q_2 = 1$, and so no two distinct Sylow q-subgroups of G have any nonidentity elements in common. Since there are p^2 Sylow q-subgroups, each containing $q-1$ nonidentity elements, this accounts for $p^2(q-1)$ elements of G.

The elements we have counted so far all have order q (and in fact, these are all the elements of G of order q, although we do not need to know that).

Let X be the set of as yet uncounted elements (including the identity). Then $|X| = |G| - p^2(q-1) = p^2q - p^2(q-1) = p^2$, and every element of G with order unequal to q lies in X.

Now let $S \in \mathrm{Syl}_p(G)$. Then no element of S has order q and therefore $S \subseteq X$. Since $|S| = p^2 = |X|$, we conclude that $S = X$, and therefore S is unique and hence normal.   ∎

We give one further result of this type.

(5.15) THEOREM. *Let $|G| = p^3q$, where p and q are primes. Then either G has a normal Sylow p- or q-subgroup or $p = 2, q = 3$ and $|G| = 24$.*

Proof. We may assume that $p \neq q$ and $n_q(G) > 1$, and so we need to consider the three possible values p, p^2 and p^3 for $n_q(G)$.

We may also assume that $n_p(G) > 1$ and thus $n_p(G) = q$ and $q \equiv 1 \bmod p$. Therefore, $q > p$ and we cannot have $p \equiv 1 \bmod q$. This eliminates the possibility that $n_q(G) = p$.

Now suppose $n_q(G) = p^3$. Counting elements as in the previous proof, we find $p^3(q-1)$ elements of order q, and so there is a set X containing all elements of order different from q, with $|X| = |G| - p^3(q-1) = p^3$. It follows that an arbitrary Sylow p-subgroup of G must equal X and thus is unique and hence normal. This contradicts $n_p(G) > 1$.

Finally, suppose $n_q(G) = p^2$. Then $p^2 \equiv 1 \bmod q$ and thus $q \mid (p^2 - 1)$. Thus $q \mid (p+1)(p-1)$ and since q is prime, either $q \mid (p+1)$ or $q \mid (p-1)$. Since $q > p$, the latter is impossible and thus $q \mid (p+1)$ and so $p < q \leq p+1$. This forces $q = p + 1$.

Since 2 and 3 are the only pair of consecutive primes, we have $p = 2, q = 3$ and $|G| = 24$.   ∎

The obvious question at this point is whether there actually exists an exceptional group of order 24 for which neither a Sylow 2- nor 3-subgroup is normal. Theorem 5.15, of course, does not answer this question.

It often happens in finite group theory that one can prove a general result and obtain a small list of exceptional cases where the proof breaks down. It seems remarkable how often it occurs that the exceptions really exist; they are not merely gaps in the proof. In particular, it turns out that the symmetric group Sym(4) of order 24 is an example (and is the only one up to isomorphism) that shows that the exceptional case in Theorem 5.15 actually can occur.

Suppose we wish to show that no group of some specific order can be simple. As we shall see shortly, the only simple p-groups are those of prime order and so we consider only groups whose order is not a prime power.

(5.16) LEMMA. *Suppose $|G| = p^a m$, where $a > 0$, $m > 1$ and $p \nmid m$. If G is simple, then $n = n_p(G)$ satisfies all of the following:*

a. *n divides m.*
b. $n \equiv 1 \mod p$.
c. $|G|$ *divides* $n!$.

Proof.   Conclusions (a) and (b) hold by Corollary 5.9 and Theorem 5.11, respectively.

Now let $H = \mathbf{N}_G(S)$, where $S \in \mathrm{Syl}_p(G)$. Thus $|G : H| = n$ by Corollary 5.9. Since G is simple and $1 < S < G$, we know that S is not normal, and so $n > 1$. By the simplicity of G and Theorem 4.4, we deduce that $|G|$ divides $n!$.   ∎

We now use a variety of techniques to work some nonsimplicity examples with varying degrees of difficulty.

(5.17) EXAMPLE.   *If* $|G| = 1,000,000 = 2^6 \cdot 5^6$, *then G is not simple.*

Proof.   Apply Lemma 5.16 with $p = 5$. The only divisors n of 2^6 that are congruent to 1 mod 5 are 1 and 16, and neither is large enough so that $|G|$ divides $n!$, since even 16! is divisible by 5 only to the third power.   ∎

Our next example is quite a bit more difficult and constitutes a review of almost everything we have done so far. It also uses results given in the problem sections of previous chapters.

(5.18) EXAMPLE.   *If* $|G| = 2376 = 2^3 \cdot 3^3 \cdot 11$, *then G is not simple.*

Proof.   We assume that G is simple and derive a contradiction. Applying Lemma 5.16 with $p = 11$, we systematically check the divisors $2^i \cdot 3^j$ of $2^3 \cdot 3^3$, and we find that the only possibility for $n_{11}(G)$ is $2^2 \cdot 3 = 12$.

Our strategy now will be to find a subgroup H such that $n = |G : H|$ satisfies $1 < n < 11$. Since in that case $|G|$ will not divide $n!$, we will be done according to Theorem 4.4. Among the techniques we have accumulated for producing subgroups are taking normalizers, centralizers and Sylow subgroups, and we shall use all of these.

Let $S \in \mathrm{Syl}_{11}(G)$ and put $N = \mathbf{N}_G(S)$. Thus $|G : N| = n_{11}(G) = 2^2 \cdot 3$ and $|N| = 2 \cdot 3^2 \cdot 11$. Now let $C = \mathbf{C}_G(S)$. By Problem 3.9, which is the "N/C theorem," we know that N/C is isomorphic to a subgroup of $\mathrm{Aut}(S)$, and by Theorem 2.11, we know that $|\mathrm{Aut}(S)| = \varphi(11) = 10$. It follows that $|N : C|$ divides 10, and therefore 3^2 divides $|C|$.

Let $P \in \mathrm{Syl}_3(C)$ so that $|P| = 3^2$, and let $H = \mathbf{N}_G(P)$. Since G is simple, we have $H < G$, and we will be done if we can show that $|G : H| < 11$. We need to show that $|H|$ is large, and we do this by finding subgroups that we can prove lie inside H.

Now $P \subseteq C = \mathbf{C}_G(S)$ and thus $S \subseteq \mathbf{C}_G(P) \subseteq \mathbf{N}_G(P) = H$, and so $|H|$ is divisible by 11.

Next, by the Sylow D-theorem, there exists $Q \in \mathrm{Syl}_3(G)$ with $P \subseteq Q$, and we have $|Q : P| = 3$. By Corollary 4.5 (since Q is a 3-group), we have $P \triangleleft Q$, and so $Q \subseteq \mathbf{N}_G(P) = |H|$. It follows that $3^3 = |Q|$ divides $|H|$.

At this point we know that $3^3 \cdot 11$ divides $|H|$, and thus $|G : H| \leq 2^3 < 11$, as required. ∎

Our final example uses the second statement of the Sylow counting theorem (5.11), which we have not yet exploited.

(5.19) EXAMPLE. *If $|G| = 8000 = 2^6 \cdot 5^3$, then G is not simple.*

Proof. Assume G is simple and apply Lemma 5.16 with $p = 5$ to obtain $n_5(G) = 16$. Since $16 \not\equiv 1 \bmod 5^2$, we conclude by Theorem 5.11 that there exist $S, T \in \mathrm{Syl}_5(G)$ with $S \neq T$ and $5^2 > |S : S \cap T|$. It follows that $|S : S \cap T| = 5$.

Now by Corollary 4.5, we have $S \cap T \lhd S$ and similarly, $S \cap T \lhd T$. Let $H = \mathbf{N}_G(S \cap T)$ and observe that $H < G$ since $1 < S \cap T$ and G is simple.

We have $S \subseteq H$ and $T \subseteq H$, and it follows by Lagrange's theorem that $S, T \in \mathrm{Syl}_5(H)$. Therefore, $n_5(H) > 1$. However, $n_5(H)$ divides $|H|$, and thus it must divide 2^6. It follows that $n_5(H) = 2^4$ (since it must be congruent to 1 mod 5), and therefore 2^4 divides $|H|$.

We now know that $2^4 \cdot 5^3$ divides $|H|$ and hence $|G : H| \leq 2^2$. This is a contradiction by the "$n!$ theorem" (4.4). ∎

5C

Since the Sylow E-theorem guarantees the existence of an assortment of p-subgroups in finite groups, it is reasonable to develop some of the properties of these groups. There is a tremendous amount known about finite p-groups, and we shall consider only a few of the most important facts.

(5.20) LEMMA. *Let the finite p-group P act on the finite set Ω, and let*

$$\Omega_0 = \{\alpha \in \Omega \mid \alpha \cdot x = \alpha \text{ for all } x \in P\}.$$

Then $|\Omega| \equiv |\Omega_0| \bmod p$.

Proof. The elements of the "fixed-point"set Ω_0 are precisely those elements of Ω that lie in "trivial" (size 1) orbits. It follows that $\Omega - \Omega_0$ is a union of nontrivial orbits. Each orbit has p-power size by the FCP, since P is a p-group, and thus each of the orbits that constitute $\Omega - \Omega_0$ has size divisible by p. The result follows. ∎

(5.21) THEOREM. *Suppose that $1 < N \lhd P$, where P is a finite p-group. Then $N \cap \mathbf{Z}(P) > 1$. In particular, a nontrivial finite p-group has a nontrivial center.*

Proof. Since $N \lhd P$, we see that P acts by conjugation on N. The set of fixed points in this action is precisely $N \cap \mathbf{Z}(P)$, and hence $|N \cap \mathbf{Z}(P)| \equiv |N| \bmod p$. Since N is a nontrivial p-group, we have $|N| \equiv 0 \bmod p$, and so $|N \cap \mathbf{Z}(P)| \equiv 0 \bmod p$ and $N \cap \mathbf{Z}(P) \neq 1$. Therefore $N \cap \mathbf{Z}(P) > 1$, as claimed. The last statement is immediate when we take $N = P$. ∎

(5.22) COROLLARY.   *If P is a finite simple p-group, then $|P| = p$.*

Proof.   We have $1 < \mathbf{Z}(P) \triangleleft P$, and thus $\mathbf{Z}(P) = P$ and P is abelian. Since P is simple, this forces $|P| = p$.   ■

(5.23) COROLLARY.   *Let P be a finite nontrivial p-group. Then P has a subgroup of index p, and every such subgroup is normal.*

Proof.   Since $P > 1$, we can choose a "maximal normal subgroup" N. In other words, $N \triangleleft P, N < P$ and there does not exist any subgroup $M \triangleleft P$ such that $N < M < P$. By the Correspondence theorem (3.7), it follows that P/N is simple, and so by Corollary 5.22, $|P : N| = p$. The last assertion follows by Corollary 4.5.   ■

Given a finite p-group P, we can apply Corollary 5.23 repeatedly to find a subgroup of index p in a subgroup of index p in a subgroup of index p, and so on. It follows that P has subgroups of every possible p-power order $\leq |P|$. Combining this with the Sylow E-theorem, we obtain the following corollary.

(5.24) COROLLARY.   *Let $|G|$ be finite and suppose p^e divides $|G|$, where p is prime. Then G has a subgroup of order p^e.*   ■

As we have seen, the fact that finite nontrivial p-groups have nontrivial centers is quite powerful. Another useful fact is that "normalizers grow" in p-groups.

(5.25) THEOREM.   *Let $H < P$, where P is a finite p-group. Then $\mathbf{N}_P(H) > H$.*

Proof.   Certainly, $\mathbf{Z}(P) \subseteq \mathbf{N}_P(H)$, and so if $\mathbf{Z}(P) \not\subseteq H$, we have $\mathbf{N}_P(H) > H$, as required. We suppose then that $\mathbf{Z}(P) \subseteq H$. Now $H/\mathbf{Z}(P) < P/\mathbf{Z}(P)$, and since $\mathbf{Z}(P) > 1$, we see that $P/\mathbf{Z}(P)$ is a p-group strictly smaller than P.

If we work by induction on $|P|$, the inductive hypothesis yields that

$$\mathbf{N}_{P/\mathbf{Z}(P)}(H/\mathbf{Z}(P)) > H/\mathbf{Z}(P).$$

By the Correspondence theorem (or, more precisely, Corollary 3.8), we can write

$$\mathbf{N}_{P/\mathbf{Z}(P)}(H/\mathbf{Z}(P)) = M/\mathbf{Z}(P)$$

for some subgroup M with $\mathbf{Z}(P) \subseteq M \subseteq P$.

We have $H/\mathbf{Z}(P) \triangleleft M/\mathbf{Z}(P)$ and thus $H \triangleleft M$. Also, $H/\mathbf{Z}(P) < M/\mathbf{Z}(P)$ and this gives $H < M$. Combining these facts, we get $H < M \subseteq \mathbf{N}_P(H)$, as required.   ■

(5.26) COROLLARY.   *Let P be a finite p-group. Then every maximal subgroup of P is normal and has index p.*

Proof.   Let $H < P$ be a maximal subgroup. By Theorem 5.25, $\mathbf{N}_P(H) > H$ and thus $\mathbf{N}_P(H) = P$ and $H \triangleleft P$. Since H is a maximal subgroup of P, the

Correspondence theorem tells us that P/H has no nontrivial proper subgroups. It therefore has prime order. ∎

We mention that infinite p-groups do not, in general, enjoy the properties we have been discussing here. In particular, they can have trivial centers and proper self-normalizing subgroups.

We close our discussion of p-groups with an application of the theory to another nonsimplicity theorem.

(5.27) THEOREM. *Let $|G| = p^a q$, where p and q are primes and $a > 0$. Then G is not simple.*

Proof. We may assume that $q \neq p$ and that $n_p(G) = q$. We consider first the situation where every pair of Sylow p-subgroups has a trivial intersection. In that case, we can count elements and conclude that there are exactly $q(p^a - 1)$ nonidentity elements in the union of all of the Sylow p-subgroups of G. The set X of as yet uncounted elements contains every Sylow q-subgroup and satisfies $|X| = |G| - q(p^a - 1) = q$. It follows that $n_q(G) = 1$ and G is not simple.

We can now suppose that there exist $S, T \in \mathrm{Syl}_p(G)$ with $S \neq T$ and $S \cap T > 1$. Choose such S and T so that $S \cap T$ is as large as possible, and let $N = \mathbf{N}_G(S \cap T)$. Since $S \cap T < S$ and $S \cap T < T$, Theorem 5.25 yields that $S \cap N > S \cap T$ and $T \cap N > S \cap T$.

If N is a p-group, then by the Sylow D-theorem, $N \subseteq P$ for some choice of $P \in \mathrm{Syl}_p(G)$, and we have

$$S \cap P \supseteq S \cap N > S \cap T.$$

By the choice of S and T, the Sylow subgroups S and P are not distinct and so $S = P$. Also, $T = P$ by exactly similar reasoning, and this contradicts the fact that $S \neq T$.

We conclude that N is not a p-group and hence $q \mid |N|$. Let $Q \in \mathrm{Syl}_q(N)$, so that $|Q| = q$. Now $SQ = G$ by Problem 2.11 or Corollary 4.17, and thus if $g \in G$, we can write $g = xy$ with $x \in S$ and $y \in Q$. It follows that

$$S^g = S^{xy} = (S^x)^y = S^y \supseteq (S \cap T)^y = S \cap T,$$

where the last equality holds since $y \in Q \subseteq N = \mathbf{N}_G(S \cap T)$. We now have

$$1 < S \cap T \subseteq \bigcap_{g \in G} S^g = \mathrm{core}_G(S) \triangleleft G,$$

and G is not simple. ∎

Much more is true than is stated in Theorem 5.27. A celebrated result of W. Burnside asserts that if $|G| = p^a q^b$, where p and q are primes, then G cannot be simple unless it has prime order. There is no known proof of this "$p^a q^b$-theorem" along the lines we have been discussing in this chapter. Burnside's proof was one of the early triumphs of character theory, a powerful tool for the study of finite groups involving certain functions (called "characters") from a group into the complex numbers. It was only relatively recently that a character-free proof of Burnside's theorem was found. In Chapter 15 we present some of the basic theorems of character theory, and we give Burnside's proof of the $p^a q^b$-theorem in Chapter 29.

5D

In the previous proof, we were able to get information about $\mathrm{core}_G(S)$ (for $S \in \mathrm{Syl}_p(G)$) by considering a maximal Sylow intersection. We close this chapter with some results about minimal Sylow intersections and their connection with $\mathrm{core}_G(S)$. We know that $\mathrm{core}_G(S)$, which is the largest normal subgroup of G contained in S, is the intersection of all of the conjugates of S, that is, of all Sylow p-subgroups. The question is whether $\mathrm{core}_G(S)$ is already the intersection of just two Sylow p-subgroups of G. We shall see that this is true if S is abelian, although it does not hold in general. We mention that the standard notation for the intersection of all of the Sylow subgroups of G is $\mathbf{O}_p(G)$. It is easy to see that this is the unique largest p-subgroup normal in the finite group G.

(5.28) THEOREM (Brodkey). *Let G be finite and assume $S \in \mathrm{Syl}_p(G)$ is abelian. Then there exists $T \in \mathrm{Syl}_p(G)$ with $S \cap T = \mathrm{core}_G(S)$.*

There is a result along the lines of Theorem 5.28 that does work in general, and we state and prove that first.

(5.29) THEOREM. *Let G be finite and let $S \in \mathrm{Syl}_p(G)$. Choose $T \in \mathrm{Syl}_p(G)$ such that $T \cap S$ is minimal, and suppose $N \subseteq T \cap S$ satisfies $N \triangleleft S$ and $N \triangleleft T$. Then $N \subseteq \mathrm{core}_G(S)$.*

Proof. Since

$$\mathrm{core}_G(S) = \bigcap_{g \in G} S^g = \bigcap \mathrm{Syl}_p(G),$$

we need to show that $N \subseteq P$ for all $P \in \mathrm{Syl}_p(G)$. Fix some $P \in \mathrm{Syl}_p(G)$.

Let $M = \mathbf{N}_G(N)$ and note that $T \subseteq M$, and so $T \in \mathrm{Syl}_p(M)$ and we have $(P \cap M)^m \subseteq T$ for some $m \in M$ by Theorem 5.8. Thus $P^m \cap M \subseteq T$, and since $S \subseteq M$, we conclude that

$$P^m \cap S = P^m \cap M \cap S \subseteq T \cap S.$$

By the choice of T, we know that $S \cap T$ is minimal among intersections of two Sylow p-subgroups of G, and thus we cannot have $P^m \cap S < T \cap S$. It follows that $P^m \cap S = T \cap S \supseteq N$, and so $N \subseteq P^m$. Finally, since $m \in M = \mathbf{N}_G(N)$, this yields

$$N = N^{m^{-1}} \subseteq P,$$

as required.   ∎

Proof of Theorem 5.28.   Choose $T \in \mathrm{Syl}_p(G)$ so that $S \cap T$ is minimal and let $N = S \cap T$. Now $S \cong T$ since S and T are conjugate by the Sylow C-theorem, and so both S and T are abelian and $N \lhd S$ and $N \lhd T$.

By Theorem 5.29, $N \subseteq \mathrm{core}_G(S)$. However, since T is one of the conjugates of S, we have $\mathrm{core}_G(S) \subseteq S \cap T = N$. It follows that $N = \mathrm{core}_G(S) \lhd G$. ∎

Problems

5.1 Let $S \in \mathrm{Syl}_p(G)$ and $N \lhd G$. Show that $S \cap N \in \mathrm{Syl}_p(N)$. In particular, if N is a p-group, then $N \subseteq S$.

  NOTE: By Problem 5.1, every normal p-subgroup of G is contained in $\mathrm{core}_G(S)$, which is therefore the unique largest normal p-subgroup of G. As we mentioned earlier, it is denoted $\mathbf{O}_p(G)$.

5.2 Let G be finite and suppose $\varphi : G \to H$ is a surjective homomorphism.

  a. If $P \in \mathrm{Syl}_p(G)$, show that $\varphi(P) \in \mathrm{Syl}_p(H)$.
  b. If $Q \in \mathrm{Syl}_p(H)$, show that $Q = \varphi(P)$ for some $P \in \mathrm{Syl}_p(G)$.
  c. Show that $n_p(H) \leq n_p(G)$.

5.3 Let $H \subseteq G$ with G finite. Show that $n_p(H) \leq n_p(G)$.

5.4 Let G be finite and suppose $H \subseteq G$ satisfies the condition that $\mathbf{C}_G(x) \subseteq H$ for all $x \in H - \{1\}$. Show that $\gcd(|H|, |G : H|) = 1$.

  HINT: Choose $P \in \mathrm{Syl}_p(H)$ and show that $P \in \mathrm{Syl}_p(G)$.

  NOTE: A subgroup of G with relatively prime order and index is called a *Hall subgroup* of G.

5.5 Let $H \subseteq G$ with G finite, and suppose $P \in \mathrm{Syl}_p(H)$. If $\mathbf{N}_G(P) \subseteq H$, show that $P \in \mathrm{Syl}_p(G)$.

5.6 Let G be finite with $G > 1$, and suppose $P \subseteq \mathrm{Aut}(G)$ is a p-subgroup. Show that there exists some nontrivial Sylow q-subgroup Q of G (for some prime q) such that $\sigma(Q) = Q$ for all $\sigma \in P$.

  HINT: Consider separately the cases $p \mid |G|$ and $p \nmid |G|$.

5.7 Let $|G| = p^2 q^2$ where $p > q$ are primes. If $|G| \neq 36$, show that G has a normal Sylow p-subgroup.

5.8 Let $P \in \mathrm{Syl}_p(G)$. Show that

$$\mathbf{N}_G(\mathbf{N}_G(P)) = \mathbf{N}_G(P).$$

5.9 If $|G| = pqr$ where p, q and r are primes, show that G is not simple.

5.10 If $|G| \leq 100$ and G is nonabelian and simple, show that $|G| = 60$.

  NOTE: There really does exist a simple group of order 60.

5.11 If G is simple of order 60, show that G is isomorphic to a subgroup of Sym(5). Conclude that if G_1 and G_2 are both simple of order 60, then $G_1 \cong G_2$.

HINT: Show that G has a subgroup of index 5. If $n_2(G) = 5$, this is easy, so assume that $n_2(G) = 15$. Show that there exists an intersection $D = S \cap T$ of two Sylow 2-subgroups with $|D| = 2$. Consider $\mathbf{N}_G(D)$.

5.12 If $|G| = 280$, show that G is not simple.

5.13 Given primes p and q with q dividing $p - 1$, construct a nonabelian group of order pq as follows. Let P have order p and let $Q \subseteq \mathrm{Aut}(P)$ have order q. Let $G \subseteq \mathrm{Sym}(P)$ be the set of all maps $\varphi_{a,\sigma}$ defined by

$$(x)\varphi_{a,\sigma} = (xa)\sigma$$

for $a \in P$ and $\sigma \in Q$. Show that G is a nonabelian group of order pq.

5.14 Let $|G| = p(p + 1)$, where p is prime. Show that G has either a normal subgroup of order p or one of order $p + 1$.

HINT: If $n_p(G) > 1$, let $x \in G$ with $o(x) \neq 1, p$. Show that $|\mathbf{C}_G(x)| = p + 1$. Count elements.

5.15 A subgroup $X \subseteq G$ is said to be a *trivial intersection set*, or *T.I.-set*, if for each $g \in G$, either $X^g = X$ or $X^g \cap X = 1$. Let $P \in \mathrm{Syl}_p(G)$ and suppose P is a T.I.-set in G. Let $H \subseteq G$ and $Q \in \mathrm{Syl}_p(H)$. Show that Q is a T.I.-set in H.

5.16 Let G be finite. Write $f_n(G) = |\{x \in G \mid x^n = 1\}|$ for positive integers n. A theorem of Frobenius asserts that if $n \big| |G|$, then $n \big| f_n(G)$. Prove this theorem in the case that n is prime.

HINT: Let $S \in \mathrm{Syl}_n(G)$ and $\Omega = \{x \in G \mid x^n = 1\}$. Let S act on Ω by conjugation and note that $\Omega \cap \mathbf{C}_G(S)$ is a subgroup of S.

5.17 Let $|G| = p^a(kp+1)$, where p is prime and $0 < k \leq p+1$. Let $S \in \mathrm{Syl}_p(G)$. Show that either S is a maximal subgroup of G or $S \triangleleft G$.

HINT: If $r \equiv 1 \equiv s \bmod p$ and $r > 1, s > 1$, then $rs \geq (p + 1)^2$.

5.18 Let $|P| = p^2$, where p is prime. Prove that P is abelian.

5.19 Let G be a finite group in which every Sylow subgroup is normal.

a. If $P \in \mathrm{Syl}_p(G)$, show that $\mathbf{Z}(P) \subseteq \mathbf{Z}(G)$.
b. If $1 < N \triangleleft G$, show that $N \cap \mathbf{Z}(G) > 1$.

NOTE: A group satisfying the hypothesis of Problem 5.19 is said to be *nilpotent*. It is also possible to define infinite nilpotent groups, and we do so in Chapter 8. Part (b) of this problem remains true even in the infinite case.

5.20 Let P be a finite p-group, and suppose $A \triangleleft P$ is maximal among abelian normal subgroups of P. Show that $A = \mathbf{C}_P(A)$.

HINT: Write $C = \mathbf{C}_P(A)$ and assume that $C > A$. Show that C/A contains a subgroup B/A normal in P/A and of order p. Prove that B is abelian.

CHAPTER SIX

Permutation Groups

6A

The goal of this chapter is to study a particular family of groups and certain subgroups. Recall that if X is any nonempty set, then $\mathrm{Sym}(X)$ denotes the symmetric group on X, the group of all permutations of this set. We shall assume throughout this chapter that X is finite. Of course, the group $\mathrm{Sym}(X)$ is determined up to isomorphism by the cardinality $|X| = n$ of X, and so we lose little if we limit our attention to sets of the form $X = \{1, 2, 3, \ldots, n\}$. In this case, one often writes S_n to denote $\mathrm{Sym}(X)$. Note that $|\mathrm{Sym}(X)| = n!$ if $|X| = n$. We remind the reader that here, as throughout this book, function composition is interpreted left to right. If $\sigma, \tau \in \mathrm{Sym}(X)$, therefore, the element $\sigma\tau$ is the result of doing σ first and then τ.

We begin with a description of the so-called cycle notation for elements of $\mathrm{Sym}(X)$. Let $\alpha_1, \alpha_2, \ldots, \alpha_m$ be any m distinct elements of X. We write $g = (\alpha_1, \alpha_2, \ldots, \alpha_m)$ to denote the permutation defined by

1. $\alpha_i g = \alpha_{i+1}$ for $1 \le i < m$,
2. $\alpha_m g = \alpha_1$ and
3. $\beta g = \beta$ for $\beta \in X - \{\alpha_1, \ldots, \alpha_m\}$.

A permutation of this form is called an *m-cycle* (or simply a *cycle*). Note that the identity is the only 1-cycle and that an m-cycle for $m > 1$ can be written in precisely m different ways, since any of the α_i can occur as the leftmost entry. (It follows that if $|X| = n$ and $m > 1$, then the number of m-cycles in $\mathrm{Sym}(X)$ is exactly $n(n-1)(n-2)\cdots(n-m+1)/m$.)

Two permutations are *disjoint* if there is no element moved by both of them. The cycles $(\alpha_1, \ldots, \alpha_m)$ and $(\beta_1, \ldots, \beta_k)$ with $m, k \ge 2$ are disjoint, therefore, precisely when none of the α_i is equal to any of the β_j. It should be clear that disjoint permutations necessarily commute.

The following lemma is essentially a triviality and we omit the proof.

(6.1) LEMMA. *Let $g \in \text{Sym}(X)$ and let $Y \subseteq X$ be an orbit for the action of $\langle g \rangle$ on X. Let $\alpha \in Y$ and write $|Y| = m$. Then*

a. *$\alpha, \alpha g, \ldots, \alpha g^{m-1}$ are all distinct,*
b. *$\alpha g^m = \alpha$ and*
c. *$Y = \{\alpha g^i \mid 0 \le i < m\}$.*   ∎

In the situation of Lemma 6.1, we write g_Y to denote the permutation that fixes each element of $X - Y$ and acts on the set Y exactly as g does. It follows that g_Y is the m-cycle $(\alpha, \alpha g, \ldots, \alpha g^{m-1})$. Note that the cycles g_Y are disjoint for distinct orbits Y for $\langle g \rangle$.

(6.2) THEOREM. *Let $g \in \text{Sym}(X)$ and let $\mathcal{Y}$ be the set of nontrivial $\langle g \rangle$-orbits on X. Then*

$$g = \prod_{Y \in \mathcal{Y}} g_Y \,,$$

and this is the only way that g can be written as a product of disjoint nontrivial cycles.

Note that since the g_Y commute with each other, their product is well defined. Also, as is usual, an empty product is defined to be the identity.

Proof of Theorem 6.2.   Write $\pi = \prod g_Y$ and let $\alpha \in X$. If g fixes α, then α lies in no $Y \in \mathcal{Y}$, and therefore every g_Y fixes α and $\alpha \pi = \alpha = \alpha g$. If g moves α, however, then $\alpha \in Y$ for some $Y \in \mathcal{Y}$. Then $\alpha g_Y = \alpha g$, and both α and αg are fixed by all other factors of π. In this case, too, $\alpha \pi = \alpha g$ and thus $\pi = g$.

Now suppose $g = \prod c$, where c runs over some set C of disjoint nontrivial cycles. If $Y \in \mathcal{Y}$, choose $\alpha \in Y$. Since g moves α, so too must some $c \in C$. All the other members of C therefore act trivially on the entire $\langle c \rangle$-orbit containing α, and thus g acts on this orbit exactly as c does. It follows that the $\langle c \rangle$-orbit containing α is Y and that $c = g_Y$. This shows that $\{g_Y \mid Y \in \mathcal{Y}\} \subseteq C$.

If $c \in C$, then c moves some $\alpha \in X$ and so g moves α also, and $\alpha \in Y$ for some $Y \in \mathcal{Y}$. Since c and g_Y both lie in C and move α, we conclude that $c = g_Y$.   ∎

Note that there is no uniqueness asserted if a permutation is written as a product of cycles that are not disjoint. For instance, let us compute $g = (2, 3, 1)(3, 5)(2, 1, 7, 4)(3, 4)$ in S_8. Remembering to work from left to right, we have the following:

$$g : \begin{pmatrix} 1 & 2 & 3 & 4 & 5 & 6 & 7 & 8 \\ \downarrow & \downarrow & \downarrow & \downarrow & \downarrow & \downarrow & \downarrow & \downarrow \\ 1 & 5 & 7 & 2 & 4 & 6 & 3 & 8 \end{pmatrix} .$$

We can use the idea of the proof of Theorem 6.2 to write g as a product of disjoint cycles. The first moved point is 2 and g carries $2 \mapsto 5 \mapsto 4 \mapsto 2$. The next unused moved point is 3 and g takes $3 \mapsto 7 \mapsto 3$. All remaining points are fixed, and so $g = (2, 5, 4)(3, 7)$.

By the *cycle structure* of an element g, we mean the data giving the lengths of disjoint cycles whose product is g and the number of cycles of each length. More precisely, if $g \in \text{Sym}(X)$, the *cycle structure function* of g is the function s, defined on positive integers, such that $s(m)$ is the number of orbits of $\langle g \rangle$ on X that have size m. Thus, $s(1)$ is the number of fixed points of g, and for $m \geq 2$, $s(m)$ is the number of m-cycles that appear in the disjoint cycle decomposition for g. A common shorthand is to write, for instance, that g has cycle structure $1^3 \cdot 3^2 \cdot 5$ to mean that $s(1) = 3$, $s(3) = 2$, $s(5) = 1$ and all other values of s are zero. In this case, $|X| = 3(1) + 2(3) + 5 = 14$, and g can be written as a product of three disjoint cycles. As we shall see, a great deal of information about an element of $\text{Sym}(X)$ can be recovered from its cycle structure.

(6.3) LEMMA. *Let $g \in \text{Sym}(X)$. Then $o(g)$ is the least common multiple of the sizes of the cycles that appear in its disjoint cycle decomposition.*

Proof.   Write $g = \prod c$, where c runs over C, a set of disjoint cycles. Since the elements of C commute pairwise, we have $g^n = \prod c^n$ for integers n. If $g^n = 1$ and $c \in C$, we claim that $c^n = 1$. To see this, let $\alpha \in X$. If c fixes α, then certainly c^n also fixes α. If C moves α, however, then all $b \in C$ with $b \neq c$ fix α, and thus b^n fixes α for all such b. It follows that $\alpha = \alpha g^n = \alpha c^n$, and therefore c^n fixes every $\alpha \in X$. Thus $c^n = 1$, as claimed.

Since an m-cycle clearly has order m, we see that $g^n = 1$ iff n is a multiple of the sizes of every cycle in C. The result follows.   ∎

Although what we have done with symmetric groups so far is relatively trivial, there are interesting applications. In Example 5.18, for instance, we showed that there is no simple group of order $2376 = 2^3 \cdot 3^3 \cdot 11$. We can now simplify that proof considerably. If G is simple of this order, we showed that $n_{11}(G) = 12$ and that $9 \big| |C_G(S)|$, where $S \in \text{Syl}_{11}(G)$. It follows from the latter that G contains elements of order 3 and 11 that commute, and so G contains an element of order 33. On the other hand, since $|G : N_G(S)| = 12$, the action of G on the right cosets of $N_G(S)$ induces a nontrivial homomorphism of G into the symmetric group S_{12}. Since G is simple, this homomorphism has a trivial kernel, and so G is isomorphic to a subgroup of S_{12}, which therefore must contain some element g of order 33. What can the cycle structure of g be? Since the least common multiple of the cycle sizes is 33, there must be either a 33-cycle or both a 3-cycle and an 11-cycle. Since only 12 points are being permuted, neither possibility can occur. This contradiction shows that G cannot be simple.

Next, we discuss the conjugacy of elements in $\text{Sym}(X)$.

(6.4) LEMMA. *Let $g = (\alpha_1, \alpha_2, \ldots, \alpha_m)$ be an m-cycle in $\text{Sym}(X)$. If $h \in \text{Sym}(X)$, then g^h is an m-cycle. We have*

$$g^h = (\alpha_1 h, \ldots, \alpha_m h).$$

Proof.   Note that the $\alpha_i h$ are all distinct, and so $(\alpha_1 h, \ldots, \alpha_m h)$ really is an m-cycle. We need to show that $(\alpha_i h) g^h = \alpha_{i+1} h$ for $1 \leq i < m$, that $(\alpha_m h) g^h = \alpha_1 h$ and that $\beta g^h = \beta$ if β is not of the form $\alpha_i h$ for any i. Since $hg^h = gh$, the first two conditions are immediate, and if β is not of the form $\alpha_i h$, then βh^{-1} is fixed by g, and so $\beta g^h = \beta h^{-1}gh = \beta h^{-1}h = \beta$, as required.   ∎

(6.5) THEOREM.   *Two elements of* $\mathrm{Sym}(X)$ *are conjugate iff they have the same cycle structure.*

Proof.   Given $g \in \mathrm{Sym}(X)$, write $g = \prod c$, where c runs over a set of disjoint nontrivial cycles. If $h \in \mathrm{Sym}(X)$, then $g^h = \prod c^h$, and it follows by Lemma 6.4 that the c^h are disjoint nontrivial cycles and that c^h has the same size as c. Therefore, g and g^h have the same cycle structure.

Conversely, suppose $u, v \in \mathrm{Sym}(X)$ have the same cycle structure. We can then write

$$u = \prod_{i=1}^{t} c_i \quad \text{and} \quad v = \prod_{i=1}^{t} d_i ,$$

where the c_i are disjoint nontrivial cycles as are the d_i, and they are numbered so that c_i and d_i have equal size m_i. For each c_i and d_i, choose a particular starting point of the cycle so that for each j with $1 \leq j \leq m_i$, we can speak unambiguously of the jth entry in c_i and in d_i. Also, u and v each have exactly $|X| - \sum m_i$ fixed points, and we choose a numbering for these so that we can refer to the jth fixed point of u or v when $1 \leq j \leq |X| - \sum m_i$.

We now construct a map h on X as follows. For $\alpha \in X$, either α is the jth entry in some c_i for some j, or α is the jth fixed point of u for some j. Define αh to be either the jth entry in d_i or the jth fixed point of v, respectively. It should now be clear that $h \in \mathrm{Sym}(X)$ and $u^h = v$.   ∎

For example, if $u = (1\ 3)(2\ 4\ 5\ 6)$ and $v = (3\ 7)(1\ 8\ 2\ 4)$ in S_8, then there are 16 possible choices for h following the recipe in the preceding proof. One of these is $h = (1\ 3\ 7\ 5\ 2)(4\ 8\ 6)$. The reader should check that $u^h = v$.

We can now compute the conjugacy classes of S_n fairly easily. If $n = 5$, for instance, we see that there are two classes of involutions, with cycle structures $1^3 \cdot 2$ and $1 \cdot 2^2$; there is one class of elements of order 3, with cycle structure $1^2 \cdot 3$; one class each of elements of orders 4, 5 and 6, with cycle structures $1 \cdot 4$, 5 and $2 \cdot 3$, respectively; and of course, the identity, which has cycle structure 1^5.

In general, the number of classes of S_n is exactly $p(n)$, the number of "partitions" of n. This is the number of ways n can be written as a sum of positive integers (not counting order).

(6.6) COROLLARY.   *If* $n \geq 3$, *then* $\mathbf{Z}(S_n) = 1$.

Proof.   Let $z \in \mathbf{Z}(S_n)$, so that $z = z^g$ for all $g \in S_n$. If $z \neq 1$, then by renumbering the points if necessary, we can write $z = (1, 2, \ldots)\ldots$, and we have $z^{(2,3)} = (1, 3, \ldots)\ldots$. Since z carries 1 to 2 but $z^{(2,3)}$ carries 1 to 3, we see that $z \neq z^{(2,3)}$. This contradiction proves that $z = 1$.   ∎

As we shall see in Section 6C, not only does S_n have a trivial center, but it has hardly any nontrivial normal subgroups at all.

6B

The 2-cycles in $\mathrm{Sym}(X)$ are of special importance, as we shall see. These elements (which constitute a conjugacy class) are called *transpositions*. Of course, the nonempty products of disjoint transpositions are precisely the involutions of $\mathrm{Sym}(X)$, but in fact, every element of $\mathrm{Sym}(X)$ can be written as a product of transpositions.

(6.7) LEMMA. *Every element of* $\mathrm{Sym}(X)$ *is a product of transpositions, and an* m-*cycle is a product of* $m - 1$ *transpositions.*

Proof.   Since every element of $\mathrm{Sym}(X)$ is a product of cycles, it suffices to prove the last statement. For that, it is enough to observe that

$$(\alpha_1, \alpha_2, \ldots, \alpha_m) = (\alpha_1, \alpha_2)(\alpha_1, \alpha_3)(\alpha_1, \alpha_4) \cdots (\alpha_1, \alpha_m). \qquad \blacksquare$$

We say that a permutation $g \in \mathrm{Sym}(X)$ is *even* if it can be written as a product of an even number of transpositions, and g is *odd* if it is a product of some odd number of transpositions. By Lemma 6.7, an m-cycle is even if m is odd and it is odd if m is even. Also, Lemma 6.7 tells us that every element $g \in \mathrm{Sym}(X)$ is either even or odd. It is far from obvious, however, that no permutation can be both even and odd.

(6.8) THEOREM.   *No element of* $\mathrm{Sym}(X)$ *can be written both as a product of an even number of transpositions and as a product of an odd number of transpositions.*

To prove this crucial result, we define an invariant $c(g)$ for $g \in \mathrm{Sym}(X)$ by setting $c(g)$ to be the number of $\langle g \rangle$-orbits into which X is decomposed. Thus $c(1) = |X|$, and in general, $c(g) = \sum s(m)$, where s is the cycle structure function for g, and we sum over all positive integers m.

(6.9) LEMMA.   *Let* $g \in \mathrm{Sym}(X)$ *and let* $t = (\alpha, \beta)$, *a transposition in* $\mathrm{Sym}(X)$. *Then*

    a. $c(gt) = c(g) + 1$ *if* α *and* β *lie in the same* $\langle g \rangle$-*orbit and*
    b. $c(gt) = c(g) - 1$ *if* α *and* β *lie in different* $\langle g \rangle$-*orbits.*

Proof.   Suppose α and β lie in the same $\langle g \rangle$-orbit Λ. We can then write

$$g_\Lambda = (\alpha_1, \alpha_2, \ldots, \alpha_m)$$

with $\alpha_1 = \alpha$ and $\alpha_k = \beta$ for some k such that $2 \le k \le m$. We compute

$$g_\Lambda t = (\alpha_1, \ldots, \alpha_m)(\alpha_1, \alpha_k) = (\alpha_1, \ldots, \alpha_{k-1})(\alpha_k, \alpha_{k+1}, \ldots, \alpha_m) \,.$$

Since g and gt act identically on every $\langle g \rangle$-orbit other than Λ, all of these are also $\langle gt \rangle$-orbits. On Λ, however, gt and $g_\Lambda t$ act identically, and we see that Λ decomposes into two $\langle gt \rangle$-orbits. It follows that $c(gt) = c(g) + 1$.

Now suppose that $\Lambda \neq \Delta$ are $\langle g \rangle$-orbits (possibly of size 1) and that $\alpha \in \Lambda$ and $\beta \in \Delta$. We can write

$$g_\Lambda = (\alpha_1, \ldots, \alpha_m) \quad \text{and} \quad g_\Delta = (\beta_1, \ldots, \beta_n)$$

with $\alpha_1 = \alpha$ and $\beta_1 = \beta$. Then

$$g_\Lambda g_\Delta t = (\alpha_1, \ldots, \alpha_m)(\beta_1, \ldots, \beta_n)(\alpha_1, \beta_1) = (\alpha_1, \ldots, \alpha_m, \beta_1, \ldots, \beta_n) \, .$$

The actions of g and gt are identical on all $\langle g \rangle$-orbits other than Λ and Δ. On $\Lambda \cup \Delta$, the action of gt is the same as that of $g_\Lambda g_\Delta t$, and it follows that $\Lambda \cup \Delta$ is a $\langle gt \rangle$-orbit and $c(gt) = g(g) - 1$.   ∎

Proof of Theorem 6.8.   By Lemma 6.9, we have $c(gt) \equiv c(g) + 1 \mod 2$ for transpositions t. By repeated application of this, we have

$$c(t_1 t_2 \cdots t_m) \equiv c(1) + m = |X| + m \quad \mod 2 \, ,$$

where all t_i are transpositions. It follows that a permutation $g \in \mathrm{Sym}(X)$ is even iff $c(g) \equiv |X| \mod 2$ and g is odd iff $c(g) \equiv |X| + 1 \mod 2$. It is not possible for both of these congruences to hold simultaneously.   ∎

(6.10) COROLLARY.   *Let $|X| > 1$. Then the set of even permutations in $\mathrm{Sym}(X)$ forms a normal subgroup of index 2.*

Proof.   It is clear that a product of two even permutations is even, and so the set A of even permutations of X is closed under multiplication and is certainly nonempty. The inverse of $g = t_1 t_2 \cdots t_m$ is $t_m t_{m-1} \cdots t_1$ if the t_i are transpositions, and so g and g^{-1} have the same parity. It follows that A is closed under inverses and under conjugation by arbitrary elements $g \in \mathrm{Sym}(X)$. Thus, A is a normal subgroup of $\mathrm{Sym}(X)$.

Let $t \in \mathrm{Sym}(X)$ be any transposition. If $g \in \mathrm{Sym}(X)$ is odd, then $gt \in A$ and so $g \in At$. It follows that $\mathrm{Sym}(X) = A \cup At$, and since $t \notin A$, we have $|\mathrm{Sym}(X) : A| = 2$.   ∎

The subgroup of $\mathrm{Sym}(X)$ consisting of even permutations is called the *alternating group* on X and is denoted $\mathrm{Alt}(X)$. If $X = \{1, 2, \ldots, n\}$, we usually write $\mathrm{Alt}(X) = A_n$.

As an application of Corollary 6.10, we have the following nonsimplicity criterion.

(6.11) COROLLARY.   *Let G act on a finite set Ω, and assume that some element $g \in G$ induces an odd permutation on Ω. Then there exists $A \triangleleft G$ with $|G : A| = 2$ and $g \notin A$.*

Proof.   By Lemma 4.2, we have a homomorphism $\theta : G \to \mathrm{Sym}(\Omega)$ for which $\theta(g) \notin \mathrm{Alt}(\Omega)$. Composition of θ with the canonical homomorphism $\mathrm{Sym}(\Omega) \to \mathrm{Sym}(\Omega)/\mathrm{Alt}(\Omega)$ yields a homomorphism from G onto a group of order 2. The element g is not in the kernel A of this homomorphism, and the result follows.   ∎

This gives us a useful result concerning groups of specific orders.

(6.12) COROLLARY.   *Let $|G| = 2m$, where m is odd. Then G has a normal subgroup of order m. In particular, if $m > 1$, then G is not simple.*

Proof.   Let $g \in G$ have order 2. (Such an element necessarily exists by Cauchy's theorem.) Let G act on itself by right multiplication and note that since $g \neq 1$, the element g has no fixed points in this action. Since $g^2 = 1$, the cycle structure of the permutation induced by g consists of only 2-cycles and 1-cycles, but since there are no fixed points, we have only 2-cycles. It follows that this permutation is a product of m disjoint transpositions and hence is odd. The result now follows by Corollary 6.11.   ∎

To obtain other consequences of Corollary 6.11, we need an efficient method of establishing the parity of a permutation.

(6.13) LEMMA.   *A permutation $g \in \mathrm{Sym}(X)$ is even iff the number of even length cycles in its cycle structure is even.*

Proof.   Write $g = \prod c$, a product of nontrivial disjoint cycles. Let $m(c)$ denote the length of the cycle c, and recall that by Lemma 6.7, c can be written as a product of $m(c) - 1$ transpositions. It follows that g is even iff there are an even number of cycles c for which $m(c) - 1$ is odd. This happens iff the number of even $m(c)$ is even.   ∎

(6.14) EXAMPLE.   *If $|G| = 120 = 2^3 \cdot 3 \cdot 5$, then G is not simple.*

Proof.   If G is simple, then necessarily $n_5(G) = 6$ and G acts on the six right cosets of a Sylow 5-normalizer. This induces a homomorphism from G into S_6, and by Corollary 6.11, the image actually lies in the alternating subgroup A_6.

The usual arguments (as in Chapter 5) yield that $n_3(G) = 10$, and so $|N_G(T)| = 12$, where $T \in \mathrm{Syl}_3(G)$. Since $|\mathrm{Aut}(T)| = 2$, Problem 3.9 gives $|N_G(T) : C_G(T)| \leq 2$, and hence $|C_G(T)|$ is even. It follows that G contains an element of order 6.

By the assumed simplicity of G, our homomorphism of G into A_6 is an isomorphism (into) and hence A_6 contains an element g of order 6. The cycle structure of g must include either a 6-cycle or a 2-cycle and a 3-cycle. Since we have only six points to work with, the only possible cycle structures are 6 and $2 \cdot 3$. In each of these, there is just one even cycle, and so g is odd by Lemma 6.13. This is a contradiction.   ∎

As another application of Lemma 6.9, we can show that an m-cycle cannot be written as a product of fewer than $m - 1$ transpositions. More precisely, we have the following corollary.

(6.15) COROLLARY. *Let $g \in \mathrm{Sym}(X)$ be an m-cycle, and suppose $g = t_1 t_2 \cdots t_k$, where the t_i are transpositions. Then $k \geq m - 1$.*

Proof. By Lemma 6.9, we have

$$c(g) = c(t_1 t_2 \cdots t_k) \geq c(1) - k = |X| - k.$$

Since g is an m-cycle, we see that $c(g) = |X| - m + 1$, and this yields $k \geq m - 1$. ∎

As we shall see in Problem 6.9, Corollary 6.15 can be strengthened. In fact, there are at least $m - 1$ *different* transpositions among the t_i.

We give one further result about writing permutations in terms of transpositions.

(6.16) LEMMA. *The group S_n is generated by the $n - 1$ transpositions $(1, 2)$, $(2, 3)$, $(3, 4)$, ..., $(n - 1, n)$.*

Proof. By Lemma 6.7, it suffices to show that an arbitrary transposition (i, j) lies in the group generated by the given transpositions. We may clearly assume that $j \geq i + 2$. By Lemma 6.4 we compute that

$$(i, i + 1)^{(i+1,i+2)(i+2,i+3)\cdots(j-1,j)} = (i, j),$$

and this completes the proof. ∎

6C

We have given a number of different conditions sufficient to guarantee that a finite group is not simple, but the only simple groups we have seen so far are the relatively uninteresting cyclic groups of prime order. This situation will now be remedied.

(6.17) THEOREM. *The alternating group A_n is simple for $n \geq 5$.*

We need an easy lemma about transitive group actions. (Note that the natural action of A_n on the set $\{1, 2, \ldots, n\}$ is always transitive for $n \geq 3$.)

(6.18) LEMMA. *Let G act transitively on some set Ω. Then the point stabilizers G_α for elements $\alpha \in \Omega$ form a conjugacy class of subgroups of G.*

Proof. Let $\alpha \in \Omega$ and $g \in G$. We claim that $(G_\alpha)^g = G_\beta$, where $\beta = \alpha \cdot g$. To see this we must show that $x \in G$ fixes α iff x^g fixes β. Now

$$\beta \cdot x^g = (\alpha \cdot g) \cdot x^g = \alpha \cdot (xg) = (\alpha \cdot x) \cdot g,$$

and this equals $\beta = \alpha \cdot g$ iff $\alpha \cdot x = \alpha$, as claimed.

It follows that every conjugate of G_α is also a point stabilizer. Also, if α and β are any two elements of Ω, then since G acts transitively, we can find $g \in G$ such that $\alpha \cdot g = \beta$, and thus $(G_\alpha)^g = G_\beta$. Thus, every two point stabilizers are conjugate. ∎

Proof of Theorem 6.17.   We begin by considering the case $n = 5$. The cycle structures of the elements of A_5 are 1^5, $1 \cdot 2^2$, $1^2 \cdot 3$ and 5, and it is easy to count that the numbers of elements with these structures are 1, 15, 20 and 24, respectively. (As a check, note that these numbers total $60 = (1/2)5! = |A_5|$.) Now let $N \triangleleft A_5$ be a proper normal subgroup. We will prove that A_5 is simple by showing that $N = 1$.

Suppose that $3 \mid |N|$. Then N contains a Sylow 3-subgroup of A_5, and hence by the Sylow C-theorem, it contains all Sylow 3-subgroups. It follows that N contains all 20 elements of order 3. Thus $|N| > 20$, and since $|N|$ divides 60, we must have $|N| = 30$. Similarly, if we assume that $5 \mid |N|$, we deduce that N contains all 24 elements of order 5 and $|N| = 30$.

By the previous paragraph, if $|N|$ is divisible by either 3 or 5, then $|N| = 30$ and N contains 20 elements of order 3 and 24 elements of order 5. Since this is impossible, we deduce that N is a 2-group. If $|N| = 4$, then N would be the unique Sylow 2-subgroup of A_5 and so would contain all 15 elements of order 2. This is impossible, and so we consider the case where $|N| = 2$. Let $x \in N$ have order 2, and note that x has a fixed point. Thus $N = \{1, x\}$ is contained in some point stabilizer G_α, where $\alpha \in \{1, 2, 3, 4, 5\}$. Since N is normal, it follows by Lemma 6.18 that N is contained in every point stabilizer. This is a contradiction, however, since the element x clearly does not fix every point, and we have $|N| = 1$, as desired.

We now use induction on n to prove the simplicity of A_n for $n \geq 6$. Suppose $N \triangleleft A_n$ is proper and we know that A_{n-1} is simple. Let $H \subseteq A_n$ be the stabilizer of any one of the points $\alpha \in \{1, 2, \ldots, n\}$. Then $H \cong A_{n-1}$ and so is simple. Since $N \cap H \triangleleft H$, we have either $N \cap H = 1$ or $N \cap H = H$. If the latter holds, then $H \subseteq N$, and since N is normal, it follows that $N \supseteq H^g$ for all $g \in A_n$. By Lemma 6.18, the conjugates H^g are exactly all of the point stabilizers in A_n, and it follows in this case that N contains every element of A_n that fixes any point of $\{1, \ldots, n\}$. In particular, N contains every product of two transpositions and so is all of A_n, a contradiction.

We now know that the identity is the only element of N that fixes any points. If $N > 1$, let $1 \neq x \in N$. Either the cycle structure of x contains an m-cycle for $m \geq 3$ or it consists entirely of 2-cycles. By renumbering the points, therefore, we may assume that x has either the form $(1, 2)(3, 4) \cdots$ or $(1, 2, 3, \ldots) \cdots$. Let $y = x^{(3,5,6)} \in N$. Then $y = (1, 2)(5, 4) \cdots$ or $y = (1, 2, 5, \ldots) \cdots$, and in either case $x \neq y$ and yet xy^{-1} fixes the point 1. Since $xy^{-1} \in N$, this is a contradiction and proves the theorem. ∎

We mention that $|A_4| = 12$ and so A_4 is definitely not simple. In fact, the subgroup K consisting of the identity and the three elements with cycle structure 2^2 is a normal subgroup of A_4 often called the *Klein* group.

It is amusing that the fact that A_6 is simple can be used to provide a nonsimplicity proof: an easier proof of Example 6.14. If G is simple of order 120, then $n_5(G) = 6$, and this yields an isomorphic copy $G_0 \subseteq A_6$ of G. However, $|A_6 : G_0| = 3$, and this contradicts the simplicity of A_6 by the $n!$-theorem.

(6.19) COROLLARY. *For $n \geq 5$, the only normal subgroups of S_n are 1, A_n and S_n.*

Proof. Let $N \lhd S_n$. Then $N \cap A_n \lhd A_n$, and so either $A_n \subseteq N$ or $A_n \cap N = 1$. In the former case, of course, we must have $N = A_n$ or $N = S_n$. What remains is to show that if $A_n \cap N = 1$, then $N = 1$.

We certainly have $|N| \leq |S_n : A_n| = 2$ in this case. If $|N| = 2$, then N is central in S_n, and this cannot happen by Corollary 6.6. ∎

6D

Finally, we consider the automorphisms of S_n.

(6.20) THEOREM. *Let $n \neq 6$. Then every automorphism of S_n is inner.*

Proof. We may suppose that $n \geq 2$. Suppose $t \in S_n$ has order 2. Its cycle structure, therefore, consists of k 2-cycles and $n - 2k$ fixed points for some integer k with $1 \leq k \leq n/2$. We compute the size of the conjugacy class of t in terms of k and n. We need to choose k disjoint transpositions. There are $n(n-1)/2$ ways to select the first one and $(n-2)(n-3)/2$ possibilities for the next. Continuing in this way, we see that there are $n!/(2^k(n-2k)!)$ ways to select ordered sets of k disjoint transpositions in S_n, and it follows that

$$|\mathrm{cl}(t)| = \frac{n!}{2^k(n-2k)!k!}.$$

If $k = 1$, so that t is a transposition, we have

$$|\mathrm{cl}(t)| = n!/2(n-2)!.$$

We claim for all $k > 1$ that the size of the class of t is different from this. We will show, in other words, that the transpositions of G are the only involutions (elements of order 2) that lie in classes of size $n!/(2(n-2)!)$. What we need to establish, therefore, is that

$$2^k(n-2k)!k! \neq 2(n-2)!$$

for $1 < k \leq n/2$. (Observe that if $k = 3$ and $n = 6$, our desired inequality is false; the two quantities are equal. Since we are assuming that $n \neq 6$, however, this is not a problem.)

If $k = 2$, the inequality simplifies to the statement that $4 \neq (n-3)(n-2)$, and this is certainly true. Assume then that $k \geq 3$, so that in fact $n > 6$. We prove the inequality by showing that

$$2^{k-1}(n-2k)!k! < (n-2)!.$$

Since

$$1 \leq \binom{n-k}{k} = \frac{(n-k)!}{k!(n-2k)!},$$

we have

$$2^{k-1}(n-2k)!k! \le 2^{k-1}(n-k)! \,.$$

It therefore suffices to check that

$$2^{k-1} < (n-2)(n-3) \cdots (n-k+1) \,.$$

On the right, there are $k-2$ factors, each exceeding 4, and so the right side exceeds $4^{k-2} = 2^{2(k-2)} \ge 2^{k-1}$, as required.

Now let $\sigma \in \mathrm{Aut}(S_n)$. Then σ permutes the involutions of S_n and maps each involution to another with the same size conjugacy class (since, in fact, σ permutes the classes of S_n). It follows by our calculations that σ permutes the transpositions of S_n.

Let $s_i = (i, i+1)$ for $1 \le i \le n-1$ and let $t_i = (s_i)^\sigma$. Write $t_1 = (a_1, a_2)$. Now s_1 and s_2 do not commute, and thus t_1 and t_2 also fail to commute and t_2 is not disjoint from t_1. We can therefore choose notation so that $t_2 = (a_2, a_3)$ with $a_3 \ne a_1$. Now t_3 does not commute with t_2 and so it involves either a_2 or a_3. Since t_3 commutes with t_1, however (because s_3 commutes with s_1), we necessarily have $t_3 = (a_3, a_4)$, where $a_4 \notin \{a_1, a_2, a_3\}$. Similarly, t_4 is not disjoint from t_3 but is disjoint from t_1 and t_2, and so $t_4 = (a_4, a_5)$ with $a_5 \notin \{a_1, a_2, a_3, a_4\}$. Continuing in this way, we can write $t_i = (a_i, a_{i+1})$ where the points $a_1, a_2, \ldots, a_n$ are distinct and so are $1, 2, \ldots, n$ in some order.

Let $g \in S_n$ be the permutation defined by $ig = a_i$. Then $(s_i)^g = t_i$, and if α denotes the inner automorphism induced by g, then $\sigma\alpha^{-1}$ fixes each s_i. These transpositions, however, generate S_n by Lemma 6.16. Therefore, $\sigma\alpha^{-1}$ fixes every element of S_n and thus $\sigma\alpha^{-1} = 1$ and $\sigma = \alpha$.   ■

It is clear why the proof of Theorem 6.20 fails when $n = 6$, but that, of course, does not show that in fact S_6 has a non-inner automorphism.

(6.21) THEOREM.   *The group S_6 has an automorphism mapping the stabilizer of a point in S_6 to a transitive subgroup. No such automorphism can be inner.*

Proof.   Note that S_5 contains exactly 24 elements of order 5 and thus $n_5(S_5) = 6$. It follows that S_5 has a subgroup H of index 6, and the action of S_5 on the right cosets of H induces a homomorphism from S_5 into S_6 whose image is transitive on the six points.

The kernel of this homomorphism is a normal subgroup of S_5 with index ≥ 6. By Corollary 6.19, this kernel is trivial, and thus S_6 contains an isomorphic copy G of S_5 acting transitively on the six points. Since $|S_6 : G| = |S_6|/|S_5| = 6$, we have an action of S_6 on the six right cosets of G, and this induces a homomorphism $\sigma : S_6 \to S_6$ that carries G into a point stabilizer in S_6.

It follows that $|S_6 : \ker(\sigma)| \ge 6$, and reasoning as before (but this time applying Corollary 6.19 to S_6 instead of S_5), we conclude that $\ker(\sigma) = 1$. It follows that σ is an automorphism of S_6 and σ^{-1} has the required property. (Actually, σ does too, but this is somewhat less obvious.)

If $\tau \in \text{Inn}(S_6)$, then τ is conjugation by some $g \in S_6$. If $K \subseteq S_6$ is the stabilizer of some point $i \in \{1, \dots, 6\}$, then $K^\tau = K^g$ is the stabilizer of ig and so is not a transitive subgroup.   ∎

In fact, $|\text{Aut}(S_6) : \text{Inn}(S_6)| = 2$, but we shall not prove this. We mention that if G is a group with a trivial center and with the property that every automorphism is inner, then the natural homomorphism $G \to \text{Inn}(G)$ is an isomorphism of G onto $\text{Aut}(G)$. A group with this property is said to be *complete*. By Corollary 6.6 and Theorem 6.20, we see that S_n is complete for all $n \notin \{2, 6\}$.

Problems

6.1 Let $P \subseteq S_n$ be a subgroup of prime order and suppose $x \in S_n$ normalizes but does not centralize P. Show that x fixes at most one point in each orbit of P.

6.2 Show that there does not exist a simple group of order $760 = 2^3 \cdot 5 \cdot 19$.

   HINT: Use Problem 6.1.

6.3 Let $x \in A_n$ have cycle structure function s. Show that the conjugacy class of x in A_n is smaller than its class in S_n iff $s(k) \leq 1$ for all k and $s(k) = 0$ for even k. How many classes in A_{10} are not classes of S_{10}?

6.4 Show that $S_n = \langle (1, 2), (1, 2, 3, \dots, n) \rangle$.

6.5 Suppose $1 < m < n$ and let $G = \langle (1, 2, \dots, m), (1, 2, \dots, n) \rangle$. Show that G contains a 3-cycle.

6.6 A transitive subgroup $G \subseteq \text{Sym}(X)$ is said to be *primitive* if it is not possible to decompose X into proper subsets of size greater than 1 such that these sets are permuted by the action of G.

   As in Problem 6.5, let $G = \langle (1, 2, \dots, m), (1, 2, \dots, n) \rangle$, where $1 < m < n$. Show that G is primitive.

   NOTE: A theorem of Jordan asserts that a primitive subgroup of $\text{Sym}(X)$ that contains a 3-cycle must contain $\text{Alt}(X)$. By Problems 6.5 and 6.6, it follows that the group G of 6.6 is either A_n or S_n. (How can one determine which it is?)

6.7 Let $a \in S_n$ be an n-cycle. Show that $\mathbf{C}_{S_n}(a) = \langle a \rangle$.

   HINT: How big is the conjugacy class of a?

6.8 Let $T \subseteq S_n$ be a subset consisting of transpositions and let $G = \langle T \rangle$. Show that G has at least $n - |T|$ orbits.

6.9 Suppose $g \in S_n$ is an m-cycle and $g = t_1 t_2 \cdots t_k$, where the t_i are transpositions. Show that $\{t_i \mid 1 \leq i \leq k\}$ contains at least $m - 1$ *different* permutations.

6.10 Let G have a cyclic Sylow 2-subgroup. Show that G has a subgroup of index 2. Conclude (by induction on $|G|$) that G has a characteristic subgroup of odd order and 2-power index.

HINT: Let G act on G.

NOTE: This problem generalizes Corollary 6.12.

6.11 Let $M \subseteq H \subseteq G$, where G is finite, and assume $|G : H|$ is odd and $|H : M| = 2$. Suppose $t \in H$ is an involution that is not G-conjugate to any element of M. Show that G has a subgroup of index 2.

6.12 Suppose that p is prime and that $s, g \in S_p$, where s is a transposition and g has order p. Show that $\langle s, g \rangle = S_p$.

HINT: Show that if we suitably renumber the points and replace g by a power of itself, we are in the situation of Problem 6.4.

6.13 If $\mathbf{Z}(A_n) > 1$, show that $n = 3$.

CHAPTER SEVEN

New Groups from Old

7A

In this chapter we discuss two techniques for constructing groups. The simpler and more important of these is the direct product, often also called the "direct sum." Although we present direct products in the context of group theory, analogous constructions occur throughout algebra. We also consider the semidirect product, which is relevant only to group theory.

Let H_1 and H_2 be arbitrary groups. We can put the structure of a group on the set $H_1 \times H_2$ of ordered pairs of elements (x_1, x_2) with $x_i \in H_i$ by setting

$$(x_1, x_2)(y_1, y_2) = (x_1 y_1, x_2 y_2).$$

It is trivial to check that this makes $H_1 \times H_2$ into a group (which is called the *external direct product* of the H_i). If $G = H_1 \times H_2$, then the identity of G is $(1_1, 1_2)$, where 1_i is the identity of H_i, and inverses in G are given by the formula $(x_1, x_2)^{-1} = (x_1^{-1}, x_2^{-1})$. If the operations in H_i are written additively rather than multiplicatively, it is common to call G the *external direct sum* of the H_i and to write $G = H_1 \oplus H_2$. For instance, it should be clear that $\mathbb{C}^+ \cong \mathbb{R}^+ \oplus \mathbb{R}^+$, where $\mathbb{C}^+$ and $\mathbb{R}^+$ are the additive groups of the complex and real numbers, respectively.

If $G = H \times K$, we can define subgroups $\overline{H}, \overline{K} \subseteq G$ by setting

$$\overline{H} = \{(h, 1) \mid h \in H\} \text{ and } \overline{K} = \{(1, k) \mid k \in K\},$$

where we are following the custom of using the same symbol "1" to represent the identities of H and K (and for that matter, the identity $1 = (1, 1)$ of G, too). Note that it is entirely possible that H and K are the same group. In this case, our "bar notation" contains an ambiguity (which we ignore).

It should be obvious that

$$H \cong \overline{H} \text{ and } K \cong \overline{K}$$

and also that

$$\overline{H}\,\overline{K} = G \text{ and } \overline{H} \cap \overline{K} = 1.$$

Another trivial observation is that $\overline{H}$ and $\overline{K}$ centralize each other, and therefore

$$\overline{H} \triangleleft G \text{ and } \overline{K} \triangleleft G.$$

In general, if G is any group having two normal subgroups M and N with $MN = G$ and $M \cap N = 1$, we say that G is the *internal direct product* of its subgroups M and N. In this situation, we write

$$G = M \mathbin{\dot\times} N.$$

The additive version of this notation is

$$A = U \mathbin{\dot+} V,$$

where A is an abelian group and $U, V \subseteq A$ with $U + V = A$ and $U \cap V = 0$. In this case, of course, we say that A is the *internal direct sum* of the subgroups U and V.

Note that despite the syntactic similarity between

$$\text{``}G = M \times N\text{'' and ``}G = M \mathbin{\dot\times} N\text{,''}$$

there is a significant semantic difference. The first statement describes the *construction* of G; the groups M and N are absolutely arbitrary and can even be equal. The second statement is an assertion about a *given* group G and two of its subgroups. One can write "$M \times N$" as the name of a group, but "$M \mathbin{\dot\times} N$" should never stand alone since the symbol $\dot\times$ makes a statement about three groups. Also, except when G is trivial, it is never true that $G = M \mathbin{\dot\times} M$.

We admit to being somewhat pedantic on this point. The distinctions we have just made between internal and external direct products are often blurred in practice, and the notation $M \times N$ is commonly used for both, with the meaning usually determined by the context. To compound the confusion, some authors use the symbols $\oplus$ and $\dot+$ with exactly the opposite of our meanings.

The result of Problem 2.13(c) will be useful, and so we present it here as a lemma.

(7.1) LEMMA.   *Let M and N be normal in G with $M \cap N = 1$. Then $mn = nm$ for all $m \in M$ and $n \in N$.*

Proof.   We compute the commutator $[m, n] = m^{-1}n^{-1}mn$. We have $[m, n] = m^{-1}m^n \in M$ since $m^n \in M^n = M$. Also, $[m, n] = (n^{-1})^m n \in N$ since $N^m = N$. Therefore $[m, n] \in M \cap N = 1$, and the result follows.   ∎

The next result should clarify somewhat the connection between internal and external direct products.

(7.2) LEMMA. *If $G = H \times K$, then $G = \overline{H} \; \dot{\times} \; \overline{K}$ in the notation used previously. Also, if $\Gamma = M \; \dot{\times} \; N$, then $\Gamma \cong M \times N$.*

Proof. We have already observed that $G = \overline{H} \; \overline{K}$, that $\overline{H} \cap \overline{K} = 1$, and that $\overline{H} \lhd G$, $\overline{K} \lhd G$. This proves the first assertion. To prove the second, note that $mn = nm$ for $m \in M$ and $n \in N$ by Lemma 7.1. Map

$$\theta : M \times N \to \Gamma$$

by $\theta(m, n) = mn$.
   We have

$$\theta((m, n)(m', n')) = \theta(mm', nn') = mm'nn' = mnm'n' = \theta(m, n)\theta(m', n'),$$

so that θ is a homomorphism. Since $MN = \Gamma$, we see that θ is surjective. Also, if $(m, n) \in \ker(\theta)$ so that $mn = 1$, we have $m = n^{-1} \in M \cap N = 1$ and hence $(m, n) = (1, 1) = 1$. The proof is now complete. ∎

(7.3) LEMMA. *If $G = M \; \dot{\times} \; N$, then $M \cong G/N$ and $N \cong G/M$.*

Proof. We have

$$G/N = MN/N \cong M/(M \cap N) = M .$$

The isomorphism $G/M \cong N$ follows similarly. ∎

(7.4) COROLLARY. *If $G = M \; \dot{\times} \; N = M \; \dot{\times} \; L$, then $N \cong L$.* ∎

   In general, there is no such cancellation property that works for isomorphisms of external direct products. It is possible that $M \times N \cong M \times L$ and yet $N \not\cong L$. If we write $\mathbb{R}^+$ to denote the additive group of the real numbers, then (believe it or not) we have $\mathbb{R}^+ \cong \mathbb{R}^+ \oplus \mathbb{R}^+$. It follows that if we write 0 to denote the trivial group, we have

$$\mathbb{R}^+ \oplus 0 \cong \mathbb{R}^+ \cong \mathbb{R}^+ \oplus \mathbb{R}^+,$$

and so cancellation fails. To see why $\mathbb{R}^+ \cong \mathbb{R}^+ \oplus \mathbb{R}^+$, we must appeal to a bit of transfinite linear algebra. Each of these objects may be viewed a vector space over $\mathbb{Q}$, and in both cases the dimensions are equal to the cardinality of the continuum. Any bijection between bases for these two spaces extends to a linear transformation that is a vector space isomorphism and hence is an isomorphism of abelian groups. We shall see later that such a phenomenon cannot happen for finite groups, where cancellation does work for external direct products.
   We can extend the notions of direct product (both internal and external) to situations where there are more than two factors. Let $\{H_\alpha \mid \alpha \in I\}$ be any indexed collection of groups, where the index set I may be finite or infinite. It is convenient to replace the notion of an ordered pair or ordered n-tuple by a function f from the index set I into the union $\bigcup H_\alpha$, with the condition that $f(\alpha) \in H_\alpha$ for all $\alpha \in I$. Let Γ be the collection of all such functions. Then Γ becomes a group under the operation that defines fg to be the function satisfying $(fg)(\alpha) = f(\alpha)g(\alpha)$. (Note

that in the case $I = \{1, 2\}$, we see that $\Gamma = H_1 \times H_2$ if we identify the pair (x_1, x_2) with the function $1 \mapsto x_1, 2 \mapsto x_2$.) We write

$$\Gamma = \underset{\alpha \in I}{\mathsf{X}} H_\alpha$$

and call Γ the *unrestricted external direct product* of the H_α. As in the case where $|I| = 2$, we define $\overline{H}_\alpha \subseteq \Gamma$ by

$$\overline{H}_\alpha = \{f \in \Gamma \mid f(\beta) = 1 \text{ whenever } \alpha \neq \beta \in I\}.$$

If $|I| < \infty$, then Γ is the group generated by the subgroups $\overline{H}_\alpha$, but this is not true when I is infinite. In that case, $\langle \overline{H}_\alpha \mid \alpha \in I \rangle$ is the group of those functions $f \in \Gamma$ that have nonidentity values for at most finitely many indices α. (This is so because one can take only finite products of elements in a group.)   This subgroup of Γ, where the functions are restricted to having identity values "almost everywhere," is called the *restricted external direct product* of the H_α. Some authors use the words "product" and "sum" to distinguish between what we have called the unrestricted and restricted direct products, respectively, and sometimes the restricted direct product is called a "coproduct" and is denoted $\coprod$. There seems to be no completely standard nomenclature and notation that clearly distinguishes these products.

   We are primarily concerned with the case where there are just finitely many factors. Then the restricted and unrestricted external direct products coincide and we need not distinguish them. Also, in this case we can (and shall) use the "n-tuple" notation in place of functions. We write, therefore, $(x_1, x_2, \ldots, x_n)$ to denote the element of $\mathsf{X}\, H_i$ corresponding to the function $f : i \mapsto x_i$.

   It is easy to see that if $M, N \triangleleft G$, then $G = M \mathbin{\dot{\times}} N$ iff every element of G is uniquely of the form $g = mn$ with $m \in M$ and $n \in N$. (In fact, this follows from Theorem 7.5 below.)   The definition of an internal direct product can be extended to the case where there may be more than two factors by means of this alternative characterization.

   Let $M_i \triangleleft G$ for $1 \leq i \leq n$, and suppose $G = M_1 M_2 \cdots M_n = \prod_{i=1}^{n} M_i$. We say that this product is *direct* or that G is the *internal direct product* of the subgroups M_i provided that for every $g \in G$, there is a unique choice of elements $x_i \in M_i$ such that $g = x_1 x_2 \cdots x_n$. We write

$$G = \overset{n}{\underset{i=1}{\boxed{\cdot}\!\prod}} M_i$$

in this case, using the dot to signify the uniqueness.

(7.5) THEOREM.   *Let $M_i \triangleleft G$ for $1 \leq i \leq n$ and assume that $\prod M_i = G$. Then the following are equivalent*:

   i. $\overset{\cdot}{\prod} M_i = G$.

   ii.   *For each i, $M_i \cap \prod_{j \neq i} M_j = 1$.*

   iii.   *For $2 \leq i \leq n$, $M_i \cap \prod_{j=1}^{i-1} M_i = 1$.*

Note that condition (ii) depends only on the set of subgroups M_i and not on their numbering, and so the order in which the M_i appear is irrelevant to the question of the directness of the product. Condition (iii) is usually easier to check than (ii) and often is easier with some orderings of the M_i than with others. It is important to remember that the condition $M_i \cap M_j = 1$ for $j \ne i$ is *not* sufficient to prove directness when $n > 2$, but it is a consequence of directness.

Proof of Theorem 7.5.   Assume condition (i) so that the product is direct, and suppose $g \in M_i \cap \prod_{j \ne i} M_j$ for some particular subscript i. Since $g \in \prod_{j \ne i} M_j$, we can write $g = x_1 x_2 \cdots x_n$ for suitable elements $x_j \in M_j$, where $x_i = 1$. Since $g \in M_i$, however, we can also write $g = y_1 y_2 \cdots y_n$, with $y_j \in M_j$, by setting $y_j = 1$ for $j \ne i$ and $y_i = g$. The uniqueness in the definition of $\prod$ implies that $x_j = y_j$ for all j. In particular, $1 = x_i = y_i = g$, and (ii) is proved.

If we assume condition (ii), then (iii) follows trivially, and so we assume (iii) and prove (i). Given $g \in G = \prod M_i$, we can certainly write $g = x_1 x_2 \cdots x_n$ with $x_i \in M_i$. Suppose also $g = y_1 y_2 \cdots y_n$ with $y_i \in M_i$. To establish (i), we must show that $x_i = y_i$ for all i. If this is false, let i be as large as possible such that $x_i \ne y_i$. If $i < n$, cancel $x_{i+1} \cdots x_n = y_{i+1} \cdots y_n$ from the two expressions for g so that in any case, we get

$$x_1 x_2 \cdots x_i = y_1 y_2 \cdots y_i .$$

Then
$$1 \ne x_i y_i^{-1} = (x_1 x_2 \cdots x_{i-1})^{-1} (y_1 y_2 \cdots y_{i-1}) \in M_i \cap \prod_{j=1}^{i-1} M_j ,$$

and this contradicts (iii).   ∎

The following is the extension of Lemma 7.2 to the case of more than two subgroups.

(7.6) LEMMA.   *Let* $G = \prod_{i=1}^{n} N_i$. *Then* $G \cong \underset{i=1}{\overset{n}{\times}} N_i$.

Proof.   Let $E = \underset{i=1}{\overset{n}{\times}} N_i$ and map $\theta : E \to G$ by

$$\theta((x_1, x_2, \ldots, x_n)) = x_1 x_2 \cdots x_n .$$

Then θ is surjective since $G = \prod N_i$ and it is injective since this product is direct. What remains is to show that θ is a homomorphism.

Let $f = (x_1, \ldots, x_n)$ and $g = (y_1, \ldots, y_n)$ be elements of E and compute that

$$\theta(fg) = x_1 y_1 x_2 y_2 \cdots x_n y_n .$$

Now x_i and y_j commute if $i > j$ since $N_i, N_j \triangleleft G$ and $N_i \cap N_j = 1$. We can therefore move x_i to the left in the above product, passing over y_j with $j < i$. Doing this with $x_2, x_3, \ldots, x_n$ in turn, we see that

$$\theta(fg) = x_1 x_2 \cdots x_n y_1 y_2 \cdots y_n = \theta(f)\theta(g) ,$$

as required.   ∎

It is sometimes useful to be able to compute the center of a group expressed as an internal direct product.

(7.7) LEMMA. *Suppose that* $G = \prod_{i=1}^{n} N_i$. *Then* $\mathbf{Z}(G) = \prod_{i=1}^{n} \mathbf{Z}(N_i)$.

Proof.   Write $Z_i = \mathbf{Z}(N_i)$. Since N_i and N_j are normal and $N_i \cap N_j = 1$ for $i \neq j$, we see that $N_j \subseteq \mathbf{C}_G(N_i) \subseteq \mathbf{C}_G(Z_i)$. Since also $N_i \subseteq \mathbf{C}_G(Z_i)$, we have

$$G = \prod N_j \subseteq \mathbf{C}_G(Z_i)$$

and $Z_i \subseteq \mathbf{Z}(G)$ for each subscript i. Thus $\prod Z_i \subseteq \mathbf{Z}(G)$.

For the reverse containment, let $z \in \mathbf{Z}(G)$ and write $z = z_1 z_2 \cdots z_n$, with $z_i \in N_i$. If $g \in G$, then

$$z = z^g = z_1^g z_2^g \cdots z_n^g .$$

However, $z_i^g \in N_i$ since $N_i \triangleleft G$. The uniqueness in the definition of an internal direct product now yields $z_i = z_i^g$ for all i, and thus $z_i \in \mathbf{Z}(G)$ and, in particular, $z_i \in Z_i$. Therefore $z \in \prod Z_i$ and we have $\mathbf{Z}(G) = \prod Z_i$. This product is direct, however, because uniqueness is inherited from the fact that $G = \prod N_i$. ∎

We give the following result describing minimal normal subgroups as an example of how direct products come up in practice.

(7.8) THEOREM.   *Let N be a finite minimal normal subgroup of G. Then there exist subgroups $S_i \subseteq N$ such that*

a. $N = \prod S_i$,
b. *each S_i is simple, and*
c. *the S_i are all G-conjugate.*

Furthermore, if N is nonabelian, then

d. *the S_i constitute a full G-conjugacy class of subgroups.*

Proof.   Let S be any minimal normal subgroup of N. Since $N \triangleleft G$, it follows for all $g \in G$ that S^g is contained in N, and in fact, S^g is a minimal normal subgroup of N. Now $\langle S^g \mid g \in G \rangle \triangleleft G$ and this subgroup is contained in N. By the minimality of N, therefore, we have $\langle S^g \mid g \in G \rangle = N$, and we can choose conjugates $S_1, S_2, \ldots, S_n$ of S that together generate N and such that no smaller set generates. In particular, (c) is satisfied. Also, since each $S_i \triangleleft N$, we have $N = \prod S_i$.

To prove (a) we verify condition (ii) of Theorem 7.5. Given i, let

$$D = S_i \cap \prod_{j \neq i} S_j ,$$

and note that $D \triangleleft N$ and, of course, $D \subseteq S_i$. Since S_i is minimal normal in N, we conclude that either $D = 1$ or $D = S_i$. In the latter situation, $S_i = D \subseteq \prod_{j \neq i} S_j$

and thus

$$\prod_{j \neq i} S_j = \prod_j S_j = N \,,$$

and this contradicts the choice of $S_1, S_2, \ldots, S_n$ as being minimally sufficient to generate N. We conclude, therefore, that $D = 1$ and (a) follows.

To prove (b), we let $1 < M \vartriangleleft S_i$ and show that $M = S_i$. Since $S_i \cap S_j = 1$ for $i \neq j$, we have

$$S_j \subseteq \mathbf{C}_G(S_i) \subseteq \mathbf{N}_G(M) \,.$$

Clearly also $S_i \subseteq \mathbf{N}_G(M)$, and so

$$N = \prod S_i \subseteq \mathbf{N}_G(M)$$

and $M \vartriangleleft N$. Since S_i is minimal normal in N, we have $M = S_i$, as required.

Finally, assume that N is nonabelian. Since $\mathbf{Z}(N) \vartriangleleft G$ and N is minimal normal in G and is nonabelian, it follows that $\mathbf{Z}(N) = 1$. Now let T be any G-conjugate of S. Thus T is minimal normal N and so if $T \neq S_i$, we have $T \cap S_i = 1$ and $S_i \subseteq \mathbf{C}_G(T)$. It follows that if $T \notin \{S_1, S_2, \ldots, S_n\}$, then

$$N = \prod S_i \subseteq \mathbf{C}_G(T)$$

and $T \subseteq \mathbf{Z}(N)$. This is a contradiction. ∎

7B

A finite group G is said to be *nilpotent* if for every prime p, a Sylow p-subgroup of G is normal (or, equivalently, is unique). Note that finite abelian groups and finite p-groups are nilpotent. Although we study nilpotent groups in greater detail later, we give one result now as an application of direct products.

(7.9) THEOREM. *Let G be a finite nilpotent group and let π be the set of prime divisors of $|G|$. For each $p \in \pi$, let S_p be the unique Sylow p-subgroup of G. Then*

$$G = \prod_{p \in \pi} \cdot \, S_p \,.$$

Proof. Since $S_p \vartriangleleft G$ for each $p \in \pi$, we see that $\prod S_p$ is a group with order divisible by $|S_p|$ for all p. By the definition of Sylow subgroups, we have $|G| = \prod_{p \in \pi} |S_p|$, and it follows that $\prod S_p = G$.

To prove that this product is direct, it suffices by Theorem 7.5 to show for each $q \in \pi$ that

$$S_q \cap \prod_{p \neq q} S_p = 1 \,.$$

It is easy to see, however, that $\left| \prod_{p \neq q} S_p \right|$ divides $\prod_{p \neq q} |S_p|$ and so is not divisible by q. (This follows, for instance, by Corollary 4.17.) The intersection is therefore trivial and we are done. ∎

Note that although Theorem 7.9, as stated, is *descriptive* of nilpotent groups, Lemma 7.6 can be used to reinterpret it as a *constructive* result that provides a recipe for building finite nilpotent groups. By Lemma 7.6 and Theorem 7.9, every finite nilpotent group can be constructed (up to isomorphism) by taking the external direct product of some finite collection of finite p-groups for various primes p. It is easy to see that every such external direct product really is nilpotent, and so what we have done is to reduce the problem of finding all finite nilpotent groups to that of finding all finite p-groups. We mention that only in theory can we refer to this as a "reduction," since there is a tremendous variety of different p-groups and it seems hopeless to attempt to construct all of them in any useful way.

If we consider the class of finite abelian groups, however, instead of the more general nilpotent groups, we can prove a descriptive result that also serves as a highly useful constructive result.

(7.10) THEOREM (Fundamental, of abelian groups). *Let G be finite and abelian. Then*

$$G = \overset{n}{\underset{i=1}{\cdot\!\prod}} C_i,$$

where the C_i are subgroups that are cyclic p-groups for various primes p.

First we need a general lemma.

(7.11) LEMMA. *Let $G = \overset{n}{\underset{i=1}{\cdot\!\prod}} N_i$ and suppose each N_i decomposes as a direct product*

$$N_i = \overset{m_i}{\underset{j=1}{\cdot\!\prod}} M_{ij}.$$

Then

$$G = \underset{i,j}{\cdot\!\prod} M_{ij}.$$

Proof. It should be clear that $G = \prod M_{ij}$. Since the different N_i centralize each other, we see that each $M_{ij} \triangleleft G$, and so it suffices to show that if $\prod x_{ij} = \prod y_{ij}$ with $x_{ij}, y_{ij} \in M_{ij}$, then $x_{ij} = y_{ij}$ for all i, j. To see this, write

$$u_i = \prod_j x_{ij} \quad \text{and} \quad v_i = \prod_j y_{ij}.$$

Then $\prod u_i = \prod v_i$ and hence $u_i = v_i$ for all i by the directness of $\cdot\!\prod N_i$. For each i, then, $\prod_j x_{ij} = \prod_j y_{ij}$ and so $x_{ij} = y_{ij}$ for all j by the directness of $\cdot\!\prod_j M_{ij}$. ∎

The following is the key step in the proof of the Fundamental theorem of abelian groups.

(7.12) THEOREM.   *Let G be a finite abelian p-group and let $C \subseteq G$ be a cyclic subgroup with maximum possible order. Then $G = C \times B$ for some subgroup $B \subseteq G$.*

Proof.   Since we can take $B = 1$ if $C = G$, we can assume $C < G$ and we choose $x \in G - C$ of smallest possible order. Since $x \neq 1$, we see that $o(x^p) < o(x)$ and hence $x^p \in C$. If x^p generates C, then $|\langle x \rangle| = p|C|$ and this contradicts the choice of C. Therefore x^p is a nongenerator of the cyclic p-group C and it follows that x^p is a pth power in C, and we can write $x^p = y^p$ for some $y \in C$. Now $xy^{-1} \notin C$ and $(xy^{-1})^p = x^p(y^p)^{-1} = 1$. By the choice of x, we have $o(x) \leq o(xy^{-1}) = p$ and so $o(x) = p$.

Now let $X = \langle x \rangle$ and use overbars to denote the canonical homomorphism $G \rightarrow G/X = \overline{G}$. Since $|X| = p$, we have $C \cap X = 1$, and the map $^-$ is an isomorphism from C to $\overline{C}$. Thus $\overline{C}$ is cyclic with order equal to $|C|$. If $\overline{G}$ has a cyclic subgroup $\langle \overline{g} \rangle$ with order larger than $|\overline{C}|$, then

$$|\langle g \rangle| = o(g) \geq o(\overline{g}) = |\langle \overline{g} \rangle| > |\overline{C}| = |C|,$$

and this contradicts the choice of C. It follows that $\overline{C}$ is a cyclic subgroup of maximum possible order in $\overline{G}$. Since $|\overline{G}| < |G|$, we may work by induction on $|G|$ and conclude that $\overline{C}$ is a direct factor of $\overline{G}$. Since every subgroup of $\overline{G}$ has the form $\overline{B}$ for some $B \subseteq G$ with $B \supseteq X$, we can find $B \supseteq X$ such that $\overline{G} = \overline{C} \times \overline{B}$. Thus $\overline{G} = \overline{C} \, \overline{B} = \overline{CB}$ and hence $CB = G$. Also $\overline{C} \cap \overline{B} = \overline{1}$ and so $C \cap B \subseteq X$. Then $C \cap B \subseteq C \cap X = 1$, and this proves that $G = C \times B$, as required.   ∎

Proof of Theorem 7.10.   Work by induction on $|G|$. If we can write $G = A \times B$ with $A < G$ and $B < G$, then each of A and B is a direct product of cyclic p-groups by the inductive hypothesis, and the theorem holds by Lemma 7.11.

We may therefore assume that G is directly indecomposable. Since G is nilpotent (being, in fact, abelian), it follows by Theorem 7.9 that G is a p-group. Using the indecomposability of G once again, we conclude by Theorem 7.12 that G is itself cyclic, and nothing more needs to be proved.   ∎

The following corollary is the "constructive" formulation of the fundamental theorem of abelian groups.

(7.13) COROLLARY.   *Every finite abelian group is isomorphic to an external direct product of cyclic p-groups.*   ∎

Suppose, for instance, we wish to construct all abelian groups of order $500 = 2^2 \cdot 5^3$. (We mean, of course, all possible isomorphism types of such groups.) Writing Z_n to denote the cyclic group of order n, we see that the only possibilities are

$$Z_2 \times Z_2 \times Z_5 \times Z_5 \times Z_5, \quad Z_4 \times Z_5 \times Z_5 \times Z_5,$$
$$Z_2 \times Z_2 \times Z_5 \times Z_{25}, \quad Z_4 \times Z_5 \times Z_{25},$$
$$Z_2 \times Z_2 \times Z_{125}, \quad Z_4 \times Z_{125}.$$

But where in this list is Z_{500}? In fact, $Z_{500} \cong Z_4 \times Z_{125}$. In general, we have the following corollary.

(7.14) COROLLARY.   *Let G be finite and abelian.   Then the following are equivalent:*

i.   *G has a cyclic Sylow p-subgroup for each prime p dividing $|G|$.*
ii.   *G is cyclic.*
iii.   *G has just one subgroup of order p for each prime p dividing $|G|$.*

Proof.   Assume condition (i) and let x_p be a generator for the Sylow p-subgroup of G for prime divisors p of G. Let $g = \prod x_p$ and note that for positive integers n, we have $g^n = \prod (x_p)^n$. Taking $n = |G|/p$, where p is one of the primes dividing G, we have $(x_q)^n = 1$ for $q \neq p$. It follows that $g^n = (x_p)^n$, and this is not the identity since $o(x_p)$ is the full p-part of $|G|$ and this does not divide n. It follows that $o(g)$ cannot be a proper divisor of $|G|$. Thus $o(g) = |G|$ and G is cyclic, proving (ii).

That (ii) implies (iii) follows from Theorem 2.9. Assuming (iii), we let $P \in \operatorname{Syl}_p(G)$ with $p \,\big|\, |G|$ and use Theorem 7.10 to write $P = \overset{n}{\underset{i=1}{\big|\cdot\big|}} C_i$, where each C_i is a nontrivial cyclic p-group. Each C_i contains a subgroup of order p, and these are distinct since $C_i \cap C_j = 1$ if $i \neq j$. By (iii) it follows that $n = 1$ and so P is cyclic, proving (i).   ■

We could also have used the theory of direct products to prove that (i) implies (ii) in the above proof. If C is a cyclic group of order $|G|$, we want to show that $G \cong C$. Writing G_p and C_p to denote the unique Sylow p-subgroups of G and C, respectively, we know that G_p and C_p are cyclic with equal orders, and thus $G_p \cong C_p$ for all p and so their external direct products are isomorphic. This yields

$$G = \big|\cdot\big| G_p \cong \underset{}{\times} G_p \cong \underset{}{\times} C_p \cong \big|\cdot\big| C_p = C$$

by two applications each of Lemma 7.6 and Theorem 7.9.

Since $Z_4 \times Z_{125}$ clearly has cyclic Sylow subgroups, it is cyclic by Corollary 7.14, and thus $Z_{500} \cong Z_4 \times Z_{125}$. This suggests the question of what other nonobvious isomorphisms might exist. Are the six groups of order 500 that we listed really all nonisomorphic? In fact, there are no isomorphisms among these groups; there are six isomorphism classes of abelian groups of order 500. More generally, we have the following theorem.

(7.15) THEOREM.   *Let $n_1, n_2, \ldots, n_r$ and $m_1, m_2, \ldots, m_s$ be nontrivial prime powers and assume that*

$$Z_{n_1} \times Z_{n_2} \times \cdots \times Z_{n_r} \cong Z_{m_1} \times Z_{m_2} \times \cdots \times Z_{m_s}.$$

Then $r = s$ and, after renumbering if necessary, we have $n_i = m_i$ for $1 \leq i \leq r$.

We do not prove Theorem 7.15 here. A more general result on the uniqueness of direct product decompositions is the Krull-Schmidt theorem, which we present in Chapter 10 (and, in fact, Corollary 10.16 is essentially Theorem 7.15). Another approach to Theorem 7.15 is to observe that one can easily reduce to the case where all of the m_i and n_j are powers of the same prime. In that case, an argument given in Problem 7.16 can serve to complete the proof.

It should be pointed out that the fundamental theorem of abelian groups (7.10) can be considerably generalized. It is not really necessary, for instance, to assume that the group is finite; the result also holds for finitely generated infinite abelian groups (except that some of the cyclic direct factors C_i are infinite in that case). More general still is a result on modules over principal ideal domains. (This result, which we decided not to present, asserts that every finitely generated module over a PID decomposes as a direct sum of cyclic submodules. If we take the ring to be $\mathbb{Z}$, the ring of integers, we recover the case of abelian groups.)

We close our discussion of the fundamental theorem of abelian groups with an application. Let G be a finite abelian group and write $\hat{G}$ to denote the set of homomorphisms from G into the multiplicative group $\mathbb{C}^\times = \mathbb{C} - \{0\}$ of the complex numbers. If $\lambda, \mu \in \hat{G}$, we define the product

$$\lambda\mu : G \to \mathbb{C}^\times$$

by $(\lambda\mu)(g) = \lambda(g)\mu(g)$, and we note that $\lambda\mu \in \hat{G}$. It is easy to check that this makes $\hat{G}$ into a group. (Note that $1 \in \hat{G}$ is the constant function, mapping every element of G to $1 \in \mathbb{C}$.) The group $\hat{G}$ we have just constructed is called the *dual group* of G.

(7.16) THEOREM.   *Let G be finite and abelian. Then $\hat{G} \cong G$.*

Proof.   Write $G = C_1 \dot{\times} \cdots \dot{\times} C_n$, where each C_i is cyclic. We construct a map

$$\theta : \hat{G} \to \bigtimes \hat{C}_i$$

by setting $\theta(\lambda) = (\lambda_1, \lambda_2, \ldots, \lambda_n)$, where $\lambda_i \in \hat{C}_i$ is the restriction of λ to C_i. If $\lambda \in \ker(\theta)$, then each $\lambda_i = 1$, and this says that $C_i \subseteq \ker(\lambda)$. It follows that $G = \prod C_i \subseteq \ker(\lambda)$ and $\lambda = 1$. Thus $\ker(\theta) = 1$.

Next we show that θ is surjective. Given arbitrary $\lambda_i \in \hat{C}_i$ for $1 \leq i \leq n$, we must produce $\lambda \in \hat{G}$ such that for each i, the restriction of λ to C_i is λ_i. We define λ as follows. For each $x \in G$, we can (uniquely) write $x = x_1 x_2 \cdots x_n$, with $x_i \in C_i$. We set

$$\lambda(x) = \prod \lambda_i(x_i) \in \mathbb{C}^\times .$$

This is well defined by the uniqueness of the x_i, and λ is easily seen to be an element of $\hat{G}$ having the correct restrictions to the subgroups C_i. Therefore θ is surjective and we have shown that

$$\hat{G} \cong \bigtimes \hat{C}_i .$$

Our next object is to show that $\hat{C}_i \cong C_i$. From this, it will follow that

$$\hat{G} \cong \bigtimes \hat{C}_i \cong \bigtimes C_i \cong \prod \cdot C_i = G ,$$

and the proof will be complete. Effectively, we have reduced the problem to proving that $\hat{G} \cong G$ for cyclic groups, and so we may now assume that G is cyclic.

Write $G = \langle g \rangle$ and $|G| = n$, and let $E \subseteq \mathbb{C}^\times$ be the group of nth roots of unity in $\mathbb{C}$. Then E is cyclic of order n (generated by $e^{2\pi i/n}$), and there exists an isomorphism $\lambda : G \to E$. Of course, $\lambda \in \hat{G}$ and we have

$$\lambda^n(x) = \lambda(x)^n = \lambda(x^n) = 1 \,.$$

If $0 < m < n$, however, then

$$\lambda^m(g) = \lambda(g^m) \neq 1 \,.$$

It follows that $o(\lambda) = n$ and so $G \cong \langle \lambda \rangle$. What remains is to show that $\langle \lambda \rangle = \hat{G}$. To see this, let $\mu \in \hat{G}$. Now $\lambda(g)$ is a generator of E, and since $\mu(g)^n = \mu(g^n) = 1$, we have $\mu(g) \in E$ and so $\mu(g) = \lambda(g)^m$ for some integer m. Then $\mu(g) = \lambda^m(g)$, and it follows that μ and λ^m agree on all of G. Thus $\mu = \lambda^m \in \hat{G}$, as required.   ∎

Note that the construction of the isomorphism $G \cong \hat{G}$ in the preceding proof depends on choices made in its construction. The subgroups C_i are generally not uniquely determined, and even in the case where G is cyclic, there is no obvious natural isomorphism between G and $\hat{G}$ other than that constructed when generators are chosen for these groups. In general, there does not exist a natural or canonical isomorphism between G and $\hat{G}$.

7C

We close this chapter with a discussion of what is called the "extension problem" for groups. Given two groups N and H, the problem is to find all (up to isomorphism) groups G such that G has a normal subgroup M with

$$M \cong N \quad \text{and} \quad G/M \cong H \,.$$

Such a group G is called an *extension of N by H*. (Unfortunately, some authors interchange the two prepositions.) At least one such extension always exists since we can take $G = N \times H$ and let $M = \overline{N}$. Depending on N and H, there may or may not exist other extensions.

Given N and H as before, G is a *split* extension of N by H if there exists $M \triangleleft G$ with $M \cong N$ and $G/M \cong H$ and there also exists $K \subseteq G$ with $KM = G$ and $K \cap M = 1$. (Note that in this case, $K \cong KM/M = G/M \cong H$.) In this situation, we say that G *splits over M* or that M is *complemented by K* in G. The direct product extension $G = N \times H$ is certainly split since the normal subgroup $\overline{H}$ complements $\overline{N}$. Conversely, if G splits over M and there exists a *normal* complement K, then $G = M \dot{\times} K \cong N \times H$. Our goal is to show how to construct split extensions where the complement is not necessarily normal.

Suppose we have a split extension. Let $M \triangleleft G$ and let K complement M. Then K acts by conjugation on M and each element of K induces an automorphism of M. If $K \triangleleft G$, then K centralizes M by Lemma 7.1, and so this action is trivial. Conversely, if the conjugacy action of K on M is trivial, then $K \subseteq \mathbf{C}(M)$. It follows that $M \subseteq \mathbf{C}(K) \subseteq \mathbf{N}(K)$ and $K \triangleleft G$. If we wish to construct split extensions that are not merely direct products, we must consider nontrivial actions.

Given groups N and H, suppose we have some action of H on the set of elements of N. We say this is an *action via automorphisms* if for every $x, y \in N$ and $h \in H$, we have

$$(xy) \cdot h = (x \cdot h)(y \cdot h).$$

This condition, of course, simply asserts that the map $N \to N$ induced by each element of H is a homomorphism. By the general properties of actions, however, this map is a permutation of N, and so it is actually an automorphism of N. In fact, in this situation, the homomorphism $H \to \mathrm{Sym}(N)$ of Lemma 4.2 maps H into $\mathrm{Aut}(N)$. (Conversely, any homomorphism $\theta : H \to \mathrm{Aut}(N)$ defines an action via automorphisms of H on N by the formula $n \cdot h = n\theta(h)$.)

Some examples of actions via automorphisms are the action by conjugation of any group G on a normal subgroup, the natural action of $\mathrm{Aut}(G)$ on G, and of course, the trivial action. It is customary to abandon the dot notation and write n^h in place of $n \cdot h$ when H acts via automorphisms on N. This does present some problems, however, since we may not be using a conjugation action, but usually such ambiguities are resolved by the context.

The point of the next theorem is that whenever H acts via automorphisms on N, we can build a group G containing isomorphic copies of N and H such that the original, abstract action becomes a conjugation action within G.

(7.17) THEOREM. *Let H act via automorphisms on N. Then there exists a group G with subgroups $\overline{N}$ and $\overline{H}$ and isomorphisms $N \cong \overline{N}$ and $H \cong \overline{H}$ such that*

a. $\overline{H}\,\overline{N} = G$,
b. $\overline{H} \cap \overline{N} = 1$,
c. $\overline{N} \triangleleft G$, *and*
d. $(\overline{n})^{\overline{h}} = \overline{n^h}$ *for all $n \in N$ and $h \in H$.*

Perhaps the conclusion expressed in (d) needs some explanation. We use overbars to denote each of the given isomorphisms $N \to \overline{N}$ and $H \to \overline{H}$. The element $\overline{n}^{\overline{h}} \in \overline{N}$ is simply the conjugate of $\overline{n}$ by $\overline{h}$ in G, whereas $n^h \in N$ is the result of the action of h on n. Usually N and H are identified with $\overline{N}$ and $\overline{H}$ via the maps $^-$, and so what (d) says is that the original action of H on N is the conjugation action in G. It is not hard to show that conditions (a) through (d) of Theorem 7.17 uniquely determine the group G, which is called the *semidirect product of N by H* and is denoted $N \rtimes H$. This result, therefore, shows how to construct all possible split extensions (up to isomorphism). Remember that the semidirect product requires three ingredients: the groups N and H and an action via automorphisms of H on N. The notation, unfortunately, obscures the third ingredient.

Proof of Theorem 7.17.   The usual proof constructs G as a set of ordered pairs and defines a multiplication that makes this a group. This procedure makes checking the group axioms necessary, and this is somewhat tedious. We instead construct G as a subgroup of $\text{Sym}(\Omega)$, where Ω is the *set* $H \times N$.

We define actions of N and of H on Ω according to the formulas

$$(k, m) \cdot h = (kh, m^h) \quad \text{and} \quad (k, m) \cdot n = (k, mn),$$

where $(k, m) \in \Omega = H \times N$ and $h \in H$ and $n \in N$. (It is immediate that these are actions and that no nonidentity element of H or N acts trivially.) By Lemma 4.2, we therefore have isomorphisms of H and N into $\text{Sym}(\Omega)$, and we denote the images by $\overline{H}$ and $\overline{N}$, respectively. We calculate that

$$(k, m)(\overline{n})^{\overline{h}} = (((k, m) \cdot h^{-1}) \cdot n) \cdot h = (k, ((m^{h^{-1}})n)^h).$$

Since

$$((m^{h^{-1}})n)^h = mn^h$$

(because the action of H on N is via automorphisms), we have

$$(k, m)(\overline{n})^{\overline{h}} = (k, m)\overline{n^h}$$

and thus

$$\overline{n}^{\overline{h}} = \overline{n^h}.$$

It follows from this that $\overline{H} \subseteq \mathbf{N}(\overline{N})$, and so $G = \overline{H}\,\overline{N}$ is a group. All that remains is to show that $\overline{H} \cap \overline{N} = 1$. To see this, suppose $\overline{h} = x = \overline{n}$. Then

$$(kh, m^h) = (k, m)x = (k, mn),$$

from which it follows that $h = 1$. Therefore $x = \overline{h} = 1$, as required.   ∎

Problems

◁ **7.1** Let U and V be nonabelian simple groups. Show that $G = U \times V$ has precisely four different normal subgroups.

✗ **7.2** A subgroup $D \subseteq G = M \times N$ is a *diagonal* subgroup if

$$D \cap M = 1 = D \cap N \quad \text{and} \quad DM = G = DN.$$

Show that G has a diagonal subgroup iff $M \cong N$.

✗ **7.3** Let $M_1, M_2, \ldots, M_n \triangleleft G$ and write $D = \bigcap M_i$. Show that G/D is isomorphic to a subgroup of $\mathsf{X}(G/M_i)$.

7.4 Let $\mathcal{X}$ be a nonempty collection of groups such that:

  i.   If $U \in \mathcal{X}$ and $V \cong U$ then $V \in \mathcal{X}$.
  ii. If $U \in \mathcal{X}$ and $V \subseteq U$ then $V \in \mathcal{X}$.
  iii. If $U, V \in \mathcal{X}$, then $U \times V \in \mathcal{X}$.

Show that every finite group has a unique normal subgroup N minimal with the property that $G/N \in \mathcal{X}$.

NOTE: Some examples of collections $\mathcal{X}$ to which Problem 7.4 can be applied are: abelian groups, p-groups (for a fixed prime p), groups whose commutator subgroup is a p-group, and groups with a normal Sylow p-subgroup.

7.5 Let S be a finite collection of minimal normal subgroups of a group G, and let $N = \langle M \mid M \in S \rangle$.

  a. Show that N is the direct product of some subcollection of S.
  b. If S consists entirely of nonabelian groups, show that N is the direct product of all of them.

7.6 The *socle* of a group is the product of all of its minimal normal subgroups. Show that the socle of a finite group is a direct product of simple groups.

7.7 Suppose $G = M \dot\times N$ and let D be a diagonal subgroup as in Problem 7.2. If M (and therefore N) is simple, show that D is a maximal subgroup of G.

7.8 Let $M \triangleleft U$ and $N \triangleleft V$ and assume $U/M \cong V/N$. Show that there exists a group G with normal subgroups $M_0 \cong M$ and $N_0 \cong N$ so that $G/N_0 \cong V$, $G/M_0 \cong U$, $G/(M_0 N_0) \cong U/M$, and $M_0 \cap N_0 = 1$.

HINT: Find G as a subgroup of $U \times V$.

7.9 Let $M \subseteq \mathbf{Z}(U)$ and $N \subseteq \mathbf{Z}(V)$ and assume $M \cong N$. Show that there exists a group G with normal subgroups K and L such that $K \cong U$, $L \cong V$, $KL = G$, $[K, L] = 1$, and $K \cap L \cong M$.

NOTE: The group G is said to be a *central product* of U and V *identifying* M and N.

7.10 Let G act on G via conjugation and construct the corresponding semidirect product $G \rtimes G$. Show that $G \rtimes G \cong G \times G$.

7.11 Show that a finite nonabelian p-group cannot split over its center.

7.12 Let H act on N via automorphisms and assume no nonidentity element of H fixes any nonidentity element of N. Construct $G = N \rtimes H$ and identify N and H with the corresponding subgroups of G.

  a. Show that $H \cap H^g = 1$ for all $g \in G - H$.
  b. If G is finite, show that $G = N \cup \bigcup_{g \in G} H^g$.

7.13 Let P be a finite p-group that acts on a finite group G and assume (after appropriate identifications) that P is a maximal subgroup of $G \rtimes P$. Show that G is an abelian q-group for some prime q.

HINT: Show that P stabilizes some Sylow subgroup of G.

7.14 An abelian group A is *divisible* if for every $a \in A$ and positive integer n, there exists $b \in A$ with $b^n = a$. Let G be abelian and let $A \subseteq G$ be divisible with $|G : A| < \infty$. Show that $G = A \times B$ for some $B \subseteq G$.

HINT: If $A < G$, show that there exists $U \subseteq G$ with $U \cap A = 1 < U$.

NOTE: The hypothesis that the index is finite is not really essential. In this case, however, an easier proof is available.

7.15 Let G be abelian. A subgroup $A \subseteq G$ is *pure* if whenever $g^n \in A$ with $g \in G$ and $n \in \mathbb{Z}$, there exists $b \in A$ with $b^n = g^n$.

a. If G splits over A, show that A is pure in G.
b. Let A be pure in G and U/A pure in G/A. Show that U is pure in G.
c. Let A be pure in G with $|G : A| < \infty$. Show that G splits over A.

7.16 Fix a prime p and let

$$G = Z_{p^{e_1}} \times Z_{p^{e_2}} \times \cdots \times Z_{p^{e_r}},$$

where $e_i \geq 1$. The object here is to show how the e_i can be determined if we know G only up to isomorphism. We define a function f on the natural numbers by setting

$$f(e) = |\{i \mid e_i = e\}|.$$

Also, if $U_e = \{x \in G \mid x^{p^e} = 1\}$, define the function g so that $|U_e| = p^{g(e)}$. Clearly, g is determined by the isomorphism class of G, and the problem is to show that g determines f. Specifically, show that

$$f(e) = 2g(e) - g(e-1) - g(e+1)$$

for all $e \geq 1$.

7.17 Let $m_1, m_2, \ldots, m_n$ be pairwise relatively prime integers with product m. Show that

$$Z_m \cong \bigtimes_{i=1}^{n} Z_{m_i}.$$

(The notation Z_k represents a cyclic group of order k.)

CHAPTER EIGHT

Solvable and Nilpotent Groups

8A

In Chapter 3 we mentioned that finite simple groups are the "bricks" from which, in some sense, all finite groups are built. We know that these simple groups come in two fundamentally different varieties: the cyclic groups of prime order and the comparatively rare and very much more complex nonabelian finite simple groups. In order to use this "building block" approach to prove general theorems about finite groups, one needs detailed information about the structure of all finite simple groups. Because of the recently completed simple group classification, this has become feasible, and indeed a number of striking theorems have been proved and many old problems have been solved by this method. To give one example of a result proved using the classification, we recall that by Corollary 4.16, if G is finite and $H < G$, then there exists some element $g \in G$ that is not conjugate to any element of H. The new result, proved by W. Kantor, is that in fact, the element g can be chosen to have prime power order. There is no known proof independent of the simple group classification for this seemingly mild extension of Corollary 4.16. (Kantor's result has applications in number theory.)

The simple group classification is, of course, irrelevant for the study of groups all of whose simple building blocks are abelian (and hence of prime order). This class of groups, the finite "solvable" groups, has been the object of a great deal of attention, and there are a number of interesting results that hold for groups of this type but do not hold in general.

The word "solvable" as applied to these groups refers to the work of E. Galois in the early nineteenth century. Galois showed that given any polynomial with rational coefficients, there is associated with it a certain finite group that is now called the "Galois group" of the polynomial. Galois defined precisely what it means for a polynomial to be "solvable" and found a necessary and sufficient condition for the solvability of a polynomial in terms of properties of the associated Galois group. It should not surprise the reader that this polynomial solvability criterion is precisely that the group be solvable. We discuss this in considerable detail and prove Galois's theorem in Chapter 22. For now, we mention only that the well-known result that polynomials of degree 5 are not generally solvable is a consequence of the fact that the symmetric group S_5 is not solvable (since A_5 is simple).

As we said, the finite solvable groups are precisely those for which all of the (necessarily simple) composition factors have prime order. For our "official" definition, however, we use the following, which applies for infinite groups as well.

(8.1) DEFINITION.   A group G is *solvable* if there exists a finite collection of normal subgroups $G_0, G_1, \ldots, G_n$ such that

$$1 = G_0 \subseteq G_1 \subseteq \cdots \subseteq G_n = G$$

and G_{i+1}/G_i is abelian for $0 \le i < n$.

In particular, note that if G is abelian, then the subgroups 1 and G meet the requirements of Definition 8.1, and so G is solvable. An equivalent, but easier to use, characterization of solvability is in terms of the "derived series." Recall that the commutator subgroup G' of G is the subgroup generated by all commutators $[x, y] = x^{-1}y^{-1}xy$ for $x, y \in G$. This subgroup, also called the *derived* subgroup, is the unique smallest normal subgroup of G such that the corresponding factor group is abelian. We can go on to define $G'' = (G')'$ and $G''' = (G'')'$ and so on, but since this notation rapidly gets unwieldy, we usually write $G^{(0)} = G$, $G^{(1)} = G'$, $G^{(2)} = G''$, and in general, $G^{(n)} = (G^{(n-1)})'$ for $n > 0$. We have $G = G^{(0)} \supseteq G^{(1)} \supseteq \cdots$, and these groups constitute the *derived series* of G.

(8.2) LEMMA.   *Let $\varphi : G \to H$ be a surjective homomorphism. Then $\varphi(G^{(n)}) = H^{(n)}$ for every $n \ge 0$. Also, $G^{(n)}$ is characteristic in G for each $n \ge 0$.*

Proof.   We have $\varphi([x, y]) = [\varphi(x), \varphi(y)]$, and since $\varphi(G) = H$, we see that φ maps the set of commutators in G onto those in H. It follows that $\varphi(G') = H'$, and repeated application of this argument yields that $\varphi(G^{(n)}) = H^{(n)}$, as required.

That the terms of the derived series of G are characteristic of course follows from the fact that they are canonically defined. Alternatively, this also follows from the first part of the lemma when we take $H = G$ and $\varphi \in \text{Aut}(G)$.   ∎

We can now replace the unspecified normal series in Definition 8.1 with a particular canonical series.

(8.3) THEOREM.   *A group G is solvable iff $G^{(n)} = 1$ for some n. Also, subgroups and factor groups of solvable groups are solvable.*

Proof.   Suppose G is solvable so that we have subgroups $G_i \lhd G$ with

$$1 = G_0 \subseteq G_1 \subseteq \cdots \subseteq G_n = G$$

and such that G_{i+1}/G_i is abelian for $0 \le i < n$. Thus $(G_{i+1})' \subseteq G_i$ for all

i, and in particular, $G' \subseteq G_{n-1}$. Then $G'' = (G')' \subseteq (G_{n-1})' \subseteq G_{n-2}$, and continuing like this, we obtain $G^{(k)} \subseteq G_{n-k}$ for $0 \leq k \leq n$. We conclude that $G^{(n)} \subseteq G_0 = 1$, as desired.

Conversely, if $G^{(n)} = 1$, then the series

$$1 = G^{(n)} \subseteq G^{(n-1)} \subseteq \cdots \subseteq G^{(1)} \subseteq G^{(0)} = G$$

shows that G is solvable by Definition 8.1, since each $G^{(k)} \triangleleft G$.

Now if $H \subseteq G$, then $H' \subseteq G'$ and in general, $H^{(k)} \subseteq G^{(k)}$. If G is solvable, then for some n we have $H^{(n)} \subseteq G^{(n)} = 1$, and H is solvable. Finally, if $N \triangleleft G$, then application of Lemma 8.2 to the canonical homomorphism $\varphi : G \to G/N$ yields that $(G/N)^{(k)} = \varphi(G^{(k)})$, and so if $G^{(n)} = 1$, we have $(G/N)^{(n)} = 1$. ∎

If G is solvable, its *derived length* $\mathrm{dl}(G)$ is the smallest integer n such that $G^{(n)} = 1$. The groups with derived length 1, for example, are exactly the nontrivial abelian groups.

Generally, groups built from solvable groups are themselves solvable.

(8.4) COROLLARY. *Let $N \triangleleft G$ and suppose that both N and G/N are solvable. Then G is solvable and*

$$\mathrm{dl}(G) \leq \mathrm{dl}(N) + \mathrm{dl}(G/N) .$$

Proof. Let $\mathrm{dl}(N) = n$ and $\mathrm{dl}(G/N) = m$. Since the canonical homomorphism $G \to G/N$ maps $G^{(m)}$ to $(G/N)^{(m)} = 1$, we see that $G^{(m)} \subseteq N$. Thus $G^{(m+n)} = (G^{(m)})^{(n)} \subseteq N^{(n)} = 1$, and the result follows. ∎

To justify our assertion that the finite solvable groups are precisely those groups whose simple "building blocks" have prime order, we prove the following corollary.

(8.5) COROLLARY. *Let*

$$1 = N_0 \triangleleft N_1 \triangleleft \cdots \triangleleft N_n = G$$

be a composition series for G. (Recall that this means that each factor N_{i+1}/N_i is a simple group for $0 \leq i < m$.) Then G is solvable iff the composition factors N_{i+1}/N_i all have prime order.

Proof. First assume that the composition factors have prime order. Despite the fact that these factors N_{i+1}/N_i are abelian, we cannot appeal directly to Definition 8.1 because we are not assuming that the subgroups N_i are all normal in G. Nevertheless, we can say that $G' \subseteq N_{n-1}$ and thus $G'' \subseteq N_{n-2}$, and so on. It follows that $G^{(n)} = 1$ and G is solvable.

Conversely, assume G is solvable. Then each factor N_{i+1}/N_i is solvable by Theorem 8.3. Since these factors are simple, it suffices to observe that a solvable simple group must have prime order. This is clear, since if G is solvable and nontrivial, then $G' < G$. If G is also simple, then $G' = 1$ and G is abelian, and we know that simple abelian groups have prime order. ∎

In general, even if G is solvable, it is not possible to find a series of normal subgroups of G, each of prime index in the next. The finite groups for which there is such a series are the *supersolvable* groups. We defer further exploration of these groups to the problems at the end of this chapter.

8B

A *chief* series in a group G is a series of normal subgroups

N/N_{i-1} : chief factor G $$1 = N_0 \subseteq N_1 \subseteq \cdots \subseteq N_n = G,$$

for which each factor N_{i+1}/N_i is a minimal nontrivial normal subgroup of G/N_i. Equivalently, the series is "saturated" in the sense that for each i with $0 \le i < n$, there exists no subgroup $M \triangleleft G$ with $N_i < M < N_{i+1}$. (We stress that the distinction between a composition series and a chief series is that in the latter we consider only subgroups normal in the whole group G.) It should be clear that every finite group has at least one chief series, and in fact, every normal subgroup of a finite group is a term in some chief series.

We have seen that the composition factors of finite solvable groups are of prime order. We now investigate the chief factors.

(8.6) LEMMA.   *Let N be a minimal normal subgroup of some group G, and assume that N is finite and solvable. Then N is an elementary abelian p-group for some prime p. (This means that N is abelian and $x^p = 1$ for all $x \in N$.)*

Proof.   Since $N > 1$ and is solvable, we have $N' < N$. Since N' char N, we see that $N' \triangleleft G$ and so $N' = 1$ by the minimality of N. It follows that N is abelian. Let p be a prime divisor of $|N|$ and write $A = \{x \in N \mid x^p = 1\}$. Then $1 < A$, and since N is abelian, A is a subgroup and, in fact, A char N. Thus $A \triangleleft G$ and it follows that $A = N$.   ∎

(8.7) COROLLARY.   *All chief factors of a finite solvable group are elementary abelian p-groups for various primes p.*   ∎

For the remainder of our discussion of solvable groups, we limit our attention to finite groups and make repeated use of Corollary 8.7. First we consider maximal subgroups.

(8.8) THEOREM.   *Let $M < G$ be a maximal proper subgroup of the finite solvable group G. Then $|G : M|$ is a prime power.*

Proof.   Let L be maximal among normal subgroups of G contained in M. (In other words, $L = \mathrm{core}_G(M)$.) (6.7) Since $L < G$, we can let K/L be a chief factor of G. (We choose K, therefore, so that K/L is minimal normal in G/L.) Since $K > L$ and $K \triangleleft G$, we have $K \not\subseteq M$ and so $KM > M$. Therefore $KM = G$ and $|G : M| = |K : K \cap M|$, which divides $|K : L|$, a prime power by Corollary 8.7.   ∎

The smallest nonsolvable group is the alternating group A_5, and it has three conjugacy classes of maximal subgroups with indices 5, 6, and 10. Theorem 8.8 would not hold, therefore, if we dropped the solvability hypothesis. It is interesting, however, that the simple group of order 168 also has exactly three conjugacy classes of maximal subgroups, and these have indices 7, 7, and 8, which are all prime powers, although the group is not solvable.

8C

Our next goal is somewhat more ambitious.

(8.9) THEOREM (P. Hall).   *Let G be finite and solvable and let π be any set of prime numbers. Then there exists a subgroup $H \subseteq G$ with order divisible only by primes in π and with index divisible by none of these primes.*

A subgroup $H \subseteq G$ satisfying the conclusion of Theorem 8.9 is called a *Hall π-subgroup* of G. Note that if $\pi = \{p\}$, a singleton set, then a Hall π-subgroup is precisely a Sylow p-subgroup, and in this case, of course, the solvability in Theorem 8.9 would not need to be assumed. Note also that A_4, of order 12, is a Hall $\{2, 3\}$-subgroup of A_5 but that A_5 has no Hall $\{2, 5\}$- or $\{3, 5\}$-subgroup.

To prove Theorem 8.9, we need a few preliminary results. The first of these is the so-called Frattini argument, which is general and quite useful.

(8.10) LEMMA (Frattini argument).   *Let $N \lhd G$, where N is finite, and let $P \in \mathrm{Syl}_p(N)$. Then $G = \mathbf{N}_G(P)N$.*

Proof.   Let $g \in G$ and note that $P^g \subseteq N^g = N$. Since $|P^g| = |P|$, we conclude that $P^g \in \mathrm{Syl}_p(N)$ and so $P^{gn} = P$ for some element $n \in N$. Then $gn \in \mathbf{N}_G(P)$ and hence $g \in \mathbf{N}_G(P)n^{-1} \subseteq \mathbf{N}_G(P)N$. Since $g \in G$ was arbitrary, the result follows. ■

The next result is actually true without the solvability hypothesis (and in that generality it is called the Schur-Zassenhaus theorem). For our purposes, however, the solvable case is sufficient (and it is much easier to prove).

(8.11) THEOREM.   *Let $M \lhd G$, where G is solvable, and assume that $\gcd(|M|, |G : M|) = 1$. Then G splits over M. (In other words, there exists $H \subseteq G$ with $MH = G$ and $M \cap H = 1$.)*

Proof.   Use induction on $|G|$. Since we could take $H = 1$ if $M = G$, we may assume that $M < G$. Thus $|G| > 1$ and our induction is initialized. Let N/M be a chief factor of G. Then N/M is a p-group for some prime p by Corollary 8.7, and we choose $P \in \mathrm{Syl}_p(N)$. Note that this forces $PM = N$.

Write $X = \mathbf{N}_G(P)$. By the Frattini argument (8.10), we have $G = XN = XPM = XM$, where the last equality holds since $P \subseteq X$.

We claim now that X satisfies the hypothesis of the theorem with respect to its normal subgroup $M \cap X$. Certainly X is solvable. We have $|X : M \cap X| = |G : M|$, and this is coprime to $|M \cap X|$ since the latter is a divisor of $|M|$. If $X < G$, the inductive hypothesis applies, and we obtain a subgroup $H \subseteq X$ such that $H(M \cap X) = X$ and $H \cap (M \cap X) = 1$. Now $G = XM = H(M \cap X)M = HM$ and $1 = (H \cap X) \cap M = H \cap M$, and so H is a complement for M in G, as required.

In the remaining case, $X = G$ and so $P \triangleleft G$, and in this situation we apply the inductive hypothesis to G/P. (Note that since $N > M$, we have $P > 1$ and so $|G/P| < |G|$.) We observe that G/P satisfies the hypotheses with respect to its normal subgroup MP/P. This follows since $|MP/P|$ divides $|M|$ and $|G/P : MP/P| = |G : MP|$, and this divides $|G : M|$.

By the inductive hypothesis, we obtain a subgroup H/P of G/P such that

$$\frac{MP}{P} \frac{H}{P} = \frac{G}{P} \quad \text{and} \quad \frac{MP}{P} \cap \frac{H}{P} = 1.$$

Thus $MPH = G$ and so $MH = G$ since $P \subseteq H$. Also, $MP \cap H = P$ and so $M \cap H \subseteq M \cap P$.

Finally, and for the first time, we actually use the assumption that $\gcd(|M|, |G : M|) = 1$. Since p divides $|G : M|$, it cannot divide $|M|$, and hence $M \cap P = 1$. Therefore $M \cap H = 1$, as required. ∎

Proof of Theorem 8.9.   We work by induction on $|G|$ and note that there is nothing to prove if $|G| = 1$. Assuming $G > 1$ then, we let M be a minimal normal subgroup of G. Since G/M is solvable and smaller than G, we can find a Hall π-subgroup H/M of G/M. Note that $|G : H| = |G/M : H/M|$ and so is not divisible by any prime in π.

By Lemma 8.6, there is a prime p such that M is a p-group. With the possible exception of p, therefore, every prime divisor of $|H|$ lies in π. If $p \in \pi$, there is nothing more to prove.

Assume now that $p \notin \pi$. Then $\gcd(|M|, |H : M|) = 1$, and Theorem 8.11 applies and yields a subgroup $K \subseteq H$ with $KM = H$ and $K \cap M = 1$. Then $|K| = |H/M|$ involves only primes in π and $|G : K| = |G : H||M|$ involves no prime of π. In this case, K is a Hall π-subgroup of G. ∎

Theorem 8.9 is analogous to the Sylow E-theorem, and the property it asserts for solvable groups is denoted E_π. The analogs C_π and D_π for the Sylow C- and D-theorems also hold for all π in solvable groups, although we do not prove them here. We also mention that if a finite group G satisfies E_π for every prime set π, then G is necessarily solvable. (All of these results were also discovered by P. Hall.)

8D

In Chapter 7 we said that a finite group is "nilpotent" if all of its Sylow subgroups are normal. We are now ready to give our "official" definition of nilpotence for not necessarily finite groups. We shall see that this agrees with the former definition in the finite case.

(8.12) DEFINITION.   A group G is *nilpotent* if there exists a finite collection of normal subgroups $G_0, G_1, \ldots, G_n$, with

$$1 = G_0 \subseteq G_1 \subseteq \cdots \subseteq G_n = G$$

and such that

$$G_{i+1}/G_i \subseteq \mathbf{Z}(G/G_i)$$

for $0 \leq i < n$.

Of course, nilpotent groups are solvable by Definition 8.1, and abelian groups are certainly nilpotent. As we will soon see, finite p-groups are nilpotent. These are perhaps the most important examples.

(8.13) LEMMA.   *Let G be finite and assume $\mathbf{Z}(G/M) > 1$ for every proper normal subgroup $M \triangleleft G$. Then G is nilpotent.*

Proof.   Define subgroups Z_i for $i \geq 0$ as follows. Let $Z_0 = 1$ and $Z_1 = \mathbf{Z}(G)$. The subgroup Z_2 is defined via the Correspondence theorem so that $Z_2/Z_1 = \mathbf{Z}(G/Z_1)$. (Note that $Z_2 \triangleleft G$.)   We continue like this, defining Z_i so that $Z_i/Z_{i-1} = \mathbf{Z}(G/Z_{i-1})$ for $i > 0$.

By hypothesis, if $Z_i < G$, then $Z_{i+1} > Z_i$, and it follows using the finiteness of G that $Z_n = G$ for some n. This proves the nilpotence of G.   ∎

(8.14) COROLLARY.   *A finite p-group is nilpotent.*

Proof.   By Theorem 5.21, nontrivial finite p-groups have nontrivial centers. The corollary is now immediate from Lemma 8.13.   ∎

A set of normal subgroups $N_i \triangleleft G$ such that

$$N_0 \subseteq N_1 \subseteq \cdots \subseteq N_n$$

is called a *central series* in G if $N_{i+1}/N_i \subseteq \mathbf{Z}(G/N_i)$ for $0 \leq i < n$. A group G is nilpotent, therefore, iff it has a finite central series containing both 1 and G. Using the idea in the proof of Lemma 8.13, we can construct a central series in any group. We set $Z_0(G) = 1$ and inductively define $Z_i(G)$ by the equation $Z_i(G)/Z_{i-1}(G) = \mathbf{Z}(G/Z_{i-1}(G))$ for $i > 0$. (Of course, these subgroups need not be distinct.) The collection $\{Z_i(G) \mid i \geq 0\}$ is called the *ascending* or *upper* central series of G. Certainly, if $Z_n(G) = G$ for some integer n, then G is nilpotent. As we shall see, the converse of this statement is true, too.

Perhaps more important is the "descending" or "lower" central series, which we are about to construct. If $H, K \subseteq G$, we write $[H, K]$ to denote the subgroup generated by all elements of the form $[h, k] = h^{-1}k^{-1}hk$, with $h \in H$ and $k \in K$. The subgroup $[H, K]$ is the *commutator* of H and K and equals 1 iff each element of H commutes with each element of K. (Note that in this notation, we can write the derived subgroup $G' = [G, G]$.) Observe that $[k, h] = [h, k]^{-1}$, from which it follows that $[H, K] = [K, H]$. It should be remembered that, in general, $[H, K]$ may contain elements other than the commutators $[h, k]$ with $h \in H$ and $k \in K$.

Unfortunately, there seems to be no universally accepted notation for the subgroups we are about to construct. We write $G^1 = G$, $G^2 = [G, G] = G'$, $G^3 = [G^2, G]$, and in general, $G^i = [G^{i-1}, G]$ for $i > 1$. Note that although $G' = G^2$, the subgroup $G'' = G^{(2)} = [G', G']$ is generally not equal to either $G^3 = [G', G]$ or $G^4 = [[[G, G], G], G]$.

If $H \subseteq K \subseteq G$, then $[H, G] \subseteq [K, G]$, and since $G^2 \subseteq G^1$, this implies that $G^3 \subseteq G^2$. In general, we have

$$G = G^1 \supseteq G^2 \supseteq G^3 \supseteq \cdots .$$

The *lower* or *descending* central series of G is the set of subgroups G^i. (Note that they are numbered in the reverse direction of the numbering in the definition of the "central series.") To verify that $\{G^i\}$ truly is a central series, we need to establish that each $G^i \triangleleft G$ and that $G^{i-1}/G^i \subseteq \mathbf{Z}(G/G^i)$ for all $i \geq 2$. The normality is immediate, since the G^i are canonically defined and hence are characteristic in G. Part (b) of the next lemma will be used to prove that the factors are central.

(8.15) LEMMA. *Let $X, Y \subseteq G$ be subgroups. Then*:

    a. $[X, Y] \subseteq X$ *iff* $Y \subseteq \mathbf{N}_G(X)$.
    b. *Suppose* $Y \triangleleft G$. *Then* $[X, G] \subseteq Y$ *iff* $XY/Y \subseteq \mathbf{Z}(G/Y)$.

Proof. To prove part (a), note that $[X, Y] \subseteq X$ iff $[x, y] \in X$ for all $x \in X$ and $y \in Y$. Since $[x, y] = x^{-1}x^y$, this happens iff $x^y \in X$ for all x and y. In other words, $[X, Y] \subseteq X$ iff $X^y \subseteq X$ for all $y \in Y$. Since Y is a *group*, this is equivalent to $Y \subseteq \mathbf{N}(X)$.

For part (b), let $\varphi : G \to G/Y$ be the canonical homomorphism. Now $\varphi(X)$ is central in $\varphi(G)$ iff

$$1 = [\varphi(X), \varphi(G)] = \varphi([X, G]),$$

and this happens iff $[X, G] \subseteq \ker(\varphi) = Y$. Since $\varphi(G) = G/Y$ and $\varphi(X) = XY/Y$, the result follows. ∎

(8.16) COROLLARY. *The lower central series of G is a central series for G.* ∎

(8.17) THEOREM. *Let G be any group and suppose $n \geq 1$. Then the following are equivalent*:

    i. $G^{n+1} = 1$.
    ii. $Z_n(G) = G$.

Furthermore, G is nilpotent iff (i) *and* (ii) *hold for some integer n.*

Proof. Certainly, if either (i) or (ii) holds, then G is nilpotent. We may therefore suppose that G is nilpotent and prove that both (i) and (ii) hold for exactly the same nonempty sets of integers n.

Let $1 = N_0 \subseteq \cdots \subseteq N_k = G$ be a central series for G. (This is possible since G is nilpotent.) Since $N_{i+1}/N_i \subseteq \mathbf{Z}(G/N_i)$, we have $[N_{i+1}, G] \subseteq N_i$ by Lemma 8.15(b). We have $G^1 = N_k$ and $G^2 = [G, G] \subseteq [N_k, G] \subseteq N_{k-1}$ and, similarly, $G^3 = [G^2, G] \subseteq [N_{k-1}, G] \subseteq N_{k-2}$. In general, $G^i \subseteq N_{k-i+1}$ and so $G^{k+1} = 1$ and (i) holds for some integer. Also (ii) implies (i) since if $Z_n(G) = G$, we can take $N_i = Z_i(G)$ and $k = n$, and we conclude that $G^{n+1} = 1$.

We now write $Z_i = Z_i(G)$ and show that $N_i \subseteq Z_i$ for all i. This certainly holds for $i = 0$. Working by induction, we may assume that $i > 0$ and $N_{i-1} \subseteq Z_{i-1}$. Now $[N_i, G] \subseteq N_{i-1} \subseteq Z_{i-1}$, and so Lemma 8.15(b) yields

$$N_i Z_{i-1}/Z_{i-1} \subseteq \mathbf{Z}(G/Z_{i-1}) = Z_i/Z_{i-1},$$

and we get $N_i \subseteq Z_i$, as claimed. It follows that $Z_k(G) = G$ and (ii) holds for some integer. Also, (i) implies (ii) since, if $G^{n+1} = 1$, we can take $N_i = G^{n-i}$ and $k = n$, and we conclude that $Z_n = G$. The proof is now complete. ∎

We mention that the smallest integer n for which $Z_n(G) = G$ (and, equivalently, $G^{n+1} = 1$) is called the *nilpotence class* of G. Abelian groups have class 1, and nonabelian groups for which $G' \subseteq \mathbf{Z}(G)$ are precisely those that are nilpotent of class 2.

(8.18) COROLLARY. *Subgroups and factor groups of nilpotent groups are nilpotent.*

Proof. If $H \subseteq G$, then $H^n \subseteq G^n$, and so if $G^n = 1$, we have $H^n = 1$ and H is nilpotent. To prove the result for factor groups, let $N \triangleleft G$ and note that if $\varphi : G \to G/N$ is the canonical homomorphism, then $\varphi(G^n) = (G/N)^n$. (The proof of this is similar to that of Lemma 8.2.) If $G^n = 1$, therefore, we have $(G/N)^{(n)} = 1$, and G/N is nilpotent. ∎

8E

For finite groups, a number of conditions are equivalent to nilpotence. One of these, item (iv) in the theorem below, was used as the temporary definition of nilpotence (for finite groups) in Chapter 7.

(8.19) THEOREM. *Let G be finite. Then the following are equivalent:*

i. *G is nilpotent.*
ii. *$\mathbf{N}_G(H) > H$ whenever $H < G$.*
iii. *Every maximal subgroup of G is normal.*
iv. *Every Sylow subgroup of G is normal.*
v. *G is isomorphic to a direct product of p-groups for various primes p.*

Proof.   Assume (i) and let $1 = N_0 \subseteq \cdots \subseteq N_n = G$ be a central series. Given $H < G$, let k be maximal such that $N_k \subseteq H$, and note that $k < n$. We prove (ii) by showing that $N_{k+1} \subseteq \mathbf{N}_G(H)$. By Lemma 8.15(a), it suffices to check that $[H, N_{k+1}] \subseteq H$. Since $\{N_i\}$ is a central series, however, we have $[H, N_{k+1}] \subseteq N_k$ by Lemma 8.15(b), and (ii) follows since $N_k \subseteq H$.

That (ii) implies (iii) is essentially trivial, since if $M < G$ is a maximal subgroup, then $\mathbf{N}_G(M) > M$ by (ii). Thus $\mathbf{N}_G(M) = G$ and $M \triangleleft G$.

Now assume (iii) and let $P \in \mathrm{Syl}_p(G)$ for some prime p. If $\mathbf{N}_G(P) < G$, choose a maximal subgroup $M \supseteq \mathbf{N}_G(P)$. Then $M \triangleleft G$ and we can apply the Frattini argument (8.10) to deduce that $G = \mathbf{N}_G(P)M$. Since $\mathbf{N}_G(P) \subseteq M < G$, however, this is a contradiction. Thus $\mathbf{N}_G(P) = G$ and $P \triangleleft G$, proving (iv).

Condition (iv) is precisely how we defined finite nilpotent groups in Chapter 7. In Theorem 7.9 we showed that such a group is the direct product of its Sylow subgroups, and so by Lemma 7.6, such a group is isomorphic to an external direct product of p-groups. This proves (v).

Finally, to prove that (v) implies (i), it suffices to show that direct products of nilpotent groups are nilpotent. This is true without assuming finiteness, and so we give this argument as the following separate lemma.   ∎

(8.20) LEMMA.   Let $G_1, G_2, \ldots, G_n$ be nilpotent groups. Then $\overset{n}{\underset{i=1}{\times}} G_i$ is nilpotent.

Proof.   If $H_i \subseteq G_i$, we view $\times H_i \subseteq \times G_i$ and observe that $[\times H_i, \times G_i] = \times [H_i, G_i]$. From this it is immediate that we can compute the lower central series of $\times G_i$ using the formula

$$\left(\overset{n}{\underset{i=1}{\times}} G_i \right)^m = \overset{n}{\underset{i=1}{\times}} (G_i)^m .$$

The result follows.   ∎

The characterization of nilpotence for finite groups in terms of Sylow subgroups is probably the most important consequence of Theorem 8.19. The following theorem is an application of this.

(8.21) THEOREM.   Let $M \triangleleft G$ and $N \triangleleft G$, where M and N are finite and nilpotent. Then MN is nilpotent.

We need a lemma for finding Sylow subgroups of groups written as products of subgroups. The following result is much more general than we really need.

(8.22) LEMMA.   Let $G = HK$ be finite and let $S \subseteq G$ be a p-subgroup. Suppose S contains a Sylow subgroup of H and also one of K. Then $S \in \mathrm{Syl}_p(G)$ and $S = (S \cap H)(S \cap K)$.

Proof.   We need to show that $|G|_p = |S|$, where we are writing the subscript p to denote the p-part of an integer. By Corollary 4.17 we have

$$|G|_p = \left(\frac{|H||K|}{|H \cap K|}\right)_p = \frac{|H|_p|K|_p}{|H \cap K|_p}.$$

Now $|H|_p \le |S \cap H|$ and $|K|_p \le |S \cap K|$ by hypothesis, and $|H \cap K|_p \ge |S \cap H \cap K|$ by Lagrange's theorem. We conclude that

$$|G|_p \le \frac{|S \cap H||S \cap K|}{|S \cap H \cap K|} = |(S \cap H)(S \cap K)| \le |S|.$$

Since $|S| \le |G|_p$ by Lagrange, we have equality throughout. In particular, $(S \cap H)(S \cap K) = S$, and the lemma follows.   ∎

Proof of Theorem 8.21.   Fix a prime p and let $P \in \mathrm{Syl}_p(M)$ and $Q \in \mathrm{Syl}_p(N)$, and note that since each of M and N is nilpotent, we have $P \lhd M$ and $Q \lhd N$. These Sylow subgroups are therefore characteristic in M and N and hence are normal in G. It follows that PQ is a normal p-subgroup of MN.

By Lemma 8.22 we have $PQ \in \mathrm{Syl}_p(MN)$, and so MN satisfies (iv) of Theorem 8.19 and hence is nilpotent.   ∎

By Theorem 8.21 every finite group G has a unique largest normal nilpotent subgroup. We use the notation $\mathbf{F}(G)$ for this subgroup, which is called the *Fitting subgroup* of G. Of course, the Fitting subgroup, being canonically determined, is characteristic in G.

Another characteristic subgroup relevant to the study of nilpotent groups is the Frattini subgroup $\Phi(G)$, defined in Problem 2.7 as the intersection of all the maximal subgroups of G.

(8.23)   THEOREM.   *Let G be finite. Then the following are equivalent*:

i.   *G is nilpotent.*
ii.   *$G/\Phi(G)$ is abelian.*
iii.   *$G/\Phi(G)$ is nilpotent.*

Proof.   First assume (i). We must show that $G' \subseteq \Phi(G)$, and so it suffices to show that $G' \subseteq M$ for every maximal subgroup M of G. Since G is nilpotent, we have $M \lhd G$ by Theorem 8.19, and by the Correspondence theorem, the identity M/M is maximal in G/M. It follows that G/M is of prime order and hence is abelian, and so $G' \subseteq M$ as desired.

That (ii) implies (iii) is trivial since every abelian group is nilpotent. We assume (iii), therefore, and prove (i). By Theorem 8.19, it suffices to show that every maximal subgroup $M \subseteq G$ is normal. We have $\Phi(G) \subseteq M$, and by the Correspondence theorem, $M/\Phi(G)$ is maximal in $G/\Phi(G)$ and hence is normal by Theorem 8.19 applied to the group $G/\Phi(G)$.   ∎

In fact, one can easily say more about the Frattini factor group $G/\Phi(G)$ of a finite nilpotent group G. Not only is it abelian, but it is also the direct product of cyclic groups of prime order. We leave a proof of this to the problems at the end of the chapter.

The following result generalizes the fact that (iii) implies (i) in Theorem 8.23.

(8.24) THEOREM.   *Let G be finite and suppose that $\Phi(G) \subseteq N \triangleleft G$. Then N is nilpotent iff $N/\Phi(G)$ is nilpotent.*

Note that if we take $N = \Phi(G)$ in Theorem 8.24, we deduce that the Frattini subgroup of a finite group is always nilpotent. In other words, $\Phi(G) \subseteq \mathbf{F}(G)$.

Proof of Theorem 8.24.   Certainly, if N is nilpotent, then so is $N/\Phi(G)$. Conversely, assuming that $N/\Phi(G)$ is nilpotent, we will show that N is nilpotent by proving that every Sylow subgroup of N is normal. Let $P \in \mathrm{Syl}_p(N)$ and note that $P\Phi(G)/\Phi(G)$ is a Sylow subgroup of $N/\Phi(G)$ and so, since $N/\Phi(G)$ is nilpotent, it is normal. It follows that $P\Phi(G)/\Phi(G)$ is characteristic in $N/\Phi(G)$ and hence is normal in $G/\Phi(G)$. We conclude that $P\Phi(G) \triangleleft G$.

Since $P \in \mathrm{Syl}_p(N)$, we have $P \in \mathrm{Syl}_p(P\Phi(G))$, and the Frattini argument yields that $G = \mathbf{N}_G(P)P\Phi(G) = \mathbf{N}_G(P)\Phi(G)$. This implies that $\mathbf{N}_G(P) = G$. (Otherwise, $\mathbf{N}_G(P)$ would be contained in some maximal subgroup $M < G$, which, by definition, would also contain $\Phi(G)$. This would contradict $\mathbf{N}(P)\Phi(G) = G$.)

We now have $P \triangleleft G$ and so $P \triangleleft N$. Since P was an arbitrary Sylow subgroup of N, we deduce that N is nilpotent, as required. ∎

The following immediate corollary provides a formula relating the Fitting and Frattini subgroups.

(8.25) COROLLARY.   *We have*

$$\mathbf{F}(G/\Phi(G)) = \mathbf{F}(G)/\Phi(G)$$

for all finite groups G. ∎

8F

A nilpotent group is certainly solvable, and it is a triviality to see that its derived length is bounded above by its nilpotence class. We will prove a much stronger inequality after we establish some preliminary results.

We need to consider "higher commutators." We write $[x, y, z]$ to denote $[[x, y], z]$, and more generally, $[x_1, \dots, x_n]$ is defined inductively for $n \geq 3$ to be $[[x_1, \dots, x_{n-1}], x_n]$. Similarly, we "left associate" for subgroups so that $[X_1, \dots, X_n]$ is defined to be $[[X_1, \dots, X_{n-1}], X_n]$. With this notation, we can write $G^n = [G, G, G, \dots, G]$, where G^n is the nth term of the lower central series and there are n G's within the brackets. We begin with an amazing identity.

(8.26) LEMMA (P. Hall).   *Let $x, y, z \in G$. Then*

$$[x, y^{-1}, z]^y [y, z^{-1}, x]^z [z, x^{-1}, y]^x = 1.$$

Proof.   Compute!   ∎

(8.27) LEMMA (Three subgroups).   *Let* $X, Y, Z \subseteq G$ *and assume*

$$[X, Y, Z] = 1 \quad \text{and} \quad [Y, Z, X] = 1.$$

Then $[Z, X, Y] = 1$.

Proof.   Let $x \in X$, $y \in Y$, and $z \in Z$. Then $[x, y^{-1}, z] = 1 = [y, z^{-1}, x]$, and by Lemma 8.26, we conclude that $[z, x^{-1}, y]^x = 1$ and so $[z, x^{-1}, y] = 1$. Therefore $[z, x^{-1}] \in \mathbf{C}_G(Y)$ for all $z \in Z$ and $x \in X$. Since these elements generate $[Z, X]$, we conclude that $[Z, X] \subseteq \mathbf{C}_G(Y)$ and hence $[Z, X, Y] = 1$.   ∎

(8.28) COROLLARY.   *Let* $X, Y, Z \subseteq G$ *and* $N \triangleleft G$. *Assume*

$$[X, Y, Z] \subseteq N \quad \text{and} \quad [Y, Z, X] \subseteq N.$$

Then $[Z, X, Y] \subseteq N$.

Proof.   Apply Lemma 8.27 in the factor group G/N.   ∎

Sometimes in the group theory literature, one reads proofs involving very complicated and unpleasant commutator calculations. Occasionally these can be replaced by an appropriate appeal to the three subgroups lemma (or to Corollary 8.28). Apparently, what is happening is that the identities being proved are essentially subsumed within P. Hall's identity (8.26). Once Lemma 8.26 is established, there is no need to redo these elementwise calculations. The following theorem is a good example of an application of the three subgroups method.

(8.29) THEOREM.   *Let* $G = G^1 \supseteq G^2 \supseteq \cdots$ *be the lower central series of an arbitrary group* G. *Then*

$$[G^i, G^j] \subseteq G^{i+j}$$

for all $i, j \geq 1$.

We mention that we chose our numbering of the terms of the lower central series so that this simple formula would hold. Some authors write "$G = G^0$" and "$[G, G] = G^1$" and so on. Although this alternative notation does have some appeal, it would complicate the statement of Theorem 8.29.

Proof of Theorem 8.29.   Induct on j. Note that by the definition of the lower central series, we have $[G^i, G] = G^{i+1}$, and so the theorem holds when $j = 1$ and we may assume that $j > 1$. We have

$$[G^i, G^j] = [G^j, G^i] = [G^{j-1}, G, G^i].$$

Since $G^{i+j} \triangleleft G$, we see by Corollary 8.28 that it suffices to show that

$$[G, G^i, G^{j-1}] \subseteq G^{i+j} \quad \text{and} \quad [G^i, G^{j-1}, G] \subseteq G^{i+j}.$$

We have $[G, G^i] = [G^i, G] = G^{i+1}$, and so

$$[G, G^i, G^{j-1}] = [G^{i+1}, G^{j-1}] \subseteq G^{i+j}$$

by the inductive hypothesis, since $j - 1 < j$. Also, $[G^i, G^{j-1}] \subseteq G^{i+j-1}$ by the inductive hypothesis, and so

$$[G^i, G^{j-1}, G] \subseteq [G^{i+j-1}, G] = G^{i+j} .$$

The proof is now complete.   ∎

We can now establish the promised inequality relating the derived length and nilpotence class of a nilpotent group.

(8.30) THEOREM.   *Let G be nilpotent with derived length d and nilpotence class c. Then*

$$d < 1 + \log_2(c + 1) .$$

Proof.   We have $G' = G^2$, and by Theorem 8.29 we get $G'' = [G^2, G^2] \subseteq G^4$ and $G''' = [G'', G''] \subseteq [G^4, G^4] \subseteq G^8$. Continuing like this, we see that $G^{(k)} \subseteq G^{2^k}$ for all $k \ge 0$.

Now $1 < G^{(d-1)} \subseteq G^{2^{d-1}}$ and yet $G^{c+1} = 1$. This gives $c + 1 > 2^{d-1}$, and the result follows.   ∎

(8.31) COROLLARY.   *If G is nilpotent with nilpotence class $c \le 3$, then G' is abelian.*

Proof.   By Theorem 8.30 we have $dl(G) < 3$ and thus $G'' = 1$.   ∎

We close this chapter with one further application of the three subgroups lemma.

(8.32) THEOREM.   *Let H act faithfully via automorphisms on some group G. Suppose that $K \subseteq G$ and that H fixes each element of K and stabilizes each right coset of K in G. Then H is abelian.*

Proof.   Let $\Gamma = G \rtimes H$, the semidirect product, and identify G and H with subgroups of Γ so that $G \triangleleft \Gamma$ and the original action of H on G is the conjugation action within Γ.

Since the right coset Kg is H-invariant, we have $g^h \in Kg$ and so $g^h g^{-1} \in K$ for all $g \in G$ and $h \in H$. We have

$$g^h g^{-1} = h^{-1} g h g^{-1} = [h, g^{-1}]$$

and hence $[G, H] = [H, G] \subseteq K$. We are assuming that H centralizes K, however, and this gives $[G, H, H] \subseteq [K, H] = 1$ and also $[H, G, H] = 1$. By the three subgroups lemma, we conclude that $[H, H, G] = 1$. In other words, $[H', G] = 1$ and $H' \in \mathbf{C}_\Gamma(G)$. Since the action of H on G is faithful, it follows that $H' = 1$, and so H is abelian.   ∎

Problems

8.1 Let G be finite and $P \in \mathrm{Syl}_p(G)$. Suppose $\mathbf{N}_G(P) \subseteq H \subseteq G$. Show that $\mathbf{N}_G(H) = H$.

8.2 Let G be finite and solvable and suppose $M < G$ is a maximal subgroup and $\mathrm{core}_G(M) = 1$. Let N be a minimal normal subgroup of G. Show the following:

    a. $NM = G$ and $N \cap M = 1$.
    b. $\mathbf{C}_M(N) = 1$.
    c. $N = \mathbf{C}_G(N)$.
    d. N is the unique minimal normal subgroup of G.

8.3 Let G be finite. Show that G is nilpotent iff $xy = yx$ whenever $x, y \in G$ have relatively prime orders.

(proper)

8.4 Let G be finite. Show that G has a unique largest solvable normal subgroup.

8.5 A group is *perfect* if it is its own derived subgroup. (In particular, note that nonabelian simple groups are perfect.) Let $N \triangleleft G$ and assume that G/N is perfect. If $\varphi : G \to S$ is a homomorphism, where S is solvable, show that $\varphi(G) = \varphi(N)$.

$F = 6 \to$ contradict.

8.6 Let G be finite, solvable, and nonabelian. Show that $\mathbf{F}(G) > \mathbf{Z}(G)$. $\left(Z(6) \nleq F(6) \right)$

8.7 Let G be finite and solvable. Show that $\mathbf{F}(G) \supseteq \mathbf{C}_G(\mathbf{F}(G))$.

    HINT: Apply Problem 8.6 to $\mathbf{C}_G(\mathbf{F}(G))$.

8.8 Let G be finite and suppose $G = HK$ for subgroups $H, K \subseteq G$. Fix a prime p and show that there exist $U \in \mathrm{Syl}_p(H)$ and $V \in \mathrm{Syl}_p(K)$ such that $UV \in \mathrm{Syl}_p(G)$.

8.9 Let G be finite and suppose $N \triangleleft G$ and G/N is nilpotent. Show that there exists a nilpotent subgroup $H \subseteq G$ such that $NH = G$.

    HINT: Consider separately the cases $N \subseteq \Phi(G)$ and $N \nsubseteq \Phi(G)$.

8.10 A group G is *supersolvable* if there exist normal subgroups N_i with

$$1 = N_0 \subseteq N_1 \subseteq \cdots \subseteq N_n = G$$

and such that N_{i+1}/N_i is cyclic for $0 \le i < n$. Show that a finite nilpotent group is necessarily supersolvable.

8.11 Let G be finite and supersolvable and suppose M is a maximal subgroup of G. Show that $|G : M|$ is a prime number.

    HINT: Find $K, L \triangleleft G$ with $L \subseteq M$ and $MK = G$ and K/L cyclic.

8.12 Suppose that every maximal subgroup of some finite group G has prime index and let p be the largest prime divisor of $|G|$.

a. Show that G has a normal Sylow p-subgroup.

b. Show that G is solvable.

HINT: If $P \in \text{Syl}_p(G)$ and $\mathbf{N}_G(P) \subseteq H \subseteq G$, compute $|G : H|$ mod p.

NOTE: In fact, G is necessarily supersolvable, but this is quite hard to prove.

8.13 Let $H, K \subseteq G$. Show that $[H, K] \lhd \langle H, K \rangle$.

8.14 Let H act faithfully on G via automorphisms and suppose $K \lhd G$ and $K \supseteq \mathbf{C}_G(K)$. If H fixes the elements of K, show that H is abelian.

HINT: Let $\Gamma = G \rtimes H$ and note that $H \subseteq \mathbf{C}_\Gamma(K) \lhd \Gamma$. Deduce that $[G, H] \subseteq \mathbf{C}_G(K) \subseteq K$.

8.15 Let G be nilpotent (and possibly infinite) and suppose $1 < N \lhd G$. Show that $N \cap \mathbf{Z}(G) > 1$.

HINT: Consider $[N, G, G, \ldots, G]$.

NOTE: Compare this with Problem 5.19.

8.16 Show that a minimal normal subgroup of a supersolvable group is cyclic.

HINT: Let M be the minimal normal subgroup and consider two cases according to whether or not $M \subseteq N_1$ in the notation of Problem 8.10. Induct on the number n.

8.17 Let P be a finite p-group. Show that $\Phi(P)$ is the unique normal subgroup of P minimal such that the corresponding factor group is elementary abelian.

CHAPTER NINE

Transfer

9A

It is useful to have a selection of theorems available that can be used to show that certain groups are not simple. Results of this sort make assertions about the structures of simple groups and help us to understand these mysterious objects. A good way to prove that a group G is not simple is to construct a homomorphism from G to some other group, and to show that the kernel of this homomorphism is both proper and nontrivial. We seek, therefore, techniques for constructing homomorphisms.

One such technique we have already exploited is that of group actions: whenever G acts on a set, there is a corresponding homomorphism from G into an appropriate symmetric group. There are two other important general methods for obtaining homomorphisms. One of these is the theory of linear representations, in which one considers homomorphisms from a group into the group of invertible $n \times n$ matrices over some field F. Perhaps the most important case of this is where $F = \mathbb{C}$, the complex numbers. (W. Burnside's proof that groups of order $p^a q^b$ are solvable, where p and q are primes, provides an excellent example of the use of complex representation theory to prove a nonsimplicity result.)

The third general technique for constructing homomorphisms is transfer theory, which was also pioneered by Burnside. In this case, all of the target groups for the homomorphisms we will define are abelian, and so if we wish to use transfer to prove that some group G is not simple, the only hope for success is when G has a proper normal subgroup with an abelian factor group. In other words, transfer theory provides tools for establishing that $G' < G$.

Suppose $H \subseteq G$ and choose a right transversal T for H in G. In other words, T is a set of representatives for the right cosets of H in G and each such coset contains a unique element of T. Now G acts on the set of right cosets of H by right multiplication, and since the elements of T can be viewed as labels for these cosets, this defines a corresponding action of G on T. For $t \in T$ and $g \in G$, we write $t \cdot g$ to denote the unique element of T that lies in the coset Htg.

(9.1) LEMMA. *Let $H \subseteq G$ and suppose that T is a right transversal. If $t \in T$ and $x, y \in G$, then*

a. $t \cdot 1 = t$,
b. $(t \cdot x) \cdot y = t \cdot (xy)$, *and*
c. $tx(t \cdot x)^{-1} \in H$.

Proof.   Statements (a) and (b) merely say that "$\cdot$" is an action of G on T, and this is clear since right multiplication defines an action on $\{Ht \mid t \in T\}$. Statement (c) is immediate from the fact that $t \cdot x \in Htx$. ∎

Suppose $H \subseteq G$ and let T be as before, and assume $|G : H| < \infty$. (Note that $|T| = |G : H|$.) Since $tg(t \cdot g)^{-1} \in H$ for each $t \in T$ and $g \in G$, we can define a map $\pi : G \to H$ by setting

$$\pi(g) = \prod_{t \in T} tg(t \cdot g)^{-1}.$$

Since H need not be abelian and we have not specified any particular order in which to carry out the multiplication, the map π is, of course, not uniquely determined. This ambiguity of definition proves to be only a minor inconvenience, however, since we intend to compose π with the canonical homomorphism $H \to H/M$, where $M \lhd H$ with H/M abelian. The coset (element) $M\pi(g) \in H/M$ is uniquely determined by g, H, M, and T.

(9.2) DEFINITION.   Let $H \subseteq G$ have finite index and suppose $M \lhd H$ with H/M abelian. The *transfer* from G to H/M is the map $v : G \to H/M$ given by

$$v(g) = M\pi(g),$$

where

$$\pi(g) = \prod_{t \in T} tg(t \cdot g)^{-1}$$

and T is a right transversal for H in G.

As we have defined it, the transfer appears to depend on the transversal T (but not on the ordering of the elements of T involved in computing $\pi(g)$). In fact, the transfer is independent of the choice of transversal. Although we do not need to use it, we begin with a proof of this independence.

It is convenient to write $x \equiv y \bmod M$ if $x, y \in H$ with $Mx = My$.

(9.3) THEOREM.   *Let S and T be right transversals for H in G and let $M \lhd H$ with H/M abelian. Assume $|G : H| < \infty$. Then for $g \in G$, we have*

$$\prod_{t \in T} tg(t \cdot g)^{-1} \equiv \prod_{s \in S} sg(s \cdot g)^{-1} \bmod M,$$

and so the transfer map $G \to H/M$ is independent of the transversal used to calculate it.

Proof.   For each $t \in T$, there is a unique element $h_t \in H$ such that $h_t t \in S$, and as t runs over T, the elements $h_t t$ run over S. Furthermore, since

$$H(h_t t)g = H(t \cdot g),$$

we see that the unique element of S in this coset is $h_{t \cdot g}(t \cdot g)$. It follows that the "$\cdot$" operations on S and on T are related by the formula

$$(h_t t) \cdot g = h_{t \cdot g}(t \cdot g).$$

Therefore

$$\prod_{s \in S} sg(s \cdot g)^{-1} \equiv \prod_{t \in T} h_t tg(h_{t \cdot g}(t \cdot g))^{-1}$$

$$\equiv \prod_{t \in T} h_t tg(t \cdot g)^{-1} h_{t \cdot g}^{-1}$$

$$\equiv \prod_{t \in T} tg(t \cdot g)^{-1} \prod_{t \in T} h_t \prod_{t \in T} h_{t \cdot g}^{-1} \mod M.$$

Since $t \cdot g$ runs over T as t does, we have

$$\left(\prod_{t \in T} h_t \right)^{-1} \equiv \prod_{t \in T} h_{t \cdot g}^{-1} \mod M,$$

and the result follows. Throughout this calculation, of course, we are taking advantage of the independence on the order of the factors of the various products (mod M), provided each factor lies in H. ■

Next we establish the fact that makes the transfer interesting.

(9.4) THEOREM.   *The transfer map $v : G \to H/M$ is a homomorphism.*

Proof.   Let $x, y \in G$. We show that $\pi(xy) \equiv \pi(x)\pi(y) \mod M$. Since $t \cdot x$ runs over T as t does, we have

$$\pi(y) = \prod_{t \in T} ty(t \cdot y)^{-1} \equiv \prod_{t \in T} (t \cdot x)y((t \cdot x) \cdot y)^{-1}$$

$$\equiv \prod_{t \in T} (t \cdot x)y(t \cdot xy)^{-1} \mod M.$$

Therefore

$$\pi(x)\pi(y) \equiv \prod_{t \in T} tx(t \cdot x)^{-1} \prod_{t \in T} (t \cdot x)y(t \cdot xy)^{-1}$$

$$\equiv \prod_{t \in T} tx(t \cdot x)^{-1}(t \cdot x)y(t \cdot xy)^{-1}$$

$$\equiv \prod_{t \in T} txy(t \cdot xy)^{-1} \mod M.$$

Again, we have freely rearranged factors that are elements of H. The result now follows.   ∎

Our major applications of the transfer homomorphism depend on a result we call the "Transfer evaluation lemma," which provides a useful computational tool. Before we present this result, however, we observe that there is one situation where the computation of the transfer $v(g)$ is especially easy: where $g \in \mathbf{Z}(G)$. We give an application that exploits this.

(9.5) THEOREM.   *Let G be finite and suppose a Sylow p-subgroup of G is abelian. Then p does not divide $|\mathbf{Z}(G) \cap G'|$.*

Proof.   Let $P \in \text{Syl}_p(G)$. Since P is abelian, we can take $M = 1$ and consider the transfer homomorphism $v : G \to P$. Let T be a right transversal for P in G, and note that if $z \in \mathbf{Z}(G) \cap P$ and $t \in T$, then $Ptz = Pzt = Pt$ and so $t \cdot z = t$ and $tz(t \cdot z)^{-1} = z$. It follows that

$$v(z) = \prod_{t \in T} z = z^{|G:P|} \,,$$

and since z is a p-element and $p \nmid |G : P|$, we conclude that $z = 1$ if $v(z) = 1$. Thus $\mathbf{Z}(G) \cap P \cap \ker(v) = 1$.

Since $v(G)$ is abelian, we have $G' \subseteq \ker(v)$ and hence $P \cap \mathbf{Z}(G) \cap G' = 1$. Because $\mathbf{Z}(G) \cap G'$ is a normal subgroup of G, however, we have

$$1 = P \cap (\mathbf{Z}(G) \cap G') \in \text{Syl}_p(\mathbf{Z}(G) \cap G') \,,$$

and the result follows.   ∎

Theorem 9.5 has applications in the theory of central extensions. Given a finite group G, we consider surjective homomorphisms $\varphi : \Gamma \to G$ such that $\ker(\varphi) \subseteq \mathbf{Z}(\Gamma) \cap \Gamma'$. (We shall not attempt to explain why anyone would be interested in such a thing.) It turns out that there are only finitely many different isomorphism classes of (necessarily abelian) groups M that can occur as $\ker(\varphi)$. As was proved by I. Schur, all of these are homomorphic images of the largest one, which is unique up to isomorphism. This largest possible kernel is called the *Schur multiplier* of G and is denoted $M(G)$. Corollary 9.6 below says that if G has a cyclic Sylow p-subgroup for some prime p, then the order of $M(G)$ is not divisible by p. Since, for example, the only prime for which A_5 has a noncyclic Sylow subgroup is 2, it follows that $M(A_5)$ is a 2-group. In fact, $|M(A_5)| = 2$.

(9.6) COROLLARY.   *Let G have a cyclic Sylow p-subgroup. If $G \cong \Gamma/M$, where Γ is finite and $M \subseteq \mathbf{Z}(\Gamma) \cap \Gamma'$, then p does not divide $|M|$.*

Proof.   Let $P \in \text{Syl}_p(\Gamma)$ and note that $P/(P \cap M) \cong PM/M \in \text{Syl}_p(\Gamma/M)$, and so it is cyclic. Since $P \cap M \subseteq \mathbf{Z}(P)$, it follows that P is abelian and Theorem 9.5 applies.   ∎

9B

Most of our applications of the transfer depend on the following lemma.

(9.7) LEMMA (Transfer evaluation). *Let $M \lhd H \subseteq G$ with $|G : H| < \infty$ and H/M abelian, and let T be a right transversal for H in G. Then for each $g \in G$, there exists a subset $T_0 \subseteq T$ and positive integers n_t for $t \in T_0$ such that*

a. $\sum n_t = |G : H|$,
b. $tg^{n_t}t^{-1} \in H$ *for all $t \in T_0$, and*
c. $\pi(g) \equiv \prod_{t \in T_0} tg^{n_t}t^{-1}$ *mod M.*

Also, if $o(g) < \infty$, then

d. *each n_t divides $o(g)$.*

Proof.   The cyclic group $\langle g \rangle$ acts on T via "$\cdot$" and decomposes T into orbits. Let T_0 be a set of representatives for these orbits and let n_t denote the size of the $\langle g \rangle$-orbit containing t. Statements (a) and (d) should now be clear.
For $t \in T_0$, the permutation induced by g on the $\langle g \rangle$-orbit containing t is an n_t-cycle, and so the elements of this orbit are

$$t, t \cdot g, t \cdot g^2, \ldots, t \cdot g^{n_t - 1},$$

and we have

$$t \cdot g^{n_t} = t.$$

Therefore

$$t \in Htg^{n_t},$$

and (b) follows.
The contribution to the product $\pi(g)$ corresponding to elements of T in the orbit containing t is

$$\prod_{i=0}^{n_t - 1} (t \cdot g^i)g(t \cdot g^{i+1})^{-1} = tg^{n_t}t^{-1},$$

and (c) follows.   ∎

An immediate application of Lemma 9.7 is the following corollary.

(9.8) COROLLARY (Schur).   *Let $|G : \mathbf{Z}(G)| = m < \infty$. Then the map $g \mapsto g^m$ is a homomorphism from G into $\mathbf{Z}(G)$.*

Proof.   In fact, we will show that this map is the transfer $v : G \to \mathbf{Z}(G)$. By the Transfer evaluation lemma, we have for $g \in G$ that

$$v(g) = \pi(g) = \prod_{t \in T_0} tg^{n_t}t^{-1},$$

where $tg^{n_t}t^{-1} \in \mathbf{Z}(G)$. It follows that $tg^{n_t}t^{-1} = g^{n_t}$ and

$$v(g) = \prod_{t \in T_0} g^{n_t} = g^{\sum n_t} = g^m \,,$$

as required.   ∎

More can be said when the hypotheses of Schur's theorem (9.8) are satisfied. If T is a transversal for $\mathbf{Z}(G)$ in G, an easy calculation shows that every commutator in G actually has the form $[s, t]$ for elements $s, t \in T$. In particular, there are only finitely many commutators in G, and so the derived subgroup G' is finitely generated. Also, since $\mathbf{Z}(G)$ is abelian, we see that G' lies in the kernel of the map of Corollary 9.8, and hence the mth power of every element of G' is the identity. As we see in Problem 9.12, this information about G' is sufficient to guarantee that G' is finite when $|G : \mathbf{Z}(G)| < \infty$.

9C

Our principal applications of transfer in finite groups involve the transfer of G into P/P', where $P \in \mathrm{Syl}_p(G)$. In order to prove nonsimplicity theorems, we would like to be able to compute the kernel of the transfer homomorphism $v : G \to P/P'$ and to find conditions sufficient to guarantee that this kernel is proper. Since we are mapping into a p-group, it follows that $v(G) = v(P)$, and so it suffices to compute $P \cap \ker(v)$. This turns out to be the "focal" subgroup of P, which we are about to define.

(9.9) DEFINITION.   Let $H \subseteq G$. Then the *focal subgroup* of H in G is

$$\mathrm{Foc}_G(H) = \langle x^{-1}y \mid x, y \in H \text{ and } x, y \text{ are } G\text{-conjugate}\rangle.$$

Note that if $x, h \in H$, then x and $y = x^h$ are certainly G-conjugate elements of H. It follows that $[x, h] = x^{-1}x^h = x^{-1}y \in \mathrm{Foc}_G(H)$, and we see that $H' \subseteq \mathrm{Foc}_G(H)$. This containment can be proper, however, since there are often elements $x, y \in H$ that are G-conjugate but not H-conjugate. We use the term "fusion" to describe this phenomenon. Specifically, we say that two classes of H are *fused* in G if both are contained in the same G-class. To say that there is *no fusion* in H means that any two elements of H that are G-conjugate are already H-conjugate. In that case, we see that $\mathrm{Foc}_G(H) = H'$.

We usually use the next result when the subgroup $H \subseteq G$ is Sylow. The proof is identical for the more general situation of Hall subgroups, however, and so we state and prove it in that generality.

(9.10) THEOREM (Focal subgroup).   *Let G be finite. Suppose $H \subseteq G$ is a Hall subgroup and let $v : G \to H/H'$ be the transfer map. Then*

$$\mathrm{Foc}_G(H) = H \cap G' = H \cap \ker(v) \,.$$

Proof. If $x, y \in H$ with $y = x^g$ for some $g \in G$, then $x^{-1}y = [x, g] \in G'$ and hence $\mathrm{Foc}_G(H) \subseteq H \cap G'$. Also, $G' \subseteq \ker(v)$ since $v(G)$ is abelian, and hence $H \cap G' \subseteq H \cap \ker(v)$. To complete the proof of the theorem, therefore, we let $g \in H \cap \ker(v)$ and show that $g \in \mathrm{Foc}_G(H)$.

Using the Transfer evaluation lemma (9.7) and the assumption that $g \in H$, we have

$$\pi(g) \equiv \prod_{t \in T_0} tg^{n_t}t^{-1} \equiv g^m \prod_{t \in T_0} g^{-n_t}tg^{n_t}t^{-1} \mod H',$$

where $m = \sum n_t = |G : H|$. Since $g \in H$ and $tg^{n_t}t^{-1} \in H$ by Lemma 9.7, each factor

$$g^{-n_t}tg^{n_t}t^{-1}$$

lies in $\mathrm{Foc}_G(H)$ by the definition of the focal subgroup.

We are assuming that $g \in \ker(v)$. Thus $\pi(g) \in H' \subseteq \mathrm{Foc}_G(H)$ and we conclude that $g^m \in \mathrm{Foc}_G(H)$. However, $m = |G : H|$ is relatively prime to $o(g)$ since $g \in H$, which is a Hall subgroup. Thus

$$g \in \langle g^m \rangle \subseteq \mathrm{Foc}_G(H),$$

and this completes the proof. ∎

(9.11) COROLLARY. *Let $P \in \mathrm{Syl}_p(G)$ and suppose there is no fusion in P. Then $G' \cap P = P'$.*

Proof. Since there is no fusion, we have $\mathrm{Foc}_G(P) = P'$. ∎

In order to compute the focal subgroup and to use results like Corollary 9.11, we need to be able to obtain information about fusion. The following consequence of the Sylow C-theorem is useful for this purpose.

(9.12) LEMMA (Burnside). *Let $P \in \mathrm{Syl}_p(G)$ and suppose $x, y \in \mathbf{C}_G(P)$ are conjugate in G. Then x and y are conjugate in $\mathbf{N}_G(P)$.*

Proof. Write $y = x^g$ for some $g \in G$, and note that $P \subseteq \mathbf{C}_G(y)$. Also $P \subseteq \mathbf{C}_G(x)$ and so

$$P^g \subseteq \mathbf{C}_G(x)^g = \mathbf{C}_G(x^g) = \mathbf{C}_G(y).$$

Therefore P and P^g are Sylow p-subgroups of $\mathbf{C}_G(y)$, and hence there exists $c \in \mathbf{C}_G(y)$ such that $P^{gc} = P$. Then $gc \in \mathbf{N}_G(P)$ and $x^{gc} = y^c = y$, as required. ∎

A subgroup N of a finite group G is said to be a *normal p-complement* in G (where p is prime) if it is a normal subgroup having index a power of p and order not divisible by p. In other words, it is a normal subgroup whose index is equal to the order of a Sylow p-subgroup of G. Corollary 9.11 and Lemma 9.12 can be combined to give a very useful sufficient condition for a group to have a normal p-complement.

(9.13) THEOREM (Burnside). *Let $P \in \mathrm{Syl}_p(G)$ and suppose $P \subseteq \mathbf{Z}(\mathbf{N}_G(P))$. Then G has a normal p-complement.*

Proof.   Suppose $x, y \in P$ are conjugate in G. Since P is abelian, we have $x, y \in \mathbf{C}_G(P)$, and hence by Lemma 9.12, we deduce that $y = x^n$ for some element $n \in \mathbf{N}_G(P)$. Since $P \subseteq \mathbf{Z}(\mathbf{N}_G(P))$, however, we have $x^n = x$ and $x^{-1}y = 1$. We conclude that $\mathrm{Foc}_G(P) = 1$ and hence $P \cap \ker(v) = 1$ by Theorem 9.10, where $v : G \to P$ is the transfer map. It follows that $p \nmid |\ker(v)|$. Since $|G : \ker(v)| = |v(G)|$ is a p-power, we see that $\ker(v)$ is a normal p-complement for G, as required.   ∎

As an application of Burnside's theorem, we give the following corollary.

(9.14) COROLLARY. *Suppose all Sylow subgroups of G are cyclic (for all primes). Then G is solvable.*

Proof.   Let $P \in \mathrm{Syl}_p(G)$, where p is the smallest prime divisor of $|G|$. By the N/C-Theorem (Problem 3.9), we know that $|\mathbf{N}_G(P) : \mathbf{C}_G(P)|$ divides $|\mathrm{Aut}(P)|$. Since P is cyclic, we have $|\mathrm{Aut}(P)| = \varphi(|P|)$, where φ is Euler's function. We can write $\varphi(|P|) = (p - 1)p^{a-1}$, where $p^a = |P|$, and it follows that $|\mathbf{N}_G(P) : \mathbf{C}_G(P)|$ is divisible by no prime larger than p. Since $P \subseteq \mathbf{C}_G(P)$, this index is not divisible by p, and by the choice of p, it certainly is not divisible by any prime smaller than p. It follows that $|\mathbf{N}_G(P) : \mathbf{C}_G(P)| = 1$ and thus $P \subseteq \mathbf{Z}(\mathbf{N}_G(P))$. By Burnside's theorem, therefore, G has a normal p-complement N.

Now $N < G$ and so, working by induction on $|G|$, we may assume that N is solvable. Since G/N is a p-group, it too is solvable, and the result follows.   ∎

Burnside's theorem is also useful for proving that certain numbers do not occur as the orders of simple groups. For instance, if $|G| = 12{,}100 = 2^2 \cdot 5^2 \cdot 11^2$, then the only possibilities for $n_{11}(G) = |\mathrm{Syl}_{11}(G)|$ are 1 and 100. In the first case, a Sylow 11-subgroup is normal and in the second, $P = \mathbf{N}_G(P)$, where $P \in \mathrm{Syl}_{11}(G)$. Since P is abelian, Burnside's theorem applies here, and so in neither case can G be simple.

The arguments used to prove Burnside's theorem give some information when there is an abelian Sylow p-subgroup, even when it is not central in its normalizer.

(9.15) COROLLARY. *Let $P \in \mathrm{Syl}_p(G)$ and assume that P is abelian. Let $N = \mathbf{N}_G(P)$. Then $G' \cap P = N' \cap P$.*

Proof.   By Burnside's lemma (9.12), we know that if $x, y \in P$ are conjugate in G, then they are conjugate in N. It follows that $\mathrm{Foc}_G(P) = \mathrm{Foc}_N(P)$, and the result follows by the Focal subgroup theorem (9.10).   ∎

Results like Corollary 9.15 allow us to work with N in place of G in order to establish the nonsimplicity of G. If it can be shown, for instance, that N has a nontrivial p-group as a homomorphic image, then $P \not\subseteq N'$ and therefore $P \not\subseteq G'$ by Corollary 9.15, and hence $G' < G$.

Corollary 9.15 can be combined with Theorem 9.5 to get a useful consequence.

(9.16) COROLLARY. *Let* $P \in \mathrm{Syl}_p(G)$ *be abelian and write* $N = \mathbf{N}_G(P)$. *Then* $\mathbf{Z}(N) \cap P \cap G' = 1$.

Proof. Application of Theorem 9.5 to N yields that $\mathbf{Z}(N) \cap P \cap N' = 1$. Since $P \cap N' = P \cap G'$, the result follows.   ∎

As an example of how Corollary 9.16 can be used, we give the following application.

(9.17) COROLLARY. *Let* G *be a finite simple group having an abelian Sylow 2-subgroup of order 8. Then* G *contains no element of order 4.*

Proof. Let $P \in \mathrm{Syl}_2(G)$ and $N = \mathbf{N}_G(P)$. If G has an element of order 4, then P either is cyclic or is isomorphic to the direct product of a cyclic group of order 2 with one of order 4. In either case, P has a characteristic subgroup Z of order 2. (Either $Z = \{x^4 \mid x \in P\}$ or $Z = \{x^2 \mid x \in P\}$.)

Since $|Z| = 2$ and $Z \triangleleft N$, we have $Z \subseteq P \cap \mathbf{Z}(N)$, and so by Corollary 9.16, we see that $Z \cap G' = 1$ and hence $G' < G$. This is a contradiction since G is simple, but does not have prime order.   ∎

We mention that the assumption that the Sylow 2-subgroup of G has order 8 in Corollary 9.17 is not really necessary. It does not appear, however, that the more general result can be proved without the use of much deeper methods.

It is natural to ask whether Corollary 9.15 would remain true if the condition that P is abelian were relaxed. If G is simple of order 168, let $P \in \mathrm{Syl}_2(G)$ and $N = \mathbf{N}_G(P)$. Then $N = P$ and $G' = G$, and so $N' \cap P = P' < P = G' \cap P$, and the conclusion of Corollary 9.15 fails in this case. Nevertheless, a recent result of T. Yoshida extends Corollary 9.15 and asserts that $G' \cap P = N' \cap P$ unless a certain group W of order p^{p+1} occurs as a homomorphic image of P. (This group is the "wreath product" of two cyclic groups of order p.) Since the nilpotence class of W is p, it follows that if $P \in \mathrm{Syl}_p(G)$ has nilpotence class smaller than p, then $N' \cap P = G' \cap P$, where, as usual, $N = \mathbf{N}_G(P)$.

9D

We close this chapter with a theorem of Frobenius that gives several necessary and sufficient conditions for a finite group to have a normal p-complement. Unlike our earlier results, Frobenius's theorem applies even if the Sylow p-subgroup is nonabelian.

(9.18) THEOREM (Frobenius). *Let* $P \in \mathrm{Syl}_p(G)$. *Then the following are equivalent*:

i.   *G has a normal p-complement.*
ii. $\mathbf{N}_G(U)$ *has a normal p-complement for all p-subgroups* $U \subseteq G$ *with* $U > 1$.
iii. $\mathbf{N}_G(U)/\mathbf{C}_G(U)$ *is a p-group for all p-subgroups* $U \subseteq G$.
iv. *There is no fusion in* P.

Several comments on Frobenius's theorem are appropriate. The restriction that $U > 1$ in condition (ii) is not essential, but the fact that (ii) implies (i) would be of no interest without it. A subgroup of the form $\mathbf{N}_G(U)$, where $U > 1$ is a p-subgroup, is said to be *p-local* in G. Essentially, (ii) $\Rightarrow$ (i) says that the existence of a normal p-complement is determined "locally."

Condition (iii) says that whenever an element of G with order not divisible by p normalizes a p-subgroup of G, it centralizes it. In other words, these elements, the *p-regular* elements, do not act nontrivially on p-subgroups in G.

It should be remarked that the implication (ii) $\Rightarrow$ (iii) can (and will) be proved separately for each p-subgroup $U \subseteq G$. When just a single subgroup U is considered, however, the reverse implication is actually false. In other words, (iii) must be assumed for all U in order to conclude (ii).

The proofs that (i) implies (ii) and that (ii) implies (iii) are essentially trivial. The work in proving Frobenius's theorem comes in showing that (iii) implies (iv) and that (iv) implies (i), and we prove these separately. We remark, however, that it is easy to get a direct proof that (i) implies (iv). We begin by showing that (iv) implies (i), which is the only point where transfer is used.

(9.19) THEOREM. *Let* $P \in \mathrm{Syl}_p(G)$ *and assume that there is no fusion in* P. *Then G has a normal p-complement.*

Proof. Let $K \triangleleft G$ be minimal such that G/K is a p-group. (Of course, it may happen that $K = G$.) We will show that $p \nmid |K|$ to complete the proof.

We observe first that $p \nmid |K : K'|$; otherwise K/K' would have a proper characteristic subgroup of p-power index. This would correspond to a normal subgroup of G smaller than K and having p-power index in G, thereby contradicting the definition of K.

Let $Q = P \cap K$ and note that $Q \in \mathrm{Syl}_p(K)$. By the previous paragraph, we have $Q \subseteq K'$, and so $Q = Q \cap K' = \mathrm{Foc}_K(Q)$ by the Focal subgroup theorem. On the other hand, if $x, y \in Q$ are K-conjugate, then these are elements of P that are G-conjugate, and so by hypothesis $y = x^u$ for some element $u \in P$. Therefore $x^{-1}y = [x, u] \in [Q, P]$, and so $Q = \mathrm{Foc}_K(Q) \subseteq [Q, P]$. It follows that

$$Q \subseteq [Q, P, P, \ldots, P] \subseteq P^n$$

for all positive integers n. Since P is nilpotent, however, we know that some term P^n of its lower central series is trivial. This forces $Q = 1$ and $p \nmid |K|$, as desired. ∎

Since $G/v(G)$ is always abelian when v is a transfer map, it seems somewhat paradoxical that the proof of Theorem 9.19 used transfer to produce a normal p-complement N of G, where G/N may be nonabelian.

The next result is the key to the proof that (iii) $\Rightarrow$ (iv) in Frobenius's theorem 9.18.

(9.20) LEMMA.   *Let G be finite and assume $\mathbf{N}_G(U)/\mathbf{C}_G(U)$ is a p-group for each p-subgroup $U \subseteq G$. Let $S, T \in \mathrm{Syl}_p(G)$ and write $D = S \cap T$. Then $T = S^c$ for some element $c \in \mathbf{C}_G(D)$.*

Proof.   Assume the result is false and choose the two Sylow subgroups S and T with $D = S \cap T$ as large as possible such that T is not of the form S^c with $c \in \mathbf{C}_G(D)$. Note that $S \neq T$. As illustrated in Figure 9.1, we construct some additional Sylow subgroups.

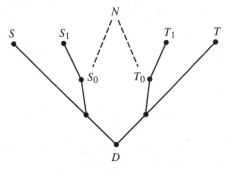

D

Figure 9.1

Let $N = \mathbf{N}_G(D)$ and choose $S_0 \in \mathrm{Syl}_p(N)$ with $S_0 \supseteq N \cap S$ and $T_0 \in \mathrm{Syl}_p(N)$ with $T_0 \supseteq N \cap T$. Also, choose $S_1 \in \mathrm{Syl}_p(G)$ with $S_1 \supseteq S_0$. Finally, choose $n \in N$ with $T_0 = S_0^n$ and set $T_1 = S_1^n$ so that $T_1 \in \mathrm{Syl}_p(G)$ and $T_1 \supseteq T_0$.

Since $S \neq T$, we have $D < S$ and hence

$$S \cap S_1 \supseteq S \cap S_0 \supseteq S \cap N = \mathbf{N}_S(D) > D,$$

since by Theorem 5.25, normalizers grow in p-groups. Similarly, $T_1 \cap T > D$, and so by the choice of S and T, we have

$$S_1 = S^a \quad \text{and} \quad T = T_1^b,$$

where $a \in \mathbf{C}_G(S \cap S_1) \subseteq \mathbf{C}_G(D)$ and $b \in \mathbf{C}_G(T_1 \cap T) \subseteq \mathbf{C}_G(D)$.

By hypothesis, $N/\mathbf{C}_G(D)$ is a p-group, and hence $N = S_0\mathbf{C}_G(D)$ since $S_0 \in \mathrm{Syl}_p(N)$. Therefore $n = sc$, where $s \in S_0 \subseteq S_1$ and $c \in \mathbf{C}_G(D)$, and we have $T_1 = S_1^{sc} = S_1^c$. It follows that $S^{acb} = S_1^{cb} = T_1^b = T$, and since $acb \in \mathbf{C}_G(D)$, we have a contradiction to our choice of S and T. ∎

(9.21) COROLLARY.   *Let $P \in \mathrm{Syl}_p(G)$ and assume that $\mathbf{N}_G(U)/\mathbf{C}_G(U)$ is a p-group for all p-subgroups $U \subseteq G$. Then there is no fusion in P.*

Proof.   Let $x, y \in P$ with $y = x^g$ for some $g \in G$. Since $y = x^g \in P^g$, we have $y \in P \cap P^g$ and Lemma 9.20 yields an element $c \in \mathbf{C}_G(y)$ with $P^{gc} = P$. Then $gc \in \mathbf{N}_G(P)$, and since $\mathbf{N}_G(P)/\mathbf{C}_G(P)$ is a p-group, we can write $gc = ua$, where $u \in P$ and $a \in \mathbf{C}_G(P)$. Now $x^u \in P$ and so $a \in \mathbf{C}_G(x^u)$, and we have

$$x^u = x^{ua} = x^{gc} = y^c = y.$$

Since $u \in P$, we have shown that x and y are P-conjugate, as required.   ■

Proof of Theorem 9.18.   Assume (i) and let N be a normal p-complement in G. If $H \subseteq G$ is any subgroup, we will show that H has a normal p-complement, and this will prove (ii). In fact, $H \cap N$ is a normal subgroup of H with order not divisible by p and $|H : H \cap N| = |HN : N|$ divides $|G : N|$, which is a p-power. It follows that $H \cap N$ is a normal p-complement in H.

Now assume (ii) and let $U \subseteq G$ be a p-subgroup. If $U = 1$, then certainly $\mathbf{N}_G(U)/\mathbf{C}_G(U)$ is a p-group, and so we may assume that $U > 1$. Let M be a normal p-complement for $\mathbf{N}_G(U)$. Then $M \cap U = 1$, and since both M and U are normal subgroups of $\mathbf{N}_G(U)$, we have $M \subseteq \mathbf{C}_G(U)$. Thus $|\mathbf{N}_G(U) : \mathbf{C}_G(U)|$ divides $|G : M|$ and is therefore a p-power. This proves (iii).

That (iii) implies (iv) is Corollary 9.21, and that (iv) implies (i) is Theorem 9.19.   ■

As an application of Frobenius's theorem, we offer the following corollary.

(9.22) COROLLARY.   *Let $|G| = p^a m$, where p is prime and $p \nmid m$. Suppose that $\gcd(m, p^e - 1) = 1$ for all integers e with $1 \le e \le a$. Then G has a normal p-complement.*

Proof.   By Frobenius's theorem, if G does not have a normal p-complement, then there exists a p-subgroup $U \subseteq G$ such that $\mathbf{N}_G(U)/\mathbf{C}_G(U)$ is not a p-group. It follows that $\text{Aut}(U)$ contains an element σ of prime order $q \mid m$.

Let $V = \{u \in U \mid u\sigma = u\}$. Then V is a proper subgroup of U and all elements of $U - V$ lie in orbits of size q under $\langle \sigma \rangle$. Therefore q divides $|U| - |V| = |V|(|U : V| - 1)$, and it follows that q divides $|U : V| - 1$. Since $|U : V| = p^e$ for some exponent e with $1 \le e \le a$, this is a contradiction.   ■

(9.23) COROLLARY.   *If G is simple with $|G| = 8m$, then one of 2, 3, or 7 must divide m.*   ■

We close this chapter by mentioning a powerful theorem of J. Thompson that strengthens Frobenius's theorem in the case where $p \ne 2$. Thompson's result is that in order to prove that G has a normal p-complement (for odd p), it suffices to check that $\mathbf{N}_G(U)$ has a normal p-complement for just two particular p-subgroups $U \subseteq G$. The two subgroups whose normalizers need to be checked are characteristic in a Sylow p-subgroup P of G; they are the center $\mathbf{Z}(P)$ and the *Thompson subgroup* $\mathbf{J}(P)$, defined to be the subgroup generated by all elementary abelian subgroups of P of largest possible order. (In fact, it suffices to check the smaller group $\mathbf{C}_G(\mathbf{Z}(P))$ in place of $\mathbf{N}_G(\mathbf{Z}(P))$).

Problems

9.1 Let $P \in \mathrm{Syl}_p(G)$ and assume $P \subseteq \mathbf{Z}(G)$. Show that $G = P \dot\times K$ for some subgroup $K \subseteq G$.

9.2 Let $A \lhd G$ be abelian and let $v : G \to A$ be the transfer homomorphism.
 a. Show that $v(A) \subseteq \mathbf{Z}(G)$.
 b. Now assume $\gcd(|A|, |G : A|) = 1$. Show that $v(G) = v(A) = A \cap \mathbf{Z}(G)$.
 c. In the situation of part (b), show that $G = v(G) \dot\times \ker(v)$.

9.3 Let $P \in \mathrm{Syl}_p(G)$ be cyclic. Show that if $p \mid |G'|$, then $p \nmid |G : G'|$.

 HINT: First do the case where $P \lhd G$. In that situation, use Problem 9.2.

9.4 Let $P \in \mathrm{Syl}_p(G)$ and $N = \mathbf{N}_G(P)$. Suppose $z \in \mathbf{Z}(N) \cap P$ and $z \notin P'$. Show that $z \notin G'$.

 HINT: Use the Transfer evaluation lemma. Note that if $tz^n t^{-1} \in P$, then $tz^n t^{-1} = z^n$ by Lemma 9.12.

9.5 Let Q be a nonabelian 2-group that contains just one involution, and let $P = Q \times C$, where C is a nontrivial cyclic 2-group. Show that P is not isomorphic to a Sylow 2-subgroup of a simple group.

9.6 Suppose that G is finite and simple and that some maximal subgroup M of G is nilpotent. Show that M is a p-group for some prime p.

 HINT: Suppose p and q are distinct prime divisors of $|M|$, and let $P \in \mathrm{Syl}_p(M)$ and $Q \in \mathrm{Syl}_q(M)$. Apply Lemma 9.12 to Q to compute fusion in P.

9.7 Suppose that $P \in \mathrm{Syl}_p(G)$ and that $M, N \lhd P$. Show that if M and N are conjugate in G, then they are conjugate in $\mathbf{N}_G(P)$.

9.8 Let $P \in \mathrm{Syl}_p(G)$ and let $W \subseteq P$. We say that W is *weakly closed* in P with respect to G if, whenever $W^g \subseteq P$ for $g \in G$, we have $W^g = W$. Show that if $\mathbf{Z}(P)$ is normal in every Sylow p-subgroup that contains it, then $\mathbf{Z}(P)$ is weakly closed in P with respect to G.

9.9 Let $P \in \mathrm{Syl}_p(G)$ and $W \subseteq \mathbf{Z}(P)$. Assume W is weakly closed in P with respect to G and let $N = \mathbf{N}_G(W)$.
 a. Show that if $x, y \in P$ are G-conjugate, then they are N-conjugate.
 b. Show that $G' \cap P = N' \cap P$.

 NOTE: The combined results of Problems 9.8 and 9.9 are from Grün, who called groups *p-normal* if they satisfy the condition of Problem 9.8 (that the center of a Sylow p-subgroup is normal in every Sylow p-subgroup that contains it).

9.10 Let G be finite and simple and suppose $P \subseteq G$ is a maximal subgroup that is a nonidentity p-group. Show that the nilpotence class of P is ≥ 3.

HINT: Note that $P \in \text{Syl}_p(G)$. If P has class ≤ 2, show that $\mathbf{Z}(P)$ cannot be contained in any Sylow p-subgroup of G other than P itself. Use Problems 9.8 and 9.9.

NOTE: A nice application of Thompson's strengthening of Frobenius's theorem is that the situation of Problem 9.10 cannot happen if $p > 2$.

9.11 **(Schreier)** Let $H \subseteq G$ be a subgroup of finite index, where G is finitely generated. Show that H is finitely generated. In fact, if T is a right transversal for H in G and X is a generating set for G, define

$$Y = \{tx(t \cdot x)^{-1} \mid t \in T \text{ and } x \in X\}$$

and show that Y generates H.

HINT: Let $K = \langle Y \rangle$ and show that $\{Kt \mid t \in T\}$ is the full set of right cosets of K in G. Do this by proving that these cosets are permuted among themselves by right multiplication by elements of G.

9.12 **(Schur)** Suppose $|G : \mathbf{Z}(G)| < \infty$. Show that $|G'| < \infty$.

HINT: Use Corollary 9.8 and the discussion following it. Apply Problem 9.11 to G' to show that $G' \cap \mathbf{Z}(G)$ is finite.

9.13 Let G be finite and suppose that $H \subseteq G$ is a Frobenius complement. (This means that $H \cap H^g = 1$ whenever $g \in G - H$.)

 a. Show that there is no fusion in H.
 b. Show that H is a Hall π-subgroup of G for some set π of primes.
 c. If $H' < H$, show that $|G : G'|$ is divisible by some prime in π.
 d. If H is solvable, show that there exists $N \triangleleft G$ such that $NH = G$ and $N \cap H = 1$.

HINT: Use Problem 5.4 to prove part (b). For (d), let $N \triangleleft G$ be minimal such that $|G : N|$ is a π-number, and show that $|N|$ is divisible by no prime in π. To do this, observe that $N \cap H$ is a Frobenius complement in N.

NOTE: A theorem of Frobenius asserts that the conclusion of part (d) holds even without the assumption that H is solvable. The only known proofs of Frobenius's theorem depend on representation theory. (See Chapter 15.)

CHAPTER TEN

Operator Groups and Unique Decompositions

10A

The subject matter of this chapter is intermediate between the "pure" group theory we have been discussing up to now and module theory, which will constitute an important part of our study of rings. We give definitions and prove theorems that have applications both in group theory and in module theory.

(10.1) DEFINITION. Let X be an arbitrary (possibly empty) set and let G be a group. We say that G is an *X-group* (or *group with operator set X*) provided that for each $x \in X$ and $g \in G$, there is defined an element $g^x \in G$ such that if $g, h \in G$, then $(gh)^x = g^x h^x$.

Essentially, X is just a set of endomorphisms of G, in other words, group homomorphisms from G into itself. We do not require, however, that distinct elements of X determine distinct endomorphisms of G. (In fact, we pay hardly any attention at all to individual elements of X.)

For the purpose of motivating the definition, perhaps the best example of an X-group is a vector space V over some field F. We consider the (additive) group structure of V as an F-group, where the scalar multiplications by the elements of F provide the required endomorphisms. (Note that if $V \neq 0$ in this situation, then distinct elements of F give distinct endomorphisms.)

When we study groups we are usually interested in all subgroups and all homomorphisms, but in vector space theory we limit our attention to subspaces and linear transformations. In general, attaching a set of operators to a group provides a way to focus on a restricted family of subgroups and homomorphisms. (Since every group is an X-group with $X = \varnothing$, however, we can use the language of X-groups even when we do not wish to limit our attention to some privileged collection of subgroups. The results of this chapter, therefore, apply to ordinary groups, too.)

Let G be an X-group. A subgroup $H \subseteq G$ is an *X-subgroup* if for every $x \in X$ and $h \in H$, we have $h^x \in H$. In this situation, we also say that H *admits* X. In the vector space example, the F-subgroups of V are precisely the subspaces of V. It was our custom when discussing pure group theory to have it understood when we wrote "$H \subseteq G$" that H is a subgroup of G and not merely a subset (unless the contrary was clear from the context). Similarly, if G is an X-group and we write "$H \subseteq G$" or "$H \lhd G$," our default assumption is that H is an X-subgroup. Note that if G is an X-group and $H \subseteq G$ admits X, then H is automatically an X-group itself, and any X-subgroup of H is an X-subgroup of G.

If G is an arbitrary group, we have seen that we can view G as an operator group simply by taking $X = \varnothing$. In that case, of course, the notions of "subgroup" and "X-subgroup" coincide. We can also view G as an operator group in a different way. Taking $X = G$, we let G act on G by conjugation. Observe that with this construction, the G-subgroups of G are precisely the normal subgroups.

Suppose G and H are X-groups. A homomorphism $\varphi : G \to H$ is an *X-homomorphism* if $\varphi(g^x) = \varphi(g)^x$ for $x \in X$ and $g \in G$. (In the vector space example, of course, the F-homomorphisms between two F-spaces are precisely the linear transformations.) In general, we write $\mathrm{Hom}_X(G, H)$ to denote the set of all X-homomorphisms from G to H.

The results of Chapter 3 on homomorphisms between groups apply without essential change to the situation of X-homomorphisms between X-groups. For instance, kernels and images of X-homomorphisms are X-subgroups, and if $\varphi : G \to H$ is a surjective X-homomorphism, then there is a bijective correspondence between the set of X-subgroups of G containing $\ker(\varphi)$ and the set of all X-subgroups of H. (In fact, the bijection given in the Correspondence theorem (3.7) makes X-subgroups correspond to X-subgroups.)

To make the analogy with our previous work complete, we need to mention that if G is an X-group and $N \lhd G$ (and N admits X), then G/N becomes an X-group in a fairly natural way. We define $(Ng)^x = Ng^x$ for $g \in G$ and $x \in X$, and we observe that this is well defined since if $Ng = Nh$, then $h = ng$ for some $n \in N$, and thus $h^x = n^x g^x$. However, $n^x \in N$ because N admits X, and so $Nh^x = Ng^x$.

At this point, the reader will not be surprised to learn that if $\varphi \in \mathrm{Hom}_X(G, H)$ is surjective and $N = \ker(\varphi)$, then G/N and H are X-isomorphic X-groups. (In other words, an isomorphism exists that is, in fact, an X-homomorphism.)

We do not propose to give proofs for all the routine facts about X-subgroups, X-factor groups, and X-homomorphisms we use. Everything works with at most trivial changes from the case where $X = \varnothing$.

10B

Although we have alluded to composition series previously, we have not yet proved any significant theorems concerning them. The major result in this subject is the Jordan-Hölder theorem, which we now intend to prove in the general context of operator groups. By varying our operator set X, we obtain results that are useful in different ways in group theory and we can also get applications to module theory.

The "basic" version of this theorem is what results when X is empty. We begin with the appropriate operator group definition.

(10.2) DEFINITION. Let G be an X-group. An *X-series* for G is a collection of X-subgroups $H_i \subseteq G$ for $0 \le i \le n$ such that

a. $H_0 = 1$ and $H_n = G$ and
b. $H_i \lhd H_{i+1}$ for $0 \le i < n$.

This X-series is an *X-composition series* for G if, in addition,

c. H_i is maximal among proper normal X-subgroups of H_{i+1} for $0 \le i < n$.

Note that all *composition factors* H_{i+1}/H_i for the X-composition series

$$1 = H_0 \lhd H_1 \lhd \cdots \lhd H_n = G$$

are X-simple X-groups. (In other words, they have no nontrivial proper normal X-subgroups.) Conversely, any X-series with X-simple factors is an X-composition series. A finite X-group clearly has an X-composition series, but an arbitrary X-group may not. If $X = \varnothing$ and G is abelian, for instance, a composition series exists iff G is finite. (This is so because each composition factor must have prime order.) If V is a nonzero finite dimensional vector space over the real numbers $\mathbb{R}$, therefore, then V has no $\varnothing$-composition series, but it does have an $\mathbb{R}$-composition series. (In fact, any nested collection of vector subspaces with dimensions increasing by 1, starting at 0 and going to V, is an $\mathbb{R}$-composition series.)

Situations where X-composition series are especially important for finite groups G are when $X = \varnothing$, in which case the series is simply called a *composition series*, and when $X = G$, acting by conjugation, where X-subgroups are normal subgroups and an X-composition series is called a *chief series*. In particular, results about X-composition series give information about both composition series and chief series for finite groups.

The existence of an X-composition series in an X-group G is a type of "finiteness condition." (A finiteness condition is a property enjoyed by all finite groups and, when it holds for an infinite group, it says that the group is not "too badly" infinite.) Another example of a finiteness condition for groups is the property of being finitely generated. In Chapter 11 we discuss some other finiteness conditions related to the existence of composition series and to being finitely generated. For now, however, we concentrate on composition series.

The following result is useful for constructing new X-composition series from old ones.

(10.3) LEMMA. *Let G be an X-group and suppose that $N, U, V \subseteq G$ are X-subgroups with $N \lhd G$ and $U \lhd V$. Suppose further that V/U is X-simple. Then*

a. *$UN \lhd VN$ and VN/UN is either trivial or X-simple, and*
b. *$(U \cap N) \lhd (V \cap N)$ and $(V \cap N)/(U \cap N)$ is either trivial or X-simple.*

Proof.   For (a), note that $V \subseteq \mathbf{N}_G(UN)$. Since, of course, $N \subseteq \mathbf{N}_G(UN)$, we deduce that $UN \triangleleft VN$, as claimed. Since $VN = V(UN)$, the X-version of the Diamond theorem (3.6) yields that

$$\frac{VN}{UN} \cong \frac{V(UN)}{UN} \cong \frac{V}{(V \cap UN)} .$$

Since $U \subseteq V \cap UN \triangleleft V$ and V/U is X-simple, it follows that either $V \cap UN = V$ or $V \cap UN = U$. We conclude that VN/UN either is trivial or is isomorphic to V/U and hence is X-simple.

The proof of (b) is similar. Since $V \subseteq \mathbf{N}_G(U \cap N)$, we certainly have $U \cap N \triangleleft V \cap N$. The Diamond theorem yields that

$$\frac{V \cap N}{U \cap N} = \frac{V \cap N}{(V \cap N) \cap U} \cong \frac{(V \cap N)U}{U} .$$

Since $U \subseteq U(V \cap N) \triangleleft V$ and V/U is X-simple, we deduce that either $U(V \cap N) = V$ or $U(V \cap N) = U$. The result now follows.   ∎

(10.4) COROLLARY.   *Suppose the X-group G has an X-composition series and let $N \triangleleft G$ be an X-subgroup. Then N has an X-composition series and N is a member of some X-composition series for G.*

Observe that Corollary 10.4 is entirely consistent with the idea that having a composition series is a kind of finiteness condition. Certainly, a normal subgroup of a not-too-big group should also be not too big.

Proof of Corollary 10.4.   Let $1 = H_0 \triangleleft H_1 \triangleleft \cdots \triangleleft H_n = G$ be the given X-composition series for G and write $K_i = N \cap H_i$, so that $1 = K_0 \triangleleft K_1 \triangleleft \cdots \triangleleft K_n = N$ is an X-series for N. By Lemma 10.3(b), each of the factors K_{i+1}/K_i is either trivial or X-simple. It follows that if we simply delete K_{i+1} from our X-series whenever $K_{i+1} = K_i$, what remains is an X-composition series for N.

To prove the second statement, we define $L_i = NH_i$ for $0 \leq i \leq n$. Then $N = L_0 \triangleleft L_1 \triangleleft \cdots \triangleleft L_n = G$, and we see that the K_i and the L_i together constitute an X-series for G containing N. By Lemma 10.3(a), we know that each L_{i+1}/L_i is either trivial or X-simple, and it follows that by deleting repeats as in the previous paragraph, we obtain an X-composition series for G.   ∎

(10.5) THEOREM (Jordan-Hölder).   *Suppose*

$$\mathcal{H} : 1 = H_0 \triangleleft \cdots \triangleleft H_h = G$$

and

$$\mathcal{K} : 1 = K_0 \triangleleft \cdots \triangleleft K_k = G$$

are two X-composition series for an X-group G. Then $h = k$ and the lists of composition factors H_{i+1}/H_i and K_{i+1}/K_i for $0 \leq i < h$ are (up to X-isomorphism) just rearrangements of one another.

The last assertion says that if S is any X-simple X-group, then the number of i for which $H_{i+1}/H_i \cong S$ is equal to the number of j for which $K_{j+1}/K_j \cong S$, although the two sets of indices may not be the same.

Proof of Theorem 10.5.   If $\mathcal{X}$ and $\mathcal{Y}$ are two X-composition series for the same X-group, we shall (for the purposes of this proof only) use the notation $\mathcal{X} \equiv \mathcal{Y}$ to mean that $\mathcal{X}$ and $\mathcal{Y}$ have equal lengths and isomorphic factors (up to rearrangement, but counting multiplicity). It is easy to see that this is an equivalence relation. We work by induction on the length h of $\mathcal{H}$ to show that $\mathcal{H} \equiv \mathcal{K}$. Note that if $h = 1$, then G is X-simple and in that case, $\mathcal{H} = \mathcal{K}$.

We label the penultimate terms of $\mathcal{H}$ and $\mathcal{K}$. Write $H = H_{h-1}$ and $K = K_{k-1}$, and let $\mathcal{H}_0$ and $\mathcal{K}_0$ denote the X-composition series for H and K obtained by deleting the term G from each of $\mathcal{H}$ and $\mathcal{K}$.

If it happens that $H = K$, then since the length of $\mathcal{H}_0$ is $h - 1$, the inductive hypothesis applied in H yields that $\mathcal{H}_0 \equiv \mathcal{K}_0$. From this it follows that $\mathcal{H} \equiv \mathcal{K}$. This is because each of these series has just one extra factor, and it is the same factor G/H for both of them.

Now suppose $H \neq K$ and write $N = H \cap K \triangleleft G$. By Corollary 10.4, we know that N has some X-composition series $\mathcal{N}$ (of length n, say). Since G/H and G/K are X-simple, it follows that $HK = G$ and that $K/N \cong G/H$ and $H/N \cong G/K$ are also X-simple. We can therefore form two new X-composition series $\mathcal{U}$ and $\mathcal{V}$ for G by appending $N \triangleleft H \triangleleft G$ and $N \triangleleft K \triangleleft G$, respectively, to $\mathcal{N}$. Truncating these, we get series $\mathcal{U}_0$ and $\mathcal{V}_0$ for H and K.

By the inductive hypothesis applied in H, we have $\mathcal{H}_0 \equiv \mathcal{U}_0$ and hence $\mathcal{H} \equiv \mathcal{U}$. Therefore $h = n + 2$, and so the length of $\mathcal{V}_0$ is $n + 1 = h - 1$, and the inductive hypothesis applies in K, yielding $\mathcal{V}_0 \equiv \mathcal{K}_0$. It follows that $\mathcal{V} \equiv \mathcal{K}$.

Finally, since $G/H \cong K/N$ and $G/K \cong H/N$, we see that $\mathcal{U} \equiv \mathcal{V}$, and the result follows.   ∎

As an application to finite groups, we prove a result about finite supersolvable groups. The general definition of supersolvability was given in the problems at the end of Chapter 8. For our purposes here, however, we say that a finite group G is supersolvable if there exists a chief series whose factors have prime order. (The reader should check that for finite groups, this "definition" agrees with the one given previously.)

(10.6) THEOREM.   *Let G be a finite supersolvable group. Then G' is nilpotent.*

Proof.   Recall that a chief series for a group is exactly an X-composition series, where X is G, acting by conjugation. Since G is finite and supersolvable, we know that G has some G-composition series with prime order factors, and hence by the Jordan-Hölder theorem, every G-composition series for G has prime order factors.

By Corollary 10.4 we know that G has a G-composition series with G' as one of its members, and we let $1 = H_0 \subseteq H_1 \subseteq \cdots \subseteq H_n = G$ be such a series, where $H_m = G'$. Recall that all $H_i \triangleleft G$ since this is a chief series.

Now G acts by conjugation on H_{i+1}/H_i, which, being of prime order, has an abelian automorphism group by Problem 2.19. It follows that G' is in the kernel of this action for each i, and in particular, $H_{i+1}/H_i \subseteq \mathbf{Z}(G'/H_i)$ for $0 \le i < m$. The series $1 = H_0 \subseteq \cdots \subseteq H_m = G'$ is therefore a central series for G', and we conclude that G' is nilpotent. ∎

If G is an X-group that has an X-composition series $1 = H_0 \lhd \cdots \lhd H_n = G$, we say that G has *finite composition length* and we write $\ell(G) = n$. (Note that by the Jordan-Hölder theorem, the length $\ell(G)$ is uniquely determined.) If G has no X-composition series, we say that it has *infinite length*. The order of a group provides a "handle" we can use to prove theorems for finite groups (for instance, by induction), and similarly, the length of an X-group often provides such a handle when the composition length is finite, even if the group happens to be infinite.

Note that if V is a finite dimensional vector space over some field F and we view V as an F-group, then the composition length $\ell(V)$ is equal to $\dim_F(V)$. Many of the familiar facts about dimension in vector spaces also hold for composition length, but the reader is warned that it is not true in general that if $H \subseteq G$, then $\ell(H) \le \ell(G)$.

(10.7) LEMMA.   *Let G be an X-group of finite length and suppose $N \lhd G$ is an X-subgroup. Then $\ell(G) = \ell(N) + \ell(G/N)$.*

Proof.   This is immediate since, by Corollary 10.4, we know that N is part of an X-composition series for G. ∎

(10.8) COROLLARY.   *In the situation of Lemma 10.7, if $N > 1$, then $\ell(G/N) < \ell(G)$ and if $N < G$, then $\ell(N) < \ell(G)$.*

Proof.   Since only the identity X-group can have length zero, both statements follow from Lemma 10.7. ∎

10C

We now consider decompositions of X-groups as direct products of X-subgroups. A nonidentity X-group is said to be *indecomposable* if it is not the (internal) direct product of two nontrivial X-subgroups. For example, taking $X = \varnothing$, a cyclic group C of order p^a is indecomposable if p is prime and $a > 0$. To see this, suppose $C = U \times V$, where each of U and V is nontrivial. Then each factor would contain a subgroup of order p, and these would be distinct since $U \cap V = 1$. This contradicts the fact that the cyclic group C has just one subgroup of order p. By the fundamental theorem of abelian groups, the nontrivial cyclic p-groups are the only finite abelian groups that are indecomposable.

(10.9) LEMMA.   *Let G be a nontrivial X-group of finite length. Then G is the direct product of some (finite) collection of indecomposable X-subgroups. (If G is itself indecomposable, this collection consists of the single group G.)*

Proof.   There is nothing to prove if G is indecomposable and so we may assume that $G = H \overset{.}{\times} K$, where H and K are nontrivial X-subgroups. Each is therefore proper, and so by Corollary 10.8, we have $\ell(H) < \ell(G)$ and $\ell(K) < \ell(G)$. Working now by induction on the length, we can write each of H and K as direct products of indecomposable X-subgroups, and the result follows by Lemma 7.11.   ■

At this point, the natural question is: How unique is the direct product decomposition in Lemma 10.9? It is easy to see that the factors are not always uniquely determined as subgroups, but are they at least unique up to isomorphism?

(10.10) THEOREM (Krull-Schmidt).   *Let G be an X-group of finite length and suppose*

$$\prod_{i=1}^{n} H_i = G = \prod_{j=1}^{m} K_j,$$

where the X-subgroups H_i and K_j are indecomposable. Then $m = n$ and, after renumbering if necessary, H_i and K_i are X-isomorphic for $1 \leq i \leq n$.

To prove the Krull-Schmidt theorem, we need to discuss endomorphisms of X-groups. We write $\text{End}_X(G) = \text{Hom}_X(G, G)$ (sometimes suppressing the subscript X) and observe that the composition of functions defines an associative "multiplication" on $\text{End}(G)$. (Recall that our standard convention is that $\alpha\beta$ means "α first, then β.") An element $\varphi \in \text{End}(G)$ is *nilpotent* if φ^n is the trivial endomorphism (so that $\varphi^n(G) = 1$) for some integer n. Also, $\varphi \in \text{End}(G)$ is *normal* if the image under φ of every normal X-subgroup of G is normal. (Note that images of X-subgroups of G under X-endomorphisms are automatically X-subgroups.) In the important case where G is abelian, of course, every endomorphism is normal.

(10.11) THEOREM (Fitting).   *Let G be an X-group of finite length and let $\varphi \in \text{End}_X(G)$ be normal. Then $G = H \overset{.}{\times} K$, where H and K are X-subgroups, each mapped into itself by φ, and where the restriction of φ to K is nilpotent in $\text{End}(K)$ and the restriction of φ to H is an automorphism of H. Either of H or K may be trivial.*

Proof.   Since φ is normal, we have $\varphi(G) \triangleleft G$ and thus $\varphi^2(G) \triangleleft G$ and, by induction, $\varphi^n(G) \triangleleft G$. By Lemma 10.7 we have $\ell(G) \geq \ell(\varphi(G)) \geq \ell(\varphi^2(G)) \geq \cdots$. Since each of these numbers is nonnegative, we can choose n so that $\ell(\varphi^n(G)) = \ell(\varphi^{n+1}(G))$, and we write $H = \varphi^n(G)$. Thus $\varphi(H) = \varphi^{n+1}(G)$, and we have $\ell(H) = \ell(\varphi(H))$. Since $\varphi(H) \triangleleft H$, we deduce by Corollary 10.8 that $H = \varphi(H)$. Also, $\varphi(H) \cong H/N$, where $N = H \cap \ker(\varphi)$, and by Corollary 10.8 again, we see that $N = 1$, and so the restriction of φ to H is an automorphism of H.

Let $K = \ker(\varphi^n)$. We have $\varphi^n(\varphi(K)) = \varphi(\varphi^n(K)) = \varphi(1) = 1$, and so $\varphi(K) \subseteq \ker(\varphi^n) = K$. Thus φ maps K into itself and the restriction of φ to K lies in $\text{End}(K)$. This restriction is clearly nilpotent since $\varphi^n(K) = 1$.

Certainly $K \lhd G$, and so what remains is to show that $H \cap K = 1$ and $HK = G$. Since the restriction of φ to H is an automorphism, so also is the restriction of φ^n, and hence $H \cap K = H \cap \ker(\varphi^n) = 1$. Finally, $\varphi^n(G) = H = \varphi^n(H) = \varphi^n(HK)$, and since both G and HK contain $\ker(\varphi^n)$, it follows by the Correspondence theorem (3.7) that $HK = G$.   ∎

(10.12) COROLLARY. *Let G be an X-group that is indecomposable and of finite length. Then every normal endomorphism of G is either nilpotent or an automorphism.*   ∎

We need to do a little technical work with endomorphisms. If G is any group and $\alpha, \beta : G \to G$ are any two maps, then we define $\alpha + \beta : G \to G$ to be the map given by $(\alpha + \beta)(g) = \alpha(g)\beta(g)$. Even if both α and β are endomorphisms, it is not in general true that $\alpha + \beta$ is an endomorphism (although this does hold if G is abelian). Also, of course, if G is nonabelian, it is not true in general that $\alpha + \beta = \beta + \alpha$. (This makes our notation rather nonstandard, since the $+$ symbol is almost never used for noncommutative operations.) We do, however, have distributive laws.

If $\gamma : G \to G$ is any map, then $\gamma(\alpha + \beta) = \gamma\alpha + \gamma\beta$, and if γ is an endomorphism, then $(\alpha + \beta)\gamma = \alpha\gamma + \beta\gamma$. (Recall that the function on the left is done first.)

(10.13) LEMMA. *Let $\alpha, \beta \in \mathrm{End}_X(G)$ and assume that their images, $\alpha(G)$ and $\beta(G)$, centralize each other. Then $\alpha + \beta \in \mathrm{End}_X(G)$. Also, if $\alpha + \beta$ is surjective, then each of α and β is normal.*

Proof.   If $g, h \in G$, then

$$(\alpha + \beta)(gh) = \alpha(gh)\beta(gh) = \alpha(g)\alpha(h)\beta(g)\beta(h)$$
$$= \alpha(g)\beta(g)\alpha(h)\beta(h)$$
$$= ((\alpha + \beta)(g))((\alpha + \beta)(h)),$$

and so $\alpha + \beta$ is an endomorphism of G. It is trivial to check that it is an X-endomorphism.

To prove the second statement, assume that $\alpha + \beta$ is surjective and let $N \lhd G$. Then $\alpha(N) \lhd \alpha(G)$ and $\beta(G) \subseteq \mathbf{C}_G(\alpha(G)) \subseteq \mathbf{C}_G(\alpha(N))$. Since $\alpha + \beta$ maps onto G, we have $G = \alpha(G)\beta(G) \subseteq \mathbf{N}_G(\alpha(N))$, and so $\alpha(N) \lhd G$ and α is a normal endomorphism. A similar proof works for β.   ∎

(10.14) LEMMA. *Let G be an indecomposable X-group of finite length and let $\alpha_1, \alpha_2, \ldots, \alpha_n \in \mathrm{End}_X(G)$. Assume that the subgroups $\alpha_i(G)$ and $\alpha_j(G)$ centralize each other for all $i \neq j$. If $\alpha_1 + \alpha_2 + \cdots + \alpha_n$ is an automorphism of G, then at least one of the α_i is an automorphism, too.*

Proof.   The result is trivial if $n = 1$, and so we start by considering the first interesting case, where $n = 2$. Write $\alpha_1 + \alpha_2 = \sigma \in \mathrm{Aut}(G)$ and note that σ is an X-automorphism. If we put $\beta_i = \alpha_i \sigma^{-1}$, we have $\beta_i \in \mathrm{End}_X(G)$ and $\beta_1 + \beta_2 = 1$, the identity map on G.

The groups $\beta_1(G)$ and $\beta_2(G)$ are the images of $\alpha_1(G)$ and $\alpha_2(G)$ under σ^{-1}, and so they centralize each other. Lemma 10.13 implies that β_1 and β_2 are normal, and so by Fitting's lemma (or, more precisely, by Corollary 10.12), each β_i is either nilpotent or an automorphism. If either β_1 or β_2 is an automorphism, then since $\alpha_i = \beta_i \sigma$, we are done. We are left with the case, therefore, that β_1 and β_2 are each nilpotent.

Since β_1 is nilpotent and $G > 1$, we see that $\ker(\beta_1) > 1$, and we can choose g with $1 \neq g \in \ker(\beta_1)$. Now $\beta_1 + \beta_2 = 1$, and this yields

$$g = (\beta_1 + \beta_2)(g) = \beta_1(g)\beta_2(g) = \beta_2(g) \, ,$$

and therefore $(\beta_2)^m(g) = g$ for all $m \geq 0$. On the other hand, β_2 is nilpotent and so $(\beta_2)^m(g) = 1$ for some m. This is a contradiction and completes the proof when $n = 2$.

If $n \geq 3$, we use induction on n. Write $\alpha = \alpha_1 + \alpha_2$ and note that $\alpha \in \text{End}_X(G)$ by Lemma 10.13. Also, $\alpha(G) \subseteq \alpha_1(G)\alpha_2(G)$, and this centralizes $\alpha_j(G)$ for $j \geq 3$. Since $\alpha + \alpha_3 + \alpha_4 + \cdots + \alpha_n$ is an automorphism, the inductive hypothesis yields that either α_j is an automorphism for some $j \geq 3$ or α is an automorphism. In the latter situation, we conclude that one of α_1 or α_2 is an automorphism by appeal to the case where $n = 2$. ∎

The following result, which is of some independent interest, is the key step in the proof of the Krull-Schmidt theorem.

(10.15) THEOREM (Factor replacement).   *Let G be an X-group of finite length with normal X-subgroups H, N, and K_i such that*

$$H \mathbin{\dot\times} N = G = \prod_{i=1}^{n} K_i,$$

where H and the K_i are indecomposable. Then for some subscript i, we have $G = K_i \mathbin{\dot\times} N$ and $H \cong K_i$.

Proof.   The direct product decompositions give us "projection maps," as follows. Since each element $g \in G$ is uniquely of the form $g = hn$, with $h \in H$ and $n \in N$, we can define a map $\sigma : G \to H$ by $(g)\sigma = h$. It is routine to see that σ is a homomorphism and, in fact, is an X-homomorphism, and we view $\sigma \in \text{End}_X(G)$. Similarly, we can define the X-homomorphisms $\tau_i : G \to K_i$ by setting $(g)\tau_i = k_i$, where $g = \prod k_i$ and $k_i \in K_i$. (Note that we have elected to write the functions on the right in this proof, since we are working in a context where function composition is relevant.)

Observe that $\sum \tau_i = 1$, the identity map on G, and thus

$$\sum (\tau_i \sigma) = \left(\sum \tau_i \right) \sigma = \sigma \, .$$

Now let α_i denote the restriction to H of $\tau_i \sigma$. Since the restriction of σ to H is the identity map, we see that $\sum \alpha_i = 1$.

We wish to apply Lemma 10.14 to the maps $\alpha_i \in \mathrm{End}_X(H)$, and so we need to check that the subgroups $(H)\alpha_i$ centralize each other. Since

$$[(H)\alpha_i, (H)\alpha_j] = [(H)\tau_i\sigma, (H)\tau_j\sigma] = ([(H)\tau_i, (H)\tau_j])\sigma \subseteq ([K_i, K_j])\sigma = 1$$

if $i \neq j$, Lemma 10.14 does apply and we conclude that some α_i (say, α_1) is an automorphism of H.

Write $U = (H)\tau_1$ and note that $U \triangleleft K_1$ since $H \triangleleft G$ and $(G)\tau_1 = K_1$. It follows that $U \triangleleft G$ since all K_j with $j > 1$ centralize K_1 and thus normalize U.

We claim that $G = U \dot\times N$. First, if $u \in U \cap N$, we can write $u = (h)\tau_1$ with $h \in H$. Because $u \in N$, we have $1 = (u)\sigma = (h)\tau_1\sigma = (h)\alpha_1$, and thus $h = 1$ since α_1 is injective. Therefore $U \cap N = 1$ and we need to show that $UN = G$. Note that $(U)\sigma = (H)\tau_1\sigma = (H)\alpha_1 = H$ since α_1 is surjective. Thus $(UN)\sigma = H = (G)\sigma$, and we conclude that $UN = G$ by the Correspondence theorem, since $N = \ker(\sigma)$ is contained in both UN and G.

Since $G = H \dot\times N$ and also $G = U \dot\times N$, we have $U \cong G/N \cong H$. Also, $U \subseteq K_1$ and the proof will be complete if we can show that $U = K_1$. Since $UN = G$ and $U \subseteq K_1$, Dedekind's lemma (Problem 2.9) implies that $K_1 = U(N \cap K_1)$. Since $U \cap N = 1$, we have $U \cap (N \cap K_1) = 1$ and hence $K_1 = U \dot\times (N \cap K_1)$. We are assuming that K_1 is indecomposable, however. Note that U is nontrivial since $U \cong H$, and we are assuming that H is indecomposable and hence is nontrivial by the definition of indecomposability. It follows that $U = K_1$, as required. ∎

Proof of Theorem 10.10.   We have

$$\prod_{i=1}^{n} \cdot\, H_i = G = \prod_{j=1}^{m} \cdot\, K_j,$$

and so we can apply the factor replacement theorem (10.15) when we let $H = H_1$ and $N = \prod_{i=2}^{n} \cdot\, H_i$. Renumbering the K_j if necessary, we conclude that $H_1 \cong K_1$ and

$$K_1 \dot\times \prod_{i=2}^{n} \cdot\, H_i = G = \prod_{j=1}^{m} \cdot\, K_j.$$

We apply factor replacement again, this time with $H = H_2$ and $N = K_1 \dot\times \prod_{i=3}^{n} \cdot\, H_i$. This yields some $K_j \cong H_2$ with $K_j \cap N = 1$, so that $j \neq 1$. We may assume that $j = 2$ and write

$$K_1 \dot\times K_2 \dot\times \prod_{i=3}^{n} \cdot\, H_i = G = \prod_{j=1}^{m} \cdot\, K_j.$$

We apply Theorem 10.15 again and again, successively replacing each H_i by some isomorphic K_j, which necessarily intersects trivially with, and hence is distinct from, all of the previously chosen K_j. When we have done this n times, appropriately renumbering the K_j, we have $H_i \cong K_i$ for $1 \le i \le n$. We also have $G = \prod_{i=1}^{n} K_i$, and this forces $n = m$.   ∎

(10.16) COROLLARY.   *Let G be a nontrivial finite abelian group. Then there exists a unique list of prime powers q_i with $1 < q_1 \le q_2 \le \cdots \le q_n$ such that G is the direct product of cyclic subgroups C_i of order q_i.*

Proof.   The existence of the q_i is precisely the content of the fundamental theorem of abelian groups (7.10). The uniqueness is immediate from the Krull-Schmidt theorem, given that nontrivial cyclic p-groups are indecomposable.   ∎

Another interesting application of the Krull-Schmidt theorem is the following "cancellation" lemma.

(10.17) COROLLARY.   *Let G, H, and K be X-groups of finite length and suppose*
$$G \times H \cong G \times K .$$
Then $H \cong K$.

To prove Corollary 10.17 we introduce the *summand multiplicity function*. This function s_U, where U is any indecomposable X-group, is defined on the collection of all X-groups of finite length as follows. If G is an X-group of finite length, we can apply Lemma 10.9 and write

$$G = \prod_{i=1}^{m} U_i$$

for indecomposable X-subgroups U_i. We now let $s_U(G)$ be the number of subscripts i such that $U_i \cong U$. By the Krull-Schmidt theorem, this is a well-defined number associated with the X-group G, and it is independent of the choice of the particular subgroups U_i.

It should be clear that if $G_1 \cong G_2$, then $s_U(G_1) = s_U(G_2)$ for every choice of U. Conversely, one can recover G (up to X-isomorphism) from a knowledge of all the integers $s_U(G)$, where U runs over all X-isomorphism types that are indecomposable X-groups. To see this, note that G is isomorphic to the external direct product of the various X-groups U, each taken $s_U(G)$ times. (Although there may well be infinitely many different isomorphism classes of indecomposable X-groups, there are clearly only finitely many with $s_U(G) > 0$.)

We should note, however, that this procedure for determining isomorphism of finite length X-groups by comparing the values of the summand multiplicity functions s_U for indecomposable X-groups U does not provide a practical computational algorithm. It is essentially impossible to determine all U and, besides, the evaluation of $s_U(G)$ can be quite difficult for particular groups U and G.

Because of the intimate connections between internal and external direct products, it is essentially trivial to see that $s_U(G \times H) = s_U(G) + s_U(H)$ for any X-groups G and H of finite length and any indecomposable U. We omit a formal proof.

Proof of Corollary 10.17.   Let U be any indecomposable X-group. Then

$$s_U(H) = s_U(G \times H) - s_U(G) = s_U(G \times K) - s_U(G) = s_U(K).$$

Since this holds for all U, we conclude that $H \cong K$.   ∎

We stress that cancellation does not hold in general without the finite length assumption. For instance, let H be any finite group and let G be the unrestricted direct product of infinitely many copies of H. Clearly then,

$$H \times G \cong G \cong 1 \times G,$$

and yet, of course, H need not be trivial.

We close this chapter with another corollary of the Krull-Schmidt theorem, which is also a sort of cancellation result.

(10.18) COROLLARY.   *Let H and K be finite length X-groups and suppose*

$$H \times H \times \cdots \times H \cong K \times K \times \cdots \times K,$$

where each product has n factors. Then $H \cong K$.

Proof.   Let U be any indecomposable X-group and note that

$$ns_U(H) = s_U(H \times \cdots \times H) = s_U(K \times \cdots \times K) = ns_U(K).$$

Thus $s_U(H) = s_U(K)$, and since this holds for all U, we have $H \cong K$.   ∎

Problems

10.1   If G is an X-group, show that G' is an X-subgroup of G.

10.2   Let $G = D_8 \times Z_2$, where D_8 is dihedral of order 8 and Z_2 is cyclic of order 2. Let $X = \{x\}$ be a singleton set. Show how to make G into an X-group so that $\mathbf{Z}(G)$ is not an X-subgroup.

10.3   If G is an X-group of finite length, show that G is solvable iff every X-composition factor of G is abelian.

10.4   Let S be any collection of finite simple groups. We say that a finite group K is an S-*group* if every composition factor of K is isomorphic to a member of S. If G is an arbitrary finite group, show that G has a unique largest normal subgroup that is an S-group and a unique smallest normal subgroup for which the factor group is an S-group.

10.5 Let G be finite and fix some prime p. Show that the following are equivalent:

    i.   Every chief factor of G either is a p-group or is of order not divisible by p.

    ii. Every composition factor of G either is a p-group or is of order not divisible by p.

    NOTE: A finite group satisfying these equivalent conditions is said to be *p-solvable*.

10.6 Let G be finite. Show that G is solvable iff G is p-solvable for all primes p.

10.7 A chief factor K/L of a finite group G is said to be *central* if $K/L \subseteq \mathbf{Z}(G/L)$. Show that all chief series for G contain equal numbers of central factors.

10.8 Let G be finite. Show that G has a normal p-complement iff every chief factor of G that has order divisible by p is central.

    HINT: To show "if," consider $N \triangleleft G$ maximal such that G/N does not have a normal p-complement. Use results from Chapter 9.

10.9 Let G be an X-group of finite length and suppose

$$G = \prod_{i=1}^{n} \cdot H_i,$$

where each X-subgroup H_i is indecomposable. Suppose $\mathbf{Z}(G) = 1$. Show that if $G = N \;\dot{\times}\; M$, then N and M are products of some of the H_i.

    HINT: Note that $\mathbf{C}_G(H_i) = \prod_{j \neq i} H_j$. Use the factor replacement theorem.

CHAPTER ELEVEN

Module Theory without Rings

11A

In Chapter 12 we will begin our study of rings and their associated modules. As will be apparent then, module theory is a crucial part of ring theory, and that makes the title of this chapter somewhat paradoxical. If R is a ring (whatever that is), then an R-module M is an abelian group for which R is a set of operators. It is more than that, of course, since we make assumptions that relate the action of R on M to the internal structure of R. (This is analogous to the situation of a group acting on a set, where the permutations effected on the set by the group elements are not arbitrary, but instead reflect the structure of the group.) Some of the basic definitions and facts in module theory, however, ignore the structure of the ring and are really concerned only with the properties of the module as an abelian R-group. It is this part of module theory with which we deal in this chapter. (Actually, much of what we do can be made to work in the nonabelian case too, but because our real interest is module theory, we do not pursue that.)

To avoid confusion, we do not refer to the objects of study here as "modules," since that word is usually reserved for the situation where we have a ring and additional structure. Our attention is directed mostly to abelian X-groups, which we usually write additively, so that the symbol "0" means either the identity element or the trivial X-subgroup. For additively written groups, it is customary to write the action of X multiplicatively rather than exponentially. In place of u^x, therefore, we write ux, where $x \in X$ and u lies in some (abelian) X-group M. We thus have $(u + v)x = ux + vx$ for $x \in X$ and $u, v \in M$.

11B

It would be too restrictive to assume that our X-groups are finite, but from time to time, we need conditions that will guarantee that they are not "too infinite." In other words, we wish to investigate useful "finiteness conditions" on abelian X-groups. We have already seen one such condition: finite composition length. We also discuss finite generation and the ascending and descending "chain conditions." It turns out that there are a number of interconnections among these four finiteness conditions.

In order to discuss the two chain conditions, it is convenient to consider a more general situation: that of "partially ordered sets." Although this material is arguably set theory rather than algebra, it is so useful for our purposes that it seems appropriate to include it.

(11.1) DEFINITION.   Let P be a set and let $\leq$ be a binary relation on P. Then P is a *partially ordered set*, or *poset*, with respect to $\leq$ provided

i. $a \leq a$ for all $a \in P$,
ii. if $a \leq b$ and $b \leq c$, then $a \leq c$, and
iii. if $a \leq b$ and $b \leq a$, then $a = b$.

Since very little is required in Definition 11.1, examples of posets abound. For instance, if S is any set, then the collection of all subsets of S is a poset with respect to set containment. Also, if G is any group, then the collections of all subgroups and of all normal subgroups form posets, also with respect to containment. Given any poset, one can "reverse the inequality" to construct a new poset, which is called the *dual* of the original poset. (In the dual of P, we have $a \leq b$ iff $b \leq a$ in P.)

Given $a \neq b$ in poset P, it may be that a and b are not comparable, that neither $a \leq b$ nor $b \leq a$. In some posets, every two elements are comparable. If this happens, the set is said to be *linearly* or *totally* ordered. Examples of this are the real numbers and the natural numbers with respect to the ordinary inequality $\leq$. An "algebra" example is the collection of all subgroups of a cyclic p-group with respect to $\subseteq$.

An *ascending chain* in a poset P is an infinite list of not necessarily distinct elements $a_1, a_2, a_3, \ldots$ of P, subscripted by the natural numbers, and such that

$$a_1 \leq a_2 \leq \cdots \leq a_n \leq \cdots .$$

Similarly, a *descending chain* in P is such a list with

$$a_1 \geq a_2 \geq \cdots \geq a_n \geq \cdots .$$

(Here, as is customary, we write $x \geq y$ to mean $y \leq x$. We also write $x > y$ or $y < x$ to mean $y \leq x$ but $y \neq x$.)

We say that P satisfies the *ACC*, or *ascending chain condition*, if every ascending chain in P is "eventually constant." This means that if $a_1 \leq a_2 \leq \cdots$ in P, then for some integer n, we have $a_n = a_{n+1} = a_{n+2} = \cdots$. (In general, the number n depends on the particular ascending chain considered.) In other words, there are at most finitely many strict inequalities between consecutive terms in every ascending chain. Similarly, the poset P satisfies the *DCC*, or *descending chain condition*, if every descending chain is eventually constant. Obviously, the poset P satisfies the DCC iff its dual satisfies the ACC.

Some examples are appropriate. If P is finite, it satisfies both the ACC and the DCC. Also, if P is any set, we can make P into a poset that satisfies both chain conditions simply by defining $\leq$ so that $a \leq b$ iff $a = b$. More interesting are the examples

$$P = \{1/n \mid n \in \mathbb{Z}, n > 0\}$$

and

$$Q = \{1 - 1/n \mid n \in \mathbb{Z}, n > 0\}.$$

It should be clear that P and Q are linearly ordered posets with respect to ordinary inequality and that P satisfies the ACC but not the DCC and Q satisfies the DCC but not the ACC.

We give one more example. Let $\mathbb{Z}$ be the additive group of the integers and let P be the poset of all subgroups of $\mathbb{Z}$. Every nontrivial subgroup of $\mathbb{Z}$ has finite index, and so if $H_1 \subseteq H_2 \subseteq \cdots$ is any ascending chain containing some nontrivial subgroup H_k, then there are only finitely many different subgroups H_m with $m \geq k$, and so the chain is eventually constant. If the given chain has no nontrivial member, then, of course, $0 = H_1 = H_2 = H_3 = \cdots$, and in this case too, the chain is eventually constant. It follows that P satisfies the ACC. It does not, however, satisfy the DCC since, for instance,

$$\langle 2 \rangle > \langle 4 \rangle > \langle 8 \rangle > \langle 16 \rangle > \cdots.$$

Before we return to algebra, we need to consider two more conditions on a poset P: the "maximal" and "minimal" conditions. If $S \subseteq P$, we say that $a \in S$ is a *maximal* element of S if there is no element $b \in S$ with $b > a$. Similarly, $a \in S$ is *minimal* if there is no $b \in S$ with $b < a$. Of course, if S is nonempty but finite, it necessarily contains both maximal and minimal elements. (There may be more than one of each, and some elements of S may be both maximal and minimal.) In general, however, nonempty subsets of a poset may fail to have maximal or minimal elements.

The poset P satisfies the *maximal condition* if every nonempty subset has a maximal element and, dually, it satisfies the *minimal condition* if every nonempty subset has a minimal element. Of course, finite posets satisfy both the maximal and minimal conditions.

(11.2) LEMMA.   *Let P be any poset. Then*

    a. *P satisfies the ACC iff it satisfies the maximal condition and*
    b. *P satisfies the DCC iff it satisfies the minimal condition.*

To prove Lemma 11.2 we need to venture more deeply into the realm of set theory and discuss the axiom of choice (which we accept as "true" and use with only the slightest hesitation).

Let S be a collection of nonempty sets. If $X \in S$, we can certainly choose an element $x \in X$. This, of course, does not require any mysterious axiom; it follows from the definition of what it means to say that X is "nonempty." Now suppose we wish to build a "machine" to do the job for us and select $x \in X$ whenever $X \in S$.

This may be an easy task. For instance, if each X consists of only some natural numbers, we can program our machine always to choose the smallest $x \in X$. This "one line program" will do the job. As another example of where we could build our machine, consider the case where the collection S of sets is finite. We can go through all $X \in S$ and make a selection in advance of $x \in X$ for each X. We can now explicitly design our machine to return our preselected choice when it is presented with any $X \in S$.

In the case where S is infinite, we cannot, in general, go through all $X \in S$ and for each one make an explicit choice of $x \in X$ (although, as we have seen, this can sometimes be easy, depending on the particular collection of sets). In essence, the axiom of choice says that a "choice machine" can always be built. In other words, there exists a *choice function* $f : S \to \bigcup S$ such that $f(X) \in X$ for all $X \in S$. The axiom tells us only that this functions exists; it gives no clue as to its construction.

We stress that we have been using the word "machine" metaphorically here. The axiom of choice definitely does not guarantee the existence of an actual algorithm that could be programmed on a real computer. On the contrary, the axiom of choice is useful precisely when no such algorithm exists. It allows us to assume the existence of a choice function even when there is no choice algorithm.

We illustrate the situation with this example: Given infinitely many pairs of shoes, an easy rule that serves to pick one shoe from each pair is to choose the right shoe. If we are given infinitely many pairs of socks, however, the axiom of choice is needed to guarantee the existence of a rule to pick one from each pair, but it does not tell us how to do it.

Proof of Lemma 11.2. Part (b) follows from (a) if we replace P by its dual poset, and so we prove only (a). If P satisfies the maximal condition and $a_1 \leq a_2 \leq \cdots$ is any ascending chain, let $S = \{a_i \mid i \geq 1\}$. Since $S \neq \varnothing$, the maximal condition guarantees the existence of a maximal element $a \in S$ and we can write $a = a_n$ for some n. Since $a_m \geq a_n$ for $m \geq n$, we have $a_m = a_n$ for these m, and the chain is eventually constant and P satisfies the ACC.

Conversely, assume P satisfies the ACC and let $S \subseteq P$ be nonempty. Suppose S has no maximal element. Then for every $a \in S$, the set

$$S(a) = \{b \in S \mid b > a\}$$

is nonempty. By the axiom of choice, there exists a choice function that selects some element $b \in S(a)$ for every $a \in S$. In other words, there exists $f : S \to S$ such that $f(a) > a$ for every $a \in S$.

Now choose $a \in S$ and define $a_n \in S$ inductively by setting $a_1 = a$, $a_2 = f(a_1)$, and in general, for $n > 1$, let $a_n = f(a_{n-1})$. Then $a_1 \leq a_2 \leq \cdots$ is an ascending chain in P in which no two consecutive terms are equal. This contradicts the ACC and completes the proof. ∎

Finally we return to algebra. Let M be an abelian X-group and consider the poset of all X-subgroups of M ordered by inclusion: $\subseteq$. We say that M is *noetherian* (after Emmy Noether) if this poset satisfies the ACC. Also, M is *artinian* (after Emil

Artin) if it satisfies the DCC. (Often, and somewhat improperly, one says that M satisfies the ACC or DCC if it is noetherian or artinian.)

Of course, if M is finite, it is both noetherian and artinian. Taking $X = \varnothing$, we see that $\mathbb{Z}$, the additive group of the integers, is noetherian but not artinian. For an example of an artinian but not noetherian group (with $X = \varnothing$), fix a prime number p and consider the multiplicative subgroup of the complex numbers consisting of all elements with order a power of p.

A finiteness condition we have already studied, finite composition length, can be viewed as "decomposing" into the two chain conditions. This is the content of our next result.

(11.3) THEOREM. *Let M be an abelian X-group. Then M has finite composition length iff M is both noetherian and artinian.*

Proof.   Suppose M is both noetherian and artinian. Let S be the set consisting of those X-subgroups of M that have finite composition length, and note that $S \neq \varnothing$ since $0 \in S$. Our goal is to show that $M \in S$.

Since M is noetherian, the poset of all X-subgroups of M satisfies the maximal condition by Lemma 11.2, and hence there exists a maximal element S in S. We may assume that $S < M$ and we let $\mathcal{T}$ be the collection of all X-subgroups of M that properly contain S. Since $M \in \mathcal{T}$, we see that $\mathcal{T}$ is not the empty set, and since M is artinian, Lemma 11.2 guarantees the existence of some minimal element $T \in \mathcal{T}$.

Now $T > S$, but there does not exist any X-subgroup U with $T > U > S$, because otherwise $U \in \mathcal{T}$ and this would contradict the minimality of T. We conclude, therefore, that T/S is X-simple, and so we can append T to the end of an X-composition series for S and get an X-composition series for T. Therefore $\ell(T) < \infty$ and hence $T \in S$. Since $T > S$, this contradicts the maximality of S.

Conversely, assume that M has finite length. If $S_1 \subseteq S_2 \subseteq \cdots$ is any ascending chain of S-subgroups, then since M is abelian, each $S_i \triangleleft M$ and so

$$\ell(S_1) \leq \ell(S_2) \leq \cdots \leq \ell(M) < \infty$$

by Lemma 10.7. It follows that the sequence $\{\ell(S_i)\}$ of integers is eventually constant, and for some integer $n \geq 1$ we have

$$\ell(S_n) = \ell(S_{n+1}) = \cdots .$$

By Corollary 10.8, this forces $S_n = S_{n+1} = \cdots$, and hence M is noetherian.

If, on the other hand, $S_1 \supseteq S_2 \supseteq \cdots$ is any descending chain of X-subgroups, then by Lemma 10.7,

$$\ell(S_1) \geq \ell(S_2) \geq \cdots ,$$

and the sequence of integers $\{\ell(S_i)\}$ must eventually be constant in this case, too. Reasoning as above and using Corollary 10.8, we have $S_n = S_{n+1} = \cdots$ for some n, and therefore M is artinian.   ∎

The real significance of the artinian condition will not become clear until we study actual modules over rings. The noetherian condition, however, is interesting even at this level of generality because of its connection with another finiteness condition: finite generation.

(11.4) THEOREM. *Let M be an abelian X-group. Then M is noetherian iff every X-subgroup of M is finitely generated.*

Perhaps we should stress here that a "generating set" for an X-group G is a subset of G not contained in any proper X-subgroup. It may well be contained in proper subgroups of G that do not admit X, in which case it would not generate G as a group but only as an X-group. For instance, any nonzero vector in a one-dimensional F-vector space generates the space as an F-group.

We need the following lemma concerning unions of subgroups. Although unions of algebraic objects seldom are interesting, nested unions, that is, unions of linearly ordered collections, are an exception.

(11.5) LEMMA. *Let G be any X-group and suppose $\mathcal{H}$ is a nonempty collection of X-subgroups, linearly ordered with respect to inclusion. Then*

a. *the union $U = \bigcup \mathcal{H}$ of all members of $\mathcal{H}$ is an X-subgroup of G, and*
b. *if U is finitely generated, then U is a member of $\mathcal{H}$.*

Proof.   To prove part (a), suppose that $u, v \in U$. Then there exist $H, K \in \mathcal{H}$ such that $u \in H$ and $v \in K$. Since $\mathcal{H}$ is linearly ordered, we may assume that $K \subseteq H$ and hence $u, v \in H$. It follows that $uv \in H \subseteq U$ and, similarly, $u^{-1} \in U$ and $u^x \in U$ for $x \in X$. This proves (a).

Now let Y be a finite generating set for U. Since $U = \bigcup \mathcal{H}$, we can choose for each $y \in Y$, some member $H_y \in \mathcal{H}$ such that $y \in H_y$. Now the finite set $\{H_y \mid y \in Y\}$ has some maximal element H, and since $\mathcal{H}$ is linearly ordered, we see that $H \supseteq H_y$ for every $y \in Y$. It follows that H contains each element of the generating set Y and so $H \supseteq U$. Since the reverse containment is trivial, we have $U = H \in \mathcal{H}$, as required. ∎

It is interesting that the converse of Lemma 11.5(b) is true, too. An X-group G is necessarily finitely generated if it cannot be written as the union of a linearly ordered collection of proper X-subgroups. This result appears (with a hint) as Problem 11.8.

Proof of Theorem 11.4.   Assume M is noetherian. To show that $N \subseteq M$ is finitely generated, we consider

$$\mathcal{S} = \{S \subseteq N \mid S \text{ is finitely generated}\},$$

and we observe that $\mathcal{S}$ is nonempty since $0 \in \mathcal{S}$. Since the poset of X-subgroups of M satisfies the maximal condition by Lemma 11.2, we can choose a maximal element $S \in \mathcal{S}$ and a finite generating set Y for S. We have $S \subseteq N$, and if this inclusion is proper, we can choose an element $n \in N$ with $n \notin S$. The

X-subgroup T generated by the finite set $Y \cup \{n\}$ thus lies in S. We have $S = \langle Y \rangle \subseteq T$, and since $n \in T$, we see that $S < T$. This contradicts the maximality of S and proves that $S = N$ and N is finitely generated.

Conversely, suppose that every X-subgroup of M is finitely generated. Let $N_1 \subseteq N_2 \subseteq \cdots$ be an ascending chain of X-subgroups of M. Our goal is to show that $N_n = N_{n+1} = \cdots$ for some integer n. Write $N = \bigcup N_i$ and note that N is an X-subgroup of M by Lemma 11.5(a). Thus N is finitely generated, and hence by Lemma 11.5(b), we see that $N = N_n$ for some integer n. For $m \geq n$, we have $N \supseteq N_m \supseteq N_n = N$ and thus $N_n = N_m$, as required.   ∎

It is trivial to see that the noetherian and artinian conditions are inherited by subgroups. If we use the Correspondence theorem, it is nearly as trivial to see that these conditions are inherited by factor groups, too. The following result, which provides a useful tool for proving that X-groups are noetherian or artinian, is somewhat of a converse to these observations.

(11.6) THEOREM.   *Let M be an abelian X-group and suppose $N \subseteq M$ is an X-subgroup.*

a. *If both N and M/N are noetherian, then so is M.*
b. *If both N and M/N are artinian, then so is M.*

The following easy result is the key to the proof of Theorem 11.6.

(11.7) LEMMA.   *Let $N, H, K \subseteq G$, where G is arbitrary. Suppose*

i. $H \subseteq K$,
ii. $H \cap N = K \cap N$, *and*
iii. $HN = KN$.

Then $H = K$.

Proof.   For $k \in K$, we show that $k \in H$. We have $k \in KN = HN$, and we can write $k = hn$ with $h \in H$ and $n \in N$. Since $h \in H \subseteq K$, we see that $n = h^{-1}k \in K \cap N = H \cap N$ and hence $h^{-1}k \in H$. The result now follows.

∎

Proof of Theorem 11.6.   Let $H_1 \subseteq H_2 \subseteq \cdots$ be an ascending chain (of X-subgroups, of course) in M. Then

$$H_1 \cap N \subseteq H_2 \cap N \subseteq \cdots$$

and

$$\frac{(H_1 + N)}{N} \subseteq \frac{(H_2 + N)}{N} \subseteq \cdots$$

are ascending chains in N and M/N, respectively, and so are eventually constant (in the situation of part (a)). We can choose an integer n, therefore, so that if $m \geq n$, we have

$$H_n \cap N = H_m \cap N$$

and

$$H_n + N = H_m + N \, .$$

By Lemma 11.5 we have $H_n = H_m$, and (a) is proved. The proof of (b) is similar and so we omit it. ∎

The following application of Theorem 11.6 is important for our study of rings and modules.

(11.8) COROLLARY. *Let M be an abelian X-group that is the sum of a finite collection of noetherian X-subgroups. Then M is noetherian. Similarly, if each of the summands is artinian, then so is M.*

Proof. We are given $M = N_1 + N_2 + \cdots + N_k$. If $k = 1$, the result is trivial and so we work by induction on k and assume $k > 1$. Let $N = N_1 + \cdots + N_{k-1}$ and note that N is noetherian (or artinian, in the situation where each N_i is artinian) by the inductive hypothesis. Also,

$$M/N = (N + N_k)/N \cong N_k/N_k \cap N$$

is a homomorphic image of N_k and so is noetherian (or artinian). The result follows by Theorem 11.6. ∎

11C

An abelian X-group M is said to be *completely reducible* if every X-subgroup $N \subseteq M$ is a direct summand. In other words, for every such N, there exists an X-subgroup $U \subseteq M$ such that $M = N \dotplus U$. (Such a subgroup U is said to be a *complement* for N in M. In general, it is by no means unique.) Note that complete reducibility is not a finiteness condition since it can easily fail even when M is finite. For instance, if $X = \varnothing$ and M is cyclic of order 4, then M is not completely reducible. If X is a field and M is any X-vector space, then M is completely reducible. (This is easy if M is finite dimensional but is, in fact, true even in the infinite-dimensional case. This follows from the characterization of complete reducibility given in the next theorem.)

(11.9) THEOREM. *An abelian X-group M is completely reducible iff M is the sum of (or, equivalently, is generated by) its simple X-subgroups.*

Note that the simple X-subgroups of an X-vector space (when X is a field) are precisely the one dimensional subspaces. Since every vector is contained in one of these, they certainly generate the whole vector space.

It is trivial to see (without appeal to Theorem 11.9) that if M is itself a simple X-group, then M is completely reducible. Because complete reducibility can be viewed as a generalization of simplicity, some authors use the adjective "semisimple" to mean what we have called "completely reducible." Unfortunately, this word also has a number of other meanings in the literature, and so we avoid its use altogether.

The full proof of Theorem 11.9 relies on Zorn's lemma, which we prove later in this chapter. If we are willing to assume a finiteness condition, however, we can prove Theorem 11.9 much more easily. As we shall see, the usual finiteness conditions coalesce for completely reducible abelian X-groups.

We begin with an easy lemma.

(11.10) LEMMA.   *Let $N \subseteq M$, where M is a completely reducible abelian X-group. Then N and M/N are completely reducible.*

Proof.   To see that N is completely reducible, let $U \subseteq N$ be an X-subgroup. Since M is completely reducible, we can write $M = U \dotplus V$. By Dedekind's lemma (see Problem 2.9), we have $N = U + (N \cap V)$, and this sum is clearly direct.

Now $N = M \dotplus L$ for some X-subgroup $L \subseteq N$. By the previous paragraph, L is completely reducible. Since $M/N \cong L$, the result follows.   ∎

(11.11) THEOREM.   *Let M be a completely reducible abelian X-group. Then the following are equivalent:*

i.   *M is noetherian.*
ii.   *M is artinian.*
iii.   *M has finite composition length.*
iv.   *M is finitely generated.*

Proof.   We first show the equivalence of conditions (i) and (ii). Assume that M is noetherian but not artinian. Since the collection of artinian X-subgroups of M is certainly nonempty, we can use the maximal condition to choose $U \subseteq M$ maximal with respect to the property of being artinian. Since M is not artinian, we have $U < M$, and so (by the maximal condition again) we can choose $V \supseteq U$ such that V is maximal among proper X-subgroups of M. Thus M/V is X-simple. Since M/U is completely reducible by Lemma 11.10, we can choose a complement S/U for V/U in M/U. Then $S/U \cong (M/U)/(V/U) \cong M/V$ is X-simple and hence is artinian. Now S/U and U are both artinian, and this implies that S is artinian by Theorem 11.6. This contradicts the maximality of U and shows that (i) implies (ii).

Now assume that M is artinian but not noetherian, and use the minimal condition to choose $U \subseteq M$ minimal with the property that M/U is noetherian. We have $U > 0$, and hence by the minimal condition again, we can choose a minimal nonzero X-subgroup $S \subseteq U$. By Lemma 11.10, let V be a complement for S in U and note that $U/V \cong S$ is X-simple and hence is noetherian. Since M/U is also noetherian, Theorem 11.6 guarantees that M/V is noetherian, and this contradicts the minimality of U.

That (iii) is equivalent to (i) and (ii) is given by Theorem 11.3 and (i) implies (iv) by Theorem 11.4. We now assume (iv) and prove (i). If $U \subseteq M$ is any X-subgroup, then $U \cong M/V$, where V is a complement for U. Since U is isomorphic to a factor group of M and the property of being finitely generated is inherited by factor groups, it follows that U is finitely generated. At this point we know that every X-subgroup of M is finitely generated, and hence M is noetherian by Theorem 11.4. The proof is now complete.   ∎

We now present a result that includes a "finite" version of Theorem 11.9. We need this and the next several results for our study of artinian rings.

(11.12) THEOREM.   *Let M be an abelian X-group. Then the following are equivalent:*

   i.   *M is completely reducible and of finite composition length.*

   ii.   *M is the direct sum of a finite collection of X-simple X-subgroups.*

   iii.   *M is the sum of a finite collection of X-simple X-subgroups.*

Proof.   We assume (i) and prove (ii). The collection S of those X-subgroups of M that can be written as direct sums of finitely many X-simple subgroups is nonempty since it contains the trivial subgroup. Now M satisfies the maximal condition by (i), and so we can choose $U \subseteq M$ maximal in S. Let V be a complement for U in M and suppose that $V > 0$. Since M satisfies the minimal condition, we can choose an X-simple subgroup $S \subseteq V$ and we note that $S \cap U = 0$. Writing $W = U + S$, we see that $W = U \dotplus S$, and hence by Lemma 7.11, we conclude that W is a direct sum of a finite collection of X-simple subgroups. Thus $W \in S$, and this is a contradiction since $W > U$. It follows that $V = 0$ and hence $U = M$ and (ii) is proved.

It is obvious that (ii) implies (iii) and so we assume (iii) and prove (i). Since X-simple abelian groups are both artinian and noetherian, we see that M is both artinian and noetherian by Corollary 11.8, and hence it has finite length. Now let $U \subseteq M$ be an X-subgroup. Using the maximal condition, we choose $V \subseteq M$ maximal with the property that $U \cap V = 0$. The sum of U and V is automatically direct, and so it suffices to show that $U + V = M$ in order to prove that V is a complement for U. Suppose, therefore, that $U + V < M$. Since M is a sum of X-simple subgroups, there must be at least one of these (say, S) not contained in $U + V$. We thus have $(U + V) \cap S = 0$, and since also $U \cap V = 0$, we see that the sum $U + V + S$ is direct by Theorem 7.5. It follows that $U \cap (V + S) = 0$, and this contradicts the maximality of V. We deduce that U is complemented in M, and (i) holds.   ∎

The following corollary gives yet another characterization of finite length, completely reducible groups.

(11.13) COROLLARY.   *Let M be an abelian X-group. Then the following are equivalent:*

   i.   *M is completely reducible and of finite length.*

   ii.   *There exists a finite collection $\{N_1, N_2, \ldots, N_k\}$ of X-subgroups of M such that $\bigcap N_i = 0$ and each M/N_i is X-simple.*

Proof.   Assuming (i), we write

$$M = S_1 \dotplus S_2 \dotplus \cdots \dotplus S_k,$$

where the S_i are X-simple. (This is possible by Theorem 11.12.) Let

$$N_i = \sum_{j \neq i} S_j$$

and note that $\bigcap N_i = 0$ because each element of M is uniquely expressible as a sum of elements of the S_j. We have that $M/N_i \cong S_i$ is X-simple, as required.

To prove the converse, construct the external direct sum

$$E = (M/N_1) \oplus (M/N_2) \oplus \cdots \oplus (M/N_k).$$

Then E is the (internal) direct sum of a finite collection of abelian X-simple X-subgroups and hence is completely reducible and of finite length. Now map $\theta : M \to E$ by setting $\theta(m)$ to be that element of E with ith coordinate equal to the coset $N_i + m$. Then θ is an X-homomorphism and $\ker(\theta) = \bigcap N_i = 0$. We conclude that M is isomorphic to an X-subgroup of E. Therefore M has finite length and it is completely reducible by Lemma 11.10.   ∎

Observe that the implication (ii) $\Rightarrow$ (i) of Corollary 11.13 fails utterly if the collection $\{N_i\}$ is not finite. Perhaps the simplest example of this is to take $X = \varnothing$ and $M = \mathbb{Z}$, the additive group of the integers. We take the N_i to be the subgroups of prime index for all primes.

We now examine a little more closely the completely reducible X-groups with finiteness condition that were the subject of Theorem 11.12 and Corollary 11.13. If M is one of these groups, we can write $M = S_1 \dotplus \cdots \dotplus S_k$, where the S_i are X-simple. The subgroups S_i are not (in general) uniquely determined, although by the Krull-Schmidt theorem (10.10) it follows that the number of S_i in any given isomorphism class is an invariant of M and is independent of the particular direct sum decomposition. (In fact, the more elementary Jordan-Hölder theorem (10.5) can be used to obtain this conclusion, too.) It is an important (and somewhat surprising) fact that the X-subgroups of M obtained by summing all of those S_i in a single isomorphism class are uniquely determined by M, independently of the particular direct sum decomposition. In order to prove this, we begin with a definition.

Let S be any simple abelian X-group. If M is an arbitrary abelian X-group, we consider the X-subgroup of M generated by all of the X-subgroups of M that happen to be X-isomorphic to S. This subgroup, called the S-*isotypic component* of M, depends, of course, only on the isomorphism class of S and is trivial if M contains no isomorphic copy of S. Note that by the Jordan-Hölder theorem, if M has finite composition length, then there are at most finitely many isomorphism classes of simple X-groups S for which the S-isotypic component of M is nontrivial.

(11.14) THEOREM.   *Let M be an abelian X-group and suppose that $M = S_1 \dotplus \cdots \dotplus S_k$, where each S_i is X-simple. If S is any X-simple abelian*

X-group, then the S-isotypic component of M is equal to the sum of all those summands S_i that are isomorphic to S.

Proof.   We may assume that the S_i have been numbered so that for some integer r with $0 \leq r \leq k$, we have that $S_i \cong S$ for $i \leq r$ and $S_i \not\cong S$ for $i > r$. We let U be the S-isotypic component of M and $V = \sum_{i=1}^{r} S_i$. (Note that if no $S_i \cong S$, then $r = 0$ and $V = 0$.) By the definition of U, it is clear that $V \subseteq U$. To prove the reverse containment, we must show that if $T \subseteq M$ and $T \cong S$, then $T \subseteq V$.

For $1 \leq i \leq k$, let $\pi_i : M \to S_i$ be the projection map, so that π_i is an X-homomorphism and $m = \sum \pi_i(m)$ for all $m \in M$. Since S_i is abelian and X-simple and $\pi_i(T)$ is an X-subgroup, we see that for each i, either $\pi_i(T) = 0$ or $\pi_i(T) = S_i$. In the latter case, since T is X-simple, we have $S_i \cong T \cong S$ and so $i \leq r$. Therefore $\pi_i(T) = 0$ for $i > r$ and hence $\pi_i(T) \subseteq V$ for all i. We conclude that $T \subseteq \sum \pi_i(T) \subseteq V$, as required.   ∎

(11.15) COROLLARY.   *Let M be a completely reducible abelian X-group and assume that M is noetherian (or, equivalently, artinian, finitely generated, or of finite composition length). Then M is the direct sum of its nonzero S-isotypic components as S runs over a set of representatives for·the isomorphism classes of X-simple abelian X-groups.*

Proof.   Write $M = S_1 + \cdots + S_k$, where the S_i are simple. By Corollary 11.15, we see that as S varies, the S-isotypic component of M is a partial sum of this direct sum, and the result follows.   ∎

(11.16) COROLLARY.   *Let M be as above and let S and T be X-simple abelian X-groups. Let U and V be the corresponding isotypic components of M. If $U \cap V \neq 0$, then $S \cong T$ and $U = V$.*   ∎

11D

To prove Theorem 11.9 (with no finiteness assumption), we need to return to set theory and obtain Zorn's lemma. We also use Zorn's lemma frequently in ring theory in later chapters of this book.

Our proof of Zorn's lemma uses the axiom of choice. It is a fact that if one were to assume the statement of Zorn's lemma as an "axiom," then the axiom of choice could be proved as a "theorem." It follows that there is no hope of finding a proof of Zorn's lemma that does not depend ultimately on the axiom of choice, since it is known that this axiom definitely is not a consequence of the rest of set theory.

From the logician's point of view, the axiom of choice and Zorn's lemma are "equivalent," and one might ask why we bother to give the following somewhat technical proof. Why not simply assume Zorn's lemma, just as we assumed the axiom of choice? The point, as this author sees it, is that although these assertions are "logically equivalent," they are not interchangeable. Logically, the fact that $2 + 2 = 4$ is equivalent to the fact that finite groups always contain Sylow p-

subgroups: both statements are true. We suspect, however, that the reader of this book would feel cheated if instead of proving Sylow's theorem, we had said that no proof is necessary since $2 + 2 = 4$. In the same way, we feel that it is appropriate to offer a proof of Zorn's lemma.

(11.17) THEOREM (Zorn's lemma).   *Let P be a nonempty poset and assume that for every nonempty linearly ordered subset $L \subseteq P$, there exists an element $u \in P$ such that $u \geq x$ for every element $x \in L$. Then P contains a maximal element.*

Before we plunge into the proof of Zorn's lemma, a few remarks about the statement are appropriate. The element u of Theorem 11.17 is said to be an *upper bound* for the linearly ordered subset L. There is no requirement that the upper bound u be an element of L, and so L need not have a maximal element. Note that if P itself were linearly ordered, the theorem would be trivial, since an upper bound for P would be a maximal element of P in that case. Equally trivial is the case where P is finite or, more generally, where P satisfies the ACC.

Proof of Theorem 11.17.   Working by contradiction, we assume that P has no maximal element. If $L \subseteq P$ is nonempty and linearly ordered, it has an upper bound $u \in P$, and since u is not maximal in P, there exists $v \in P$ with $v > u$. Then $v > x$ for every $x \in L$ and v is a "strict" upper bound for L. (Note that if L is empty, then every element of P is a strict upper bound.) Since the set of strict upper bounds is nonempty for every linearly ordered subset of P, the axiom of choice guarantees the existence of a choice function f that assigns to each linearly ordered subset L a strict upper bound for L. Thus $f(L) \in P$ and $f(L) > x$ for all $x \in L$.

We now restrict our attention to those linearly ordered subsets of P that happen to satisfy the minimal condition. Such a subset S is said to be *well ordered*. It has the property that if $\varnothing \neq X \subseteq S$, then X contains a minimal element (which is necessarily unique in this situation since S and therefore X are linearly ordered). If S is linearly ordered and $s \in S$, we introduce the notation

$$S^s = \{x \in S \mid x < s\}$$

and we say that S^s is an *initial segment* of S. We say that a subset $S \subseteq P$ is *admissible* if it is well ordered and $s = f(S^s)$ for every $s \in S$. (In other words, the well-ordered set S is admissible if each initial segment determines the "next" element of S via the function f.) Note that the empty set is admissible, and if S is any admissible set, then so is $S \cup \{v\}$, where $v = f(S)$.

Let S and T be unequal admissible subsets of P. We intend to show that either $S = T^t$ for some $t \in T$ or $T = S^s$ for some $s \in S$. By symmetry, since $S \neq T$, we may assume that $T \not\subseteq S$ and so, using the fact that T is well ordered, we can choose $t \in T$ minimal such that $t \notin S$. Then $T^t \subseteq S$ and we claim that we have equality here.

Suppose $T^t < S$ and choose $s \in S$ minimal such that $s \notin T^t$. We thus have $S^s \subseteq T^t$. Certainly $T \nsubseteq S^s$, and so we can choose $u \in T$ minimal with $u \notin S^s$ and hence

$$T^u \subseteq S^s \subseteq T^t \subseteq S.$$

We claim that $S^s = T^u$. By the containments in the previous paragraph, this is certainly true if $t = u$. We cannot have $t < u$ since that would imply that $t \in T^u \subseteq S$, and this contradicts our choice of t. The remaining possibility is that $t > u$ and so $u \in T^t \subseteq S$. We know that $u \notin S^s$, however, and hence $u \geq s$. Also in this case, S^u is defined and we have $S^s \subseteq S^u \cap T \subseteq T^u$. We conclude that $S^s = T^u$, as claimed.

Since S and T are admissible, we have $s = f(S^s) = f(T^u) = u$. In the previous paragraph, we saw that $u \leq t$. We cannot have $u = t$ because then $t = u = s \in S$, a contradiction. Therefore $t > u$ and hence $s = u \in T^t$, which is also a contradiction. This proves $T^t = S$, as desired. In particular, we have now shown that the collection of admissible subsets is linearly ordered by inclusion.

Next, let A be the union of all admissible subsets of P. We shall show that A is itself admissible. We check first that A is linearly ordered. To see this, let $a, b \in A$ and choose admissible sets S and T such that $a \in S$ and $b \in T$. Since the collection of admissible subsets is linearly ordered, we may assume that $S \subseteq T$ and hence $a, b \in T$. Since T is linearly ordered, we conclude that a and b are comparable, as required.

Now let $a \in A$ and choose an admissible set S such that $a \in S$. We will show that $A^a = S^a$. Since $S \subseteq A$, it is obvious that $S^a \subseteq A^a$, and so we work to prove the reverse containment. Let $b \in A^a$ and note that $b < a$ so that if $b \in S$, we have $b \in S^a$, as desired. We may assume, therefore, that $b \notin S$ and we work for a contradiction. Choose an admissible T with $b \in T$. Then $T \nsubseteq S$ and hence $S \subseteq T$, and in fact $S = T^t$ for some member $t \in T$. Since $b \notin S$, we have $b \geq t > a$, where the second inequality follows since $a \in S = T^t$. This is a contradiction and proves that $A^a = S^a$, as claimed.

To show that A is well ordered, let $X \subseteq A$ be any nonempty subset and let S be admissible with $X \cap S$ nonempty. Let x be a minimal element of $X \cap S$. Then $A^x = S^x$ contains no elements of X and hence x is minimal in X, as desired.

To complete the proof that A is admissible, let $a \in A$. We must show that $a = f(A^a)$. We know, however, that $A^a = S^a$ for some admissible set S, and so we have $a = f(S^a) = f(A^a)$, as required.

Since A is admissible, we know that $A \cup \{v\}$ is admissible, where $v = f(A)$. This is a contradiction, however, since $v \notin A$ and yet A is supposed to contain all admissible subsets of P. The proof is now complete.   ∎

The following lemma, which is the key to the proof of Theorem 11.9, is typical of applications of Zorn's lemma to algebra.

(11.18) LEMMA. *Let G be an X-group and suppose $Y \subseteq G$ is a subset consisting of nonidentity elements. Let S be the collection of all X-subgroups $H \subseteq G$ such* *that $H \cap Y = \varnothing$. Then S is nonempty and each element of S is contained in some maximal element of S.*

Proof.   Note that S contains the trivial subgroup of G, and so it is nonempty. Given $H \in S$, write

$$\mathcal{P} = \{U \in S \mid U \supseteq H\}$$

and partially order $\mathcal{P}$ by inclusion. To find a maximal element of $\mathcal{P}$, it suffices to show that it satisfies the hypotheses of Zorn's lemma.

Suppose that $\mathcal{L}$ is a nonempty linearly ordered subset of $\mathcal{P}$. By Lemma 11.5(a), the union $U = \bigcup \mathcal{L}$ is an X-subgroup of G. Since each member of $\mathcal{L}$ intersects trivially with Y, we see that $U \cap Y = \varnothing$ and hence $U \in S$. Clearly, $U \supseteq H$ and hence $U \in \mathcal{P}$. Of course, we have $U \supseteq L$ whenever $L \in \mathcal{L}$, and so we have produced an upper bound for $\mathcal{L}$ in the poset $\mathcal{P}$. Zorn's lemma thus applies.   ∎

Proof of Theorem 11.9.   First assume that M is completely reducible and let N be the X-subgroup generated by all the simple X-subgroups. (In other words, N is the *socle* of M. At this point, it is not obvious that M has any simple X-subgroups, in which case we have $N = 0$.)

We wish to show that $N = M$, and so we assume the contrary and choose $m \in M - N$. Using Lemma 11.18 and taking $Y = \{m\}$, we can choose $U \supseteq N$ maximal among X-subgroups that do not contain m, and we note that $U < M$. Let $U \subseteq V < M$ with $V > U$ if possible, and choose a complement $W \subseteq M$ for V, using complete reducibility.

Now $W \not\subseteq U$ and so $U + W > U$ and hence $m \in U + W$ by the maximality of U. We can thus write $m = u + w$ with $u \in U$ and $w \in W$. If $V > U$, then $m \in V$ and so $w = m - u \in V$. Thus $w \in V \cap W = 0$ and $m = u \in U$, a contradiction. This shows that it is not possible to choose $V > U$, and hence $W \cong M/V = M/U$ is X-simple. Since $W \not\subseteq U$, this contradicts the fact that U contains the socle of M.

Conversely now we assume that M is its own socle, and we prove that it is completely reducible. Let U be an arbitrary X-subgroup of M. Our goal is to produce a complement for U in M.

Using Lemma 11.18 and taking $Y = U - \{0\}$ and $H = 0$, we can choose $V \subseteq M$ maximal such that $U \cap V = 0$. We complete the argument as in the proof of Theorem 11.12. Our task is to show that $U + V = M$, and so we suppose that $U + V < M$. Since M is generated by its X-simple subgroups, there must be at least one of these (say, S) with $S \not\subseteq U + V$. In particular, since S is X-simple and abelian, we have $(U + V) \cap S = 0$. Since also $U \cap V = 0$, the sum $U + V + S$ is direct and hence $U \cap (V + S) = 0$. Because $V + S > V$, this contradicts the maximality of V. It follows that $U + V = M$, and the proof is complete.   ∎

Problems

11.1   Let G be a solvable group whose poset of subgroups satisfies both chain conditions. Show that G is finite.

11.2 Let L be a linearly ordered set and suppose that $X \subseteq L$. (We allow $X = \varnothing$.) We say that X is a *cut* for L if $x < y$ whenever $x \in X$ and $y \in L - X$. Recall also that X is an *initial segment* of L if it has the form $X = L^a = \{x \in L \mid x < a\}$ for some $a \in L$.

a. Show that every initial segment of L is a cut.

b. Show that L is well ordered (satisfies the minimal condition) iff every proper cut in L is an initial segment.

11.3 Let P be any poset. Show that there exists a maximal linearly ordered subset in P.

HINT: Apply Zorn's lemma to the poset $\mathcal{L}$ of all linearly ordered subsets of P.

11.4 If P is a poset, a subset $A \subseteq P$ is called an *antichain* in P if, whenever $x, y \in A$ with $x \leq y$, we have $x = y$. Show that P is linearly ordered iff every maximal antichain in P has just one element.

11.5 Let M be an artinian abelian X group and assume that the intersection of all of the maximal X-subgroups of M is zero. Show that M is completely reducible.

NOTE: As the example $M = \mathbb{Z}$ shows (with $X = \varnothing$), the problem becomes false if "noetherian" replaces "artinian."

11.6 Let M and N be abelian X-groups and let $\varphi : M \to N$ be an X-homomorphism. Show that for each simple abelian X-group S, the S-isotypic component of M is mapped by φ into the S-isotypic component of N.

11.7 Let S be a simple abelian X-group. Show that for some prime number p, every nonzero element of S has order p.

11.8 Let G be a (not necessarily abelian) X-group with the property that whenever $\mathcal{H}$ is a linearly ordered (by inclusion) collection of proper X-subgroups of G, then $\bigcup \mathcal{H} < G$. Show that G is finitely generated.

HINT: Let $\mathcal{S}$ be the set of all of those X-subgroups $H \subseteq G$ for which there is no finite subset $Y \subseteq G$ with $G = \langle H \cup Y \rangle$. Note that there can be no maximal elements in $\mathcal{S}$. Use Zorn's lemma to derive a contradiction if $\mathcal{S} \neq \varnothing$.

11.9 Let $V \neq 0$ be a (not necessarily finitely generated) vector space and let $\mathcal{P}$ be the set of all linearly independent subsets of V, ordered by containment. Show that $\mathcal{P}$ has a maximal element and that this set spans V (and so is a basis). Recall that only finite linear combinations are ever considered.

11.10 Let M be a noetherian abelian X-group. Show without using Zorn's lemma that Theorem 11.9 holds for M.

11.11 Let G be a finite solvable group and write $F = \mathbf{F}(G)$, the Fitting subgroup.

a. If M is a maximal subgroup of G not containing F, show that $F \cap M$ is normal in G and that $F/(F \cap M)$ is abelian and G-simple, where G acts by conjugation.

b. Assume that the Frattini subgroup $\Phi(G) = 1$. Show that F is abelian and is completely reducible as a G-group.

11.12 Let G be finite and solvable and have trivial center and trivial Frattini subgroup. Show that $\mathbf{F}(G) \subseteq G'$.

HINT: Show that there exists $A \triangleleft G$ such that $A \cap G' = 1$ and $\mathbf{F}(G) \subseteq AG'$.

11.13 **(Clifford)** Let G be a finite group acting by automorphisms on a finite abelian group A, and assume that A is completely reducible as a G-group. If $N \triangleleft G$, show that A is completely reducible when viewed as an N-group.

HINT: It suffices to do the case where A is G-simple. Observe that if $B \subseteq A$ is an N-simple N-subgroup, then so is B^g for all $g \in G$.

CHAPTER TWELVE

Rings, Ideals, and Modules

12A

In some sense, the symmetric group on a set Ω can be thought of as the fundamental example of a group. Not only does $\mathrm{Sym}(\Omega)$ serve to motivate the axioms that define groups, but also, by Cayley's theorem, every group appears as a subgroup of some $\mathrm{Sym}(\Omega)$. Similarly, for ring theory, the fundamental example is $\mathrm{End}(A)$, the set of all endomorphisms of an abelian group A. Multiplication within $\mathrm{End}(A)$ is defined by function composition. (The reader is reminded of our convention that if f and g are functions, we *always* write fg to denote the function constructed by doing first f and then g. Thus $(x)(fg) = (xf)g$ or, equivalently, $(fg)(x) = g(f(x))$. Because the former formula looks more natural, we generally write our functions on the right in situations where function composition is important.)

Because A is abelian, we have at our disposal a second operation on $\mathrm{End}(A)$. If r and s are endomorphisms of A, we define $r + s : A \to A$ by

$$(x)(r + s) = (x)r + (x)s$$

for all $x \in A$. (Of course, the $+$ on the right side of this equation is the operation in the group A.) Observe that if $r, s \in \mathrm{End}(A)$, then $r + s \in \mathrm{End}(A)$, but that the proof of this requires the commutativity of A. It should be clear the $\mathrm{End}(A)$ satisfies the conditions in the following definition.

(12.1) DEFINITION. Let R be a set that has two binary operations: addition and multiplication. (These are denoted by $+$ and by juxtaposition, as usual.) Then R is a *ring* provided that

    i. R is an abelian group with respect to $+$,
    ii. $(rs)t = r(st)$,
    iii. $r(s + t) = rs + rt$, and
    iv. $(s + t)r = sr + tr$ for all $r, s, t \in R$.
    v.   There is an element of R, usually denoted 1, such that $1r = r = r1$ for all $r \in R$.

The multiplicative identity 1 of R is called the *unity* element of R. (It is easily seen to be unique, justifying our use of the word "the.") Many authors do not require the existence of a unity for a ring, and they call the object in Definition 12.1 a "ring with 1" or "ring with unit." There certainly are times when it is convenient to drop condition (v) and consider unitless "rings," but in this book, we assume axiom (v) unless the contrary is made explicit. Of course, the unity of End(A) is the identity map on A.

The reader should note that the condition that addition be commutative in Definition 12.1(i) is actually redundant, since for $r, s \in R$, we have

$$(r + s)(1 + 1) = r(1 + 1) + s(1 + 1) = r + r + s + s$$

and also

$$(r + s)(1 + 1) = (r + s)1 + (r + s)1 = r + s + r + s.$$

Comparison of these two computations yields that $r + s = s + r$.

If R is a ring, we use the standard notation of abelian groups and write 0 for the additive identity, $-r$ for the additive inverse of $r \in R$, and $r - s$ for $r + (-s)$.

Before proceeding with our study of ring theory, we discuss a few examples. Perhaps the most familiar examples of rings are the integers $\mathbb{Z}$ and the sets $\mathbb{Q}$, $\mathbb{R}$, and $\mathbb{C}$ of rational, real, and complex numbers, respectively, all with respect to the usual addition and multiplication. All of these are *commutative* rings, since multiplication is commutative in each of them.

We can build new rings from old by using matrices. If R is any ring, we write $M_n(R)$ to denote the ring of all $n \times n$ matrices with entries in R. The operations in $M_n(R)$ are the usual ones of matrix theory: If $A = (a_{ij})$ and $B = (b_{ij})$, then the (i, j)-entry of $A + B$ is $a_{ij} + b_{ij}$ and the (i, j)-entry of AB is

$$\sum_{k=1}^{n} a_{ik}b_{kj}.$$

(Note that if the coefficient ring R is not commutative, the formula for the (i, j)-entry of the product AB is sensitive to the order of the factors in each term of the sum.) We omit the computations necessary to show that $M_n(R)$ really is a ring for each choice of coefficient ring R and integer $n \geq 1$. Observe that if R is any ring with more than one element and if $n \geq 2$, then $M_n(R)$ is necessarily noncommutative.

We can also build new rings from old by taking external direct sums and defining both addition and multiplication componentwise. Note that the unity element of the direct sum

$$R = R_1 \oplus \cdots \oplus R_n$$

is the n-tuple $(1, 1, \ldots, 1)$, where the ith entry is the unity of R_i. (If there are infinitely many terms in the sum, therefore, we must take the unrestricted direct sum in order for the resulting object to be a ring with unit.)

We defer until later discussion of another important technique for building new rings from old: the construction of polynomial rings.

If R is a ring, a subset $S \subseteq R$ is a *subring* if S is a ring with the operations inherited from R. Equivalently, S is an additive subgroup of R that is closed under multiplication and happens to contain some element that behaves as a unity for S.

Although it is true for rings, as it is for groups, that the multiplicative identity is unique, it is not generally true that a subring $S \subseteq R$ has the same unity as the whole ring. Of course, if the element 1 of R happens to be in S, it is necessarily the unity of S, and in this case we say that S is a *unitary* subring of R. Examples of this are $\mathbb{Z} \subseteq \mathbb{Q} \subseteq \mathbb{R} \subseteq \mathbb{C}$ and $D \subseteq M_n(R)$, where D is the subring consisting of the diagonal matrices of $M_n(R)$.

The trivial subring $S = \{0\} \subseteq R$ provides an example of a subring that is not unitary (if R is not itself trivial). More interesting examples can be constructed using direct sums. Let $R = R_1 \oplus R_2$, where multiplication and addition are defined componentwise and both summands are nontrivial. Let $S = \{(0, x) \mid x \in R_2\}$. Then S is a subring of R that is not unitary since $(1, 1)$ is the unity of R and $(0, 1)$ is the unity of S.

We have seen that $\text{End}(A)$ is a ring whenever A is an abelian group. If X is a set and A is an abelian X-group, we write $\text{End}_X(A)$ to denote the set of X-endomorphisms of A. Then $\text{End}_X(A)$ is a subring of $\text{End}(A)$ and, in fact, it is a unitary subring, since the identity map on A is an X-homomorphism. (Note that the endomorphisms of A induced by elements of X do not generally lie in $\text{End}_X(A)$.)

Actually, $\text{End}_X(A)$ is the most general possible example of a ring. This is the content of the following ring analog of Cayley's theorem.

(12.2) THEOREM. *Let R be any ring (with unit, of course). Then R is isomorphic to a ring of the form $\text{End}_X(A)$ for some abelian X-group A. In fact, we can take $X = R$ and $A = R$ (the additive group) with the action of X on A given by left multiplication in R.*

Although presumably it is clear what it means for two rings to be isomorphic, we shall make this explicit. A map $\theta : R \to S$ is a *ring homomorphism* if $\theta(x + y) = \theta(x) + \theta(y)$ and $\theta(xy) = \theta(x)\theta(y)$ for all $x, y \in R$. If, in addition, θ is a bijection of R onto S, then it is a *ring isomorphism*, and in this case we say that R and S are *isomorphic*. Of course, two isomorphic rings enjoy identical "ring theoretic" properties.

Note that if $\theta : R \to S$ is a ring homomorphism, then it is automatically a group homomorphism from the additive group of R to that of S, and so everything we know about group homomorphisms applies here. In particular, θ is injective precisely when $\ker(\theta) = 0$. Of course, $\ker(\theta) = \{r \in R \mid \theta(r) = 0\}$.

A not entirely trivial example of isomorphic rings is $\text{End}_F(V) \cong M_n(F)$, where F is a field and V is an F-vector space of dimension n. Here, $\text{End}_F(V)$ is the ring of F-linear transformations from V to itself, and we construct the isomorphism with the matrix ring $M_n(F)$ by choosing a basis for V.

Proof of Theorem 12.2.   If $a \in R$ (viewed as an abelian group) and $x \in R$ (viewed as a set), we write $a \cdot x = xa$. This "dot" action of (the set) R on (the abelian group) R makes the additive group of R into an R-group since

$$(a + b) \cdot x = x(a + b) = xa + xb = a \cdot x + b \cdot x$$

by the left distributive law.

If $r \in R$ (the ring), define the map $e_r : R \rightarrow R$ by $(a)e_r = ar$, so that $e_r \in \text{End}(R)$ by the right distributive law. In fact, we have

$$(a \cdot x)e_r = xar = (ae_r) \cdot x$$

by the associative law, and hence $e_r \in \text{End}_R(R)$. We define $\theta : R \rightarrow \text{End}_R(R)$ by $\theta(r) = e_r$, and we show that θ is a ring isomorphism.

Now $\theta(rs) = e_{rs}$, and for $a \in R$, we have

$$ae_{rs} = ars = ae_r e_s,$$

and so $\theta(rs) = e_r e_s = \theta(r)\theta(s)$. Also, $\theta(r + s) = e_{r+s}$ and

$$ae_{r+s} = a(r + s) = ar + as = ae_r + ae_s = a(e_r + e_s),$$

where the last equality is by the definition of addition in $\text{End}_R(R)$. Therefore $\theta(r+s) = e_r + e_s = \theta(r) + \theta(s)$, and θ is a ring homomorphism. If $r \in \ker(\theta)$, then $e_r = 0$ and we have $r = 1r = 1e_r = 0$, and so θ is injective.

Finally, to show that θ is surjective, let $\alpha \in \text{End}_R(R)$. Write $(1)\alpha = r$ and compute for $a \in R$ that

$$a\alpha = (a1)\alpha = (1 \cdot a)\alpha = (1\alpha) \cdot a = r \cdot a = ar = ae_r.$$

Since this holds for all $a \in R$, we have $\alpha = e_r = \theta(r)$, and θ is surjective.   ∎

The following lemma, though trivial, is useful.

(12.3) LEMMA.   *Let R be a ring and suppose $r, s \in R$. Then*

a. $0r = 0 = r0$ *and*
b. $(-r)s = -(rs) = r(-s)$.

Proof.   We have $0r + 0r = (0 + 0)r = 0r$, and so $0r = 0$. The proof that $r0 = 0$ is similar. We have

$$0 = 0s = (r + (-r))s = rs + (-r)s$$

and hence $(-r)s = -rs$. The remainder of (b) follows symmetrically.   ∎

Lemma 12.3 has numerous immediate corollaries such as $(-1)(-1) = 1$ and $r(s - t) = rs - rt$ for $r, s, t \in R$. We neither list all of these nor prove them, but they are used freely when needed. Note that facts such as $-(-r) = r$ and $r - r = 0$ are not corollaries of Lemma 12.3 and are not really ring theory at all. These are immediate from the group structure of $(R, +)$.

In group theory, the kernels of homomorphisms turned out to be very important objects: they were precisely the normal subgroups. In ring theory, too, kernels of homomorphisms are of significance. Their normality in the additive group says nothing because the group is abelian, but we do have the following lemma.

(12.4) LEMMA. *Let $\theta : R \to S$ be a ring homomorphism and let $I = \ker(\theta)$. Then*

a. *I is an additive subgroup of R.*

Also, if $a \in I$ and $r \in R$, then

b. *$ar \in I$ and*
c. *$ra \in I$.*

Proof.   Conclusion (a) is immediate since the ring homomorphism θ is a group homomorphism. For (b) we compute

$$\theta(ar) = \theta(a)\theta(r) = 0\theta(r) = 0,$$

and so $ar \in \ker(\theta) = I$. The proof of (c) is similar.   ■

A subset $I \subseteq R$ that satisfies the three conclusions of Lemma 12.4 is called an *ideal* of R. Note that each of (b) and (c) implies that I is closed under multiplication (although they say much more than that), and so an ideal that happens to contain a multiplicative identity is a subring. If $I \subseteq R$ is an ideal that contains the unity of R, then since I contains every multiple of each of its elements and $r = 1r$ for all $r \in R$, we see that $I = R$. A proper ideal, therefore, can never be a unitary subring.

We discuss a few examples of ideals. For any ring, the ring itself and the trivial subring 0 are ideals. If $R = R_1 \oplus R_2$ and we write (as we did earlier) $S = \{(0, x) \mid x \in R_2\}$, then S is an ideal (which happens to be a subring). If R is commutative, the set $aR = \{ar \mid r \in R\}$ is an ideal called the *principal* ideal generated by a. This ideal, often simply denoted (a), contains a and is contained in every other ideal that contains a. In the case $R = \mathbb{Z}$, for instance, we know that every additive subgroup has the form $n\mathbb{Z}$ for some $n \geq 0$. Therefore these principal ideals are all of the ideals of $\mathbb{Z}$. (Observe that $n\mathbb{Z}$ has no unity and so is not a subring for $n > 1$.) A commutative ring such as $\mathbb{Z}$, in which every ideal is principal, is called a *principal ideal ring*, or a *PIR*.

If R is not commutative, then aR may not be an ideal. It satisfies two of the three defining properties of an ideal, however: it is an additive subgroup and it is closed under right multiplication by arbitrary elements of R. Such an object is called a *right ideal* of R and, similarly, an additive subgroup closed under left multiplication by all elements of R is a *left ideal*. (It is interesting to note that unlike the situation in natural languages where adjectives generally restrict the noun they modify, in this case, the words "right" and "left" create a more (rather than less) general situation. Neither a left ideal nor a right ideal need be an ideal and, in fact, an ideal is precisely an object that is both a left ideal and a right ideal.)

Other examples of one-sided (that is right or left) ideals are "annihilators." It often happens in rings that a product $xy = 0$ even if neither factor is zero. Given any subset $X \subseteq R$, the *right annihilator* ra(X) is the set $\{r \in R \mid xr = 0$ for all $x \in X\}$ and the *left annihilator* la(X) is defined similarly. The reader should check that for any subset $X \subseteq R$, the right annihilator ra(X) is a right ideal, and that if X is itself a right ideal, then ra(X) is an ideal. (Remember, this means "two-sided.") If "left" replaces "right" above, everything still works, of course.

In group theory, we know not only that kernels of homomorphisms are normal subgroups, but also that every normal subgroup is the kernel of some homomorphism. This latter fact was proved by consideration of the canonical homomorphism of the group onto the factor group by the given normal subgroup. If $I \subseteq R$ is an ideal, we would like (by analogy with the group case) to define the *factor ring R/I*. Since I is, in particular, a normal subgroup of the additive group of R, we can define the factor group R/I. Its elements are the additive cosets $I + r$ for elements $r \in R$. To make R/I into a ring, we need to define multiplication, and since we want the canonical group homomorphism $R \to R/I$ to be a ring homomorphism, we are forced to define $(I + r)(I + s) = I + rs$.

(12.5) LEMMA. *Let $I \subseteq R$, where I is an ideal. Define multiplication on R/I by setting*

$$(I + r)(I + s) = I + rs \ \text{ for } \ r, s \in R.$$

This multiplication is well defined and makes R/I into a ring.

Proof.   We must prove that our product formula gives an unambiguous result, independent of the choices of coset representatives. If we have $I + r = I + r_0$ and $I + s = I + s_0$, we need to check that $I + rs = I + r_0 s_0$. To see that this holds, write $r_0 = u + r$ and $s_0 = v + s$ with $u, v \in I$. Then

$$r_0 s_0 = rv + us + uv + rs,$$

and since each of rv, us, and uv lies in I (because I is an ideal), we get $r_0 s_0 \in I + rs$ and hence $I + rs = I + r_0 s_0$.

The completely routine verification that this definition of multiplication makes R/I into a ring is omitted except for the observation that the unity of R/I is the coset $I + 1$.   ∎

If $I \subseteq R$ is an ideal, we see that the canonical group homomorphism $\pi : R \to R/I$, carrying $r \in R$ to the coset $I + r$, is a ring homomorphism with kernel I.

(12.6) COROLLARY. *The ideals of R are precisely the kernels of all ring homomorphisms from R to other rings.*   ∎

The analogy between the theories of group and ring homomorphisms goes further and we state two more such results. Their proofs are omitted, however, since they are quite routine.

(12.7) LEMMA. *Let $\theta : R \to S$ be a surjective ring homomorphism with $I = \ker(\theta)$. Then $S \cong R/I$.*   ∎

(12.8) LEMMA. *Let $\theta : R \to S$ and I be as above, so that by the Correspondence theorem, θ determines a bijection from the set of additive subgroups of R containing I onto the set of additive subgroups of S. Under this correspondence,* *ideals of R correspond to ideals of S and, similarly, right ideals correspond to right ideals and left ideals correspond to left ideals.*   ∎

Finally, we mention one way in which the analogy between group and ring homomorphisms might lead one astray. If $\theta : R \to S$ is a ring homomorphism, then $\theta(R)$ is a subring of S and $\theta(1)$ is, of course, its unity element. It may happen, however, that $\theta(R)$ is not a unitary subring of S, and so in general we cannot conclude that $\theta(1) = 1$. This clearly does hold, however, if θ is surjective.

12B

An element $u \in R$ is called a *unit* and is said to be *invertible* in R if it has both a left inverse and a right inverse. Thus u is a unit if there exist $x, y \in R$ with $ux = 1 = yu$. In this case, we have $x = 1x = yux = y1 = y$, and so this element $x = y$ is the only left or right inverse u has in R. It can thus be unambiguously denoted u^{-1}. Of course, u^{-1} is also a unit of R and $(u^{-1})^{-1} = u$. (This use of the word "unit" is standard but unfortunately is not quite consistent with the phrase "ring with unit," which also is quite common.)

In the matrix ring $M_n(F)$, where F is a field, the units, the invertible matrices, are precisely the matrices with nonzero determinant, and it follows that any matrix having either a left inverse or a right inverse is invertible. This situation is not typical since, in general, in a noncommutative ring, an element can have a one-sided inverse and yet not be invertible. For an example of this, consider the ring $R = \text{End}(A)$, where A is the abelian group of all infinite sequences $\{a_i \mid i \geq 0\}$ of integers with respect to term-by-term addition. Let $r \in R$ be the "right-shift" endomorphism so that $\{a_i\}r = \{b_i\}$, where $b_0 = 0$ and $b_i = a_{i-1}$ for $i > 0$. Similarly, let $s \in R$ denote the "left-shift" endomorphism defined by $\{a_i\}s = \{b_i\}$, where $b_i = a_{i+1}$ for $i \geq 0$. It is clear that $rs = 1$, but neither r nor s is invertible since s is not injective and r is not surjective.

The set $U(R)$ of units in the ring R is nonempty since it contains 1. It is clearly closed under the taking of inverses and it is closed under multiplication, since if u and v are units, then $v^{-1}u^{-1}$ is an inverse for uv, which is therefore a unit. It follows that $U(R)$ is a group: the *group of units* of R. We have, for instance, $U(\text{End}(A)) = \text{Aut}(A)$ and $U(M_n(F)) = GL(n, F)$, the general linear group, where F is a field. Also $U(\mathbb{Z}) = \{1, -1\}$. More interesting is the case $R = \mathbb{Z}/n\mathbb{Z}$, the ring of integers mod n, with $n \geq 1$. The reader should check that the units in this case are precisely the cosets of the form $n\mathbb{Z} + x$ for integers x with $\gcd(n, x) = 1$. (Note that the value of $\gcd(n, x)$ is constant as x runs over any coset of $n\mathbb{Z}$.) It follows that $|U(\mathbb{Z}/n\mathbb{Z})| = \varphi(n)$, where φ is Euler's function. In fact, if C is cyclic and of order n, then $\text{End}(C) \cong \mathbb{Z}/n\mathbb{Z}$ and so $U(\mathbb{Z}/n\mathbb{Z}) \cong \text{Aut}(C)$.

The element $0 \in R$ is never a unit, except in the trivial case where $R = \{0\}$. A nontrivial ring R for which every nonzero element is a unit is called a *division ring*. In other words, a ring R is a division ring iff $U(R) = R - \{0\}$, and in this case we write $R - \{0\} = R^{\times}$, the *multiplicative group* of R. A commutative division ring is a *field*, and examples of these are very familiar. In addition to $\mathbb{Q}$, $\mathbb{R}$, and $\mathbb{C}$, it is easy to find many other subfields of $\mathbb{C}$. For example, the object $\{a + b\sqrt{2} \mid a, b \in \mathbb{Q}\}$, which is denoted $\mathbb{Q}[\sqrt{2}]$, is a field. (We study fields of this type in considerable depth later.) We also have finite fields such as $\mathbb{Z}/p\mathbb{Z}$, where p is a prime number. (To see that this is a field, observe that $|U(\mathbb{Z}/p\mathbb{Z})| = \varphi(p) = p - 1$.) As we shall see later, there are also other finite fields that are not of prime order.

It is not so easy to produce concrete examples of noncommutative division rings, although, in fact, these objects abound. Perhaps the most familiar example is the ring $\mathbb{H}$ of quaternions, discovered by Hamilton. We construct $\mathbb{H}$ as a four-dimensional real vector space with basis $\{1, i, j, k\}$, and we define multiplication in $\mathbb{H}$ by giving rules for multiplying the four basis vectors and then extending the definition to all of $\mathbb{H}$. Before we give the details, we digress to consider this method of constructing rings (not necessarily division rings) in greater generality.

Suppose F is a field. An *F-algebra* is a ring A that is simultaneously a vector space over F with the same addition, and such that ring multiplication and scalar multiplication are related by the formula

$$(\alpha a)b = \alpha(ab) = a(\alpha b)$$

for all $a, b \in A$ and $\alpha \in F$. For example, if F is any field, then the matrix ring $M_n(F)$ is an F-algebra. Also, any field is an algebra over itself, and $\mathbb{C}$ is also a two-dimensional algebra over $\mathbb{R}$.

We remark that for us, since an algebra is a ring, its multiplication is associative and there is a unity element. The word "algebra" is occasionally used, however, in situations where the only assumptions on the multiplication are the two distributive laws and the compatibility with scalar multiplication described above. (Equivalently, left and right multiplication are F-linear transformations.)

Let A be an F-algebra and suppose B is an F-basis for A. If we know enough of the multiplication of A to be able to compute the product in A of any two elements of B, writing the result in terms of B, then it is easy to recover the rest of the multiplication of A. If $u, v \in A$, we can write

$$u = \sum_{x \in B} \alpha_x x \quad \text{and} \quad v = \sum_{x \in B} \beta_x x$$

with $\alpha_x, \beta_x \in F$. (In the case where the basis B is infinite, of course, we insist that all but finitely many of the coefficients α_x and β_x are zero.) Using the distributive law and the definition of an algebra, we have

$$uv = \sum_{x,y \in B} (\alpha_x \beta_y)(xy),$$

and since we are assuming that we know xy for all $x, y \in B$, this enables us to compute uv.

This technique can also be used to *construct* an algebra. Start with a field F and an arbitrary set B, and let A be the set of all "formal" F-linear combinations of the elements of B. Thus A is the set of all objects that look like

$$\sum_{x \in B} \alpha_x x,$$

where all the coefficients α_x lie in F and only finitely many of them are nonzero. Addition and scalar multiplication are defined in the obvious manner to make A into an F-space. If we identify the element $x \in B$ with the formal linear combination

$$1x + \sum_{y \neq x} 0y,$$

we can view B as a subset of and basis for A. Next invent some rule for multiplication of elements of B with results in A, and extend this to a multiplication for all of A as above.

The problem with this construction, of course, is that for an arbitrary multiplication on B, it is not very likely that the multiplication in A will turn out to be associative. The point here is that it is enough to check associativity for elements of B in order to guarantee associativity in A, and if B is reasonably small, or otherwise under good control, this may not be hard to do. We leave it to the reader to check that if the associative law holds for the basis vectors, then it holds in general.

Returning now to our construction of the quaternions $\mathbb{H}$ as an $\mathbb{R}$-algebra, we see that we need only define and check associativity of multiplication on the four element set $\{1, i, j, k\}$. As the reader has guessed, 1 multiplies as an identity. The rules invented by Hamilton for the other products are as follows: $i^2 = j^2 = k^2 = -1$; also $ij = k$, $jk = i$, $ki = j$ and $ji = -k$, $kj = -i$, $ik = -j$. (We leave the associativity check to the reader.)

We now have that $\mathbb{H}$ is a ring (in fact, an $\mathbb{R}$-algebra), and we need to establish that it is a division ring. If $u \in \mathbb{H}$, write $u = a+bi+cj+dk$ with $a, b, c, d \in \mathbb{R}$, and let $v = a - bi - cj - dk \in \mathbb{H}$. Computation shows that $uv = (a^2 + b^2 + c^2 + d^2)1$, and so if $u \neq 0$, we have $a^2 + b^2 + c^2 + d^2 \neq 0$ and thus $(1/(a^2 + b^2 + c^2 + d^2))v$ is a right inverse for u. Similarly, it is a left inverse, and thus every nonzero element of $\mathbb{H}$ is a unit and $\mathbb{H}$ is a division ring. (It is not commutative, of course, since $ij \neq ji$, for instance.)

As some justification for our assertion that noncommutative division rings abound, we mention now an easy result that we will need later.

(12.9) LEMMA (Schur).   *Let S and T be X-simple abelian X-groups for some set X of operators. If $\alpha : S \to T$ is a nonzero X-homomorphism, then α is an isomorphism. Also, $\mathrm{End}_X(S)$ is a division ring.*

Proof.   We have $\ker(\alpha) < S$ since $\alpha \neq 0$. Because $\ker(\alpha)$ is an X-subgroup of S, which is X-simple, we conclude that $\ker(\alpha) = 0$ and α is injective. Also, the image $S\alpha \neq 0$ since $\alpha \neq 0$. Because $S\alpha$ is an X-subgroup of T and T is X-simple, we see that $S\alpha = T$ and α is an isomorphism.

Now take $T = S$. If $\alpha \in \mathrm{End}_X(S)$ and $\alpha \neq 0$, we have just shown that α is invertible in $\mathrm{End}(S)$. To see that α is invertible in $\mathrm{End}_X(S)$, we need to check that $(sx)\alpha^{-1} = (s\alpha^{-1})x$ for $s \in S$ and $x \in X$. To prove this, apply α to both sides (using the fact that α is an X-endomorphism) and observe that the result is the same. Since α is injective, this yields the desired equality.   ∎

Next we investigate connections between the units and the ideal structure of a ring.

(12.10) LEMMA. *Let $I \subseteq R$ be an ideal or a one-sided ideal. If I contains a unit, then $I = R$.*

Proof.   Suppose $u \in I$ is a unit in R. For $r \in R$, we have $(ru^{-1})u = r = u(u^{-1}r)$, and so r is both a left multiple and a right multiple of u. It thus lies in I.   ∎

(12.11) COROLLARY. *Let R be a nontrivial ring. Then the following are equivalent:*

i.   *R is a division ring.*
ii. *The only right ideals of R are 0 and R.*
iii. *The only left ideals of R are 0 and R.*

Proof.   That (i) implies both (ii) and (iii) is immediate from Lemma 12.10. Now assume (ii) and let $0 \neq a \in R$. Then aR is a nonzero right ideal, and so $aR = R$ and there exists some $b \in R$ with $ab = 1$. In particular, $b \neq 0$ and we can apply the same argument to b and find a right inverse $c \in R$ for b. Then $ab = 1 = bc$ and so $a = c$ and b is a (two-sided) inverse for a, which is therefore a unit. This shows that R is a division ring and (ii) implies (i). That (iii) also implies (i) follows similarly.   ∎

(12.12) COROLLARY. *A commutative ring is a field iff 0 is maximal among its proper ideals.*   ∎

We mention that in any ring, an ideal (or a right or left ideal) that is maximal among proper ideals (or proper right or left ideals) is called a *maximal ideal* (or a *maximal right ideal* or *maximal left ideal*). When describing ideals or one-sided ideals, we always use the word "maximal" to mean "maximal proper."

By Lemma 12.8, which is the Correspondence theorem for rings, we see that $M \subseteq R$ is a maximal ideal iff 0 is maximal as an ideal of R/M. If R is commutative, therefore, then R/M is a field when M is a maximal ideal.

(12.13) LEMMA. *Let $I \subseteq R$ be a proper ideal (or right or left ideal). Then I is contained in some maximal ideal (or maximal right or left ideal, respectively).*

Proof.   Note that the right ideals, left ideals, and ideals of R are just the X-subgroups, where $X = R$ acting by right multiplication or $X = R$ acting by left multiplication or X is the union of two copies of R, one acting by right multiplication and one by left. In all three cases, Lemma 12.10 tells us that being proper is exactly equivalent to being disjoint from the subset $\{1\} \subseteq R$.

We are now set up to apply Lemma 11.18 (which is our main application of Zorn's lemma), and the result follows.   ∎

(12.14) COROLLARY. *If R is a nontrivial commutative ring, then there exists a homomorphism of R onto some field.*

Proof.   The ideal $I = 0$ is proper and so by Lemma 12.13 is contained in some maximal ideal M. Then R/M is a field and the canonical map $R \to R/M$ is the desired homomorphism.   ∎

What happens if R is not commutative? A ring R for which 0 is a maximal ideal is a *simple* ring. Division rings are clearly simple, but noncommutative simple rings need not be division rings.

(12.15) THEOREM.   *Let D be a division ring. (In particular, D may be a field.) Then the matrix ring $M_n(D)$ is simple.*

Proof.   Calculations in matrix rings are often facilitated by using the so-called *matrix units* $e_{ij} \in M_n(D)$ for $1 \le i, j \le n$. The matrix e_{ij} has (i, j)-entry equal to 1 and all other entries are zero. (This, despite its name, is certainly not a unit of $M_n(D)$, except when $n = 1$.) Observe that

$$e_{ij}e_{kl} = \begin{cases} 0 & \text{if } j \ne k \\ e_{il} & \text{if } j = k. \end{cases}$$

Suppose $I \subseteq M_n(D)$ is a nonzero ideal. We shall show that $I = M_n(D)$ by proving that I contains the identity matrix. Let $0 \ne a \in I$ and write $a_{ij} \in D$ to denote the (i, j)-entry in the matrix a. Then

$$a = \sum_{i,j} a_{ij}e_{ij} ,$$

and so $e_{ii}ae_{jj} = a_{ij}e_{ij}$ for all i, j. Since I is an ideal, we have $a_{ij}e_{ij} \in I$.

Because $a \ne 0$, we can certainly choose subscripts i and j such that $a_{ij} \ne 0$. Let $b \in D$ with $ba_{ij} = 1$ and compute that

$$e_{kk} = (be_{ki})(a_{ij}e_{ij})(e_{jk}) \in I .$$

Since $\sum e_{kk} = 1$, the identity matrix, we have $1 \in I$ and hence $I = R$, as desired.   ■

Observe that the simple ring $M_n(D)$ of Theorem 12.15 is definitely not a division ring if $n > 1$. This is clear since the matrix unit e_{ij} is nonzero and yet its square is zero if $i \ne j$.

We close this section with the remark that every finite division ring is commutative. This surprising result of M. Wedderburn is proved later.

12C

We wish now to consider the ring theory analog of a group acting on a set.

(12.16) DEFINITION.   Let M be an additively written abelian group and let R be a ring. Suppose that for each $m \in M$ and $r \in R$, there is defined an element of M denoted mr. Then M is a *right R-module* if the following conditions hold:

i. $(x + y)r = xr + yr$ for all $x, y \in M$ and $r \in R$.
ii. $x(r + s) = xr + xs$.

    iii. $x(rs) = (xr)s$ for all $x \in M$ and $r, s \in R$.
    iv. $x1 = x$ for all $x \in M$.

Just as the key example of a group action is $\text{Sym}(\Omega)$ acting on Ω, we have for rings that any abelian group A is a right $\text{End}(A)$-module with the natural action. Also, for any ring R, the ring itself and all right ideals of R are right R-modules. A ring R viewed as a right R-module is called the *regular* right R-module. When it is necessary to distinguish it from the original ring, the regular module is denoted $R^\bullet$. (Of course, R and $R^\bullet$ are merely different ways of thinking about the same set.) Many authors write R_R where we have written $R^\bullet$, and more generally, they write M_R to describe the abelian group M viewed as a right R-module.

Next observe that an arbitrary abelian group A can be viewed as a right $\mathbb{Z}$-module. Write A additively and, for $a \in H$ and $n \in \mathbb{Z}$, define an (as usual) to be the sum of n copies of a (or $-n$ copies of $-a$ if $n < 0$). Our final example of a module is probably the most familiar of all. If $R = F$, a field, then a right R-module is precisely the same thing as an F-vector space (if we agree to write scalars on the right for scalar multiplication). Definition 12.16 simply generalizes the definition of a vector space from fields to arbitrary rings.

Observe that condition (i) in Definition 12.16 is exactly the assertion that R is a set of operators for M and thus M is an R-group. If $N \subseteq M$ is an R-subgroup, then the R-groups N and M/N inherit conditions (i)–(iv) from M and so are right R-modules. They are a *submodule* and *factor module*, respectively, of M. Because right R-modules are just a certain type of group with operators, we can apply all of the terminology and results of Chapter 11. We can thus speak of simple, artinian, noetherian, finitely generated, or completely reducible right R-modules.

Conditions (ii) and (iii) of Definition 12.16 relate the operator action of R on M to the internal operations in R, and condition (iv) turns out to be equivalent to the assertion that no nonzero element $x \in M$ satisfies $xr = 0$ for all $r \in R$. Some authors do not require condition (iv) and call modules that satisfy it "unital" R-modules.

A *left R-module* is an abelian group M with an action of R written on the left, so that if $r \in R$ and $m \in M$, then the element $rm \in M$ is defined. The conditions corresponding to those in Definition 12.16 are $r(x+y) = rx+ry$, $(r+s)x = rx+sx$, $(rs)x = r(sx)$, and $1x = x$ for $r, s \in R$ and $x, y \in M$. The reader may wonder whether or not there is a real difference between a left and a right R-module beyond the merely typographic detail of the side on which we have chosen to write the ring elements. Given a left R-module M, we might try to convert it to a right R-module by defining mr for $m \in M$ and $r \in R$ to be simply rm. Conditions (i), (ii), and (iv) of Definition 12.16 certainly hold, but in general, (iii) does not since

$$(xr)s = (rx)s = s(rx) = (sr)x = x(sr),$$

and there is no reason to believe that this equals $x(rs)$. This computation shows, however, that if R is commutative, then left R-modules are essentially interchangeable with right R-modules, and nothing can be gained by distinguishing between the two types.

If R is not commutative, then left R-modules can be viewed as right modules for the *opposite ring* of R. This ring, denoted R^{op}, has the same underlying set and the same addition as R, but the product rs in R^{op} is defined to be the element sr of R. When R is commutative, of course, $R = R^{op}$ but, in general, R and R^{op} are not even isomorphic. The reader might wonder why we did not define G^{op} in our study of groups; this seems to provide an easy way to build a new group from an existing one. The reason is that for all groups, $G \cong G^{op}$ via the map $g \mapsto g^{-1}$. This is also the reason we never considered "left actions" of groups on sets. Although these would be the group analog of left R-modules, they would contribute nothing new.

Perhaps the most obvious examples of left R-modules are the ring itself with respect to left multiplication (this is the *regular* left R-module, which we denote $^\bullet R$) and all of its submodules, which are the left ideals of R. Our practice in this book is to focus on right R-modules and to use left R-modules only when necessary.

12D

A ring R is *right artinian* if the regular right R-module $R^\bullet$ is artinian. (Equivalently, the set of right ideals of R satisfies the minimal condition or the DCC.) Similarly, we define *right noetherian*, *left artinian*, and *left noetherian* rings. If R is commutative, then right ideals and left ideals are simply ideals, and so the right and left versions of each chain condition are equivalent. In this case, we just say that R is artinian or noetherian if the appropriate condition is satisfied. For example, the ring $\mathbb{Z}$ is noetherian but not artinian, since the additive group of $\mathbb{Z}$ was seen to have this property in Chapter 11, and the ideals of $\mathbb{Z}$ are precisely the additive subgroups.

(12.17) LEMMA. *Let A be a finite dimensional F-algebra for some field F. Then A is both right and left artinian and noetherian.*

The proof of Lemma 12.17 is an easy consequence of the next result.

(12.18) LEMMA. *Let A be an F-algebra. Then every right or left ideal of A is an F-subspace of A.*

Proof. Let $I \subseteq A$ be a right ideal. We must show that $\alpha x \in I$ whenever $\alpha \in F$ and $x \in I$. This follows because $\alpha x = \alpha(x1) = x(\alpha 1)$ is a right multiple of x. If I were a left ideal instead, the proof would be similar since $\alpha x = \alpha(1x) = (\alpha 1)x$. $\blacksquare$

Proof of Lemma 12.17. The posets of right ideals and of left ideals are subsets of the poset of all F-subspaces of A. Since A is finite dimensional, the latter poset satisfies both chain conditions, and the result follows. $\blacksquare$

So far, we have given no example of a right artinian ring that is not right noetherian. In fact, there is no such example, as we prove later. There are examples of rings that are artinian or noetherian on one side but not the other, but we do not take the time to construct such a ring here. (See Problem 12.24.)

In the case of vector spaces over fields, being finitely generated is clearly equivalent to being finite dimensional, and so subspaces of finitely generated vector spaces are themselves finitely generated. This generalizes to right modules over any right noetherian ring. Although we do not use this fact until much later, it is instructive to go through the proof now. We end this chapter by proving somewhat more.

(12.19) THEOREM.   *Let R be either right noetherian or right artinian and let M be a finitely generated right R-module. Then M is either noetherian or artinian, respectively, and in the noetherian case, every R-submodule of M is finitely generated.*

Proof.   If $x \in M$, then $xR = \{xr \mid r \in R\}$ is an R-submodule (called a *cyclic* submodule). If S is a generating set for M, we have

$$M = \sum_{x \in S} xR .$$

By Corollary 11.8, it suffices to show that each cyclic submodule xR is noetherian or artinian in order to prove the same about M.

Now map $\theta : R^\bullet \to xR$ by $\theta(a) = xa$. We claim that θ is a homomorphism of R-modules. (In other words, θ is an R-homomorphism of R-groups.) The map clearly respects addition and so we need only check the R-action. We have $\theta(ar) = x(ar) = (xa)r = \theta(a)r$, as required. Of course, θ maps the regular module $R^\bullet$ onto the cyclic module xR, and so xR, being a homomorphic image of $R^\bullet$, is noetherian or artinian. (The *ring* R is right noetherian and so the *module* $R^\bullet$ is noetherian, and similarly for artinian.)

Finally, if R is right noetherian, then so is M, and thus every submodule of M is finitely generated by Theorem 11.4.   ∎

Problems

12.1 Let R be a finite ring. Show that the following are equivalent:
  i. R is a division ring.
  ii. R is nontrivial and if $r, s \in R$, with $rs = 0$, then either $r = 0$ or $s = 0$.

  NOTE: A commutative ring that satisfies (ii) is called an *integral domain* or sometimes just a *domain*. The problem tells us that finite domains are fields.

12.2 Let R be a finite dimensional F-algebra, where F is a field. Show that conditions (i) and (ii) of Problem 12.1 are equivalent in this case, too.

12.3 Let $S \subseteq R$ be a subring. Show that the following are equivalent:
  i. S is a unitary subring.
  ii.   If $rS = 0$ for some $r \in R$, then $r = 0$.

12.4 Let R be a "ring" but do not assume that $1 \in R$. Let $R^* = R \oplus \mathbb{Z}$ (as abelian groups). Show how to define multiplication on R^* so that it becomes a ring with an ideal that is "ring" isomorphic to R.

NOTE: This problem shows that the theory of "rings without 1" is included within the theory of "rings and ideals."

12.5 Let R be a ring and suppose $I \subseteq R$ is the unique maximal right ideal of R.

a. Show that I is an ideal of R.
b. Show that every element $a \in R - I$ is a unit.
c. Show that I is the unique maximal left ideal of R.

HINT: For part (a), show that $aI \subseteq I$ for all $a \in R$. If this fails, then $aI = R$. Deduce from this that the right annihilator $\mathrm{ra}(a)$ contains an element of the form $1 - xa$ for some $x \in I$, and obtain a contradiction.

NOTE: A ring that satisfies the hypothesis of this problem is said to be *local*.

12.6 Let R be a ring and let I be the set of nonunits of R. Suppose that I is an additive subgroup of R. Show that I is an ideal of R and hence that R is local.

HINT: If $a \in I$, show that $ab \neq 1$ for $b \in R$. This is easy if b is a unit. Assume that $b \in I$ and $ab = 1$. Show that $a = (a - 1)(1 - b)^{-1}$ and derive a contradiction.

12.7 Fix a prime p and let

$$R = \{m/n \in \mathbb{Q} \mid m, n \in \mathbb{Z} \text{ and } p \text{ does not divide } n\}.$$

Show that R is a subring of $\mathbb{Q}$ and is local.

12.8 Let G be any group (written multiplicatively) and let F be any field. Construct the F-algebra with G as a basis and with multiplication defined by extending the original group multiplication on G to the whole algebra. This F-algebra, denoted FG, is called the *group algebra* of G over F.

a. Show that there is a unique F-linear ring homomorphism $\delta : FG \to F$ such that $\delta(g) = 1$ for all $g \in G$.
b. Let δ be as in part (a) and write $\Delta = \ker(\delta)$. Show that $\{1 - g \mid g \in G\}$ is an F-basis for Δ.

NOTE: The map δ is called the *augmentation map* and its kernel Δ is the *augmentation ideal* of the group algebra FG.

12.9 Let S be a simple ring. Show that all nonzero elements of S have equal additive order. Show that this order either is a prime number p or is infinite.

NOTE: We say that S has *characteristic p* in the former case and *characteristic 0* in the latter.

12.10 An element $r \in R$ is *nilpotent* if $r^n = 0$ for some integer $n \geq 1$. Show that if r is nilpotent, then $1 - r$ is a unit in R.

12.11 Show that in a commutative ring, the set of nilpotent elements forms an ideal, and produce an example showing that this can fail without the commutativity assumption.

12.12 Show that the group algebra FG contains a finite-dimensional ideal iff G is finite. Show that if G is finite, then there is a one-dimensional ideal.

HINT: Show that a finite-dimensional subspace of FG is contained in the linear span of some finite subset of G.

12.13 An element $e \in R$ is *idempotent* if $e^2 = e \neq 0$. If e is idempotent, then it is a triviality that eRe is a subring with e as its unity element. Show that if R is finite and contains no nonzero nilpotent elements, then for every idempotent e we have $eRe = eR$.

HINT: Consider the map $eR \to eR$ defined by $x \mapsto xe$. Show that this map must be surjective since its kernel is trivial.

12.14 Show that every division ring is isomorphic to $\text{End}_X(A)$ for some X-simple X-group A.

NOTE: This is a converse to Schur's lemma (12.9).

12.15 Let D be a division ring and let $X \subseteq D$ be any subset. We write $\mathbf{C}(X) = \{r \in D \mid xr = rx \text{ for all } x \in X\}$. Show that this *centralizer* subring is a division ring.

12.16 Show that R is ring-isomorphic to some F-algebra iff there exists a unitary subring $S \subseteq R$ such that $S \cong F$ and $S \subseteq \mathbf{Z}(R)$, the *center* of R. (Of course, $\mathbf{Z}(R) = \{z \in R \mid zr = rz \text{ for all } r \in R\}$. The center of any ring is a unitary subring.)

12.17 If R is simple, show that $\mathbf{Z}(R)$ is a field. In particular, if $R = M_n(F)$, where F is a field, show that $\mathbf{Z}(R) \cong F$.

12.18 Let A be an F-algebra and let M be a right A-module. Show that there is a unique way to give M the structure of an F-vector space such that

$$\alpha(ma) = (\alpha m)a = m(\alpha a)$$

for all $\alpha \in F, m \in M$, and $a \in A$.

12.19 Let A be a finite dimensional $\mathbb{C}$-algebra and let M be a simple right A-module. Show that if $\alpha \in \text{End}_A(M)$, then there exists $\lambda \in \mathbb{C}$ such that $m\alpha = \lambda m$ for all $m \in M$ (where λm is defined by Problem 12.18).

HINT: Show that $\dim_{\mathbb{C}}(M)$ is finite and that α is a $\mathbb{C}$-linear transformation on M. Let λ be an eigenvalue of α on M and consider $\alpha - \lambda 1 \in \text{End}_A(M)$. Use Schur's lemma (12.9).

12.20 Let A be a finite dimensional commutative $\mathbb{C}$-algebra and let M be any simple right A-module. Show that $\dim_{\mathbb{C}}(M) = 1$.

12.21 Let $X \subseteq M_n(\mathbb{C})$ be an arbitrary set of pairwise commuting matrices. Show that there is some nonzero row vector that is an eigenvector for all matrices in X.

12.22 If R is nontrivial, write R^n to denote the direct sum of n copies of the regular module $R^\bullet$ (where n is a positive integer). Assume $R^n \cong R^m$ where, of course, this is an R-isomorphism.

 a. Show that $n = m$ if R is both artinian and noetherian.
 b. Show that $n = m$ if R is commutative.

 HINT: For part (a), consider composition lengths. For part (b), show that $(R/I)^n \cong (R/I)^m$ as (R/I)-modules, when $I \subseteq R$ is an ideal. Apply this when I is a maximal ideal of R.

 NOTE: It is not true that $m = n$ in complete generality. We mention that any right R-module that is R-isomorphic to R^n is said to be *free* of *rank n*.

12.23 Let B and C be rings and write $R = B \oplus C$. Show that R is right artinian or noetherian iff each of B and C is right artinian or noetherian, respectively.

12.24 Let $B \subseteq A$ and $C \subseteq A$ be unitary subrings of a ring A, and note that we can view A as a right B-module, which we denote A_B, or as a left C-module, which we denote $_C A$. Let $R = \left\{ \begin{bmatrix} b & 0 \\ a & c \end{bmatrix} \mid a \in A, b \in B, c \in C \right\} \subseteq M_2(A)$ and let $I = \left\{ \begin{bmatrix} 0 & 0 \\ a & 0 \end{bmatrix} \mid a \in A \right\} \subseteq R$.

 a. Show that R is a ring and that I is an ideal of R.
 b. Show that $R/I \cong B \oplus C$.
 c. Show that I is noetherian or artinian as a right R-module iff A_B is noetherian or artinian.
 d. Show that I is noetherian or artinian as a left R-module iff $_C A$ is noetherian or artinian.
 e. In the case $A = B = \mathbb{R}$ and $C = \mathbb{Q}$, show that R is right noetherian and right artinian but is neither left noetherian nor left artinian.

12.25 Let F be a field and G a finite group. Show that the center $\mathbf{Z}(FG)$ is an F-subspace of the group algebra FG with dimension equal to the number of conjugacy classes in G.

12.26 Suppose that the additive group of R decomposes as an internal direct sum

$$R = I_1 \dotplus I_2 \dotplus \cdots \dotplus I_n,$$

where each I_i is an ideal of R. Show that each of these ideals is a subring of R (so that it has a unity) and that R is ring isomorphic to the (external) direct sum of these subrings.

12.27 Write $R = M_n(D)$, where D is a division ring, and for $1 \leq i \leq n$, let I_i be the set of all matrices with all rows other than the ith row equal to zero. (The ith row of an element of I_i is arbitrary.) Show that each I_i is a right ideal of R that is simple as a right R-module.

HINT: Use the matrix units $e_{ij} \in R$ as in the proof of Theorem 12.15.

12.28 Let F be an arbitrary field and let $\mathbb{H}(F)$ denote the four-dimensional F-algebra with basis $1, i, j, k$ and multiplication rules as in the definition of the quaternion division algebra $\mathbb{H}$. (Thus $\mathbb{H}(\mathbb{R}) = \mathbb{H}$.) Show that $\mathbb{H}(F)$ is a division ring iff the equation $x^2 + y^2 + z^2 = -1$ has no solution with $x, y, z \in F$.

NOTE: It is a fact that in a finite field F, the equation $x^2 + y^2 = a$ has a solution for every $a \in F$, and thus $\mathbb{H}(F)$ is not a division ring in this case. This is consistent with the theorem of Wedderburn that finite division rings are necessarily commutative.

12.29 Let $S \subseteq R$ be a subring. Show that there exists a unique subring $T \subseteq R$ maximal with the property that T contains S as a unitary subring.

12.30 Let $x, y \in R$ with $xy = 1$.
   a. If $yx \neq 1$, show that the right annihilator $\text{ra}(x) > 0$.
   b. Note that
$$0 = \text{ra}(x^0) \subseteq \text{ra}(x^1) \subseteq \text{ra}(x^2) \subseteq \cdots$$
   is an ascending chain of right ideals. Show that if $\text{ra}(x^{n-1}) < \text{ra}(x^n)$ with $n > 0$, then $\text{ra}(x^n) < \text{ra}(x^{n+1})$.
   c. If R is right noetherian, show that $yx = 1$.

   HINT: For part (b), suppose $x^n a = 0 \neq x^{n-1}a$. Consider the element ya.

CHAPTER THIRTEEN

Simple Modules and Primitive Rings

13A

We are now ready to begin in earnest our study of ring theory. Our concern in this chapter is those properties of a (not necessarily commutative) ring that are related to its collection of simple right R-modules.

If U is any right R-module and $X \subseteq U$ is a subset, we define the *annihilator*

$$\mathrm{ann}_R(X) = \{r \in R \mid xr = 0 \text{ for all } x \in X\}$$

and we write $\mathrm{ann}_R(x)$ for $\mathrm{ann}_R(X)$ when $X = \{x\}$, a singleton. We sometimes suppress the subscript and write $\mathrm{ann}(X)$ in place of $\mathrm{ann}_R(X)$.

(13.1) LEMMA.   *Let U be a right R-module and let $X \subseteq U$ be any subset. Then:*

a. $\mathrm{ann}_R(X)$ *is a right ideal of R.*
b. *If X is an R-submodule of U, then $\mathrm{ann}_R(X)$ is an ideal of R.*
c. *If U is simple and $0 \neq x \in U$, then $\mathrm{ann}_R(x)$ is a maximal right ideal of R and $U \cong R^{\bullet}/\mathrm{ann}_R(x)$.*
d. *Every maximal right ideal of R has the form $\mathrm{ann}_R(x)$ for some nonzero element x of some simple right R-module U.*

Before we give the (easy) proof of Lemma 13.1, a few comments on part (c) are appropriate. The dot in the expression $R^{\bullet}/\mathrm{ann}_R(x)$ is necessary. By part (a), we see that $\mathrm{ann}_R(x)$ is a right ideal of R; it is a submodule of $R^{\bullet}$ and the object $R^{\bullet}/\mathrm{ann}_R(x)$ is the corresponding factor module. The absence of the dot would suggest that we are referring to a factor ring of R, and this would require that the denominator be an ideal. It is not usually true, however, that $\mathrm{ann}_R(x)$ is an ideal.

The real significance of part (c) is that all simple right R-modules can be found (at least theoretically, and up to R-isomorphism) within the ring R as factor modules of $R^{\bullet}$. When we study simple right R-modules, therefore, we are really studying the ring itself.

Proof of Lemma 13.1. Note that $0 \in \text{ann}_R(X)$ since $x0 = x(0 + 0) = x0 + x0$, and so $x0 = 0$ for all $x \in X$. From this, it is immediate that $\text{ann}_R(X)$ is an additive subgroup of R. For $a \in \text{ann}_R(X)$ and $r \in R$, we have $Xar = 0r = 0$, and so $ar \in \text{ann}_R(X)$, and part (a) follows. If X is a submodule, then $Xr \subseteq X$ and thus $Xra \subseteq Xa = 0$, and part (b) holds.

For part (c), map $\theta : R^{\bullet} \to U$ by $\theta(r) = xr$, and check that θ is an R-homomorphism with $\ker(\theta) = \text{ann}_R(x)$. Since $x = x1 = \theta(1) \in \theta(R^{\bullet})$, we see that $\theta(R^{\bullet})$ is a nonzero submodule of the simple module U. It follows that $\theta(R^{\bullet}) = U$ and hence $U \cong R^{\bullet}/\ker(\theta) = R^{\bullet}/\text{ann}_R(x)$. Since U is simple, the Correspondence theorem tells us that $\text{ann}_R(x)$ is maximal among proper submodules of $R^{\bullet}$.

Finally, let M be a maximal right ideal of R and write $U = R^{\bullet}/M$, so that U is a simple right R-module. Let $x = M + 1 \in U$. Then $r \in R$ lies in $\text{ann}_R(x)$ iff $(M + 1)r = M + 0$. (Recall that the coset $M + 0 = M$ is the zero element of U.) By definition of the operator action in operator factor groups, we have $(M + 1)r = M + 1r = M + r$, and so $r \in \text{ann}_R(x)$ iff $r \in M$. Thus $\text{ann}_R(x) = M$, as required. ■

Next, suppose $\theta : R \to S$ is a surjective ring homomorphism. If U is any right S-module, we can make U into a right R-module simply by defining $ur = u(\theta(r))$ for all $r \in R$. (It is trivial to check that this works.) Since the actions of R and S induce exactly the same sets of endomorphisms of U (from the assumption that θ is surjective), it follows that a subgroup $V \subseteq U$ is an R-submodule iff it is an S-submodule. Because of this, adjectives such as "artinian," "finitely generated," "simple," and "completely reducible" either apply or fail to apply equally to U as an R-module and to U as an S-module.

Which R-modules do we get from S-modules in this way? In the above situation, if $r \in \ker(\theta)$, then $\theta(r) = 0 \in \text{ann}_S(U)$, and so $r \in \text{ann}_R(U)$ and we see that $\ker(\theta) \subseteq \text{ann}_R(U)$. Conversely, if V is any right R-module with $\ker(\theta) \subseteq \text{ann}_R(V)$, then we can "undo" the above construction and view V as a right S-module by defining $vs = vr$, where $v \in V$, $s \in S$, and $r \in R$ is any element such that $\theta(r) = s$. Since r is not uniquely determined, we observe that if also $\theta(r_0) = s$, then $r - r_0 \in \ker(\theta) \subseteq \text{ann}_R(V)$, and so $v(r - r_0) = 0$. Thus $vr = vr_0$ and vs is well defined. Because of these remarks, we can, and occasionally shall, identify right S-modules with right R-modules that are annihilated by $\ker(\theta)$.

Now we consider the question of how much we could tell about a ring R if we knew all of the simple right R-modules. Some nonzero elements of R may be indistinguishable from zero from the point of view of these modules. Such elements, together with zero, form (as we shall see) an ideal J of R. The factor ring R/J is really the only part of R "visible" to the simple modules and about which they can give information.

We should mention that a nontrivial ring R necessarily has some simple right modules. This is because, by Lemma 12.13, there exists some maximal right ideal $M \subseteq R$, and so $R^\bullet/M$ is a simple module.

13B

The *Jacobson radical* of R, denoted $J(R)$, is the intersection of all the annihilators $\text{ann}_R(U)$ for all simple right R-modules U. Since there are some set theoretic difficulties with the phrase "all ... modules" in the previous sentence, we point out that if $U \cong V$ are R-isomorphic right R-modules, then $\text{ann}_R(U) = \text{ann}_R(V)$ (since U and V both look the same to R or, for readers who do not find that explanation convincing, by an easy argument, which we omit). It follows that in the definition of the Jacobson radical, it suffices to let U run over a set of representatives for the isomorphism classes of simple right R-modules. This poses no logical difficulties since by Lemma 13.1(c) we could choose our representative set from among the modules $R^\bullet/M$ as M runs over the maximal right ideals of R, and so the Jacobson radical is essentially defined "within" R.

Observe that by Lemma 13.1, if U is a right R-module, then $\text{ann}_R(U)$ is an ideal of R. Ideals of this form for simple modules U are called *primitive* ideals and so, by definition, $J(R)$ is the intersection of all the primitive ideals of R and hence is itself an ideal.

(13.2) COROLLARY. *For any ring R, we have $J(R/J(R)) = 0$. More generally, if $\theta : R \to S$ is any surjective ring homomorphism with $\ker(\theta) \subseteq J(R)$, then $\theta(J(R)) = J(S)$.*

Proof. Note that the first statement follows from the second by taking $\theta : R \to R/J(R)$ to be the canonical homomorphism. To prove the second statement, observe that every simple right S-module can be viewed as an R-module, and as such, it is simple and so is annihilated by $J(R)$. This means that the original S-module is annihilated by $\theta(J(R))$ and hence $\theta(J(R)) \subseteq J(S)$.

Since every simple R-module is annihilated by $J(R) \supseteq \ker(\theta)$, each can be viewed as an S-module and so is annihilated by $J(S)$. It follows that $J(R)$ contains the full inverse image of $J(S)$, and the result follows. ∎

Other types of "radicals" have been defined for rings. Roughly, the idea is to focus on some kind of "bad" elements and to observe that the set of these forms an ideal $I \subseteq R$ such that if we "mod out" by I, the resulting ring will have no nonzero "bad" elements. For the Jacobson radical, "bad" means "annihilating all simple right modules." The classical situation is where R is commutative and "bad" is "nilpotent." In this case, the corresponding radical is the *nilradical* and is often denoted $\sqrt{0}$. We shall have considerably more to say about this when we study commutative rings, but for now, we mention only that the symbol $\sqrt{0}$ and the word "radical" are explained by the fact that a nilpotent element is an "nth root" of zero for some integer $n > 0$.

The Jacobson radical has a useful internal characterization.

(13.3) COROLLARY. *The Jacobson radical $J(R)$ is the intersection of all maximal right ideals of R.*

Proof. Let $M \subseteq R$ be any maximal right ideal of R. By Lemma 13.1(d), we have that $M = \mathrm{ann}_R(x)$ for some element x of some simple right R-module U. We have $J(R) \subseteq \mathrm{ann}_R(U) \subseteq \mathrm{ann}_R(x) = M$, and hence $J(R)$ is contained in every maximal right ideal.

Conversely, let $r \in R$ lie in every maximal right ideal of R. If U is any simple right R-module, we conclude by Lemma 13.1(c) that $r \in \mathrm{ann}_R(x)$ for every $x \in U$, and thus $r \in \mathrm{ann}_R(U)$. Since U is arbitrary, we conclude that $r \in J(R)$. ∎

Note that we could have defined $J(R)$ as the intersection of all maximal right ideals of R. From that point of view, however, it is not so obvious that $J(R)$ is an ideal, although it is clearly a right ideal.

Observe that Corollary 13.3 is relevant to Problem 12.5 concerning local rings. If I is the unique maximal right ideal of R, then $I = J(R)$ and so it is an ideal, and this solves part (a) of the problem.

Neither the definition of $J(R)$ nor its characterization in Corollary 13.3 proves to be computationally convenient for working with the Jacobson radical. Much more useful is the notion of quasiregularity.

An element $r \in R$ is *quasiregular* if $1 - r$ is invertible. For instance, if r is nilpotent so that $r^n = 0$ for some integer $n \geq 1$, then an easy computation shows that $1 + r + r^2 + \cdots + r^{n-1}$ is an inverse of $1 - r$. Thus every nilpotent element is quasiregular. (Of course, not every quasiregular element is nilpotent, since if R is a division ring, for instance, every element $r \in R - \{1\}$ is quasiregular.) We say that $r \in R$ is *right quasiregular* if $1 - r$ has a right inverse in R but possibly not a (two-sided) inverse. This is equivalent to the assertion that $(1 - r)R = R$.

(13.4) THEOREM. *Let R be any ring.*

a. *Every element of $J(R)$ is quasiregular.*
b. *If I is a right ideal and every element of I is right quasiregular, then $I \subseteq J(R)$.*

Proof. Let $r \in J(R)$ and note that if $(1 - r)R < R$, then by Lemma 12.13 we can find a maximal right ideal M containing $(1 - r)R$, and so $1 - r \in M$. Since also $r \in M$ by Corollary 13.3, we conclude that $1 \in M$, and this is a contradiction since no proper right ideal can contain 1. It follows that $(1 - r)R = R$, and thus r is right quasiregular and we can write $(1 - r)s = 1$ for some $s \in R$.

Now put $y = 1 - s$ so that $s = 1 - y$, and we have $(1 - r)(1 - y) = 1$, which yields $ry - r - y = 0$. Therefore $y = r(y - 1)$ and this, being a multiple of r, lies in $J(R)$. Applying the reasoning of the first paragraph to y in place of r shows that $1 - y$ has some right inverse in R. Since it also has the left inverse

$1 - r$, we conclude that these two one-sided inverses for $1 - y$ are equal, and so $(1 - r)(1 - y) = 1 = (1 - y)(1 - r)$. Therefore $(1 - y) = (1 - r)^{-1}$ and r is quasiregular, proving (a).

Now let I be as in (b). By Corollary 13.3, it suffices to show that $I \subseteq M$ for an arbitrary maximal right ideal M of R. If this is not true, then $I + M > M$, and since $I + M$ is a right ideal, the maximality of M forces $I + M = R$. We can therefore write $1 = i + m$ with $i \in I$ and $m \in M$. Thus $m = 1 - i$ and

$$M \supseteq mR = (1 - i)R = R$$

since i is right quasiregular. This contradiction completes the proof.   ∎

Some caution is needed here since Theorem 13.4 does not say (and it is not true) that *every* quasiregular element of R lies in $J(R)$. Nevertheless, Theorem 13.4 is often useful for proving that elements either are or are not in $J(R)$.

(13.5) COROLLARY. *Let $z \in R$ be nilpotent and central in R. Then $z \in J(R)$.*

Proof. If $r \in R$, then $rz = zr$ and so $(zr)^n = z^n r^n = 0$ if n is large enough. Therefore zR is a right ideal consisting of nilpotent and hence quasiregular elements. By Theorem 13.4(b), therefore, $z \in zR \subseteq J(R)$.   ∎

(13.6) COROLLARY. *Let $e \in R$ be idempotent. (Recall that this means that $e^2 = e$ and $e \neq 0$.) Then $e \notin J(R)$.*

Proof. Since $(1 - e)e = 0$ and $e \neq 0$, we see that $1 - e$ cannot be invertible. The result follows by Theorem 13.4(a).   ∎

As a further example of how Theorem 13.4 can be used computationally, let R be any ring and let $S \subseteq M_2(R)$ be the subring consisting of all lower triangular matrices. (The upper right entries of the matrices in S are zero.) Let

$$I = \left\{ \begin{bmatrix} 0 & 0 \\ r & 0 \end{bmatrix} \,\middle|\, r \in R \right\} .$$

It is an easy computation to see that I is an ideal of S and that each of its elements has square zero and so is nilpotent and hence quasiregular. Thus $I \subseteq J(S)$. In fact,

$$J(S) = \left\{ \begin{bmatrix} a & 0 \\ r & b \end{bmatrix} \,\middle|\, r \in R; \ a, b \in J(R) \right\} ,$$

as we shall see in the problems at the end of this chapter.

Theorem 13.4 gives another internal characterization of the Jacobson radical.

(13.7) COROLLARY. *The radical $J(R)$ is the unique right ideal of R maximal with the property that every element is right quasiregular.*   ∎

We could, of course, have chosen to work with left modules rather than right modules and defined the "left handed" version of the Jacobson radical as the intersection of the annihilators of all simple left R-modules. It would also be the intersection of all maximal left ideals by the "mirror image" of Corollary 13.3. (We could also describe this object as $J(R^{op})$, where R^{op} is the opposite ring of R.) It is remarkable that we get nothing new this way.

(13.8) THEOREM. *The "left-handed" version of the Jacobson radical of R is equal to $J(R)$.*

Proof. Let $L = J(R^{op})$ be the left Jacobson radical. Then L is an ideal and hence is a right ideal. By the "mirror image" of Theorem 13.4(a), every element of L is quasiregular. By Theorem 13.4(b), therefore, $L \subseteq J(R)$. The reverse inclusion follows by application of the above argument to R^{op}. ∎

(13.9) COROLLARY. *Let I be a left ideal consisting of left quasiregular elements. Then $I \subseteq J(R)$.*

Proof. Apply Theorem 13.4(b) to R^{op} and then use Theorem 13.8. ∎

Another application of Theorem 13.8 is the following corollary.

(13.10) COROLLARY. *Suppose I is the unique maximal right ideal of R. It is then also the unique maximal left ideal of R.*

Proof. By Corollary 13.3, we have that $I = J(R)$ is an ideal, and by the Correspondence theorem (12.8), we see that in the ring R/I, the trivial right ideal 0 is maximal. By Corollary 12.11, therefore, R/I is a division ring and 0 is maximal among left ideals and so I is a maximal left ideal of R by Lemma 12.8, again. Since $I = J(R)$ is contained in every maximal left ideal by Theorem 13.8 and the "mirror image" of Corollary 13.3, we are done. ∎

We need to introduce some notation. Let U be a right R-module and let $X \subseteq U$ and $Y \subseteq R$ be arbitrary additive subgroups. We then write XY to denote the additive subgroup of U generated by all elements of the form xy with $x \in X$ and $y \in Y$. Therefore an element $u \in U$ lies in XY iff u is a finite sum of the form $u = x_1 y_1 + \cdots + x_n y_n$, where $x_i \in X$ and $y_i \in Y$. An important special case of this is where $U = R^{\bullet}$, since this defines a multiplicative structure on the collection of additive subgroups of R.

As an example of the use of this notation, observe that if U is a right R-module and $V \subseteq U$ is maximal among proper submodules, then $U J(R) \subseteq V$. To see this, note that U/V is simple and so is annihilated by all elements $r \in J(R)$. Thus

$$V + ur = (V + u)r = V + 0$$

and $ur \in V$ for all $u \in U$. This gives $U J(R) \subseteq V$, as claimed. It is tempting to try to conclude from the fact that $U J(R)$ is contained in every maximal proper submodule of U that necessarily $UJ(R) < U$ whenever U is a nontrivial right R-module. This reasoning is not valid, however (and the conclusion is false), since U may not have any maximal submodules.

If $U \neq 0$ is a noetherian right R-module, then by the maximal condition, there do exist maximal proper submodules and it is true that $UJ(R) < U$. It is somewhat less trivial that the weaker finiteness condition that U be finitely generated is sufficient to obtain the same conclusion. (Note, however, that by Theorem 12.19, if R is right noetherian, then the module U is finitely generated iff it is noetherian.)

(13.11) THEOREM. *Let U be a finitely generated right R-module. If $UJ(R) = U$, then $U = 0$.*

Proof. Let $X \subseteq U$ be a finite subset minimal with the property that it generates U. Then

$$U = \sum_{x \in X} xR$$

and hence

$$U = UJ(R) = \sum_{x} xRJ(R) \subseteq \sum_{x} xJ(R).$$

If $U \neq 0$, then $X \neq \varnothing$ and we choose some element $y \in X$. By the above, we can write

$$y = \sum_{x \in X} xr_x,$$

where $r_x \in J(R)$. This gives

$$y(1 - r_y) = \sum_{x \neq y} xr_x$$

and thus

$$y = \sum_{x \neq y} xr_x(1 - r_y)^{-1}$$

since $r_y \in J(R)$ is quasiregular.

We now know that y is an element of the submodule generated by $X - \{y\}$, and so this submodule contains X and hence is all of U. This contradicts the minimality of X. ∎

Theorem 13.11 is usually stated and used in the following, essentially equivalent formulation.

(13.12) COROLLARY (Nakayama's lemma). *Let U be a finitely generated right R-module and suppose $V \subseteq U$ is a submodule such that $U = V + UJ(R)$. Then $V = U$.*

Proof. The elements of $(U/V)J(R)$ are the cosets $V + u$, where $u \in UJ(R)$, and so we see that $(U/V)J(R) = U/V$. Also U/V is finitely generated (by the images in U/V of the finite generating set for U). By Theorem 13.11, we conclude that $U/V = 0$. ∎

As an example of how Nakayama's lemma can be used, we prove the following corollary. To explain the statement, we mention that if $S \subseteq R$ is a unitary subring, then R can be viewed as a right (or left) S-module, where the module action is just multiplication in R. Note that it is essential that S be unitary for this to work.

(13.13) COROLLARY. *Let $S \subseteq R$ be a unitary subring contained in the center of R and suppose that R is finitely generated as a right S-module. Then $J(S) \subseteq J(R)$.*

Proof.   Let U be a simple right R-module and let $s \in J(S)$. Our object is to show that $Us = 0$. Because $sR = Rs$, we see that $UsR = URs \subseteq Us$, and so Us is an R-submodule of the simple R-module U. If we can show that $Us < U$, therefore, it will follow that $Us = 0$.

Since R is finitely generated as an S-module, we can write

$$R = \sum_{x \in X} xS$$

for some finite subset $X \subseteq R$. Now let $u \in U$ with $u \neq 0$. Since U is simple, we have

$$U = uR = \sum_{x \in X} uxS,$$

and thus the set $\{ux \mid x \in X\}$ is a finite generating set for U as an S-module. By Theorem 13.11, therefore, we have $Us \subseteq UJ(S) < U$.   ∎

We resume our study of the Jacobson radical in Chapter 14, where we consider right artinian rings.

13C

In studying the Jacobson radical, we considered the collection of all simple right R-modules. Now we intend to examine one of those modules in greater detail. Suppose then that U is a simple right R-module. If $r \in R$, then r induces an endomorphism of U by $u \mapsto ur$, and we write r_U to denote this map. The set $R_U = \{r_U \mid r \in R\}$ is a subring of $\text{End}(U)$ since the map $r \mapsto r_U$ is a ring homomorphism from R into $\text{End}(U)$.

Let $D = \text{End}_R(U)$. Then D is a subring of $\text{End}(U)$ and D is a division ring by Schur's lemma (12.9). If $r \in R$, then r_U commutes with every element $\alpha \in D$. To see this, note that we have

$$(u)r_U\alpha = (ur)\alpha = (u\alpha)r = (u)\alpha r_U .$$

It should be clear, in fact, that D is precisely the set of all elements of $\text{End}(U)$ that commute with all elements of R_U. We can write, therefore, $D = \mathbf{C}_{\text{End}(U)}(R_U)$, the *centralizer ring* of R_U in $\text{End}(U)$.

Which elements of $\text{End}(U)$ are induced by elements of R? Equivalently, how can we recognize the elements of R_U? Certainly, these elements must commute with every element of D, and so

$$R_U \subseteq \mathbf{C}_{\mathrm{End}(U)}(D) = \mathrm{End}_D(U).$$

The main result in this direction is the Jacobson density theorem, which guarantees that R_U is a "large" subring of $\mathrm{End}_D(U)$. (We shall see that if R is artinian, then in fact we have the equality: $R_U = \mathrm{End}_D(U)$.)

We can view U as a right D-module, but since D is a division ring (and not just a ring) we use the more suggestive linear algebra language and say that D is a "right D-vector space" (or D-space) and the elements of $\mathrm{End}_D(U)$ are referred to as "D-linear operators" on U. (Most of the familiar basic facts of linear algebra work as well over noncommutative division rings as they do in the commutative case, over fields. For instance, if $X \subseteq U$ is a D-linearly independent subset of U, then any map $X \to U$ can be extended to a D-linear operator $U \to U$.)

(13.14) THEOREM (Jacobson density).   *Let U be a simple right R-module and write $D = \mathrm{End}_R(U)$. Let α be any D-linear operator on U and let $X \subseteq U$ be any finite D-linearly independent subset. Then there exists an element $r \in R$ such that $xr = x\alpha$ for all $x \in X$.*

In other words, the arbitrary element $\alpha \in \mathrm{End}_D(U)$ is matched in its behavior on X by some element $r_U \in R_U$. It follows that α and r_U agree on every D-linear combination of the elements of X. They agree, in other words, on the entire D-subspace spanned by X. In particular, if U has finite D-dimension, then the density theorem tells us that R_U is exactly the full ring $\mathrm{End}_D(U)$ of all D-linear operators on U. As we shall see, a condition sufficient to guarantee that $\dim_D(U) < \infty$ (or, equivalently, that it is finitely generated as a D-module) is that R is right artinian.

In general, if A is any abelian operator group with operator set S, then a subset $E \subseteq \mathrm{End}_S(A)$ is said to be *dense* if for every $\alpha \in \mathrm{End}_S(A)$ and every finitely generated S-subgroup $B \subseteq A$, there exists $\beta \in E$ such that β agrees with α on B. Observe that if A itself is finitely generated as an S-group, then no proper subset of $\mathrm{End}_S(A)$ can be dense.

The conclusion of the Jacobson density theorem is that R_U is always a dense subring of $\mathrm{End}_D(U)$. The following example shows that in the infinite dimensional case, there do exist proper dense subrings of $\mathrm{End}_D(U)$. Let U be any infinite-dimensional D-space and let I be the set of all finite rank D-linear operators on U. (The rank of a linear transformation is the dimension of its image.) Note that I is a dense ideal of $\mathrm{End}_D(U)$. To build a dense ring, we need a unity element, and so we let R be the subring of $\mathrm{End}_D(U)$ generated by I and the identity operator 1. Thus R is the set of all operators of the form $n \cdot 1 + i$, where $n \in \mathbb{Z}$ and $i \in I$. It is not hard to see in this situation that U is simple as an R-module and that $D = \mathrm{End}_R(U)$. In this case, $R_U = R$ is proper in $\mathrm{End}_D(U)$.

We should explain the word "dense". It is possible to topologize $\mathrm{End}_S(A)$ in such a way that two elements are "close" if they agree on a "large" finitely generated S-subgroup of A. Then dense subsets are, in fact, dense in this topology.

We begin now to work toward a proof of the Jacobson density theorem.

(13.15) THEOREM.   *Let U be a simple right R-module and let $D = \mathrm{End}_R(U)$. Let $X \subseteq U$ be a finite subset and write $I = \mathrm{ann}_R(X)$. Suppose $u \in U$ with $uI = 0$. Then $u \in XD$, the D-span of X.*

Note that the converse of Theorem 13.15 is trivially true: every element of XD is clearly annihilated by I.

Proof of Theorem 13.15.   Observe that if $X = \varnothing$, then $I = R$ and $u = 0$, and thus u lies in $XD = 0$, as required. We may therefore suppose that $X \neq \varnothing$ and work by induction on $|X|$. Let $x \in X$ and write $Y = X - \{x\}$.

Let $J = \text{ann}_R(Y)$ and note that $I = J \cap \text{ann}_R(x)$. If $J \subseteq \text{ann}_R(x)$, then $J = I$ and so $uJ = 0$. In this case, the inductive hypothesis yields that $u \in YD \subseteq XD$ and we are done. We can thus suppose that $xJ \neq 0$. Now $J \subseteq R$ is a right ideal and thus xJ is an R-submodule of U. Since U is simple, it follows that $xJ = U$.

Next we wish to define a map $\alpha : U \to U$. Since $xJ = U$, every element of U has the form xj for some $j \in J$, and it suffices to define α for elements of this form. We put $(xj)\alpha = uj$ and check that this is well defined. Specifically, we must show that if $xj = xk$ for $j, k \in J$, then $uj = uk$. We have $x(j-k) = 0$ and so $j - k \in J \cap \text{ann}_R(x) = I$. By hypothesis, then, $u(j-k) = 0$ and hence $uj = uk$, as required.

We claim that $\alpha \in D$. It is easy to see that $\alpha \in \text{End}(U)$, and so we must check that it is an R-endomorphism. Let $r \in R$ and $z \in U$ and write $z = xj$ for some $j \in J$. Since $jr \in J$, we compute that

$$(zr)\alpha = (xjr)\alpha = ujr = (xj)\alpha r = (z)\alpha r \,.$$

This shows that $\alpha \in \text{End}_R(U) = D$, as desired.

To complete the proof, it suffices to show that $u - x\alpha \in YD$, and this follows by the inductive hypothesis if we can show that $(u - x\alpha)J = 0$. Let $j \in J$ and compute

$$(u - x\alpha)j = uj - x\alpha j = uj - (xj)\alpha = 0 \,,$$

as required.   ∎

Proof of Theorem 13.14.   The result is vacuously true if $X = \varnothing$, and so we assume X is nonempty and use induction on $|X|$. Let $x \in X$ and write $Y = X - \{x\}$. By the inductive hypothesis, there exists $s \in R$ such that $ys = y\alpha$ for all $y \in Y$.

Now we need to "adjust" s to make it work for x also, and we must do this without changing its effect on Y. Let $I = \text{ann}_R(Y)$. Then for every $i \in I$, the element $r = s + i$ satisfies $yr = y\alpha$ for all $y \in Y$. Our task is to choose $i \in I$ so that $x(s + i) = x\alpha$.

By linear independence, $x \notin YD$ and hence Theorem 13.15 tells us that $xI \neq 0$. Since I is a right ideal, however, xI is an R-submodule of the simple R-module U. Therefore $xI = U$ and we can choose $i \in I$ with $xi = x\alpha - xs$, as required.   ∎

What does the Jacobson density theorem tell us about an arbitrary ring R? If U is a simple right R-module, then the map $\theta : R \to R_U$ defined by $\theta(r) = r_U$ is a ring homomorphism of R onto R_U. (The map θ is sometimes called the *representation* of R associated with U.) We have $\ker(\theta) = \mathrm{ann}_R(U)$, and so $R/\mathrm{ann}_R(U) \cong R_U$, which is a dense ring of linear operators on some D-space for some division ring D. In other words, if $I \subseteq R$ is any primitive ideal, then R/I is isomorphic to a dense ring of linear operators.

The ring R is said to be a *primitive* ring if the trivial ideal is primitive. This means there exists some simple right R-module with trivial annihilator. (A module with trivial annihilator is said to be *faithful*.) All primitive rings, therefore, are isomorphic to dense rings of linear operators.

(13.16) COROLLARY.   *A primitive ring is isomorphic to a dense subring of the ring of D-linear operators on some D-space U for some division ring D.*   ∎

We mention that a weaker condition than primitivity for a ring is semiprimitivity. We say that R is *semiprimitive* if $J(R) = 0$. Thus R is semiprimitive precisely when the trivial ideal is an intersection of primitive ideals. In general, if "foo" is some property for ideals, we often say that "R is a foo ring" if 0 is a "foo" ideal and "R is a semifoo ring" if 0 is an intersection of "foo" ideals.

13D

We now consider artinian rings. Our goal is to classify all right artinian simple rings. As we shall see, these turn out to be exactly the full matrix rings over division rings.

(13.17) LEMMA.   *Let R be right artinian and let U be a simple right R-module and $D = \mathrm{End}_R(U)$. Then U is a finite dimensional D-space.*

Proof.   Consider the collection of right ideals of R of the form $\mathrm{ann}_R(X)$, where $X \subseteq U$ is a finite subset. By the minimal condition on right ideals of R, we can choose a finite subset $X \subseteq U$ so that $I = \mathrm{ann}_R(X)$ is minimal in this collection.

Now let $u \in U$ be arbitrary. If $uI \neq 0$, then $\mathrm{ann}_R(X \cup \{u\}) < I$ and this contradicts the minimality of I. Therefore $uI = 0$, and so by Theorem 13.15, we have $u \in XD$. Since u was arbitrary, $XD = U$ and so U is generated as a D-space by the finite set X. It follows that U is finite dimensional.   ∎

(13.18) COROLLARY (Double centralizer).   *In the situation of the previous lemma, we have*

$$R_U = \mathbf{C}_{\mathrm{End}(U)}(\mathbf{C}_{\mathrm{End}(U)}(R_U)) .$$

Equivalently, we have $R_U = \mathrm{End}_D(U)$.   ∎

By elementary linear algebra, we know that by choosing a basis in a finite dimensional vector space V over some field F, we establish a bijection between $\mathrm{End}_F(V)$, the set of linear operators on V, and the set $M_n(F)$ of $n \times n$ matrices over F, where $n = \dim_F(V)$. Furthermore, this bijection is a ring isomorphism, and so $\mathrm{End}_F(V) \cong M_n(F)$. This goes through with one additional complication in the noncommutative case: we need to take the opposite ring of D.

(13.19) LEMMA.   *Let V be an n-dimensional right D-vector space, where D is a division ring. Then*

$$\operatorname{End}_D(V) \cong M_n(D^{\mathrm{op}}).$$

Proof.   Fix a D-basis $v_1, v_2, \ldots, v_n$ for V. For $\alpha \in \operatorname{End}_D(V)$, write

$$v_i\alpha = \sum_{j=1}^{n} v_j a_{ij}$$

with $a_{ij} \in D$. Write $[\alpha]$ to denote the matrix $[a_{ij}]$ viewed as an element of $M_n(D^{\mathrm{op}})$. (We use dots to represent multiplication in D^{op} so that $a \cdot b = ba$ for $a, b \in D$.) We claim that the map $\theta : \alpha \mapsto [\alpha]$ is the desired isomorphism.

First, it is clear that θ is an additive homomorphism. If $\theta(\alpha) = 0$, then $v_i\alpha = 0$ for all i, and so $\alpha = 0$ and θ is injective. It is surjective since any map from $\{v_i\}$ into V can be extended to a D-linear transformation, and so every matrix has the form $[\alpha]$ for some $\alpha \in \operatorname{End}_D(V)$.

Now suppose that $\alpha, \beta \in \operatorname{End}_D(V)$. We need to show that $[\alpha\beta] = [\alpha][\beta]$. We have

$$(v_i)\alpha\beta = \left(\sum_{j=1}^{n} v_j a_{ij}\right)\beta = \sum_j v_j\beta a_{ij}.$$

Since

$$v_j\beta = \sum_{k=1}^{n} v_k b_{jk},$$

where the coefficients b_{jk} are the entries in $[\beta]$, we have

$$(v_i)\alpha\beta = \sum_j \sum_k v_k b_{jk} a_{ij} = \sum_k v_k \sum_j a_{ij} \cdot b_{jk}.$$

We see, therefore, that the (i, k)-entry in $[\alpha\beta]$ is equal to that in $[\alpha][\beta]$, computed in $M_n(D^{\mathrm{op}})$.   ∎

It is worth noting that $M_n(D^{\mathrm{op}}) \cong M_n(D)^{\mathrm{op}}$ via the map carrying each matrix to its transpose. It follows that $\operatorname{End}_D(V)^{\mathrm{op}} \cong M_n(D)$ in the context of the previous lemma.

Recall that by Theorem 12.15, the matrix ring $M_n(D)$ is simple when D is a division ring. The following lemma is another fact about $M_n(D)$ that we need.

(13.20) LEMMA.   *Let D be a division ring and write $R = M_n(D)$. Then the composition length of the regular R-module $R^\bullet$ is precisely n.*

Proof.   We appeal to Problem 12.27, which asserts that the set $I_k \subseteq R$ of matrices having all but the kth row zero (and the kth row arbitrary) is a simple submodule of $R^\bullet$. Let

$$U_k = \sum_{i=1}^{k} I_i,$$

so that U_k is the set of all matrices with rows numbered $k + 1$ through n being zero. It is clear that

$$0 < U_1 < U_2 < \cdots < U_n = R.$$

Also, since $U_{k-1} \cap I_k = 0$, we have

$$U_k / U_{k-1} = (U_{k-1} + I_k)/U_{k-1} \cong I_k$$

for $1 \leq k \leq n$, where we have written $U_0 = 0$. It follows that the U_i form an R-composition series for $R^\bullet$, which thus has length n.   ∎

A similar argument, working with columns rather than rows, can be used to show that the left regular $M_n(D)$-module also has composition length n when D is a division ring.

(13.21) COROLLARY.   *A full matrix ring over a division ring is right and left artinian and noetherian.*   ∎

We can now describe all simple artinian rings.

(13.22) COROLLARY.   *Let R be a ring. Then the following are equivalent:*

   i.   *R is right artinian and simple.*
  ii.   *R is right artinian and primitive.*
 iii.   *$R \cong M_n(D)$ for some integer $n \geq 1$ and division ring D.*

Proof.   If U is a simple right R-module, then $\operatorname{ann}_R(U)$ is a proper ideal. If R is a simple ring, therefore, then $\operatorname{ann}_R(U) = 0$ and R is primitive. Therefore (i) implies (ii). If R is primitive, then $R \cong R_U$ for some simple right R-module U, and $R_U = \operatorname{End}_D(U) \cong M_n(D^{\mathrm{op}})$ by Corollary 13.18 and Lemma 13.19, where $D = \operatorname{End}_R(U)$ and $n = \dim_D(U)$. Since D^{op} is a division ring, we have now shown that (ii) implies (iii).

     Now assume (iii). By Theorem 12.15, we know that R is simple, and by Corollary 13.21, it is right artinian, proving (i).   ∎

13E

By Corollary 13.22, if R is any simple right artinian ring, we can write $R \cong M_n(D)$ for some integer n and division ring D. It is natural to ask if the ring R somehow uniquely determines n and D. That n is determined is clear, since by Lemma 13.20 we see that n is the composition length of $R^\bullet$. It is also true that D is uniquely determined by R (up to isomorphism), but this is more difficult to prove (and therefore more interesting).

(13.23) THEOREM.   *Let R be a simple right artinian ring and fix a simple right R-module U. If $R \cong M_n(D)$ for some division ring D, then $D \cong \operatorname{End}_R(U)^{\mathrm{op}}$.*

We need two general lemmas.

(13.24) LEMMA. *Let $e \in R$ be an idempotent. Then eRe is a subring of R and $eRe \cong \operatorname{End}_R(eR)^{\mathrm{op}}$.*

Proof. It is clear that eRe is an additive subgroup of R that is closed under multiplication. Since e is idempotent, we see that $e \in eRe$ and that e acts as a unity for eRe.

Now map $\theta : eRe \to \operatorname{End}_R(eR)$ by $\theta(x) = \alpha_x$, where $\alpha_x : eR \to eR$ is left multiplication by $x \in eRe$. Since $x \in eR$, we have $xeR \subseteq eR$, and we see that $\alpha_x \in \operatorname{End}(eR)$. Also, if $u \in eR$, then

$$(ur)\alpha_x = xur = (u\alpha_x)r .$$

Thus α_x lies in $\operatorname{End}_R(eR)$, as desired.

It is fairly clear that the map θ is an additive homomorphism, and if $x \in \ker(\theta)$, then $\alpha_x = 0$ and so $0 = e\alpha_x = xe = x$ and θ is injective. To see that θ is surjective, let $\beta \in \operatorname{End}_R(eR)$. Then $e\beta = (e^2)\beta = (e\beta)e$, and since $e\beta \in eR$, we also have $e\beta = e(e\beta)$. Therefore $e\beta = e(e\beta)e \in eRe$. For $u \in eR$, we compute

$$u\alpha_{(e\beta)} = (e\beta)u = (eu)\beta = u\beta .$$

Thus $\theta(e\beta) = \alpha_{(e\beta)} = \beta$ and hence θ is surjective. It is thus an isomorphism of abelian groups.

To show that $\theta : eRe \to \operatorname{End}_R(eR)^{\mathrm{op}}$ is a ring isomorphism, we must check that $\theta(xy) = \theta(y)\theta(x)$ for $x, y \in eRe$. If $u \in eR$, we have

$$u\alpha_{(xy)} = xyu = (yu)\alpha_x = u\alpha_y\alpha_x ,$$

and the result follows. ∎

(13.25) LEMMA. *Let U be a faithful simple right R-module and let $I \subseteq R$ be a right ideal that is simple as a right R-module. Then $U \cong I$.*

Note that the condition on I is really that I is a *minimal right ideal*; it is minimal in the set of nonzero right ideals. In general, a ring may fail to have any minimal right ideals (for instance, $\mathbb{Z}$), although a right artinian ring necessarily has such ideals. What Lemma 13.25 tells us is that if R has a minimal right ideal, then any two faithful simple right R-modules are isomorphic.

Proof of Lemma 13.25. Since $I \not\subseteq 0 = \operatorname{ann}_R(U)$, we can find $u \in U$ with $uI \neq 0$. Now map $\theta : I \to U$ by $\theta(i) = ui$, and check that θ is a homomorphism of right R-modules. Since $\theta \neq 0$, Schur's lemma (12.9) applies; it tells us that θ is an isomorphism. ∎

Proof of Theorem 13.23. It is no loss to assume that $R = M_n(D)$. Let $I \subseteq R$ be the set of matrices with all rows other than the first equal to zero. Then I is a simple submodule of $R^{\bullet}$ by Problem 12.27, and $I = e_{11}R$, where e_{11} is the "matrix unit" with the $(1, 1)$-entry equal to one and all other entries zero.

Since R is simple, we see that $\text{ann}_R(U) = 0$, and so U is faithful and Lemma 13.25 applies. We conclude that $I \cong U$ and thus $\text{End}_R(I) \cong \text{End}_R(U)$. Also, by Lemma 13.24 we have $\text{End}_R(I)^{\text{op}} \cong e_{11}Re_{11}$ since e_{11} is an idempotent. Finally, $e_{11}Re_{11}$ is exactly the set of all matrices with the $(1, 1)$-entry arbitrary and all other entries zero. Thus $e_{11}Re_{11} \cong D$. Combining these isomorphisms, we get

$$D \cong e_{11}Re_{11} \cong \text{End}_R(I)^{\text{op}} = \text{End}_R(U)^{\text{op}}.$$

The proof is now complete. ∎

(13.26) COROLLARY. *Let D and E be division rings and suppose*

$$M_m(D) \cong M_n(E)$$

for integers m and n. Then $m = n$ and $D \cong E$. ∎

We close this chapter with a few remarks about the fact that we had to use opposite rings in several of our theorems. Some authors avoid this (or, rather, disguise it) by composing maps sometimes from left to right (as we do) and sometimes from right to left. There is no way to avoid the problem totally, since there do exist division rings that are not isomorphic to their opposite rings. Unfortunately, the construction of an example of this phenomenon would carry us too far from our path and so, with apologies to the reader, we will not provide an example.

Problems

13.1 Show that if $I \subseteq R$ is a maximal ideal, then $I = \text{ann}_R(U)$ for some simple right R-module U.

NOTE: In other words, maximal ideals are primitive.

13.2 If R is commutative, show that every primitive ideal is maximal.

13.3 A ring R is *von Neumann regular* if for every $a \in R$, there exists $b \in R$ such that $a = aba$. Show that if R is von Neumann regular, then it is semiprimitive. (In other words, $J(R) = 0$.)

13.4 Let D be a division ring and write $R = M_n(D)$. Let V be the set of $1 \times n$ "row vectors" over D.

a. Show that V is a simple R-module with the natural action of R on V.
b. Let $E = \text{End}_R(V)$. Show that E is exactly the set of left scalar multiplications by elements of D.

13.5 Let R be any ring and let S be the subring of $M_2(R)$ consisting of all matrices with the $(1, 2)$-entry equal to zero. Let

$$s = \begin{bmatrix} a & 0 \\ r & b \end{bmatrix} \quad \text{with } a, b, r \in R.$$

   a. If $s \in J(S)$, show that $a, b \in J(R)$.
   b. If a and b are invertible in R, show that s is invertible in S.
   c. If $a, b \in J(R)$, show that $s \in J(S)$.

13.6 Let G be a finite p-group and F a field of characteristic p. Construct the group algebra $A = FG$. Show that $J(A)$ has codimension one in A. (In other words, $\dim(J(A)) = |G| - 1$.)

   HINT: Work by induction on $|G|$. If $|G| > 1$, let $Z \subseteq \mathbf{Z}(G)$ with $|Z| = p$, and let $1 \neq z \in Z$. Show that $A/(1 - z)A \cong F(G/Z)$ and that $(1 - z)A \subseteq J(A)$.

   NOTE: The augmentation ideal of A (see the note following Problem 12.8) has codimension one and so is a maximal right ideal and thus contains (and therefore equals) $J(A)$. Therefore $1 - g \in J(A)$ for all $g \in G$. Observe that each element $1 - g$ is nilpotent and so is clearly quasiregular.

13.7 Fix a prime p and construct the ring

$$L_p = \{m/n \in \mathbb{Q} \mid m, n \in \mathbb{Z} \text{ and } p \nmid n\}.$$

   Compute $J(L_p)$.

   HINT: Which elements of L_p are not units?

13.8 Let $S \subseteq R$ be a unitary subring. For each of the following statements, give either a proof or a counterexample.
   a. $J(S) \subseteq J(R)$.
   b. $J(R) \cap S \subseteq J(S)$.

13.9 Let $e \in R$ be an idempotent and recall that eRe is a subring of R with unity e.

   a. Let U be a simple right R-module and observe that Ue is an eRe-submodule of U. Show that either $Ue = 0$ or Ue is simple.
   b. Show that $J(eRe) \subseteq J(R)$.

13.10 In the situation of Problem 13.9, show that $J(eRe) = eJ(R)e$.

   HINT: If $x \in eRe$ and $(1 - x)y = 1$, compute $(e - x)(eye)$.

13.11 Let $V \neq 0$ be a right D-space, where D is a division ring. Let $R = \operatorname{End}_D(V)$. Show that R is simple iff $\dim_D(V) < \infty$.

   NOTE: In any case, R is primitive.

13.12 Let R be primitive.
   a. Show that all simple submodules of $R^\bullet$ (if there are any) are faithful and isomorphic to one another.
   b. Let $I \subseteq R$ be a nonzero ideal. Show that I contains every minimal right ideal of R.

NOTE: If R is right artinian, then there necessarily exist minimal right ideals, but in that case, part (b) is not very interesting since, by Lemma 13.20, we have $I = R$.

13.13 Suppose that $S \subseteq R$ is a subring and that $S + J(R) = R$. Show that S is a unitary subring.

13.14 Let $e \in R$ be an idempotent and suppose eRe is a division ring. Show that eR has an R-submodule U such that eR/U is simple and $U^2 = 0$. (Of course, the possibility $U = 0$ is allowed.)

13.15 Let U be a finitely generated right R-module and let $J = J(R)$. Let $X \subseteq U$ be a subset whose image in U/UJ generates U/UJ. Show that X generates U.

NOTE: The module U/UJ may be viewed either as an R-module or as an (R/J)-module because J annihilates U/UJ. If the image of X generates U/UJ in either point of view, it also generates in the other.

13.16 Show that the canonical homomorphism $R \to R/J(R)$ maps the unit group $U(R)$ onto the unit group $U(R/J(R))$.

13.17 Let A be a simple finite dimensional $\mathbb{C}$-algebra. Show that $A \cong M_n(\mathbb{C})$ for some integer $n \geq 1$.

HINT: See Problem 12.19.

CHAPTER FOURTEEN

Artinian Rings and Projective Modules

14A

Recall that a ring R is said to be *right artinian* if its regular module $R^\bullet$ is an artinian module. As we shall see, it turns out that the assumption that a ring is right artinian is just strong enough to enable us to deduce fairly detailed structural information about the ring. (Of course, we could obtain similar results with left artinian rings, but we have made the choice to work with right modules and right ideals.) Fortunately, it often happens in applications of ring theory (to finite groups, for instance) that the rings involved are actually finite dimensional algebras, and so in particular, they are artinian and the theory we develop in this chapter can be used.

If $I, J \subseteq R$ are additive subgroups, we have defined IJ to be the additive subgroup generated by all products uv with $u \in I$ and $v \in J$. In particular, we can define the powers I^n of I by repeated application of this definition. Using the distributive law, we easily see that I^n is the additive subgroup generated by all elements $u_1 u_2 u_3 \cdots u_n$, where each $u_i \in I$. We say that I is *nilpotent* if $I^n = 0$ for some $n \geq 1$. (Note that $I^n = 0$ iff every product of n elements of I is zero.)

An additive subgroup $I \subseteq R$ is said to be *nil* if each element is nilpotent. Thus I is nil if for each element $u \in I$, there exists $n \geq 1$, depending on u, such that $u^n = 0$. Of course, if I is nilpotent, it is certainly nil, but in general, the condition "nil" is weaker than "nilpotent" in two distinct ways. First, if I is nil, then there need not be any global upper bound on the exponent required to annihilate the various elements of I, and second, the condition that I is nil does not force products of different elements of I to vanish.

We have seen that if $u^n = 0$, then $1 - u$ is invertible in R, and so nilpotent elements of R are quasiregular and we have the following corollary.

(14.1) COROLLARY. *If I is any nil right or left ideal of R, then $I \subseteq J(R)$.*

Proof. Apply Theorem 13.4(b) and Corollary 13.9. ∎

There is a very strong converse to this for artinian rings.

(14.2) THEOREM. *Let R be right artinian. Then $J(R)$ is nilpotent.*

Proof. Write $J = J(R)$ and observe that we have $J \supseteq J^2 \supseteq J^3 \supseteq \cdots$. Since these powers of J are certainly right ideals, the artinian hypothesis tells us that $J^n = J^{n+1}$ for some integer $n \geq 1$. Writing $N = J^n$, we see that $N^2 = N$, and our goal is to show that $N = 0$.

Assume $N \neq 0$. Then $N^2 = N \neq 0$, and so the set of all right ideals K of R such that $KN \neq 0$ is nonempty. By the minimal condition, therefore, we can choose K to be a minimal element of this set. Choose $k \in K$ with $kN \neq 0$ and observe that kN is a right ideal. Also, $kN \subseteq K$ and

$$(kN)N = kN^2 = kN \neq 0,$$

and so by the minimality of K, we conclude that $kN = K$. We can therefore choose $u \in N$ such that $ku = k$ and hence $k(1 - u) = 0$. However, since $u \in N \subseteq J$, we know that u is quasiregular and thus $1 - u$ is invertible. Since $k(1 - u) = 0$, this yields $k = 0$ and so $kN = 0$, a contradiction. ∎

(14.3) COROLLARY. *Let R be right artinian and let $I \subseteq R$ be any right or left ideal. Then the following are equivalent:*

   i.   *I is nil.*
   ii. *$I \subseteq J(R)$.*
   iii. *I is nilpotent.* ∎

Recall that a ring R for which $J(R) = 0$ is said to be "semiprimitive." We say that R is *semiprime* if the only nilpotent ideal of R is zero. (For the reader curious about the notation, we mention that a ring R is *prime* if, whenever $A, B \subseteq R$ are ideals with $AB = 0$, either $A = 0$ or $B = 0$. Also, if $P \subseteq R$ is an ideal, it is called a *prime* ideal if R/P is a prime ring. We know that R is semiprimitive iff 0 is an intersection of primitive ideals. Similarly, but not so easy to prove, R is semiprime iff 0 is an intersection of prime ideals. This is proved in the problems at the end of this chapter.)

(14.4) COROLLARY. *Let R be right artinian. Then R is semiprimitive iff it is semiprime.* ∎

In fact, Corollary 14.4 is sometimes a practical way to prove that a ring is semiprimitive. In Chapter 15, for instance, we use this method to show that $J(FG) = 0$, where FG is the group algebra of the finite group G over the field F, and the characteristic of F does not divide the order $|G|$.

The following lemma gives some conditions that are easily seen to be equivalent to being semiprime.

(14.5) LEMMA. *The following are equivalent for a ring R:*

i. R is semiprime.
ii. $I^2 \neq 0$ whenever $I \neq 0$ for ideals I of R.
iii. $I^2 \neq 0$ whenever $I \neq 0$ for right ideals I of R.
iv. $I^2 \neq 0$ whenever $I \neq 0$ for left ideals I of R.
v. $xRx \neq 0$ whenever $0 \neq x \in R$.

Proof.   That (i) implies (ii) is a triviality and that (ii) implies (i) is very nearly so, since, if $I^n = 0$ with $I \neq 0$ for an ideal I, we may assume that $I^{n-1} \neq 0$. But then $(I^{n-1})^2 = 0$, contradicting (ii).

Of course, (iii) and (iv) each imply (ii). Conversely, we assume (ii) and suppose that I is a right ideal with $I^2 = 0$. Then RI is an ideal and $(RI)^2 \subseteq RI^2 = 0$ since $IR \subseteq I$. We then have $I \subseteq RI = 0$. Thus (ii) implies (iii) and the proof that (ii) implies (iv) is similar.

If $xRx = 0$, then $(xR)^2 = 0$ and so if (iii) holds, then $xR = 0$ and hence $x = 0$, proving that (iii) implies (v). Finally, assume (v). If I is a right ideal with $I^2 = 0$, let $x \in I$. Then $xRx \in I^2 = 0$, and so $x = 0$ by (v). Thus $I = 0$, and we have shown that (v) implies (iii).   ∎

14B

We work now toward a proof of the Wedderburn-Artin theorems. These results give detailed structural information about artinian rings that are semiprimitive or, equivalently, semiprime. (Note that condition (v) of Lemma 14.5 is probably the most elementary and direct way to state the semiprimitive/semiprime hypothesis.) We say that R is a *Wedderburn ring* if it is right artinian and semiprime.

As an example, let $R = M_n(D)$, the full ring of $n \times n$ matrices over some division ring D. Then R is simple by Theorem 12.15, and so it is trivially semiprime. By Lemma 13.20, the regular module $R^\bullet$ has finite composition length and so R is right artinian.

It is not hard to see that a finite external direct sum of Wedderburn rings is again a Wedderburn ring, and thus any ring constructed as a direct sum of finitely many full matrix rings over division rings is Wedderburn. The Wedderburn-Artin theorems tell us that every Wedderburn ring is isomorphic to a ring constructed in this way. Note also that our definition is not symmetric since we assumed that R was *right* artinian. Nevertheless, since full matrix rings over division rings are both right and left artinian, it follows that we would have obtained exactly the same class of rings if we had required Wedderburn rings to be left artinian instead.

(14.6) THEOREM.   *Let R be a Wedderburn ring. Then every right ideal of R is the direct sum of a finite collection of minimal right ideals of R.*

As usual, the phrase "minimal right ideal" refers to right ideals that are minimal with the property of being nonzero. These are exactly the simple submodules of the regular module $R^\bullet$. We need a couple of lemmas.

(14.7) LEMMA.   *Let I be a minimal right ideal of an arbitrary ring R and assume that $I^2 \neq 0$. Then $I = eR$ for some idempotent $e \in I$.*

Proof.   Since $I^2 \neq 0$, choose $a \in I$ with $aI \neq 0$ and note that $aI \subseteq aR \subseteq I$ since $a \in I$. By the minimality of I, we conclude that $aI = I$ and hence $ae = a$ for some element $e \in I$.

Now $0 \neq aI = aeI$ and hence $eI \neq 0$. Since $e \in I$, we have $0 \neq eI \subseteq eR \subseteq I$, and this forces $eR = I$. It remains to show that e is idempotent.

Since $ae = a$, we have $ae^2 = (ae)e = ae$, and so $a(e^2 - e) = 0$. Let $J = \{x \in I \mid ax = 0\}$. Then J is a right ideal and $e^2 - e \in J$. Now $e \notin J$ since $ae = a \neq 0$. Since $e \in I$, we have $J < I$ and the minimality of I gives $J = 0$, and thus $e^2 - e = 0$. Since $e \neq 0$ (because $eR = I \neq 0$), we have that e is idempotent, as desired. ∎

(14.8) LEMMA (Pierce decomposition).   *Let $I = eR$, where $e^2 = e$, and suppose $I \subseteq U$ for some right ideal U of R. Then $U = I \dotplus V$ for some right ideal V with the property that $eV = 0$.*

Proof.   Let $u \in U$. We have $eu \in eR = I \subseteq U$ and hence $(1-e)u \in U$. Therefore $(1 - e)U \subseteq U$ and we write $V = (1 - e)U$. Note that $e(1 - e) = e - e^2 = 0$, and thus $eV = 0$. Certainly, V is a right ideal of R and so it suffices to show that $I + V = U$ and that $I \cap V = 0$.

For $u \in U$, we have $u = eu + (1 - e)u \in I + V$ and hence $I + V = U$, as required. If $x \in I \cap V$, then $x = er$ for some $r \in R$ and so

$$x = er = e(er) = ex \in eV = 0,$$

and we are done. ∎

Proof of Theorem 14.6.   Suppose the theorem is false and use the minimal condition to choose a right ideal $I \subseteq R$ minimal with the property that it is not a finite direct sum of minimal right ideals. In particular, $I \neq 0$, and we can use the minimal condition once again to choose a minimal right ideal $J \subseteq I$.

Since R is a Wedderburn ring, it is semiprime and so $J^2 \neq 0$, and by Lemma 14.7, we can write $J = eR$ for some idempotent $e \in J$. By Lemma 14.8 we have $I = J \dotplus V$, where V is some right ideal of R. We have $V < I$ since $J \neq 0$ and so, by the choice of I, we know that V is a finite direct sum of minimal right ideals of R. Since $I = J \dotplus V$, we see that I is such a sum also, and this contradicts the choice of I. ∎

The most important consequence of Theorem 14.6 is that R itself is a sum of minimal right ideals. By Theorem 11.9 we see from this that the regular R-module $R^\bullet$ is completely reducible.

(14.9) COROLLARY.   *Let R be a Wedderburn ring. Then every right R-module is completely reducible.*

Proof.   By Theorem 14.6 we can choose a set $\mathcal{X}$ of minimal right ideals of R such that $R = \sum \mathcal{X}$. Let M be an arbitrary right R-module, and note that for each element $m \in M$ and each $I \in \mathcal{X}$, we have an R-module homomorphism

$\theta : I \rightarrow M$ defined by $\theta(x) = mx$. Since I is a simple R-module, either $\theta(I) = 0$ or $\theta(I) \cong I$ and $\theta(I)$ is simple. In either case, $\theta(I) \subseteq S$, the socle of M. (Recall that this is the sum of all of the simple submodules of M.) We have shown that $mI \subseteq S$ for every $I \in \mathcal{X}$, and thus $mR \subseteq S$ since $R = \sum \mathcal{X}$. Because $m \in mR$, we conclude that $m \in S$ and hence, since $m \in M$ was arbitrary, we have $S = M$. Thus M is completely reducible by Theorem 11.9. ∎

(14.10) COROLLARY. *Let M be a right module for the right artinian ring R. Then M is completely reducible iff $MJ(R) = 0$.*

Proof. If M is completely reducible, it is a sum of simple submodules, and since each of these is annihilated by $J(R)$, it follows that M is annihilated, too.

Conversely, suppose $MJ(R) = 0$. Then M may be viewed as a right $R/J(R)$-module, and as such, its collection of submodules is unchanged. (See the discussion following the proof of Lemma 13.1.) In particular, it suffices to observe that M is completely reducible as an $R/J(R)$-module.

The ring $R/J(R)$ is certainly right artinian (since its right ideals all have the form $I/J(R)$ for right ideals I of R containing $J(R)$). Also, $J(R/J(R)) = 0$ by Corollary 13.2, and thus $R/J(R)$ is a Wedderburn ring. (Being semiprimitive, it is certainly semiprime.) The result now follows from Corollary 14.9. ∎

As an application of what we have done so far, we obtain the following striking result.

(14.11) THEOREM (Hopkins). *Let R be right artinian. Then R is right noetherian.*

Proof. Write $J = J(R)$ and observe that $J^n = 0$ for some $n \geq 1$ by Theorem 14.2. Each of the right R-modules J^k/J^{k+1} for $0 \leq k < n$ is annihilated by J and so is completely reducible by Corollary 14.10. Since each of them is artinian (because $R^\bullet$ is), it follows by Theorem 11.11 that each has finite composition length. Therefore $R^\bullet$ has finite composition length and hence is noetherian. ∎

Before we continue with our analysis of Wedderburn rings, we digress to show that the condition that $R^\bullet$ is completely reducible in fact characterizes these rings.

(14.12) THEOREM. *A ring R is Wedderburn iff $R^\bullet$ is completely reducible.*

Proof. Theorem 14.6 gives us the "only if" part and so we assume that $R^\bullet$ is completely reducible and we show that R is right artinian and that $J(R) = 0$.

By complete reducibility, $R^\bullet$ is the sum of all of its simple submodules, and so there is some finite collection $\mathcal{X}$ of minimal right ideals of R such that $1 \in \sum \mathcal{X}$. However, $\sum \mathcal{X}$ is a right ideal of R, and this forces $\sum \mathcal{X} = R$. It follows by Corollary 11.8 that $R^\bullet$ is artinian, as desired.

By complete reducibility again, there exists a right ideal I of R such that $R = J(R) \dotplus I$. We can write $1 = j + i$, where $j \in J(R)$ and $i \in I$, and we see that $i = 1 - j$ is invertible since j is quasiregular. Since $i \in I$, we conclude that $I = R$ and hence $J(R) = 0$.   ■

As a variation on Theorem 14.12, we give another characterization of Wedderburn rings.

(14.13) THEOREM.   *A ring R is Wedderburn iff for each right ideal $I \subseteq R$, there exists $e \in I$ such that $e^2 = e$ and $I = eR$.*

Proof.   It suffices to show that this condition on right ideals is equivalent to $R^\bullet$ being completely reducible. First, suppose that $R^\bullet$ is completely reducible, and let I be any right ideal of R. Then $R = I \dotplus J$ for some right ideal J, and we can write $1 = e + f$, where $e \in I$ and $f \in J$. If $x \in I$, then $fx \in fR \subseteq J$ and also $fx = (1 - e)x = x - ex \in I$. It follows that $fx \in I \cap J = 0$. Therefore $x = (e + f)x = ex \in eR$ and thus $I \subseteq eR$. Since also $eR \subseteq I$, we have $I = eR$. Furthermore, since $x = ex$ for all $x \in I$, we have in particular that $e = e^2$, as desired.

Conversely, if each right ideal of R satisfies the condition of the theorem, then by the Pierce decomposition (14.8), it is a direct summand of $R^\bullet$. This means that $R^\bullet$ is completely reducible.   ■

Perhaps this is a good place to mention some of the ways that our notation diverges from that of some other books. What we have called "Wedderburn" rings are often called "semisimple artinian" rings, and so we should explain what "semisimple" means for a ring R. For some authors, "semisimple" means that $J(R) = 0$, and for others, it means that $R^\bullet$ is completely reducible. (That a *module* is semisimple means that it is completely reducible.) With either meaning for a semisimple ring, we see that Wedderburn rings are precisely semisimple artinian rings.

We now resume our study of Wedderburn rings. Given a ring R, we choose a *representative set S* of simple right R-modules. This means that no two of the simple modules that make up S are isomorphic, but that every simple right R-module is isomorphic to some (unique) member of S. (Recall that we could, if we wished, choose S to consist of modules of the form $R^\bullet/M$ for some maximal right ideals M of R.)

(14.14) THEOREM (Wedderburn-Artin).   *Let R be a Wedderburn ring and let S be a representative set for the simple right R-modules. For each $S \in S$, let I_S be the sum of all of those minimal right ideals of R that are isomorphic to S. Then:*

a. *$|S|$ is finite.*
b. *Each I_S is a minimal (nonzero, two-sided) ideal of R.*
c. *R is the direct sum of the ideals I_S.*
d. *Each ideal I_S is a simple right artinian subring of R.*
e. *If $S, T \in S$ with $S \neq T$, then $T I_S = 0$.*

Proof.   By definition, I_S is the S-isotypic component of $R^{\bullet}$. Theorem 14.6 tells us that $R^{\bullet}$ is a finite direct sum of simple submodules, and so Theorem 11.14 applies. We conclude that each I_S is a sum of those summands isomorphic to S, and (c) follows.

Next we prove (e). If $I \subseteq R$ is a minimal right ideal with $I \cong S$, and if $t \in T$, then the map $\theta : I \to T$ defined by $\theta(x) = tx$ is an R-module homomorphism. Now $I \cong S \ncong T$, and hence θ cannot be an isomorphism. By Schur's lemma (12.9), we conclude that $\theta = 0$ and thus $tI = 0$. It follows that I annihilates T. Since I_S is a sum of right ideals like I, each isomorphic to S and hence annihilating T, conclusion (e) follows.

Now if $I_T = 0$ for some $T \in \mathcal{S}$, then $TI_S = 0$ for all $S \in \mathcal{S}$, and hence $TR = 0$ by (c). This, of course, is impossible, and so we have shown that $I_T \neq 0$ for all T. Since the T-isotypic component of $R^{\bullet}$ is nonzero, it follows by Theorem 11.14 that T is represented (up to isomorphism) in any particular direct sum decomposition of $R^{\bullet}$ as a sum of minimal right ideals. Since $R^{\bullet}$ has a finite decomposition of this form (by Theorem 14.6), we conclude that $\mathcal{S}$ is a finite set, proving (a).

Next we show that I_S is an ideal of R. Since I_T is a sum of modules isomorphic to T, part (e) yields that $I_T I_S = 0$ if $T \neq S$. Therefore $RI_S = I_S I_S \subseteq I_S$ by (c), and hence I_S is a left ideal. It is a right ideal by construction, and so it is a (two-sided) ideal.

To complete the proof of (b), suppose that I is an ideal of R such that $I < I_S$. We need to show that $I = 0$. Since I is proper, we can choose a minimal right ideal $J \subseteq I_S$ with $J \cong S$ and $J \nsubseteq I$. Then $J \cap I$ is a right ideal of R properly contained in J, and hence $J \cap I = 0$. However, $JI \subseteq J \cap I$ since I is a left ideal, and thus $JI = 0$. It follows that $I_S I = 0$ since I_S is a sum of modules isomorphic to J, and each is annihilated by I. Also, if $T \neq S$, then $I_T I \subseteq I_T \cap I_S = 0$, and hence $RI = 0$ and so $I = 0$, as desired.

Finally, we prove (d). We can write $1 = \sum e_T$, with $e_T \in I_T$. If $x \in I_S$, then $x = x1 = \sum x e_T$. However, for $S \neq T$, we have $x e_T \in I_S \cap I_T = 0$ and therefore $x = x e_S$. Similarly, $x = e_S x$, and so e_S is a unity element of I_S, which is therefore a subring. If $I \subseteq I_S$ is any additive subgroup, then since $I_S I_T = 0 = I_T I_S$ for $S \neq T$, we see from (c) that $RI = I_S I$ and $IR = II_S$. It follows that the ideals and right ideals of the ring I_S are, respectively, ideals and right ideals of R. Since R is right artinian, it follows that I_S is too, and since I_S is minimal as an ideal of R by (b), we see that I_S is a simple ring.   ■

(14.15) THEOREM (Wedderburn-Artin).   *A Wedderburn ring is isomorphic to a finite external direct sum of full matrix rings over division rings.*

Proof.   The ring R of Theorem 14.14 is isomorphic to the external direct sum of the rings I_S. This is so since, given elements $x_S, y_S \in I_S$, we have $(\sum x_S)(\sum y_S) = \sum x_S y_S$ because $I_S I_T \subseteq I_S \cap I_T = 0$ for $S \neq T$.

Each ring I_S, being simple and right artinian, is isomorphic to a full matrix ring over a division ring by Corollary 13.22.   ∎

We briefly consider the question of uniqueness in Theorem 14.15. We need a preliminary result.

(14.16) LEMMA.   *In the situation of Theorem 14.14, every ideal of R has the form $\sum I_S$ as S runs over some subset of $\mathcal{S}$.*

Proof.   Let $I \subseteq R$ be an ideal and let $\mathcal{S}_0 = \{S \in \mathcal{S} \mid I_S \subseteq I\}$. We certainly have $\displaystyle\sum_{S \in \mathcal{S}_0} I_S \subseteq I$, and we claim that equality holds.

Let $x \in I$ and write $x = \sum x_S$, where the sum runs over $\mathcal{S}$ and $x_S \in I_S$. We need to show that $x_S = 0$ if $S \notin \mathcal{S}_0$. Suppose then that $x_S \neq 0$. Now $x_T I_S \subseteq I_T \cap I_S = 0$ if $T \neq S$, and so $x I_S = x_S I_S$. This is nonzero since $x_S \neq 0$ and I_S is a ring (with unity). We thus have $0 < x I_S \subseteq I \cap I_S$, and by Theorem 14.14(b), this forces $I \cap I_S = I_S$. Thus $S \in \mathcal{S}_0$, as required.   ∎

(14.17) COROLLARY.   *In the notation of Theorem 14.14, if $I \subseteq R$ is an ideal that is a simple subring, then $I = I_S$ for some $S \in \mathcal{S}$.*   ∎

Now suppose that our Wedderburn ring R is isomorphic to an external direct sum of full matrix rings $M_{n_i}(D_i)$ for $1 \leq i \leq k$, where the D_i are division rings. Then R is an internal direct sum of ideals I_i, where I_i is isomorphic as a ring to $M_{n_i}(D_i)$. It follows that I_i is simple, and by Corollary 14.17, $I_i = I_{S_i}$ for some $S_i \in \mathcal{S}$. Furthermore, the S_i are distinct and exhaust $\mathcal{S}$. It follows that $k = |\mathcal{S}|$ and by Lemma 13.20 and Theorem 13.23, the integers n_i are uniquely determined and the division rings D_i are determined up to isomorphism (except for possible renumbering).

We mention one application of the Wedderburn-Artin theory and leave others for the problems at the end of the chapter.

(14.18) COROLLARY.   *Let R be right artinian and suppose R contains no nonzero nilpotent elements. Then R is isomorphic to a direct sum of division rings.*

Proof.   Certainly, R is semiprime and so it is a Wedderburn ring and hence is isomorphic to a direct sum of full matrix rings over division rings. Observe that if $n \geq 2$, the ring $M_n(D)$ contains the nonzero nilpotent matrix unit e_{12} (in the notation of Theorem 12.15). The result follows.   ∎

(14.19) COROLLARY.   *A ring R is a division ring iff it is artinian and $xy \neq 0$ whenever $x \neq 0$ and $y \neq 0$.*

Proof.   A division ring clearly has the stated properties. Conversely, a ring with these properties has no nonzero nilpotent elements and so, by Corollary 14.18, is a direct sum of division rings. It is clear from the hypothesis that R cannot be a proper direct sum.   ∎

14C

Before we continue our study of artinian rings, we need to do a little general module theory. Let R be an arbitrary ring and suppose M is a right R-module. A subset $X \subseteq M$ is called a *basis* for M if X generates M as an R-module and X is "linearly independent" over R. (This, of course, means that if

$$\sum_{x \in X} x r_x = 0,$$

where $r_x \in R$ and at most finitely many of these coefficients are nonzero, then in fact, all $r_x = 0$.) If X is a basis for M, then every element of M has a unique expression as an R-linear combination of X.

If the R-module M has a basis, we say that it is a *free* R-module. For example, all modules over fields are free. (The fact that every vector space has a basis requires a Zorn's lemma argument in the infinite-dimensional case.) For most rings R, most R-modules are nonfree. Examples of nonfree modules for $R = \mathbb{Z}$, for instance, include abelian groups that have nonzero elements of finite order.

Observe that for any ring R, the set $\{1\}$ is a basis for the regular module $R^{\bullet}$, and it follows that regular modules are always free. More generally, arbitrary restricted direct sums of regular modules are free. To see this, let Y be any set, and write F_Y to denote the restricted external direct sum of a collection of copies of $R^{\bullet}$ subscripted by the elements of Y. (Recall that the elements of F_Y can be viewed as "Y-tuples," or functions from Y to $R^{\bullet}$. Since we are dealing with a restricted direct sum, at most finitely many components can be nonzero.) For $y \in Y$, let $x_y \in F_Y$ be the element with y-component equal to one and all other components zero. It is easy to see that $X = \{x_y \mid y \in Y\}$ is a basis for F_Y, which is therefore free. Up to isomorphism, these free modules F_Y are the only free R-modules.

(14.20) LEMMA.   *A right R-module is free iff it is the internal direct sum of a (possibly infinite) collection of submodules, each isomorphic to $R^{\bullet}$.*

Proof.   If M is free with basis X, we claim that $F_X \cong M$. To construct an isomorphism from F_X to M, view the element $f \in F_X$ as a function from X to $R^{\bullet}$, and map it to $\sum_x x f(x) \in M$. This clearly defines a homomorphism of right R-modules. It is surjective since X generates M, and it is injective since X is linearly independent.

Conversely, if M is an internal direct sum of submodules isomorphic to $R^{\bullet}$, then M is free because it is isomorphic to the corresponding restricted external direct sum.   ∎

Free modules are useful primarily because it is easy to construct homomorphisms from them.

(14.21) LEMMA.   *Let F be a free right R-module with basis X, and let A be any right R-module. If $\theta : X \to A$ is an arbitrary map, then θ has a unique extension to an R-module homomorphism $\hat{\theta} : F \to A$.*

Proof.   Each element $f \in F$ has the form $f = \sum x r_x$ for some unique choice of coefficients $r_x \in R$, with all but finitely many of these coefficients zero. If $\hat{\theta}$ exists, we have

$$(f)\hat{\theta} = \sum_{x \in X}(x\hat{\theta})r_x = \sum_{x \in X}(x\theta)r_x \, ,$$

and so $\hat{\theta}$ is uniquely determined by θ.

Since the coefficients r_x are uniquely determined by f, we can use the above formula to define $\hat{\theta}$. We omit the easy computation that proves that $\hat{\theta}$ actually is a module homomorphism.   ∎

(14.22) COROLLARY.   *Let M be any right R-module. Then M is a homomorphic image of some free right R-module.*

Proof.   Choose any generating set Y for M. (We could even take $Y = M$.) Now let F_Y be the free module we constructed earlier, with basis $X = \{x_y \mid y \in Y\}$. Define the map $\theta : X \to M$ by $(x_y)\theta = y$, and extend θ to a module homomorphism $F_Y \to M$ by Lemma 14.21. This map is surjective since X generates M, and the proof is complete.   ∎

(14.23) COROLLARY.   *Let A, B, and F be right R-modules, and assume that F is free. Suppose $\pi : A \to B$ is a surjective homomorphism and $\theta : F \to B$ is an arbitrary homomorphism. Then there exists a homomorphism $\varphi : F \to A$ such that $\varphi\pi = \theta$.*

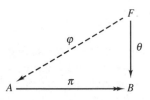

Figure 14.1

Note that the double head on the arrow representing π in Figure 14.1 is intended to indicate that π is surjective.

Proof of Corollary 14.23.   Let X be a basis for F. For each element $x \in X$, use the assumption that π is surjective to choose $a_x \in A$ such that $a_x\pi = x\theta$. By Lemma 14.21 there exists a homomorphism $\varphi : F \to A$ that extends the map taking x to a_x. Then $x\varphi\pi = a_x\pi = x\theta$, and so $\varphi\pi$ agrees with θ on all elements of X. Since X generates F, these two maps agree on F, as required.   ∎

We are now ready to give the definition toward which we have been working. We abstract the property of free modules proved in Corollary 14.23.

(14.24) DEFINITION.   A right R-module P is *projective* if, whenever A and B are right R-modules and $\pi : A \to B$ and $\theta : P \to B$ are homomorphisms with π surjective, there exists a homomorphism $\varphi : P \to A$ such that $\varphi\pi = \theta$.

(14.25) COROLLARY.   *Free modules are projective.*   ∎

The following theorem is a useful characterization of projective modules and clarifies, to some extent, the connection between free modules and projective modules. (Recall that the symbol $\oplus$ denotes external direct sums.)

(14.26) THEOREM.   *Let P be a right R-module. Then P is projective iff there exists another right R-module Q such that $P \oplus Q$ is free.*

Expressing this result in terms of internal direct sums, we can say that a module is projective iff it is isomorphic to a direct summand of some free module. We begin the proof of Theorem 14.26 with a slightly strengthened form of the "if" part.

(14.27) LEMMA.   *Let P be a projective right R-module and suppose that $P = U \dotplus V$ for submodules U and V. Then U and V are projective.*

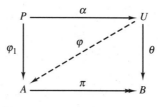

Figure 14.2

Proof.   To prove that U is projective, we suppose that we are given right R-modules A and B, a surjection $\pi : A \to B$, and a homomorphism $\theta : U \to B$. We seek a homomorphism $\varphi : U \to A$ such that $\varphi\pi = \theta$. See Figure 14.2.

Let $\alpha : P \to U$ be the projection with respect to V. In other words, $(u + v)\alpha = u$ if $u \in U$ and $v \in V$. Then $\alpha\theta$ maps P to B, and since P is projective, Definition 14.24 yields a homomorphism $\varphi_1 : P \to A$ such that $\varphi_1\pi = \alpha\theta$. If φ is the restriction of φ_1 to U, we have

$$u\varphi\pi = u\varphi_1\pi = u\alpha\theta = u\theta$$

for $u \in U$. Thus $\varphi\pi = \theta$, as required. The proof that V is projective is, of course, similar.   ∎

The next lemma, which is the key to the "only if" part of Theorem 14.26, shows how one can prove that a projective module is isomorphic to a direct summand of some other module.

(14.28) LEMMA.   *Let $\pi : M \to P$ be a surjection of right R-modules, where P is projective. Then there exists an isomorphism φ of P into M such that $\varphi\pi$ is the identity map on P. Also, $M = (P)\varphi \dotplus \ker(\pi)$, and so P is isomorphic to a direct summand of M.*

Proof.   Let $i : P \to P$ be the identity map. Since π is surjective, Definition 14.24 yields a homomorphism $\varphi : P \to M$ such that $\varphi\pi = i$.

If $x \in P$ and $x\varphi = 0$ or, even more generally, $x\varphi \in \ker(\pi)$, then $0 = (x\varphi)\pi = xi = x$ and φ is injective, as desired. This also shows that if $x\varphi \in \ker(\pi)$, then $x = 0$ and so $x\varphi = 0$. In other words, $(P)\varphi \cap \ker(\pi) = 0$.

To show that $M = (P)\varphi + \ker(\pi)$, let $m \in M$ and compute that

$$(m - m\pi\varphi)\pi = m\pi - m\pi i = 0.$$

Thus

$$m = (m\pi)\varphi + (m - m\pi\varphi) \in (P)\varphi + \ker(\pi),$$

and the proof is complete.   ∎

Proof of Theorem 14.26.   If $P \oplus Q$ is free, then it is certainly projective. Since P is isomorphic to a direct summand of this module, P is projective by Lemma 14.27.

Conversely, suppose that P is projective. By Corollary 14.22 there exists a free module F and a surjection $\pi : F \to P$. By Lemma 14.28 it follows that P is isomorphic to a direct summand of F, and thus $P \oplus Q$ is isomorphic to F for some module Q.   ∎

(14.29) COROLLARY.   *Direct sums of projective modules are projective.*

Proof.   Suppose P_1 and P_2 are projective. By Theorem 14.26 there exist modules Q_1 and Q_2 such that $P_1 \oplus Q_1$ and $P_2 \oplus Q_2$ are free. By Lemma 14.20 each of these free modules is isomorphic to a direct sum of a (possibly infinite) number of copies of the regular module. It follows that their direct sum is also a direct sum of regular modules and hence is free. We now know that $P_1 \oplus Q_1 \oplus P_2 \oplus Q_2$ is free. This module, however, is isomorphic to $(P_1 \oplus P_2) \oplus (Q_1 \oplus Q_2)$, which is therefore free. It follows that $P_1 \oplus P_2$ is projective by Theorem 14.26.   ∎

We can now give a "homological" characterization of Wedderburn rings. (Homological algebra is that part of algebra primarily concerned with homomorphisms between modules. Since projectivity is defined in terms of such maps, it is a homological concept.)

(14.30) THEOREM.   *Let R be a ring. Then the following are equivalent:*

   i.   *R is Wedderburn.*
   ii.   *Every right R-module is projective.*
   iii. *Every simple right R-module is projective.*

Proof.   Assume that R is Wedderburn and let M be an arbitrary right R-module. By Corollary 14.22 we can find a free module F and a surjection π of F onto

M. Since R is Wedderburn, F is completely reducible by Corollary 14.9, and thus we can write $F = N \,\dot{+}\, \ker(\pi)$ for some submodule $N \subseteq F$. By Lemma 14.27 we see that N is projective and we have

$$N \cong F/\ker(\pi) \cong (F)\pi = M ,$$

and so M is projective. This proves that (i) implies (ii).

That (ii) implies (iii) is trivial, and so now we assume that every simple right R-module is projective. Our goal is to prove that $R^{\bullet}$ is completely reducible, since then Theorem 14.12 implies that R is Wedderburn. Let S be the socle of $R^{\bullet}$ so that S is a right ideal of R containing every minimal right ideal. If $S < R$, then by Lemma 12.13, S is contained in some maximal right ideal $M \subseteq R$, and the module $R^{\bullet}/M$ is simple and hence projective. If $\pi : R^{\bullet} \to R^{\bullet}/M$ is the canonical homomorphism, then by Lemma 14.28, since $\ker(\pi) = M$, we have $R^{\bullet} = M \,\dot{+}\, N$ for some submodule $N \cong R^{\bullet}/M$. Then N is simple but is not contained in S. This contradiction shows that $S = R^{\bullet}$, and thus $R^{\bullet}$ is completely reducible by Theorem 11.9.   ∎

14D

We now wish to consider the projective modules for an artinian ring that is not necessarily semiprime. If P is projective and $P = U \,\dot{+}\, V$, then each of U and V is also projective, and this suggests that we might wish to focus attention on the projective indecomposable right R-modules for the right artinian ring R. (Recall that a module is *indecomposable* if it is nonzero and it is not of the form $U \,\dot{+}\, V$ for nonzero submodules U and V.)

If R is Wedderburn, then since every R-module is completely reducible, the indecomposable R-modules are precisely the simple R-modules, and these are all projective. For general right artinian rings, the situation is more complex, but it is still true that there is a natural connection between simple modules and projective indecomposable modules.

We say that a right ideal of R is *minimally potent* if it is a minimal member of the set of those right ideals of R that are not nilpotent. Of course, these exist for right artinian rings, and they are simple for Wedderburn rings.

(14.31) THEOREM.   *Let R be right artinian. Then:*

a. *Every minimally potent right ideal of R is a projective indecomposable R-module.*
b. *The regular module $R^{\bullet}$ is a direct sum of minimally potent right ideals I_i of R.*
c. *Every projective indecomposable right R-module is module-isomorphic to one (or more) of the minimally potent right ideals I_i of (b).*

We need a few preliminary results.

(14.32) LEMMA. *Let R be right artinian and suppose that $I \subseteq R$ is a right ideal. If $I = aI$ for some $a \in I$, then $I = eR$, where $e \in I$ and $e^2 = e$.*

Proof. By Hopkins's theorem (14.11), the module $R^\bullet$ is noetherian as well as artinian, and so it has finite composition length by Theorem 11.3. If $\theta : I \to I$ is an R-module homomorphism, then computing composition lengths, we get

$$\ell(I) = \ell(\ker(\theta)) + \ell(I/\ker(\theta)) = \ell(\ker(\theta)) + \ell(\theta(I)),$$

and it follows that θ is injective iff it is surjective.

The map $x \mapsto ax$ is an R-module homomorphism $I \to I$ that is surjective by hypothesis, and hence it is injective and we conclude that $ax \neq 0$ if $0 \neq x \in I$.

Since $a \in I = aI$, we can write $a = ae$ for some $e \in I$, and thus $ae = ae^2$ and $a(e - e^2) = 0$. We conclude that $e - e^2 = 0$ and $e = e^2$.

Since $e \in I$, we have $eR \subseteq I$ and it suffices to show that the map $x \mapsto ex$ maps I onto I. Since this is an R-homomorphism $I \to I$, it is enough to show that it is injective, and so we assume $ex = 0$ with $x \in I$. Then $ax = aex = 0$, and we conclude that $x = 0$, as desired. ∎

(14.33) COROLLARY. *Let R be right artinian and suppose $I \subseteq R$ is a minimally potent right ideal. Then $I = eR$ for some idempotent $e \in I$.*

Proof. The ideal I is not nilpotent, and thus $I^2 \not\subseteq J(R)$ and we can choose $a \in I$ with $aI \not\subseteq J(R)$. It follows that the right ideal aI is not nilpotent. Since $a \in I$, we have $aI \subseteq I$, and the minimality of I yields that $aI = I$. The result now follows by Lemma 14.32. ∎

(14.34) LEMMA. *Let I be a right ideal of the right artinian ring R. The following are equivalent:*

   i.   *I is minimally potent.*
   ii. *$I \cap J(R)$ is the unique maximal R-submodule of I.*
   iii. *$I \not\subseteq J(R)$ and I has a unique maximal submodule.*

Of course, when we speak of maximal submodules here, we mean maximal *proper* submodules. Note that a submodule Y of a module X is a unique maximal submodule precisely when Y is proper and contains every other proper submodule of X.

Proof of Lemma 14.34. Assuming (i), we have that $I \not\subseteq J(R)$ since I is not nilpotent. Thus $I \cap J(R) < I$. Also, if $U < I$ is a right ideal, then U is nilpotent and so $U \subseteq J(R)$. Thus $U \subseteq I \cap J(R)$ and (ii) holds.

That (ii) implies (iii) is trivial, and so we assume (iii) and prove (i). Since $I \not\subseteq J(R)$, it is not nilpotent, and thus I contains some minimally potent right ideal which, by Corollary 14.33, has the form eR for some idempotent e. To complete the proof, we show that $I = eR$. If this is not the case, then $eR < I$ and so $eR \subseteq M$, where M is the unique maximal submodule of I. By the Pierce decomposition (14.8), we can write $I = eR \dotplus V$ for some right ideal V. Since

$eR \neq 0$, we have $V < I$ and thus $V \subseteq M$. This implies that $I = eR + V \subseteq M$, and this is a contradiction. $\blacksquare$

(14.35) COROLLARY. *Let R be right artinian and assume that $R = \sum \mathcal{X}$, where $\mathcal{X}$ is a collection of minimally potent right ideals. If S is any simple right R-module, then $S \cong I/M$ for some member I of $\mathcal{X}$, where $M = I \cap J(R)$ is the unique maximal proper submodule of I.*

Proof. If $0 \neq s \in S$, then the map $\theta : R \to S$ defined by $\theta(r) = sr$ is a surjective module homomorphism $R^{\bullet} \to S$. It follows that $\theta(I) \neq 0$ for some $I \in \mathcal{X}$, and thus $\theta(I) = S$ since S is simple. Let $N = I \cap \ker(\theta)$. Then $I/N \cong S$, and since S is simple, we see that N is a maximal submodule of I and hence $N = M$ by Lemma 14.34. $\blacksquare$

(14.36) COROLLARY. *Let R be right artinian and suppose that I is a minimal potent right ideal. Let S be a simple right R-module that is a homomorphic image of I, and let P be any projective right R-module that has S as a homomorphic image. Then there exists a homomorphism of P onto I.*

Proof. Let $\pi : I \to S$ and $\theta : P \to S$ be surjections. Since P is projective, there exists a homomorphism $\varphi : P \to I$ such that $\varphi\pi = \theta$. Then $(P)\varphi \subseteq I$ is a submodule not contained in $\ker(\pi)$ since $(P)\varphi\pi = (P)\theta = S \neq 0$. Since S is simple, we see that $\ker(\pi)$ is a maximal submodule of I and thus it contains every proper submodule of I by Lemma 14.34. We conclude that $(P)\varphi$ is all of I, as required. $\blacksquare$

We can now put the pieces together.

Proof of Theorem 14.31. Let $I \subseteq R$ be a minimally potent right ideal. By Corollary 14.33, $I = eR$ for some idempotent e, and thus if $I \subseteq U$, where U is any right ideal of R, we have $U = I \dotplus V$ for some right ideal $V \subseteq U$ by the Pierce decomposition (14.8). In particular, I is a direct summand of the free module $R^{\bullet}$ and so is projective by Corollary 14.29. It is indecomposable, since by Lemma 14.34, it has a unique maximal submodule. This proves (a).

To prove (b), consider the collection $\mathcal{X}$ of right ideals $U \subseteq R$ with the property that there exist minimally potent right ideals $I_1, I_2, \ldots, I_m$ such that their sum S is direct and $R^{\bullet} = S \dotplus U$. Note that R lies in $\mathcal{X}$ (taking $m = 0$), and so we can choose a minimal member U of $\mathcal{X}$. We wish to show that $U = 0$, since that forces

$$R^{\bullet} = S = I_1 \dotplus I_2 \dotplus \cdots \dotplus I_m ,$$

as required.

Assume that $U > 0$ and write $1 = s + u$ with $s \in S$ and $u \in U$. Then for $x \in U$, we have $sx = x - ux \in U$. Since also $sx \in S$, we conclude that $sx = 0$ and $ux = x$. Thus $uU = U$ and hence $U^n = U \neq 0$ for all positive integers n. It follows that U is not nilpotent, and so it contains a minimally potent right ideal I. By the first paragraph of the proof, we have $U = I \dotplus V$ for some right ideal V, and so $R^{\bullet} = (S \dotplus I) \dotplus V$. Thus $V \in \mathcal{X}$, and this contradicts the minimality of U. Part (b) is now proved.

Now write $R^{\bullet} = \sum I_i$, where the I_i are minimally potent right ideals. (By (b), this can be done so that the sum is direct.) Let P be a projective indecomposable right R-module. To prove (c), we must show that $P \cong I_i$ for some subscript i. We show first that P has some simple module as a homomorphic image. To see this, note that $P J(R)^n = 0$ for some integer n but that $P \neq 0$. It follows that $P J(R) < P$. The nonzero R-module $P/P J(R)$ is annihilated by $J(R)$, and so it is completely reducible by Corollary 14.10. Since it has a simple submodule, it follows that $P/P J(R)$ also has a simple homomorphic image, and this establishes our claim.

Let S be a simple homomorphic image of P. By Corollary 14.35, S is also a homomorphic image of I_i for some subscript i, and so by Corollary 14.36, there is a homomorphism of P onto I_i. Since I_i is projective by (a), we conclude by Lemma 14.28 that P has a direct sum decomposition with one summand isomorphic to I_i. Since P is indecomposable, however, we have $P \cong I_i$. ∎

(14.37) COROLLARY. *Let R be right artinian.*

    a. *If P is a projective indecomposable R-module, then P has a unique maximal proper submodule $M(P)$.*

    b. *Every simple R-module S is isomorphic to $P/M(P)$ for some projective indecomposable module P, and P is unique up to isomorphism.*

Proof. Part (a) is clear from Theorem 14.31(c) and Lemma 14.34. Existence in (b) follows from Theorem 14.31(b) and Corollary 14.35, which give us that $S \cong I/M(I)$ for some minimally potent right ideal I of R. (Note that $M(I) = I \cap J(R)$.) If also $S \cong P/M(P)$, then we need to show that $P \cong I$. This follows (as in the proof of Theorem 14.31) if we use Corollary 14.36 to get a homomorphism of P onto I. The projectivity of I and the indecomposability of P force this to be an isomorphism. ∎

Note that Corollary 14.37(b) can be neatly paraphrased if we write $[X]$ to denote the isomorphism class of a module X. It says that the map $[P] \mapsto [P/M(P)]$ is a bijection from the set of isomorphism classes of projective indecomposable right R-modules onto the set of isomorphism classes of simple right R-modules. Note that the latter set is also in bijective correspondence with the set of isomorphism classes of simple right $R/J(R)$-modules, and so is finite.

14E

We conclude this chapter with material that does not concern artinian rings but is closely related to the material with which we began. We observed that a nil one-sided ideal of any ring R is contained in $J(R)$ and so is nilpotent if R is right artinian. We can generalize this to the case where we replace "artinian" by the weaker hypothesis "noetherian." (Note that by Hopkins's theorem, right artinian rings are right noetherian, and so the next theorem actually does include our earlier result that nil implies nilpotent for one-sided ideals of right artinan rings.)

(14.38) THEOREM (Levitsky).   *Let R be right noetherian. Then any nil one-sided ideal of R is nilpotent.*

Proof.   Suppose L is a nonzero nil left ideal of R. For each nonzero element $x \in L$, consider the right annihilator ra$(x) = \{r \in R \mid xr = 0\}$. This is a right ideal of R, and so we can choose nonzero $u \in L$ such that ra(u) is maximal in this set of right annihilators. For $r \in R$, we have ra$(ru) \supseteq$ ra(u) and $ru \in L$. If $ru \neq 0$, therefore, we have ra$(ru) =$ ra(u).

   We claim for every choice of $s \in R$ that $usu = 0$. We may suppose $su \neq 0$, but since su lies in the nil left ideal L, we have $(su)^n = 0$ for some integer $n > 1$ with $(su)^{n-1} \neq 0$. We can write $(su)^{n-1} = ru$ with $r = (su)^{n-2}s$. We have

$$su \in \text{ra}((su)^{n-1}) = \text{ra}(ru) = \text{ra}(u) \,,$$

so that $usu = 0$, as claimed. It follows that $(RuR)^2 = 0$ and thus R contains a nonzero nilpotent (two-sided) ideal.

   Next let $N \subseteq R$ be maximal among nilpotent ideals of R. We will show that R/N contains no nonzero nilpotent ideals and therefore, by the first part of the proof, R/N contains no nonzero nil left ideal. If U/N is a nilpotent ideal of R/N (where $U \subseteq R$ is an ideal), we have $(U/N)^m = (N/N)$ for some $m \geq 1$. Therefore $U^m \subseteq N$, and since N is nilpotent, this forces U to be nilpotent and thus $U = N$ by the maximality of N. Thus $U/N = 0$, as desired.

   Now let I be any nil one-sided ideal of R. If $a \in I$ and I is a left ideal, then of course Ra is nil. If I is a right ideal, then for $r \in R$, we have $(ra)^{n+1} = r(ar)^n a$. Since $ar \in I$, which is nil, we have $(ar)^n = 0$ for sufficiently large n, and thus $(ra)^{n+1} = 0$. In this case too, Ra is a nil left ideal.

   Let $\pi : R \to R/N$ be the canonical ring homomorphism. Then $\pi(Ra)$ is a nil left ideal of R/N and hence is zero. Thus $a \in Ra \subseteq \ker(\pi) = N$. Since $a \in I$ was arbitrary, we have $I \subseteq N$ and therefore I is nilpotent.   ∎

Problems _____

14.1   Let R be a commutative ring and suppose that $X \subseteq R$ is a finite subset consisting of nilpotent elements. Prove that X is contained in a nilpotent ideal of R.

14.2   If $I \subseteq R$, we write

$$\text{ra}(I) = \{x \in R \mid Ix = 0\}$$

and

$$\text{la}(I) = \{x \in R \mid xI = 0\}$$

to denote the right and left annihilators of I, respectively. If R is semiprime and I is an ideal, prove that ra$(I) =$ la(I).

14.3 Let R be semiprime and let $0 \neq a \in R$.

    a. Show that there exists an infinite sequence $a = a_1, a_2, a_3, \ldots$ of nonzero elements of R such that $a_j \in a_i R a_i$ whenever $1 \leq i < j$.

    b. Let $S = \{a_i \mid 1 \leq i < \infty\}$ as in part (a). If $I, J \subseteq R$ are ideals with $I \cap S \neq \varnothing \neq J \cap S$, show that $IJ \cap S \neq \varnothing$.

    c. Let $P \subseteq R$ be an ideal maximal with the property that $P \cap S = \varnothing$, where S is as in part (b). Show that P is prime.

    HINT: To show that P is prime it suffices to show that if $A > P$ and $B > P$ are ideals, then $AB \not\subseteq P$.

14.4 Show that R is semiprime iff the intersection of all prime ideals of R is zero.

    HINT: If R is semiprime and $0 \neq a \in R$, use Problem 14.3 to produce a prime ideal that does not contain a. Use Zorn's lemma.

14.5 Let R_1 and R_2 be rings and let $R = R_1 \oplus R_2$, the external direct sum (with componentwise multiplication).

    a. If R_1 and R_2 are right artinian, prove that R is right artinian.

    b. If R_1 and R_2 are semiprime, prove that R is semiprime.

14.6 Prove that a right artinian ring has only finitely many maximal ideals.

14.7 Idempotents e and f of R are said to be *orthogonal* if $ef = 0 = fe$. An idempotent e is *primitive* if it is not the sum of two orthogonal idempotents. Prove that an idempotent e is primitive iff eR is an indecomposable right R-module.

14.8 If R is right artinian, show that every idempotent is a sum of pairwise orthogonal primitive idempotents.

14.9 Let U be an ideal of a right artinian ring R. Show that there exists an ideal V such that $U + V = R$ and $U \cap V \subseteq J(R)$.

14.10 Let U and V be ideals of a ring R and assume $U + V = R$ and $U \cap V \subseteq J(R)$. Suppose that $v \in V$ and that $U + v$ is invertible in R/U. Show that there exists $u \in U$ such that $u + v$ is invertible in R.

14.11 Let R be artinian and let $I \subseteq R$ be a proper ideal. Show that the canonical homomorphism $R \to R/I$ maps the invertible elements of R onto the invertible elements of R/I.

    HINT: Use the previous two problems.

14.12 Let R be artinian and let $K > J(R)$ be a right ideal of R. Let I be minimal among right ideals of R such that $K = I + J(R)$. Show that $I = eR$ for some idempotent $e \in R$.

    HINT: Apply Theorem 14.13 to the right ideal $K/J(R)$ of $R/J(R)$ and choose $a \in I$ mapping to an idempotent generator of $K/J(R)$. Show that $I = aI$ and use Lemma 14.32.

14.13 Let R be right artinian. Let S be a simple right R-module and let $I \subseteq R$ be a minimally potent right ideal of R. Show that $SI \neq 0$ iff there exists a module homomorphism of I onto S.

14.14 Suppose that U and V are right R-modules of finite composition length, where R is right artinian. Show that U and V have a composition factor in common iff there exists a minimal potent right ideal I such that $UI \neq 0$ and $VI \neq 0$.

14.15 Let $R = A \dotplus B$, where A and B are ideals.

   a. If I is a minimally potent right ideal of R, show that either $I \subseteq A$ or $I \subseteq B$.
   b. Show that, when viewed as right R-modules, A and B have no composition factors in common.

   HINT: For part (a), observe that $IA + IB = I$.

14.16 Let K be a field and let R be the ring of 2×2 upper triangular matrices over K. Let

$$U = \left\{ \begin{bmatrix} 0 & x \\ 0 & 0 \end{bmatrix} \,\middle|\, x \in K \right\} \quad \text{and} \quad V = \left\{ \begin{bmatrix} 0 & 0 \\ 0 & x \end{bmatrix} \,\middle|\, x \in K \right\}.$$

Show that U and V are right ideals of R that are isomorphic as R-modules but that V is minimally potent and U is not.

CHAPTER FIFTEEN

An Introduction to Character Theory

15A

Group representation theory is a synthesis of group theory and ring theory that has proved to be valuable in the study of finite groups, and so it seems appropriate to devote this last chapter of the noncommutative part of this book to at least some portion of this theory. Specifically, the classical theory of "ordinary" characters, invented by Frobenius, can be approached through a study of the complex group algebra $\mathbb{C}G$ for a finite group G. We have chosen to present a small part of character theory here, just enough to prove the following theorem of Frobenius (for which no purely group theoretic proof, independent of characters and representations, is known).

(15.1) THEOREM (Frobenius). *Let $H \subseteq G$, where G is a finite group, and suppose $H \cap H^g = 1$ whenever $g \in G$ with $g \notin H$. Then the subset $N \subseteq G$ consisting of all elements of G not conjugate to any nonidentity element of H is a normal subgroup of G. Also, $G = NH$ and $N \cap H = 1$.*

Later in this book, after we have discussed Galois theory and established some facts concerning algebraic integers, we return to character theory and use it to prove Burnside's famous theorem that a group of order $p^a q^b$, where p and q are primes, is necessarily solvable. Unlike the situation with Frobenius's theorem, however, Burnside's theorem has now been proved without characters (but only with considerable difficulty).

We begin our study of group representations with a brief review of algebras in general and group algebras in particular. If F is a field, recall that an F-algebra A is both a ring and an F-vector space. The ring and vector "addition" operations are identical, and the ring and scalar "multiplications" are related by the identity

$$(\alpha a)b = \alpha(ab) = a(\alpha b)$$

for all $\alpha \in F$ and $a, b \in A$. The one-dimensional subspace $F \cdot 1 \subseteq A$ is a subring isomorphic to F. For $\alpha \in F$ and $a \in A$, we have $(\alpha 1)a = \alpha a = a(\alpha 1)$, and thus $F \cdot 1 \subseteq \mathbf{Z}(A)$ and scalar multiplication by α can be effected by ring multiplication by $\alpha 1$ on either side. It follows that every right or left ideal of A is an F-subspace, and so if A is finite dimensional, then it is automatically both right and left artinian and noetherian.

An F-algebra A is a *group algebra* if it has a basis that is closed under ring multiplication and that forms a group with respect to that multiplication. Given an arbitrary group G, the group algebra FG is constructed as the set of all formal linear combinations $\sum \alpha_g g$, where g runs over G and the coefficients α_g lie in F. The vector space structure is defined in the obvious way and multiplication is defined linearly, using the multiplication in G. The elements $1g \in FG$ (with $1 \in F$) are identified with the group elements $g \in G$, so that we do have $G \subseteq FG$ and G is a basis for FG.

(15.2) THEOREM (Maschke).   *Let G be a finite group. Then the group algebra FG is semiprimitive (has trivial Jacobson radical) iff the characteristic of F does not divide $|G|$.*

Recall that the characteristic of the field F is the common additive order of the nonzero elements of F if this order is finite, and it is zero if the nonzero elements of F all have infinite order. (See Problem 12.9.) The only possible nonzero characteristics of fields are prime numbers, and the field $F = \mathbb{Z}/p\mathbb{Z}$ shows that every prime can occur. The condition that an integer $n > 0$ is not divisible by the characteristic of F is automatically satisfied if the characteristic is zero. In general, if n is not divisible by the characteristic of F, then the equation $n\alpha = 0$ implies that $\alpha = 0$ for $\alpha \in F$.

Proof of Theorem 15.2.   Since G is finite, FG is right artinian and thus $J(FG)$ is nilpotent by Theorem 14.2. We therefore investigate nilpotent elements of FG, and our tool for doing this is a "representation" of the algebra.

If $a \in FG$, then right multiplication by a induces an F-linear transformation FG, and we write $\mathcal{R}(a)$ to denote the matrix of this transformation computed with respect to the basis G of FG. To be more precise, we fix a particular (though arbitrary) order for the elements of G and write $G = \{g_1, g_2, \ldots, g_n\}$, where $n = |G|$. Then the (i, j)-entry of the matrix $\mathcal{R}(a)$ is the coefficient of g_j in $g_i a$.

Since $((x)a)b = x(ab)$ and $x(a + b) = xa + xb$ for $x, a, b \in FG$, it follows that $\mathcal{R}$ is a ring homomorphism from FG into the full matrix algebra $M_n(F)$. In particular, if $u \in FG$ is nilpotent, then $\mathcal{R}(u)$ is a nilpotent matrix.

Now define $\rho : FG \rightarrow F$ by setting $\rho(a) = \text{tr}(\mathcal{R}(a))$, the trace of the matrix. Note that ρ is an F-linear map.

A nilpotent matrix (over a field) necessarily has trace zero. To see this, note that the minimal polynomial has the form X^m. The characteristic polynomial has degree n, the size of the matrix, and it divides a power of the minimal polynomial. It follows that the characteristic polynomial must be X^n, and since the trace of the matrix is the negative of the coefficient of X^{n-1} in the characteristic polynomial, we see that the trace is zero.

We now know that if $u \in FG$ is nilpotent, then $\rho(u) = 0$. We can also compute $\rho(u)$ in a different way, using the linearity of ρ. Writing $u = \sum \alpha_g g$ with $\alpha_g \in F$, we have $\rho(u) = \sum \alpha_g \rho(g)$. It is easy to compute $\rho(g)$ for $g \in G$, since for $g \neq 1$, we have $g_i g = g_j$ for some subscript $j \neq i$. It follows that the (i, i)-entry of $\mathcal{R}(g)$ is zero, and from this we see that $\rho(g) = 0$ when $g \neq 1$. On the other hand, $\mathcal{R}(1)$ is the identity matrix and thus $\rho(1) = n \cdot 1$. It follows when u is nilpotent that

$$0 = \rho(u) = \rho(\sum_{g \in G} \alpha_g g) = n\alpha_1 .$$

Assume now that n is not divisible by the characteristic of F. Then the condition $n\alpha_1 = 0$ implies that $\alpha_1 = 0$, and we have shown that the coefficient of the identity in any nilpotent element of FG is zero.

Now let $u \in J(FG)$ and write $u = \sum \alpha_g g$. Then $ug^{-1} \in J(FG)$ and so is nilpotent. The coefficient of the identity in this element is α_g, and hence $\alpha_g = 0$ for all $g \in G$. Thus $u = 0$, as required.

Conversely, suppose $|G|$ is divisible by the characteristic of F. Let $s \in FG$ be the sum of all of the elements of G. Since $sg = s$ for all $g \in G$, the one-dimensional subspace $F \cdot s$ is a right ideal of FG. Also $s^2 = |G|s = 0$, and hence $F \cdot s$ is nilpotent. By Corollary 14.1 we see that $F \cdot s$ is contained in $J(FG)$, which therefore is nonzero. ∎

In particular, if F has characteristic zero, for instance, $F = \mathbb{C}$ or $F = \mathbb{Q}$, then all group algebras FG for finite groups G are semiprimitive (and thus they are Wedderburn rings). It is known that $\mathbb{C}G$ is semiprimitive for every group G, even infinite groups, but it is still unknown whether or not $\mathbb{Q}G$ is always semiprimitive.

15B

At this point we reduce our generality and work with only the field $\mathbb{C}$ of complex numbers (and with only finite groups). This is sufficient for our development of character theory and for the proof of Frobenius's theorem (15.1). We hasten to add, however, that there is very much of interest that can be said about group algebras FG for other fields F, and some of that has significant applications in proving theorems about finite groups.

Recall that $GL(n, \mathbb{C})$, the "general linear" group, is the group of all invertible complex $n \times n$ matrices. If G is any finite group, then a group homomorphism $\mathcal{X} : G \to GL(n, \mathbb{C})$ is called a *representation* of G. (More properly, it is a $\mathbb{C}$-representation, but since we are limiting our attention to $\mathbb{C}$, we suppress the prefix.) The positive integer n is the *degree* of $\mathcal{X}$.

We have seen an example of a representation in the proof of Theorem 15.2. There (when $F = \mathbb{C}$) we constructed a ring homomorphism $\mathcal{R} : \mathbb{C}G \to M_n(\mathbb{C})$, where $n = |G|$. Since $\mathcal{R}(1)$ is the identity matrix, we see that $\mathcal{R}(g)$ is an invertible matrix, and thus the restriction of $\mathcal{R}$ to the group G is a representation of G whose degree is $|G|$; it is called the *regular* representation.

There is a natural way to build new representations from old ones. If $\mathcal{X}_1, \mathcal{X}_2, \ldots, \mathcal{X}_n$ are representations of G, we can form the representation $\mathcal{Y}$ (with degree equal to the sum of the degrees of the $\mathcal{X}_i$) by setting

$$\mathcal{Y}(g) = \begin{bmatrix} \mathcal{X}_1(g) & 0 & \cdots & 0 \\ 0 & \mathcal{X}_2(g) & \cdots & 0 \\ \vdots & \vdots & \ddots & \vdots \\ 0 & 0 & \cdots & \mathcal{X}_n(g) \end{bmatrix}.$$

In this situation, we say that $\mathcal{Y}$ is in *block diagonal* form and that the representations $\mathcal{X}_i$ are *constituents* of $\mathcal{Y}$.

We proved Theorem 15.2 by considering the function ρ obtained by taking the trace of $\mathcal{R}$, and we observed that $\rho(1) = |G|$ and $\rho(g) = 0$ if $1 \neq g \in G$. This function ρ, viewed as being defined on G only (rather than on all of $\mathbb{C}G$), is called the *regular character* of G. More generally, if $\mathcal{X}$ is any representation of G, then the function $\chi : G \to \mathbb{C}$ defined by $\chi(g) = \mathrm{tr}(\mathcal{X}(g))$ is called the *character* associated with or *afforded* by the representation $\mathcal{X}$. Note that in this situation, $\chi(1)$ is equal to the degree of $\mathcal{X}$. This positive integer is called the *degree* of the character. If a representation $\mathcal{Y}$ has block diagonal form with constituents $\mathcal{X}_i$, as in the previous paragraph, we observe that the character afforded by $\mathcal{Y}$ is just the sum of the characters of the $\mathcal{X}_i$.

Usually many different representations of G afford the same character. For instance, let $\mathcal{X}$ be a representation of degree n and fix any invertible $n \times n$ matrix P. The function $\mathcal{Y} : G \to GL(n, \mathbb{C})$ defined by $\mathcal{Y}(g) = P^{-1}\mathcal{X}(g)P$ is a representation, most likely different from $\mathcal{X}$, that affords the same character that $\mathcal{X}$ does. (The characters are the same because similar matrices have equal traces.) We write $\mathcal{Y} = P^{-1}\mathcal{X}P$, and we say that $\mathcal{Y}$ and $\mathcal{X}$ are *similar* representations.

What are the characters of G of degree 1? Since $GL(1, \mathbb{C})$ is essentially the same as $\mathbb{C}^\times$, the multiplicative group of $\mathbb{C}$, we see that a character of degree 1, a so called *linear* character, is just a homomorphism $G \to \mathbb{C}^\times$. There is always at least one of these: the constant function with value 1, denoted 1_G and called the *principal* character. Note that by Theorem 7.16, if G is abelian, then its number of linear characters is exactly $|G|$. (This also follows from the results we prove in this chapter.)

We write $\mathrm{Char}(G)$ to denote the set of all characters of the finite group G. If $\chi, \psi \in \mathrm{Char}(G)$, we define $\chi + \psi : G \to \mathbb{C}$ by $(\chi + \psi)(g) = \chi(g) + \psi(g)$. In fact, $\chi + \psi \in \mathrm{Char}(G)$. To see this, it suffices to observe that $\chi + \psi$ is afforded by any block diagonal representation with two constituents that afford χ and ψ, respectively.

A character $\chi \in \mathrm{Char}(G)$ is *irreducible* if it cannot be written as a sum of two characters.

(15.3) LEMMA. *Let $\chi \in \mathrm{Char}(G)$. Then χ is a sum of (not necessarily distinct) irreducible characters.*

Proof.   Since character degrees are always positive integers, if the result were false, we could choose a counterexample χ of minimum possible degree. In particular, χ is not itself irreducible, and so we can write $\chi = \alpha + \beta$, where $\alpha, \beta \in \text{Char}(G)$. Then $\chi(1) = \alpha(1) + \beta(1)$, and so the degrees $\alpha(1) < \chi(1)$ and $\beta(1) < \chi(1)$.

By the minimality of $\chi(1)$, each of α and β is a sum of irreducible characters, and thus $\chi = \alpha + \beta$ is also a sum of irreducibles. This contradiction proves the result.　　　　　　　　　　　　　　　　　　　　　　　　　　■

We write $\text{Irr}(G)$ to denote the set of irreducible characters of G. Note that every linear (degree 1) character is necessarily irreducible.

15C

A function $\alpha : G \to \mathbb{C}$ is said to be a *class function* on G if α is constant on conjugacy classes of G. The set $\text{cf}(G)$ of all class functions of G is clearly a $\mathbb{C}$-vector space with dimension equal to the number of conjugacy classes of G.

(15.4) LEMMA.   *Every character is a class function.*

Proof.   Let $\chi \in \text{Char}(G)$ and let $\mathcal{X}$ afford χ. Since the matrix $\mathcal{X}(g)$ is similar to $\mathcal{X}(h)^{-1}\mathcal{X}(g)\mathcal{X}(h) = \mathcal{X}(h^{-1}gh)$, we have

$$\chi(h^{-1}gh) = \text{tr}(\mathcal{X}(h^{-1}gh)) = \text{tr}(\mathcal{X}(g)) = \chi(g)$$

for all $h, g \in G$. Thus χ is a class function, as desired.　　　　　　　■

We need one more definition before we can state the "fundamental theorem" of character theory. We define an inner product $[\,\cdot\,,\,\cdot\,]$ on the complex vector space $\text{cf}(G)$ by setting

$$[\alpha, \beta] = \frac{1}{|G|} \sum_{g \in G} \alpha(g)\overline{\beta(g)}.$$

Note that this really is an inner product because

1. $[c_1\alpha_1 + c_2\alpha_2, \beta] = c_1[\alpha_1, \beta] + c_2[\alpha_2, \beta]$ for $c_1, c_2 \in \mathbb{C}$,
2. $[\beta, \alpha] = \overline{[\alpha, \beta]}$, and
3. $[\alpha, \alpha] > 0$ if $\alpha \neq 0$.

(15.5) THEOREM (Fundamental, of character theory).   *Let G be any finite group. Then* $\text{Irr}(G)$ *is an orthonormal basis for* $\text{cf}(G)$.

Before going on to prove the fundamental theorem, we mention a few of its consequences. We know that $\dim_{\mathbb{C}}(\text{cf}(G))$ is equal to the number k of conjugacy classes of G. By Theorem 15.5, therefore, it follows that $|\text{Irr}(G)| = k$. Since every character of G is a sum of irreducible characters and every such sum is a character, we see that to determine all characters of G, it suffices to know the values of each of the k irreducible characters. Because each of these is a class function, the characters of

G are completely determined by a $k \times k$ table that gives the values of the irreducible characters on each of the conjugacy classes of G. For example, Table 15.1 is the character table of the symmetric group S_4. We do not give the full proof that this is correct.

TABLE 15.1

Class representative:	1	(12)	(12)(34)	(1234)	(123)
Size of class:	1	6	3	6	8
χ_1:	1	1	1	1	1
χ_2:	1	−1	1	−1	1
χ_3:	2	0	2	0	−1
χ_4:	3	1	−1	−1	0
χ_5:	3	−1	−1	1	0

The first irreducible character χ_1 in the table is the constant function 1_G, the principal character of $G = S_4$. The character χ_2 is the "sign" character of S_4, having the value 1 on even permutations and −1 on odd permutations. Certainly, this function is a homomorphism $G \to \mathbb{C}^\times$, and thus it is a linear (and hence irreducible) character.

To facilitate the computation of inner products, we included in Table 15.1 a row labeled "size of class." Since the definition of the inner product requires a sum over all group elements, it can be computed by multiplying the appropriate quantity by the class size and then summing over the classes. To practice using the table, the reader may want to confirm, for example, that $[\chi_4, \chi_4] = 1$ and $[\chi_4, \chi_2] = 0$.

Now let $\psi \in \mathrm{cf}(G)$, where G is an arbitrary finite group. Since $\mathrm{Irr}(G)$ is a basis for $\mathrm{cf}(G)$, we can write

$$\psi = \sum_{\chi \in \mathrm{Irr}(G)} c_\chi \chi$$

for some complex numbers c_χ. (These coefficients can be determined by means of orthonormality, since we have the formula $c_\chi = [\psi, \chi]$. As an exercise, try this with the class function $\psi(g) = |C_G(g)|$, using the character table of S_4.) When is the class function ψ actually a character? This happens when ψ can be written as a sum of irreducible characters. It follows from the linear independence of $\mathrm{Irr}(G)$, therefore, that ψ is a character precisely when all c_χ are nonnegative integers and at least one of them is nonzero. In other words, we have the following corollary.

(15.6) COROLLARY. *Let $0 \neq \psi \in \mathrm{cf}(G)$. Then $\psi \in \mathrm{Char}(G)$ iff $[\psi, \chi]$ is a nonnegative integer for all $\chi \in \mathrm{Irr}(G)$. Also, if $\psi, \eta \in \mathrm{Char}(G)$, then $[\psi, \eta]$ is a nonnegative integer.* ∎

Suppose we start with the regular character ρ_G of G. (Recall that this character vanishes at all nonidentity group elements and that $\rho_G(1) = |G|$.) We see that

$$[\rho_G, \chi] = \frac{1}{|G|}\rho(1)\chi(1) = \chi(1),$$

which confirms by Corollary 15.6 that ρ_G is a character. Also, this gives

$$\rho_G = \sum_{\chi \in \mathrm{Irr}(G)} \chi(1)\chi .$$

(The reader should check this formula for $G = S_4$ using Table 15.1.) In particular,

$$|G| = \rho_G(1) = \sum_{\chi} \chi(1)^2 .$$

(15.7) COROLLARY.   *The sum of the squares of the degrees of the irreducible characters of G equals $|G|$.*   ∎

Let us begin now to work toward a proof of Theorem 15.5. We examine connections between characters of G and $\mathbb{C}G$-modules.

If M is any module for a finite dimensional F-algebra, then we can make M into an F-vector space by defining $m\alpha = m(\alpha \cdot 1)$, where $m \in M$, $\alpha \in F$, and 1 is the unity element of A. With this definition, we see that the endomorphisms $a_M \in \mathrm{End}(M)$ induced by elements $a \in A$ are F-linear, as are the elements of $\mathrm{End}_A(M)$. Also, if M is finitely generated as an A-module, then $\dim_F(M) < \infty$. To see this, let X be a finite generating set for M and let B be a (finite) F-basis for A. The finite set $XB \subseteq M$ then spans M over F.

In the above situation, if we fix an F-basis for M, we can write $\mathcal{X}(a)$ for $a \in A$ to denote the matrix of the linear transformation a_M induced by a on M. Then $\mathcal{X} : A \to M_n(F)$ is an F-linear ring homomorphism, where $n = \dim_F(M)$. Also, $\mathcal{X}(1)$ is the $n \times n$ identity matrix.

Now fix a finite group G and observe that the group algebra $\mathbb{C}G$ is a Wedderburn ring. (It is semiprimitive by Theorem 15.2.) Let $\mathcal{S}$ be a representative set for the simple $\mathbb{C}G$-modules and fix a $\mathbb{C}$-basis for each $S \in \mathcal{S}$. (Note that a simple module is certainly finitely generated, and thus S is finite dimensional for each $S \in \mathcal{S}$.) Let $\mathcal{X}_S : \mathbb{C}G \to M_{n_S}(\mathbb{C})$ be the corresponding $\mathbb{C}$-linear ring homomorphism, where $n_S = \dim_{\mathbb{C}}(S)$.

For each $S \in \mathcal{S}$, the restriction of $\mathcal{X}_S$ to G is a representation of G of degree n_S. (We need to observe that $\mathcal{X}_S(G)$ consists of invertible matrices; this follows because $\mathcal{X}_S(1)$ is the identity matrix.) We write χ_S to denote the character of G afforded by $\mathcal{X}_S$. In particular, we have $\chi_S(1) = n_S$.

(15.8) LEMMA.   *Let $\mathcal{X}$ be any representation of G. Then $\mathcal{X}$ is similar to a block diagonal representation with each constituent block being one of the representations $\mathcal{X}_S$.*

Proof.   Let $n = \deg(\mathcal{X})$ and extend $\mathcal{X}$ to a map $\mathbb{C}G \to M_n(\mathbb{C})$ by defining

$$\mathcal{X}\left(\sum_g \alpha_g g\right) = \sum_g \alpha_g \mathcal{X}(g) .$$

Observe that this extension is an F-linear ring homomorphism. (We omit the routine calculation that establishes this.)

Now let V be the n-dimensional row space $\mathbb{C}^n$ and make V into a $\mathbb{C}G$-module by defining $va = v\mathcal{X}(a)$ for $v \in V$ and $a \in \mathbb{C}G$. (We do not give the details here either.) Observe that the $\mathbb{C}$-space structure that V carries because it is a $\mathbb{C}G$-module is identical to its natural $\mathbb{C}$-structure as a row space.

Since the group algebra $\mathbb{C}G$ is a Wedderburn ring, Corollary 14.9 tells us that V is completely reducible. We can therefore write

$$V = V_1 \dotplus V_2 \dotplus \cdots \dotplus V_m \,,$$

where each of the submodules V_i is simple. Each V_i, therefore, is isomorphic to some member $S_i \in \mathcal{S}$. We can thus choose a basis for V_i such that, with respect to that basis, the matrix determined by the action of an element $g \in G$ on V_i is exactly the matrix $\mathcal{X}_{S_i}(g)$. Combining the bases we have just chosen for each direct summand V_i, we obtain a basis for V with respect to which V yields a representation in block diagonal form with constituent blocks, each of the form $\mathcal{X}_S$ for various $S \in \mathcal{S}$. This completes the proof.   ∎

In view of Lemma 15.8, it is appropriate to investigate the representations $\mathcal{X}_S$ for $S \in \mathcal{S}$ more closely. By the Wedderburn-Artin theorem (14.14), we can write the group algebra $\mathbb{C}G$ as a direct sum of minimal ideals I_S for $S \in \mathcal{S}$, and this gives us a unique choice of elements $e_S \in I_S$ for $S \in \mathcal{S}$ such that

$$\sum_{S \in \mathcal{S}} e_S = 1 \,.$$

Since I_T annihilates S when $S, T \in \mathcal{S}$ are different, we have $\mathcal{X}_S(I_T) = 0$ in this case, and in particular, $\mathcal{X}_S(e_T) = 0$ when $S \neq T$. It follows that $\mathcal{X}_S(e_S) = \mathcal{X}_S(1)$, the identity matrix.

We need a bit more notation. We have defined the character χ_S only on G. We need to discuss $\mathrm{tr}(\mathcal{X}_S(a))$ for elements $a \in \mathbb{C}G$ other than group elements, and so we write $\widehat{\chi}_S(a) = \mathrm{tr}(\mathcal{X}_S(a))$. Note that

$$\widehat{\chi}_S\Big(\sum \alpha_g g\Big) = \sum \alpha_g \chi_S(g) \,,$$

so that $\widehat{\chi}_S$ is simply the linear extension of χ_S to all of $\mathbb{C}G$. Observe that $\widehat{\chi}_S(e_T) = 0$ if $S \neq T$ and $\widehat{\chi}_S(e_S) = n_S$.

(15.9) THEOREM. *The characters χ_S for $S \in \mathcal{S}$ are distinct and linearly independent. They are irreducible and are all of the irreducible characters of G.*

Proof. Suppose that $\sum_{T \in \mathcal{S}} \alpha_T \chi_T$ vanishes identically on G for some coefficients $\alpha_T \in \mathbb{C}$. Then $\sum \alpha_T \widehat{\chi}_T$ vanishes identically on $\mathbb{C}G$. Evaluating at e_S yields $0 = \alpha_S n_S$, and thus $\alpha_S = 0$ for all $S \in \mathcal{S}$. It follows that the characters χ_S are distinct and linearly independent.

If χ is any character of G, then by Lemma 15.8 it follows that χ is a sum of some of the characters χ_S. In particular, if χ is irreducible, we have $\chi = \chi_S$

for some $S \in \mathcal{S}$. To see that each character χ_S is irreducible, suppose to the contrary that we can write $\chi_S = \xi + \eta$, where $\xi, \eta \in \mathrm{Char}(G)$. Each of ξ and η is a sum of characters χ_T for various members $T \in \mathcal{S}$. It follows that χ_S can be written as a sum of at least two characters of the form χ_T. This contradicts the linear independence and thus completes the proof. ∎

(15.10) LEMMA.   *Each of the ring homomorphisms $\mathcal{X}_S$ for $S \in \mathcal{S}$ maps $\mathbb{C}G$ onto $M_{n_S}(\mathbb{C})$.*

Proof.   Fix $S \in \mathcal{S}$ and let $D = \mathrm{End}_A(S)$. Then D is a division ring of $\mathbb{C}$-linear transformations of S that contains all of the scalar multiplications by elements $\alpha \in \mathbb{C}$. We claim that D consists entirely of scalar multiplications.

Let $\delta \in D$. Then δ has some eigenvalue $\lambda \in \mathbb{C}$. (This uses the "fundamental theorem of algebra," which asserts that every nonconstant polynomial with coefficients in $\mathbb{C}$ has a root in $\mathbb{C}$.)   Let $s \in S$ be an eigenvector for δ so that $s\delta = s\lambda$ and hence $s(\delta - \lambda \cdot 1) = 0$. However, $\delta - \lambda \cdot 1 \in D$, and it is not invertible since it annihilates the nonzero vector s. This forces $\delta - \lambda \cdot 1 = 0$ and $\delta = \lambda \cdot 1$. Therefore $D = \mathbb{C} \cdot 1$, as claimed.

Since S is simple, the Jacobson density theorem (13.13) applies. Because S is finite dimensional, we conclude that every element of $\mathrm{End}_D(S) = \mathrm{End}_\mathbb{C}(S)$ is induced by the action of some element of $\mathbb{C}G$. In matrix language, this says that $\mathcal{X}_S$ maps $\mathbb{C}G$ onto the full matrix ring $M_{n_S}(\mathbb{C})$. ∎

We know that $\mathcal{X}_S(I_T) = 0$ for $T \neq S$. Since $\mathbb{C}G = \sum I_S$, it follows by Lemma 15.10 that $\mathcal{X}_S$ maps I_S onto $M_{n_S}(\mathbb{C})$. Also, by the minimality of I_S, we must have $I_S \cap \ker(\mathcal{X}_S) = 0$, and thus $\mathcal{X}_S$ defines an isomorphism of I_S onto $M_{n_S}(\mathbb{C})$. We can use this to prove the following corollary.

(15.11) COROLLARY.   *The set $\mathrm{Irr}(G)$ forms a basis for $\mathrm{cf}(G)$.*

Proof.   By Theorem 15.9 we know that $\mathrm{Irr}(G) = \{\chi_S \mid S \in \mathcal{S}\}$ and that this set is linearly independent. Since the cardinality of this set is $|\mathcal{S}|$, it suffices to show that $\dim(\mathrm{cf}(G)) = |\mathcal{S}|$. We know, however, that $\dim(\mathrm{cf}(G)) = k$, the number of conjugacy classes of G. Our task is therefore to show that $|\mathcal{S}| = k$.

An element $a = \sum \alpha_g g \in \mathbb{C}G$ is central in $\mathbb{C}G$ precisely when $g^{-1}ag = a$ for all elements $g \in G$. This happens iff the coefficients $\alpha_x = \alpha_y$ whenever x and y are conjugate elements of G. In other words, an element $a \in \mathbb{C}G$ lies in $\mathbf{Z}(\mathbb{C}G)$ iff a is a linear combination of the class sums of G, the elements of $\mathbb{C}G$ obtained by summing the elements in the various classes of G. Since these class sums are clearly linearly independent (because the classes are disjoint), it follows that $\dim(\mathbf{Z}(\mathbb{C}G)) = k$.

To complete the proof we show that $\dim(\mathbf{Z}(\mathbb{C}G)) = |\mathcal{S}|$. To see this, recall that we have $\mathbb{C}G = \sum I_S$. Since this sum is direct, we see that $\mathbf{Z}(\mathbb{C}G)$ is the direct sum of the centers of the subrings I_S. It suffices, therefore, to show that $\dim(\mathbf{Z}(I_S)) = 1$. We know that I_S is isomorphic to the full ring of $n_S \times n_S$ matrices over $\mathbb{C}$. The center of a full matrix ring, however, is exactly the ring of scalar matrices $\alpha \cdot 1$. It thus has dimension 1, as required. ∎

We have obtained enough information now, on our way toward the proof of the fundamental theorem of character theory, to give a direct proof of Corollary 15.7. (The assertion of Corollary 15.7 was that $|G| = \sum \chi(1)^2$, where χ runs over Irr(G).) Certainly, dim($\mathbb{C}G$) = $|G|$, and thus the sum of the dimensions of the direct summands I_S for $S \in \mathcal{S}$ is $|G|$. Since I_S is isomorphic to the full matrix ring $M_{n_S}(\mathbb{C})$, its dimension is n_S^2. The result now follows.

To complete the proof of the fundamental theorem (15.5), we need to establish the orthonormality of the characters χ_S for $S \in \mathcal{S}$. For this purpose, it is useful to compute explicitly the elements e_S for $S \in \mathcal{S}$.

(15.12) LEMMA.   *We have*

$$e_S = \frac{\chi_S(1)}{|G|} \sum_{g \in G} \chi_S(g^{-1})g \,.$$

Proof.   For $a \in \mathbb{C}G$, write $\hat{\rho}(a) = \text{tr}(\mathcal{R}(a))$, where $\mathcal{R}(a)$ is the matrix corresponding to the linear transformation on $\mathbb{C}G$ induced by right multiplication by a. (We can use any basis we please for $\mathbb{C}G$.) Since $\hat{\rho}$ is the linear extension of the regular character ρ of G to all of $\mathbb{C}G$, and since $\rho(g) = 0$ for $1 \neq g \in G$ while $\rho(1) = |G|$, we see that if $a = \sum \alpha_g g$, then $\hat{\rho}(ag^{-1}) = |G|\alpha_g$. It follows that

$$a = \frac{1}{|G|} \sum_{g \in G} \hat{\rho}(ag^{-1})g \,.$$

To prove the theorem, therefore, it suffices to show that

$$\hat{\rho}(e_S g^{-1}) = \chi_S(1)\chi_S(g^{-1}) \,.$$

Our first task is to express ρ in terms of the irreducible characters χ_S. (By the calculation we did just before the statement of Corollary 15.7, we know that the correct formula is $\rho = \sum \chi_S(1)\chi_S$, but because we obtained that formula from the fundamental theorem, we need to find an independent proof now.)

We know that we can write $\rho = \sum m_S \chi_S$ for some choice of coefficients m_S. (In fact, since ρ is a character and the χ_S are the irreducible characters, we deduce that the m_S are all nonnegative integers.) It follows that

$$\hat{\rho} = \sum_{S \in \mathcal{S}} m_S \widehat{\chi}_S \,.$$

Evaluating at e_T gives

$$\hat{\rho}(e_T) = m_T n_T \,.$$

Also, $\hat{\rho}(e_T)$ is the trace of the linear transformation of $\mathbb{C}G$ induced by right multiplication by e_T (computed using any convenient basis). Since e_T annihilates the direct summand I_S for $S \neq T$ and e_T is the identity on I_T, it follows that $\hat{\rho}(e_T) = \dim(I_T) = n_T^2$. Thus $m_T n_T = n_T^2$ and so $m_T = n_T = \chi_T(1)$. Therefore $\hat{\rho} = \sum \chi_S(1)\widehat{\chi}_S$, as expected.

Now

$$\widehat{\chi}_T(esg^{-1}) = \text{tr}(\mathcal{X}_T(esg^{-1})) = \text{tr}(\mathcal{X}_T(es)\mathcal{X}_T(g^{-1})).$$

Since $\mathcal{X}_T(e_S) = 0$ if $S \neq T$ and $\mathcal{X}_S(e_S)$ is the identity matrix, we have

$$\widehat{\chi}_T(esg^{-1}) = \begin{cases} 0 & \text{if } S \neq T \\ \chi_S(g^{-1}) & \text{if } S = T, \end{cases}$$

and so

$$\hat{\rho}(esg^{-1}) = \chi_S(1)\chi_S(g^{-1}),$$

as desired.   ∎

The following lemma will help us to compute $\chi(g^{-1})$ for characters χ of G, and it will also be useful when we apply character theory to finding normal subgroups.

(15.13) LEMMA.   *Let $\mathcal{Y}$ be any representation of G and suppose $g \in G$. Then $\mathcal{Y}$ is similar to some representation $\mathcal{X}$ such that $\mathcal{X}(g)$ is a diagonal matrix.*

Proof.   We prove the result first in the case where G is abelian. By Lemma 15.8 we know that $\mathcal{Y}$ is similar to a representation $\mathcal{X}$ in block diagonal form, where each constituent block is a representation of the form $\mathcal{X}_S$ with $S \in \mathcal{S}$.

Since G is abelian, the image of $\mathbb{C}G$ under $\mathcal{X}_S$ must be a commutative ring. We know, however, that this image is the full matrix algebra $M_{n_S}(\mathbb{C})$, and it follows that $n_S = 1$. The matrices $\mathcal{X}(x)$ are therefore actually diagonal, and not just block diagonal, for elements $x \in G$.

Now in the general case, let $H = \langle g \rangle \subseteq G$ and observe that the restriction of $\mathcal{Y}$ to H is a representation of H. Since H is abelian, the first part of the proof guarantees the existence of an invertible matrix P such that $P^{-1}\mathcal{Y}(g)P$ is diagonal. The representation $\mathcal{X} = P^{-1}\mathcal{Y}P$ therefore has the desired properties.   ∎

What are the entries in the diagonal matrix $\mathcal{X}(g)$ of Lemma 15.13? Writing $\mathcal{X}(g) = \text{diag}(\epsilon_1, \epsilon_2, \ldots, \epsilon_f)$, where f is the degree of $\mathcal{X}$, we see that

$$\mathcal{X}(g^m) = \text{diag}(\epsilon_1^m, \epsilon_2^m, \ldots, \epsilon_f^m)$$

for all integers m. In particular, if we take m to be the order of g, then $\mathcal{X}(g^m)$ is the identity matrix, and we deduce that $\epsilon_i^m = 1$. In other words, the entries ϵ_i are *roots of unity*, that is, complex numbers with some power equal to 1. It follows that $|\epsilon_i| = 1$ and hence $\epsilon_i^{-1} = \overline{\epsilon_i}$, the complex conjugate.

(15.14) COROLLARY.   *Let $\chi \in \text{Char}(G)$ and $g \in G$. Then $\chi(g^{-1}) = \overline{\chi(g)}$, the complex conjugate of $\chi(g)$.*

Proof.   By Lemma 15.13 we know that χ is afforded by some representation $\mathcal{X}$ such that $\mathcal{X}(g)$ is a diagonal matrix. We let $\epsilon_1, \epsilon_2, \ldots, \epsilon_f$ be the diagonal entries of $\mathcal{X}(g)$, and we observe that the diagonal entries of $\mathcal{X}(g^{-1})$ are the reciprocals of these complex numbers. By the above discussion, these reciprocals are the complex conjugates of the ϵ_i, and we have

$$\chi(g^{-1}) = \sum \overline{\epsilon_i} = \overline{\chi(g)},$$

as required.   ∎

(15.15) COROLLARY.   *We have*

$$e_S = \frac{\chi_S(1)}{|G|} \sum_{g \in G} \overline{\chi_S(g)} g \, .$$

Proof.   Apply Corollary 15.14 to the formula of Lemma 15.12.   ∎

We are finally ready to prove the orthonormality of $\mathrm{Irr}(G)$ and thereby complete the proof of Theorem 15.5.

(15.16) THEOREM.   *The irreducible characters of G form an orthonormal set.*

Proof.   Using the previous notation, we let $S, T \in \mathcal{S}$. In view of Theorem 15.9, it suffices to show that $[\chi_S, \chi_S] = 1$ and that $[\chi_S, \chi_T] = 0$ if S and T are different. We compute the coefficient of 1 in $e_S e_T$. The coefficients of g^{-1} in e_S and of g in e_T are, respectively,

$$\frac{1}{|G|} \chi_S(1) \chi_S(g) \quad \text{and} \quad \frac{1}{|G|} \chi_T(1) \overline{\chi_T(g)} \, .$$

It follows that the coefficient α of 1 in $e_S e_T$ is given by

$$\alpha = \frac{\chi_S(1)\chi_T(1)}{|G|^2} \sum_{g \in G} \chi_S(g)\overline{\chi_T(g)} = \frac{\chi_S(1)\chi_T(1)}{|G|}[\chi_S, \chi_T] \, .$$

Since $e_S e_T = 0$ if $S \neq T$, we deduce that $\alpha = 0$ in this case, and thus $[\chi_S, \chi_T] = 0$, as desired. On the other hand, e_S is the unity element of I_S and thus $(e_S)^2 = e_S$. In the case where $S = T$, therefore, we see that α is the coefficient of 1 in e_S, and so $\alpha = \chi_S(1)^2/|G|$. Therefore $[\chi_S, \chi_S] = 1$.   ∎

15D

How can we use characters to prove theorems about groups? One key to this is the following result, which produces normal subgroups from character theoretic information.

(15.17) LEMMA.   *Let $\chi \in \mathrm{Char}(G)$. Then $|\chi(g)| \leq \chi(1)$ for all $g \in G$. Also,*

$$K = \{g \in G \mid \chi(g) = \chi(1)\}$$

and

$$Z = \{g \in G \mid |\chi(g)| = \chi(1)\}$$

*are normal subgroups of G and Z/K ⊆ **Z**(G/K).*

Proof.   Let $\mathcal{X}$ be a representation of G that affords χ. We will show that $K = \ker(\mathcal{X})$, so that K is a normal subgroup. Certainly, if $g \in \ker(\mathcal{X})$, then $\mathcal{X}(g) = \mathcal{X}(1)$ and hence $\chi(g) = \chi(1)$ and $g \in K$. Now let $g \in G$ be arbitrary. By Lemma 15.13, $\mathcal{X}(g)$ is similar to a diagonal matrix with diagonal entries $\epsilon_1, \epsilon_2, \ldots, \epsilon_n$, where $n = \chi(1)$. As we have seen, these entries ϵ_i are roots of unity and thus $|\epsilon_i| = 1$. By the triangle inequality we see that

$$|\chi(g)| = |\epsilon_1 + \cdots + \epsilon_n| \le n = \chi(1),$$

and equality holds precisely when all ϵ_i are equal.

If $|\chi(g)| = \chi(1)$, therefore, $\mathcal{X}(g)$ is similar to, and hence equal to, the scalar matrix ϵI, where I is the $n \times n$ identity matrix and ϵ is the common value of the ϵ_i. In particular, if $g \in K$, we have $\epsilon = 1$, and hence $g \in \ker(\mathcal{X})$, as required.

We have seen that Z is exactly the set of elements of G such that $\mathcal{X}(g)$ is a scalar matrix. Note that the scalar matrices in $\mathcal{X}(G)$ form a central subgroup and that Z is the full inverse image in G of this (necessarily normal) subgroup of $\mathcal{X}(G)$. It follows that $Z \lhd G$, and since $K = \ker(\mathcal{X})$, we have $Z/K \subseteq \mathbf{Z}(G/K)$.   ∎

We write $\ker(\chi)$ and $\mathbf{Z}(\chi)$, respectively, to denote the subgroups K and Z of Lemma 15.17. In fact, every normal subgroup of G is the kernel of some (not necessarily irreducible) character. It follows that, theoretically at least, all normal subgroups of G can be found from a knowledge of the characters of G. We see from this that the character table of G (which describes all $\chi \in \mathrm{Irr}(G)$) has embedded within it considerable information concerning the normal subgroup structure of G.

Except for the following observation, which we need, we leave the proof of all of this to the problems at the end of the chapter.

(15.18) COROLLARY.   *We have*

$$\bigcap_{\chi \in \mathrm{Irr}(G)} \ker(\chi) = 1.$$

Proof.   We know that the regular character $\rho = \sum \chi(1)\chi$, where the sum is over $\chi \in \mathrm{Irr}(G)$. It follows that if $g \in \bigcap \ker(\chi)$, we have $\rho(g) = \sum \chi(1)^2 \ne 0$. Thus $g = 1$.   ∎

In order to prove Theorem 15.1 of Frobenius, we need more information relating the character theory of a group G to that of its subgroups. If $\chi \in \mathrm{Char}(G)$ and $H \subseteq G$, we write χ_H to denote the function $H \to \mathbb{C}$ obtained by restricting χ. Since the restriction of a representation of G to H clearly yields a representation of

H, it follows that $\chi_H \in \text{Char}(H)$. There is no reason, of course, to believe that in general χ_H will be irreducible even if $\chi \in \text{Irr}(G)$.

Next we describe how to get characters of G from characters of subgroups $H \subseteq G$. We first work with class functions. If $\varphi \in \text{cf}(H)$, define $\varphi^\circ : G \to \mathbb{C}$ by

$$\varphi^\circ(g) = \begin{cases} \varphi(g) & \text{if } g \in H \\ 0 & \text{if } g \notin H. \end{cases}$$

In general, φ° is not constant on classes of G, and so we define a new function that repairs this defect. We define the *induced* class function $\varphi^G : G \to \mathbb{C}$ by the formula

$$\varphi^G(g) = \frac{1}{|H|} \sum_{x \in G} \varphi^\circ(xgx^{-1}).$$

It should be clear that $\varphi^G(g)$ depends on only the conjugacy class of g, and so we do not give a formal proof that $\varphi^G \in \text{cf}(G)$. Note that if $H = G$, then $\varphi^\circ = \varphi$ and $\varphi(xgx^{-1}) = \varphi(g)$ for all $x \in G$. It follows in this case that $\varphi^G = \varphi$. In general, we have $\varphi^G(1) = |G : H|\varphi(1)$.

(15.19) LEMMA (Frobenius reciprocity). *Let $H \subseteq G$ and suppose $\varphi \in \text{cf}(H)$ and $\theta \in \text{cf}(G)$. Then*

$$[\varphi^G, \theta] = [\varphi, \theta_H],$$

where, of course, the second inner product is computed in $\text{cf}(H)$.

Proof.   We have

$$\begin{aligned}
[\varphi^G, \theta] &= \frac{1}{|G|} \sum_{g \in G} \varphi^G(g)\overline{\theta(g)} = \frac{1}{|G||H|} \sum_{g \in G} \sum_{x \in G} \varphi^\circ(xgx^{-1})\overline{\theta(g)} \\
&= \frac{1}{|G||H|} \sum_{y \in G} \sum_{x \in G} \varphi^\circ(y)\overline{\theta(x^{-1}yx)} \\
&= \frac{1}{|G||H|} \sum_{y \in H} \sum_{x \in G} \varphi(y)\overline{\theta(y)} \\
&= \frac{1}{|H|} \sum_{y \in H} \varphi(y)\overline{\theta(y)} \\
&= [\varphi, \theta_H],
\end{aligned}$$

as desired.   ∎

(15.20) COROLLARY.   *Let $H \subseteq G$. If $\varphi \in \text{Char}(H)$, then $\varphi^G \in \text{Char}(G)$.*

Proof.   By Corollary 15.6 it suffices to show that $[\varphi^G, \chi]$ is a nonnegative integer for all $\chi \in \text{Irr}(H)$. (Certainly, $\varphi^G \neq 0$ since $\varphi^G(1) = |G : H|\varphi(1) \neq 0$.)

Frobenius reciprocity (15.19) tells us that $[\varphi^G, \chi] = [\varphi, \chi_H]$, and this latter quantity is certainly a nonnegative integer (by Corollary 15.6) since both φ and χ_H are characters of H.   ∎

15E

Recall that Theorem 15.1 concerns a subgroup $H \subseteq G$ with the property that $H^g \cap H = 1$ whenever $g \in G - H$. We call such a subgroup $H \subseteq G$ a *Frobenius complement* in G. (Usually this phrase is defined so as to require $1 < H < G$, but we do not exclude the two trivial cases $H = 1$ and $H = G$.)

(15.21) LEMMA. *Let $H \subseteq G$ be a Frobenius complement. Suppose $\varphi, \theta \in \mathrm{cf}(H)$ and $\varphi(1) = 0$. Then*

  a. $(\varphi^G)_H = \varphi$ *and*
  b. $[\theta, \varphi] = [\theta^G, \varphi^G]$.

Proof. Let $h \in H$. We need to show that $\varphi^G(h) = \varphi(h)$. If $h = 1$, this holds because $\varphi^G(1) = |G : H|\varphi(1) = 0$. Suppose that $h \neq 1$. Then

$$\varphi^G(h) = \frac{1}{|H|} \sum_{x \in G} \varphi^\circ(xhx^{-1}).$$

If $x \notin H$, however, then $H \cap H^x = 1$, and so $h \notin H^x$ and $xhx^{-1} \notin H$. In this case, $\varphi^\circ(xhx^{-1}) = 0$, and it follows that

$$\varphi^G(h) = \frac{1}{|H|} \sum_{x \in H} \varphi(xhx^{-1}) = \varphi(h),$$

where the latter equality holds since φ is a class function of H. This proves part (a).

For part (b), compute that $[\theta, \varphi] = [\theta, (\varphi^G)_H] = [\theta^G, \varphi^G]$ by (a) and Frobenius reciprocity. ∎

The reader may wonder how we can possibly hope to apply Lemma 15.21, since the hypothesis that $\varphi(1) = 0$ precludes the possibility that φ is a character. The trick is to let φ be a difference of two characters.

(15.22) LEMMA. *Let $H \subseteq G$ be a Frobenius complement and let $\chi \in \mathrm{Irr}(H)$. Then there exists $\chi^* \in \mathrm{Irr}(G)$ such that*

  a. $\chi_H^* = \chi$ *and*
  b. $\ker(\chi^*)$ *contains all elements of G that are not conjugate to elements of H.*

Proof. If $\chi = 1_H$, the principal character (with constant value one), we can take $\chi^* = 1_G$. (Note that $\ker(1_G) = G$, and so (b) holds in this case.) Suppose, then, that $\chi \neq 1_H$ and define $\varphi = \chi - \chi(1)1_H \in \mathrm{cf}(H)$.

Since φ is a difference of characters of H, it follows that φ^G is a difference of characters of G by Corollary 15.20, and therefore φ^G is a $\mathbb{Z}$-linear combination of $\mathrm{Irr}(G)$. Also, $[\varphi^G, 1_G] = [\varphi, 1_H] = -\chi(1)$ by Frobenius reciprocity, and thus we can write $\varphi^G = \chi^* - \chi(1)1_G$, where $\chi^* \in \mathrm{cf}(G)$ is a $\mathbb{Z}$-linear combination of nonprincipal irreducible characters of G. This yields

$$[\varphi^G, \varphi^G] = [\chi^*, \chi^*] + \chi(1)^2.$$

Now $\varphi(1) = 0$, and thus by Lemma 15.21(b) we have

$$[\varphi^G, \varphi^G] = [\varphi, \varphi] = [\chi, \chi] + \chi(1)^2 = 1 + \chi(1)^2 .$$

Comparison of these two equations yields $[\chi^*, \chi^*] = 1$.
    If we write

$$\chi^* = \sum_{\psi \in \mathrm{Irr}(G)} a_\psi \psi$$

with $a_\psi \in \mathbb{Z}$, then $1 = [\chi^*, \chi^*] = \sum a_\psi^2$, and it follows that $\chi^* = \pm \psi$ for some character $\psi \in \mathrm{Irr}(G)$.
    By Lemma 15.21(a) we have $(\varphi^G)_H = \varphi$ and thus

$$(\chi^* - \chi(1)1_G)_H = \chi - \chi(1)1_H .$$

From this we see that $(\chi^*)_H = \chi$. In particular, $\chi^*(1) = \chi(1) > 0$, and we cannot have $\chi^* = -\psi$ above. Thus $\chi^* = \psi \in \mathrm{Irr}(G)$.
    To prove (b), let $g \in G$ with g not conjugate to any element of H. By the definition of an induced class function, it follows that

$$0 = \varphi^G(g) = \chi^*(g) - \chi(1)$$

and thus

$$\chi^*(g) = \chi(1) = \chi^*(1),$$

and so $g \in \ker(\chi^*)$, completing the proof.   ∎

We can now present the proof of Frobenius's theorem.

Proof of Theorem 15.1.   Let $N \subseteq G$ be the subset defined by

$$N = \left(G - \bigcup_{g \in G} H^g \right) \cup \{1\}.$$

For each $\chi \in \mathrm{Irr}(H)$, let $\chi^* \in \mathrm{Irr}(G)$ be as in Lemma 15.22. By Lemma 15.22(b) we have $N \subseteq \ker(\chi^*)$ for all $\chi \in \mathrm{Irr}(H)$.
    Let $M = \bigcap \ker(\chi^*)$, where χ runs over $\mathrm{Irr}(H)$. Then M is a normal subgroup of G containing N. Also, since $\chi_H^* = \chi$ by Lemma 15.22(a), we have $M \cap H = \bigcap \ker(\chi) = 1$ from Corollary 15.18. Since $M \cap H = 1$ and $M \lhd G$, we have $M \cap H^g = 1$ for all $g \in G$, and thus $M \subseteq N$. It follows that $M = N$, and so N is a normal subgroup of G.
    To complete the proof, we need to show that $NH = G$. For this purpose, we compute $|N|$. We have $\mathbf{N}_G(H) = H$, and so H has $|G : H|$ distinct conjugates in G and, except for the identity element, these are disjoint. It follows that

$$|N| = |G| - |\bigcup_{g \in G}(H^g - 1)| = |G| - |G : H|(|H| - 1) = |G : H|.$$

Therefore, since $N \cap H = 1$, we have

$$|NH| = |N||H| = |G : H||H| = |G|,$$

and so $NH = G$, as required.   ∎

(15.23) COROLLARY. *Let G act transitively on some finite set Ω and assume that only the identity in G fixes as many as two elements of Ω. Let*

$$N = \{g \in G \mid g \text{ fixes no element of } \Omega\} \cup \{1\}.$$

Then N is a subgroup of G that is transitive on Ω.

Proof. Let $\alpha \in \Omega$ and let $H = G_\alpha$, the point stabilizer. If $g \in G - H$, then $\alpha \cdot g \neq \alpha$, and so every element of $H \cap H^g$ fixes both α and $\alpha \cdot g$ and hence $H \cap H^g = 1$. In other words, H is a Frobenius complement in G. Furthermore, N is exactly the set of those elements of G that are not conjugate to any nonidentity element of H. It follows by Theorem 15.1 that N is a subgroup of G and $HN = G$. In particular, since H stabilizes α, we see that the N-orbit of α is the whole G-orbit, Ω.   ∎

Problems

15.1 Let $N \triangleleft G$ and let $\psi \in \text{Char}(G/N)$. Define $\chi : G \to \mathbb{C}$ by $\chi(g) = \psi(Ng)$.
    a. Show that $\chi \in \text{Char}(G)$.
    b. Show that $\chi \in \text{Irr}(G)$ iff $\psi \in \text{Irr}(G/N)$.

15.2 Let $N \triangleleft G$ and $\chi \in \text{Char}(G)$ and assume $N \subseteq \ker(\chi)$. Show that there exists $\psi \in \text{Irr}(G/N)$ such that $\chi(g) = \psi(Ng)$ for all $g \in G$.

15.3 Let $N \triangleleft G$ and write $\mathcal{N} = \{\chi \in \text{Irr}(G) \mid N \subseteq \ker(\chi)\}$.
    a. Show that $N = \bigcap_{\chi \in \mathcal{N}} \ker(\chi)$.
    b. Show that $|G : N| = \sum_{\chi \in \mathcal{N}} \chi(1)^2$.

15.4 Show that $\mathbf{Z}(G) = \bigcap_{\chi \in \text{Irr}(G)} \mathbf{Z}(\chi)$.

    HINT: To show "$\subseteq$," use the fact that $\chi \in \text{Irr}(G)$ is afforded by a representation that maps onto $M_n(\mathbb{C})$, where $n = \chi(1)$.

15.5 Prove that G is abelian iff every character $\chi \in \text{Irr}(G)$ is linear.

15.6 Prove that the number of linear characters of a group G is equal to $|G : G'|$.

15.7 If G is abelian and $\chi \in \text{Char}(G)$, show that $[\chi, \chi] \geq \chi(1)$.

15.8 Let D be a dihedral group of order $2n$, where n is odd. Show that D has exactly $(n + 3)/2$ conjugacy classes, and conclude that $\text{Irr}(D)$ consists of two linear characters and $(n - 1)/2$ characters of degree 2.

15.9 Suppose $H \subseteq G$ and $\theta \in \text{Char}(H)$. Show that $\mathbf{Z}(\theta^G) \subseteq H$.

15.10 Suppose $H, K \subseteq G$ with $HK = G$. If $\varphi \in \text{cf}(H)$, show that

$$(\varphi^G)_K = (\varphi_{H \cap K})^K.$$

15.11 Let $H \subseteq G$ and suppose that $H\mathbf{Z}(G) = G$. If $\chi \in \text{Irr}(G)$, show that the restriction χ_H is irreducible.

15.12 Let $H \subseteq G$ with $|G : H| = n$ and suppose that $\chi \in \text{Char}(G)$.

a. Show that $[\chi, \chi] \geq [\chi_H, \chi_H]/n$.
b. If H is abelian and $\chi \in \text{Irr}(G)$, show that $\chi(1) \leq n$.
c. If equality holds in part (b), show that $H \triangleleft G$.

HINT: Use Problem 15.7 for part (b).

15.13 Suppose H is an abelian subgroup of G. Show that G has at least $|H|^2/|G|$ conjugacy classes.

PART TWO

Commutative Algebra

CHAPTER SIXTEEN

Polynomial Rings, PIDs, and UFDs

16A

The purpose of this, our first "commutative" chapter is to present some of the basic ring theoretic ideas that will be useful both for our study of field extensions and for our development of some of the deeper parts of ring theory. As usual, our use of the word "ring" includes the assumption of the existence of a unity element 1. Since we shall be working exclusively with commutative rings from now on, the commutativity assumption is not usually made explicit.

Let R be a ring. If $a \in R$ is any element, we write (a) to denote the ideal $\{ra \mid r \in R\}$, the *principal ideal* generated by a. It is trivial to check that (a) is an ideal and that it is proper iff a is not a unit (an invertible element) in R. (Note that in this commutative situation, we need not consider left or right ideals or left or right inverses.)

If every ideal of R is principal, then we say that R is a *principal ideal ring*, usually abbreviated PIR. For example, the ring $\mathbb{Z}$ of ordinary integers is a PIR since the additive group of $\mathbb{Z}$ is cyclic and so every subgroup is cyclic. (In $\mathbb{Z}$, the additive subgroups are exactly the ideals, and $(a) = \langle a \rangle$ for all $a \in \mathbb{Z}$.)

We say that R is an *integral domain* (or just a *domain*) if $ab \neq 0$ for nonzero elements $a, b \in R$. (We require also that R have some nonzero elements; we do not consider the trivial ring $R = \{0\}$ to be a domain.) Again, $\mathbb{Z}$ is an example, and we say that $\mathbb{Z}$ is a PID, a principal ideal domain. Other examples of domains and PIDs are polynomial rings over domains and, in particular, over fields. We digress to consider these important rings.

Let R be any ring and let "X" be some symbol (which is not the name of any element of R). An expression of the form

$$f(X) = a_n X^n + a_{n-1} X^{n-1} + \cdots + a_1 X + a_0$$

(where the *coefficients* a_i lie in R) is called a *polynomial over R* in the *indeterminate* X. What is this, really? We should not think of $f(X)$ as a function because it has no particular domain; we are not willing to specify in advance the set of possible values that we might choose to "plug in" for the indeterminate. (They might be elements of R, but they could also be, for instance, elements of some ring that contains R.)

In its purest form, the polynomial $f(X)$ can be thought of as an ordered pair whose first (and most important) component is the sequence $(a_0, a_1, a_2, \ldots)$ of coefficients and whose second component is the symbol "X" that names the indeterminate. We think of the coefficient sequence (a_i) as being infinite but having at most finitely many (and possibly no) nonzero terms. In other words, there exists some positive integer n such that $a_i = 0$ when $i > n$.

As an example, consider the case $R = \mathbb{Z}/(3)$, the integers mod 3. The elements of this ring are the three cosets (3), $(3)+1$, and $(3)+2$, usually denoted simply 0, 1, and 2. The polynomial $f(X) = X^3 - X + 1$ "over" R (that is, with coefficients in R) corresponds to the coefficient sequence $(1, 2, 0, 1, 0, 0, 0, \ldots)$.

If we substitute the elements of R (one at a time) for X in the formula for $f(X)$, we easily compute that $f(0) = 1$, $f(1) = 1$, and $f(2) = 1$, and so our polynomial $f(X)$ defines the same function on R as does the "constant" polynomial $g(X) = 1$ with coefficient sequence $(1, 0, 0, 0, \ldots)$. Nevertheless, the polynomials f and g have different coefficient sequences, and so they are certainly different polynomials. Note that if S is the noncommutative ring of 2×2 matrices over R, and if we identify R with the scalar matrices (the subring of S consisting of the matrices rI, where $r \in R$ and I is the identity matrix), then

$$f\left(\begin{bmatrix} 0 & 1 \\ 0 & 0 \end{bmatrix}\right) = \begin{bmatrix} 1 & 2 \\ 0 & 1 \end{bmatrix} \quad \text{and} \quad g\left(\begin{bmatrix} 0 & 1 \\ 0 & 0 \end{bmatrix}\right) = \begin{bmatrix} 1 & 0 \\ 0 & 1 \end{bmatrix},$$

and so f and g do *not* define the same function on S.

The set of all polynomials over R in the indeterminate X is denoted $R[X]$. This set can be made into a ring quite easily. Polynomials are added by adding the coefficients of like powers of X, and they are multiplied by using the rule that $X^i X^j = X^{i+j}$ and the distributive law.

We could do this more formally as follows. If $f, g \in R[X]$ have coefficient sequences (a_i) and (b_i), respectively, then $f+g$ is defined by the coefficient sequence (c_j), where $c_j = a_j + b_j$ for all $j \geq 0$. Also, fg is defined by the coefficient sequence (d_j), where

$$d_j = \sum_{i=0}^{j} a_i b_{j-i}.$$

One should observe that these sum and product sequences have just finitely many nonzero terms, and the axioms that define a ring should be checked. All of this is utterly routine, and so we omit the proof that $R[X]$ is a ring.

Next we make precise the process of evaluating a polynomial $f \in R[X]$ at some element $s \in S$ for some ring $S \supseteq R$. If our polynomial is $f(X) = \sum a_i X^i$, we clearly want to define $f(s) = \sum a_i s^i$. Unlike the sum defining $f(X)$ (which is only formal, in that nothing is actually being added), the sum defining $f(s)$ is genuinely addition in the ring S. It is crucial, therefore, that there are at most finitely many nonzero terms.

If our polynomial is $f(X) = X$, we would like to have $f(s) = s$, and this necessitates $1s = s$, where "1" denotes the unity of R. We must therefore require that the unity of R is the unity of S, in other words, that S is a *unitary overring* of R.

If $S \supseteq R$ is a unitary overring, then $f \in R[X]$ defines a function $S \to S$ by $s \mapsto f(s)$. This function is (confusingly) also called f. If we start with the polynomial $f(X) = X$, then the function f is the identity map on S.

Polynomials with all coefficients $a_i = 0$ for $i > 0$ are called *constant* polynomials since they define constant functions on every unitary overring $S \supseteq R$. Note that the subset of $R[X]$ consisting of the constant polynomials forms a subring of $R[X]$ that is naturally isomorphic to R, and it is usually identified with R. The element $1 \in R$ corresponds to the constant polynomial $1 \in R[X]$, and this is clearly the unity element of $R[X]$. In short, $R[X]$ is a unitary overring of R.

When we started this rather extended (and perhaps pedantic) discussion of the definition and trivial properties of polynomials, the symbol "X" had no particular meaning, but now it is a certain element of the ring $R[X]$; it is the polynomial with coefficient sequence $(0, 1, 0, 0, 0, \ldots)$. What do we mean by X^n? On the face of it, there are two possibilities: either X^n is the polynomial with coefficient sequence $a_n = 1$ and $a_i = 0$ for $i \neq n$, or X^n is the result of multiplying X by itself n times in the ring $R[X]$. Fortunately (but not very surprisingly), the result of the multiplication is precisely the first interpretation of X^n, and so there is no ambiguity. Similarly, if $f \in R[X]$, we can evaluate f at the element X of the unitary overring $R[X] \supseteq R$, and we find that $f(X) = f$. Another way of saying this is that if we write $f(X) = a_n X^n + a_{n-1} X^{n-1} + \cdots + a_1 X + a_0$, then we can view this formula not only as not being merely formal, but also as an expression involving the elements $a_0, a_1, \ldots, a_n$ and X of an actual ring, $R[X]$, combined using addition and multiplication in that ring.

(16.1) LEMMA. *Let $S \supseteq R$ be a unitary overring and fix $s \in S$. Then evaluation at s defines a ring homomorphism $R[X] \to S$.*

Proof. Let $f, g \in R[X]$. We need to show that

$$(f + g)(s) = f(s) + g(s) \quad \text{and} \quad (fg)(s) = f(s)g(s).$$

If the coefficient sequences for f and g are (a_i) and (b_i), respectively, we must show that

$$\sum_{j \geq 0} (a_j + b_j) s^j = \left(\sum_{j \geq 0} a_j s^j \right) + \left(\sum_{j \geq 0} b_j s^j \right)$$

and

$$\sum_{j \geq 0} \left(\sum_{i=0}^{j} a_i b_{j-1} \right) s^j = \left(\sum_{i \geq 0} a_i s^i \right) \left(\sum_{k \geq 0} b_k s^k \right).$$

The first equation is a triviality and the second is very nearly so, since the right side is equal to

$$\sum_j \left(\sum_{i+k=j} a_i b_k \right) s^{i+k} = \sum_j \left(\sum_{i=0}^{j} a_i b_{j-1} \right) s^j$$

because the coefficients b_k commute with s.   ∎

The homomorphism $R[X] \to S$ of Lemma 16.1 is called the *evaluation* map at s. Its image, $\{f(s) \mid f \in R[X]\}$, is a subring of S containing R (because $R[X]$ contains the constant polynomials), and this subring is denoted $R[s]$. (Note that this gives a second possible meaning to "$R[X]$" since we could take $S = R[X]$ and $s = X$. In this case, "evaluate at s" is the identity map and so $R[s] = S$ and both meanings of $R[X]$ agree.)

We need a few more definitions. If $f \in R[X]$ is not the zero polynomial and its coefficient sequence is (a_i), then there is some largest subscript n such that $a_n \neq 0$. This nonnegative integer is the *degree* of f and is denoted $\deg(f)$. If $\deg(f) = n$, then the (nonzero) coefficient a_n is the *leading coefficient* of f. If the leading coefficient is 1, the polynomial is said to be *monic*. (Observe that the zero polynomial has neither a degree nor a leading coefficient.)

(16.2) LEMMA.   *Let $f, g \in R[X]$ have degrees m and n and leading coefficients a and b, respectively. If $ab \neq 0$, then $\deg(fg) = m + n$ and the leading coefficient of fg is ab. In particular:*

a. *If R is a domain, then so is $R[X]$.*
b. *If f and g are monic, then so is fg.*

Proof.   From the definition of polynomial multiplication, we see that the coefficient of X^{m+n} in fg is ab and that the coefficients of all higher powers of X vanish. Everything asserted follows from this.   ∎

When we started this long digression on polynomial rings, we promised that they would provide us with examples not only of domains but also of PIDs. The key to this is the following familiar lemma.

(16.3) LEMMA (Division algorithm).   *Let $f, g \in R[X]$ with $f \neq 0$. Assume that the leading coefficient of f is a unit in R. Then there exist $r, q \in R[X]$ such that*

$$g = fq + r$$

and either $r = 0$ or $\deg(r) < \deg(f)$.

Proof.   Although easily seen to be true from the elementary process of polynomial long division, this fact deserves a more formal proof.

Write $n = \deg(f)$ and $f(X) = aX^n + \cdots$. If $g = 0$ or $\deg(g) < n$, we can take $q = 0$ and $r = g$, and there is nothing to prove. We assume, therefore, that $\deg(g) = m \geq n$, and working by induction on m, we assume that the result holds if g is replaced by any polynomial of degree less than m.

Let b be the leading coefficient of g and write

$$h(X) = ba^{-1} f(X) X^{m-n}.$$

Observe that the degree and leading coefficient of the polynomial $h(X)$ match those of g. It follows that $g - h$ involves no power of X as high as X^m, and so either $g - h = 0$ or $\deg(g - h) < m$. Our theorem thus holds for $g - h$, and we can write $g - h = fq + r$ with $r = 0$ or $\deg(r) < n$. Now $g = h + fq + r$, and since h is a multiple of f, the result follows. ∎

An important case of Lemma 16.3 is where the coefficient ring is a field. (Recall that that means that every nonzero element is a unit.)   In that case, the only requirement on f in the lemma is that $f \neq 0$.

(16.4) THEOREM.   *Let F be a field. Then $F[X]$ is a PID.*

Proof.   Let $A \subseteq F[X]$ be any ideal. To show that A is principal, we may assume that $A \neq (0)$. Then A contains some nonzero elements, and we can choose $f \in A$ with $\deg(f)$ as small as possible.

Since $f \in A$ and A is an ideal, we have $(f) \subseteq A$ and we claim that in fact $(f) = A$. To see this, let $g \in A$ and write $g = fq + r$ by Lemma 16.3, where either $r = 0$ or $\deg(r) < \deg(f)$. Now $r = g - fq$ is an element of A, and so by our choice of f, it cannot be that $\deg(r) < \deg(f)$. The only alternative, therefore, is that $r = 0$ and hence $g = fq \in (f)$, as required. Therefore $F[X]$ is a PIR.

Since the field F is surely a domain, $F[X]$ is a domain by Lemma 16.2, and we are done. ∎

We mention that the condition in Theorem 16.4 that the coefficient ring is a field is essential. For instance, $\mathbb{Z}[X]$ is not a PID. To see this, let $A = \{f \in \mathbb{Z}[X] \mid f(0) \text{ is even}\}$. It is clear that A is an ideal containing the polynomial X and the constant polynomial 2. If $A = (f)$, then $2 = fg$ for some $g \in F[X]$, and so $\deg(f) = 0$, and this forces f to be one of the constants ± 1 or ± 2. If $f = \pm 1$, then $(f) = \mathbb{Z}[X] \neq A$, and if $f = \pm 2$, then $X \notin (f)$ and again $(f) \neq A$.

16B

Now let R be any domain. A nonunit $a \in R$ is said to be *irreducible* if whenever we write $a = bc$ with $b, c \in R$, we have that either b or c is a unit. In the ring $\mathbb{Z}$, for instance, the irreducible elements are precisely the numbers $\pm p$ for prime numbers p. In $F[X]$, where F is a field, the units are the nonzero constant polynomials (the polynomials of degree 0). Over a field, therefore, the irreducible polynomials are exactly those polynomials of positive degree that cannot be factored into two positive degree polynomials.

Consider the subset of our domain R consisting of elements that either are irreducible or are products of some (finite) number of irreducible elements. We clearly have not included zero or any unit in this set. Is anything else missing? In general, there may be, but for many interesting domains, it is true that every nonzero nonunit is a product of irreducible elements. (This includes the possibility that it is itself irreducible.)

To explore this further, we introduce the notation $a|b$ for elements $a, b \in R$ to mean that a *divides* b. In other words, $b = ar$ for some element $r \in R$ or, equivalently, $b \in (a)$. Note that $a|0$ for all $a \in R$ but that $0|b$ only when $b = 0$.

(16.5) LEMMA.   *Let $a, b, x, y \in R$, where R is a domain.*

    a. *If $ax = ay$ with $a \neq 0$, then $x = y$.*
    b. *If $a|b$ and $b|a$, then $b = au$ for some unit $u \in R$.*

Proof.   If $ax = ay$, then $a(x - y) = 0$. Since R is a domain, we see that if $a \neq 0$, then $x - y = 0$ and $x = y$, proving (a). For (b), write $b = au$ and $a = bv$ for some $u, v \in R$. Substitution yields $a = auv$, and so if $a \neq 0$, we have $1 = uv$ by part (a), and (b) holds. If, on the other hand, $a = 0$, then necessarily $b = 0$ and we can take $u = 1$. ∎

We recall that a ring R is "noetherian" if its set of ideals satisfies the ascending chain condition or, equivalently, the maximal condition. (Of course, in this commutative situation, we need not worry about left or right ideals.) By Theorem 11.4 we know that R is noetherian iff every ideal is finitely generated. In particular, a PIR, in which every ideal is generated by one element, is certainly noetherian.

(16.6) LEMMA.   *Let R be a noetherian domain. Then every element of R is zero, a unit, or a product of (a finite number of) irreducible elements.*

Proof.   Let $X \subseteq R$ be the set of all nonzero nonunits that are not products of irreducible elements. Assuming that the lemma is false, we know X is nonempty, and so the set of ideals $\{(x) \mid x \in X\}$ is nonempty. By the maximal condition, this set contains a maximal element (m) with $m \in X$.

Since $m \in X$, it is not irreducible, and so we can write $m = ab$, where neither a nor b is a unit, and since $m \neq 0$, neither a nor b is zero. If both a and b were products of irreducible elements, then $m = ab$ would be such a product, too, and this contradicts $m \in X$. We may assume, therefore, that (say) a is not a product of irreducible elements and so $a \in X$.

Now $a|m$ and so $m \in (a)$ and $(m) \subseteq (a)$. By the maximality of (m), we have equality, and so $a \in (a) = (m)$ and $m|a$. By Lemma 16.5(b) it follows that $m = au$ for some unit u. Since $a \neq 0$ and $m = ab$, Lemma 16.5(a) yields that $b = u$, and this is a contradiction. ∎

Next we investigate the question of the uniqueness of the factorization of a nonzero nonunit into irreducible factors. True uniqueness of factorization fails even in the ring $\mathbb{Z}$ of ordinary integers. For example, we have $2^2 = 4 = (-2)^2$, and 2 and -2 are irreducible. We must settle for a weaker form of uniqueness: uniqueness up to unit multiples.

(16.7) DEFINITION.   A domain R is a *unique factorization domain* (abbreviated UFD) provided that

    i.   every nonzero nonunit of R is a product of irreducible elements, and

    ii.   if $a_1 a_2 \cdots a_n = b_1 b_2 \cdots b_m$, where each of the elements a_i and b_j is irreducible, then $n = m$ and (after renumbering the b_j if necessary) we have $b_i = u_i a_i$, where the u_i are units of R.

We shall show that every PID is a UFD and that if R is a UFD, then so is the polynomial ring $R[X]$. Although every noetherian domain satisfies condition (i) by Lemma 16.6, it is certainly not true that every such domain is a UFD.

To investigate when condition (ii) of Definition 16.7 holds, we introduce the notion of a "prime" element of a domain. To put this into context, we first define prime ideals. An ideal P of R (not necessarily a domain) is a *prime* ideal if $P < R$ and whenever $ab \in P$, either $a \in P$ or $b \in P$.

Observe that R is a domain iff the ideal (0) is prime and that, more generally, if A is any ideal of R, then the ring R/A is a domain iff A is prime. We know that if M is a maximal ideal of R, then R/M has no nonzero proper ideals and so is a field and hence is a domain. This shows that maximal ideals are always prime.

Now suppose that R is a domain. An element $\pi \in R$ is a *prime* element if $\pi \neq 0$ and (π) is a prime ideal. Equivalently, an element $\pi \in R$ is prime if it is a nonzero nonunit such that whenever $\pi \mid ab$ for $a, b \in R$, we have that either $\pi \mid a$ or $\pi \mid b$.

We collect a few facts about prime elements.

(16.8) LEMMA.   *Let R be a domain.*

    a.   *Every prime element of R is irreducible.*

    b.   *If $\pi \in R$ is prime and $\pi \mid b_1 b_2 \cdots b_m$, then π divides one of the factors b_i.*

    c.   *Suppose $a_1 a_2 \cdots a_n = u b_1 b_2 \cdots b_m$, where each a_i is prime, each b_j is irreducible, and u is a unit. Then $n = m$ and, after possible renumbering of the b_i, we have $b_i = u_i a_i$ for some units u_i.*

Proof.   Let $\pi \in R$ be prime. Certainly, π is a nonunit, and so to prove (a), it suffices to assume $\pi = xy$ and to show that either x or y is a unit. We certainly have $\pi \mid xy$ and so (say) $\pi \mid x$ since π is prime. Also, $x \mid \pi$, and hence by Lemma 16.5(b) we have $\pi = xu$ for some unit u. Since also $\pi = xy$ and $x \neq 0$ (because $\pi \neq 0$), Lemma 16.5(a) yields that $y = u$, a unit, as desired.

To prove (b), we work by induction on n. The case $n = 1$ is a triviality (and the case $n = 2$ is the definition of "prime"). We may assume that $\pi \nmid b_n$ and so $n > 1$. Since π is prime and it divides the product $(b_1 b_2 \cdots b_{n-1}) \cdot b_n$, we deduce that $\pi \mid b_1 b_2 \cdots b_{n-1}$, and the inductive hypothesis yields the result.

We also use induction on n to prove (c). In the case that $n = 0$, we have $1 = u b_1 b_2 \cdots b_m$, and so each b_i is a unit. Since the b_i are irreducible, they cannot be units, and this forces $m = 0 = n$, starting our induction.

Now assume that $n > 0$. Since a_1 is prime and divides $u b_1 b_2 \cdots b_m$, and since a_1 cannot divide the unit u, it follows by part (b) that a_1 divides some b_i.

By renumbering if necessary, we can write $b_1 = a_1 u_1$ for some element $u_1 \in R$. Since b_1 is irreducible and a_1 is not a unit, this forces u_1 to be a unit. We now have

$$a_1 a_2 \cdots a_n = u u_1 a_1 b_2 b_3 \cdots b_m \ .$$

We can cancel a_1 by Lemma 16.5(a), and the inductive hypothesis then yields the result.   ∎

(16.9) LEMMA.   *Let R be a PID and let $a \in R$. The following are equivalent:*

i.   *a is prime.*
ii.   *a is irreducible.*
iii.   *(a) is a maximal ideal.*

Proof.   That (i) implies (ii) follows by Lemma 16.8(a). Now assume (ii), so that a is irreducible. Since a is not a unit, we have $1 \notin (a)$ and hence $(a) < R$. To show that (a) is a maximal ideal, suppose that I is some ideal with $(a) \subseteq I \subseteq R$. We must show that either $I = (a)$ or $I = R$.

Since R is a PID, we can write $I = (b)$ for some element $b \in R$. Because $a \in (a) \subseteq I = (b)$, we have $a = bx$ for some $x \in R$, and the irreducibility of a forces either b or x to be a unit. If b is a unit, then $I = (b) = R$, and if x is a unit, then $b = ax^{-1} \in (a)$ and so $(b) \subseteq (a)$. In the latter case, $I = (b) = (a)$. This completes the proof that (ii) implies (iii).

Finally, assume (iii). Since (a) is a maximal ideal, it is certainly prime. Therefore a is a prime element and (i) holds.   ∎

(16.10) THEOREM.   *Let R be a PID. Then R is a UFD.*

Proof.   Since R is noetherian, condition (i) of Definition 16.7 holds by Lemma 16.6. To prove condition (ii) of Definition 16.7, suppose that

$$a_1 a_2 \cdots a_n = b_1 b_2 \cdots b_m \ ,$$

where each a_i and b_j is irreducible. By Lemma 16.9 each a_i is prime and so Lemma 16.8(c) applies (with $u = 1$). We conclude that $m = n$ and that (after possible renumbering) $b_i = u_i a_i$, where the u_i are units. This is the required condition.   ∎

The key to our proof that a PID is necessarily a UFD was the part of Lemma 16.9 that asserts that in a PID, every irreducible element is prime. In fact, it is easy to see that this is true in any UFD.

(16.11) LEMMA.   *Let $a \in R$ be irreducible, where R is a UFD. Then a is prime.*

Proof.   We certainly know that a is a nonzero nonunit, and so it suffices to assume that $a | xy$ for $x, y \in R$ and to show that either $a | x$ or $a | y$. We have $ab = xy$ and we may assume that neither x nor y is zero or is a unit. Also, b is nonzero and, furthermore, b is a nonunit, since otherwise we would have $a = x(yb^{-1})$

and this would contradict the irreducibility of a. (Note that neither x nor yb^{-1} is a unit.)

Each of x, y, and b can be written as a product of irreducible factors x_i, y_j, and b_k, respectively. The equation $ab = xy$ thus yields

$$ab_1b_2 \cdots b_n = x_1x_2 \cdots x_s y_1 y_2 \cdots y_t \,,$$

and since a is one of the irreducible factors on the left, there is a unit u such that au is equal to one of the x_i or one of the y_j. In particular, au divides either x or y, and hence a divides one of these elements, as required.   ∎

Because of Lemmas 16.11 and 16.8(a), the words "prime" and "irreducible" are often used interchangeably about elements of a UFD. If R is a UFD and $a \in R$ is any nonzero element, we can unambiguously discuss the "number of prime factors" of a. If a is a unit, this is zero, and otherwise it is the number of (not necessarily distinct) primes (or irreducibles) that occur in a factorization of a as in condition (i) of Definition 16.7. Also, we refer to two elements $a, b \in R$ as being *coprime* or *relatively prime* if they are not both zero and they have no common prime divisor.

16C

The converse of Theorem 16.10 is false; not every UFD is a PID. Perhaps the simplest counterexample is the ring $\mathbb{Z}[X]$, which we have already observed is not a PID. That it is a UFD follows from the following general result.

(16.12) THEOREM.   *Let R be a UFD. Then $R[X]$ is also a UFD.*

Throughout the following discussion leading to a proof of Theorem 16.12, we assume that R is a domain. We first identify the units of $R[X]$. As usual, we view R as a subring of $R[X]$ and we observe that every unit of R is still a unit in $R[X]$. Also, these are the only units in $R[X]$, since no polynomial of positive degree can possibly have an inverse. (This is because R is a domain, and so the degree of a product is the sum of the degrees of the factors.)

Next we observe that the irreducible elements of R remain irreducible in $R[X]$. To see this, observe that if $a \in R$ is irreducible, then it is a nonzero nonunit, and if $a = xy$ for $x, y \in R[X]$, then necessarily both x and y have degree 0 and hence lie in R. One of them is therefore a unit, and hence a is irreducible in $R[X]$.

There exist irreducible elements of positive degree in $R[X]$ also. To describe these, we say that a nonconstant polynomial $f \in R[X]$ is *primitive* if the only elements of R that divide f in $R[X]$ are the units of R. Since an element $a \in R$ divides $f \in R[X]$ precisely when a divides each of the coefficients of f, we see that the primitive polynomials are exactly those nonconstant polynomials for which the only common divisors of the coefficients are the units of R. In particular, every monic polynomial is primitive. Taking $R = \mathbb{Z}$, we see, for example, that the polynomials $4X^2 - 1$ and $6X^2 + 10X + 15$ are primitive, whereas $4X - 2$ is not. If R is a field, on the other hand, then every nonconstant polynomial is primitive.

(16.13) LEMMA   *Let R be a domain and suppose $f \in R[X]$ is primitive. If f cannot be factored into two polynomials of positive degree, then f is an irreducible element of $R[X]$.*

Proof.   Certainly, f is a nonzero nonunit. If $f = xy$, with $x, y \in R[X]$, then by hypothesis one of x or y (say x) has degree 0 and hence lies in R. Since f is primitive and $x \mid f$, we deduce that x is a unit.   ∎

Next we must show that if R is a UFD, then $R[X]$ has enough irreducible elements so that condition (i) of Definition 16.7 holds. In particular, we need to produce primitive polynomials.

(16.14) LEMMA.   *Let R be a UFD and suppose $f \in R[X]$ is nonconstant. Then there exists $a \in R$ and a primitive polynomial $g \in R[X]$ such that $f = ag$.*

Proof.   Write $n = n(f)$ to denote the minimum number of prime factors for the nonzero coefficients of f, and note that $n(f)$ is well defined by unique factorization. Observe that $n = 0$ precisely when f has some unit coefficient, and in this case, f is primitive.

We work by induction on n. If f is primitive (and, in particular, if $n = 0$), we can take $a = 1$ and there is nothing to prove. We may assume, therefore, that there is some nonunit $b \in R$ such that $b \mid f$, and we write $f = bh$ for some polynomial $h \in R[X]$. Since b has a prime factor, we see that $n(h) < n(f)$, and hence by the inductive hypothesis, $h = cg$ for some element $c \in R$ and some primitive polynomial g. The result now follows by taking $a = bc$.   ∎

(16.15) COROLLARY.   *Let R be a UFD. Then every nonzero nonunit of $R[X]$ is a product of irreducible elements, each of which is either a prime of R or a primitive polynomial that cannot be factored into two polynomials of positive degree.*

Proof.   Let $f \in R[X]$ be a nonzero nonunit and write $n = \deg(f)$. We prove by induction on n that f can be written as a product of irreducibles of the specified types. If $n = 0$, then $f \in R$, and hence f is a product of primes of R. Since these are irreducible in $R[X]$, we are done in this case.

Now assume $n > 0$. If $f = gh$, where g and h have positive degree, then each of g and h has degree less than n, and so by the inductive hypothesis, each is a product of irreducibles of the two types. We are done in this case, and so we may assume that f has no such factorization. If f is primitive, it is irreducible and we are done. Finally, if f is imprimitive, we can write $f = ag$, where g is primitive and $a \in R$ is a nonunit. Now g does not factor into two polynomials of positive degree and a is a product of primes of R. The proof is now complete.   ∎

To verify condition (ii) of Definition 16.7, and thereby complete the proof that $R[X]$ is a UFD when R is, it suffices by Lemma 16.8 to show that irreducible elements of each of the two types we have found are actually prime in $R[X]$.

(16.16) THEOREM.   *Let R be a UFD and suppose that $\pi \in R$ is prime. Then π is prime in R[X].*

Proof.   Suppose that $\pi \mid fg$, where $f, g \in R[X]$. We must show that either $\pi \mid f$ or $\pi \mid g$. In other words, we must show either that every coefficient of f is divisible by π or that every coefficient of g is divisible by π.

Let $\overline{R}$ be the ring $R/(\pi)$ and note that the canonical homomorphism $a \mapsto \overline{a}$ of R onto $\overline{R}$ extends naturally to a homomorphism $R[X] \to \overline{R}[X]$ by mapping

$$\sum a_i X^i \mapsto \sum \overline{a}_i X^i .$$

(We omit the trivial proof that this is, in fact, a ring homomorphism.)

Since each coefficient of fg lies in the ideal (π), we have $0 = \overline{fg} = \overline{f}\,\overline{g}$. However, since π is prime, the ideal (π) is prime and so $\overline{R}$ is a domain. It follows that $\overline{R}[X]$ is a domain and hence either $\overline{f} = 0$ or $\overline{g} = 0$. Thus, as required, either all coefficients of f or all coefficients of g are divisible by π.   ∎

To complete the proof of Theorem 16.12, we need the following fact.

(16.17) THEOREM.   *Let R be a domain. Then R is a unitary subring of some field F with the property that every element $\alpha \in F$ is of the form $\alpha = a/b$, where $a, b \in R$.*

A field F as described in Theorem 16.17 is called a *field of fractions* or *quotient field* of R. For example, the rational number field $\mathbb{Q}$ is a field of fractions for $\mathbb{Z}$. It turns out that for every domain R, there is a unique (up to isomorphism) field of fractions. The process used to construct the field of fractions of a domain, called "localization," is much more widely applicable in ring theory than merely in the proof of Theorem 16.17. We devote the final section of this chapter to the study of localization, but for now, let us assume Theorem 16.17 and complete our discussion of UFDs.

(16.18) LEMMA.   *Let R be a UFD and suppose that $f, g \in R[X]$ and that f is primitive. Let F be a field of fractions for R and suppose that $f \mid g$ in F[X]. Then $f \mid g$ in R[X].*

Proof.   Write $g = fh$, where $h \in F[X]$. Each coefficient of h has the form a/b for suitable elements $a, b \in R$ with $b \neq 0$. If we multiply h by the product of the denominators of its coefficients, we get some polynomial in $R[X]$. It follows that we can find a nonzero element $c \in R$ such that $cg = fk$, where $k \in R[X]$. Choose c with this property such that the number n of prime factors of c is as small as possible. If $n = 0$ so that c is a unit, we have $c^{-1}k \in R[X]$ and thus $f \mid g$ in $R[X]$, as required. We complete the proof by obtaining a contradiction if $n > 0$.

Suppose $\pi \in R$ is prime and $\pi \mid c$. Then $\pi \mid fk$. By Theorem 16.16, however, we know that π is prime in $R[X]$, and it follows that either $\pi \mid f$ or $\pi \mid k$. Since

f is primitive, the latter must hold, and we write $k_0 = (1/\pi)k \in R[X]$. Since $(c/\pi)g = fk_0$ and c/π has fewer than n prime factors, this is our desired contradiction.   ∎

(16.19) LEMMA (Gauss). *Let R be a UFD with field of fractions F. Suppose $0 \neq f \in R[X]$ and $f = gh$ for some $g, h \in F[X]$. Then $f = g_0 h_0$ for polynomials $g_0, h_0 \in R[X]$, where $g_0 = \alpha g$ and $h_0 = \beta h$ for suitable elements $\alpha, \beta \in F$. In particular, the degrees of g_0 and h_0 are equal to those of g and h, respectively.*

For example, consider the polynomial $f(X) = X^2 - X - 6 \in \mathbb{Z}[X]$. Imagine that we do not know how to factor this polynomial in $\mathbb{Z}[X]$, but that somehow we discover the factorization

$$f(X) = \left(\frac{2}{3}X - 2\right)\left(\frac{3}{2}X + 3\right)$$

in $\mathbb{Q}[X]$. What Gauss's lemma tells us is that there must exist some corresponding factorization in $\mathbb{Z}[X]$. In fact, if we multiply the first factor by $\alpha = 3/2$ and the second by $\beta = 2/3$, we obtain polynomials in $\mathbb{Z}[X]$ whose product is f. Note that in the situation of Lemma 16.19, we always have $\alpha\beta = 1$.

Proof of Lemma 16.19. Note that g and h are nonzero. Multiplication of g by the product b of the denominators of its coefficients yields the polynomial $bg \in R[X]$. By Lemma 16.14 we can find a primitive polynomial $g_0 \in R[X]$ and an element $a \in R$ such that $bg = ag_0$. Write $\alpha = b/a \in F$ and $\beta = a/b \in F$, and note that $f = (\alpha g)(\beta h)$ and that $\alpha g = g_0$ is a primitive polynomial in $R[X]$. Since g_0 divides f in $F[X]$, Lemma 16.18 implies that g_0 divides f in $R[X]$. Thus $h_0 = \beta h = f/g_0 \in R[X]$, and the proof is complete.   ∎

Proof of Theorem 16.12. What remains is to show that if $f \in R[X]$ is primitive and does not factor into two polynomials with positive degree, then f is prime. Let f be such a polynomial, and suppose $f \mid kl$ with $k, l \in R[X]$.

Let F be a field of fractions for R. We claim that f is irreducible in $F[X]$. If this were false, we could write $f = gh$ with $g, h \in F[X]$, each of positive degree. By Gauss's lemma (16.19), then, $f = g_0 h_0$ with $g_0, h_0 \in R[X]$, and we have $\deg(g_0) = \deg(g) > 0$ and $\deg(h_0) = \deg(h) > 0$. This contradicts the assumptions on f.

Now $F[X]$ is a PID and f is irreducible and hence prime in $F[X]$, and so f must divide one of k or l (say k) in $F[X]$. Since f is primitive, however, we deduce from Lemma 16.18 that f divides k in $R[X]$, as required.   ∎

Polynomial rings in several indeterminates over fields (or UFDs) are other examples of UFDs. If R is any ring and X and Y are two indeterminates, we can define the polynomial ring $R[X, Y]$ as the set of all formal linear combinations of "monomials" $X^i Y^j$ with coefficients from R. (We allow, of course, at most finitely many nonzero coefficients.) Polynomials in X and Y are added and multiplied in the obvious ways, making $R[X, Y]$ into a ring. Note that any $f \in R[X, Y]$ can be viewed as a polynomial in Y with coefficients from $R[X]$. For instance,

$$aX^2Y + bXY^2 + cX + dY + e = (bX)Y^2 + (aX^2 + d)Y + (cX + e).$$

It follows that $R[X, Y] \cong (R[X])[Y]$. Similarly, we can work with polynomial rings in any number of indeterminates.

(16.20) COROLLARY.   *Let R be a UFD. Then $R[X_1, X_2, \ldots, X_n]$ is a UFD.*

Proof.   Use induction on n, starting with the trivial case where $n = 0$. We have that $R[X_1, \ldots, X_n] \cong S[X_n]$, where $S = R[X_1, \ldots, X_{n-1}]$. By the inductive hypothesis, S is a UFD and the result follows by Theorem 16.12.   ∎

We should point out that polynomial rings in two or more indeterminates over a field are never PIDs. In fact, it is easy to see that the ideal generated by the two polynomials X and Y is not principal.

Before we conclude our study of UFDs, we mention another application of Gauss's lemma (16.19). This is the Eisenstein criterion for proving the irreducibility of certain polynomials in $F[X]$, where F is a field (such as $\mathbb{Q}$) that is a field of fractions of a UFD.

(16.21) THEOREM (Eisenstein criterion).   *Let R be a UFD with fraction field F and let $f \in R[X]$. Write*

$$f(X) = a_n X^n + a_{n-1} X^{n-1} + \cdots + a_1 X + a_0.$$

Then f is irreducible in $F[X]$ if there exists a prime $\pi \in R$ for which all of the following hold:

i. $\pi \nmid a_n.$
ii. $\pi \mid a_i$ for $0 \le i < n.$
iii. $\pi^2 \nmid a_0.$

For example, taking $F = \mathbb{Q}$ and $R = \mathbb{Z}$, we see that the polynomial $5X^3 + 18X + 12$ is irreducible in $\mathbb{Q}[X]$, using the prime $\pi = 3$.

Proof of Theorem 16.21.   If f is not irreducible in $F[X]$, then by Lemma 16.19 we can write $f = gh$, where $g, h \in R[X]$ and each of g and h has positive degree.

Now let $\overline{R} = R/(\pi)$ and extend the natural homomorphism $a \mapsto \overline{a}$ to a ring homomorphism $f \mapsto \overline{f}$ of $R[X] \to \overline{R}[X]$. By conditions (i) and (ii) we have $\overline{g}\overline{h} = \overline{f} = \overline{a}_n X^n \neq 0.$

We argue that $\overline{g} = \overline{b}X^k$ and $\overline{h} = \overline{c}X^l$ for some elements $b, c \in R$ and positive integers k, l. To see this, note that since π is prime in R, it follows that the ideal (π) is prime, and so $\overline{R}$ is a domain and thus is contained in some field K by Theorem 16.17. The factorization $\overline{f} = \overline{g}\overline{h}$ can be viewed as happening in the ring $K[X]$. Since (in this ring) $\overline{f}$ is a unit multiple of a power of the irreducible polynomial X, it follows by the uniqueness of factorization in the ring $K[X]$ that each of $\overline{g}$ and $\overline{h}$ is a unit multiple of powers of X, as claimed. To see that $k, l > 0$, it suffices to observe that $k = \deg(\overline{g}) \leq \deg(g)$ and similarly $l \leq \deg(h)$. Since $n = k + l \leq \deg(g) + \deg(h) = \deg(f) = n$, we have equality throughout, and $k = \deg(g) > 0$ and $l = \deg(h) > 0$.

It follows that the constant terms of $\overline{g}$ and $\overline{h}$ are zero, and thus the constant terms of g and h are each divisible by π. Since a_0, the constant term of f, is the product of these, we conclude that $\pi^2 | a_0$, contradicting (iii).   ∎

Applying the Eisenstein criterion in $\mathbb{Q}[X]$ (with $R = \mathbb{Z}$), we see immediately that polynomials of the form $X^n - p$ and $X^n + p$ are irreducible for prime numbers p. A less obvious application of the Eisenstein criterion is to show for prime numbers p that $f(X) = X^{p-1} + X^{p-2} + \cdots + X + 1$ is irreducible in $\mathbb{Q}[X]$. The trick here is to define a new polynomial by the formula $g(X) = f(X + 1)$ and to apply the Eisenstein criterion to g. We conclude that g is irreducible and deduce from that that f is irreducible, too.

We need to compute the coefficients of $g(X)$. An easy way to do this is to observe that $(X - 1) f(X) = X^p - 1$. Substituting $X + 1$ for X and recalling that substitution (evaluation) is a ring homomorphism, we conclude that

$$ Xg(X) = (X + 1)^p - 1 = \sum_{i=1}^{p} \binom{p}{i} X^i . $$

It is easy to see from this that the Eisenstein criterion does apply to g (with respect to the prime p), and g is irreducible. Now if $f = f_1 f_2$ with $f_1, f_2 \in \mathbb{Q}[X]$ of positive degree, then $g(X) = f(X + 1) = f_1(X + 1) f_2(X + 1)$, and this contradicts the irreducibility of g. Therefore f is irreducible, as claimed.

We mention that the polynomial f that we just considered is an example of a "cyclotomic" polynomial. In Chapter 20 we study these polynomials in some detail and, in particular, we prove Gauss's theorem that all cyclotomic polynomials are irreducible over $\mathbb{Q}$.

16D

We now consider the technique of "localization," which we use to prove Theorem 16.17 and which appears again in Chapters 26 and 27. Let R be an arbitrary commutative ring. A subset $M \subseteq R$ is called a *multiplicative system* in R if $1 \in M, 0 \notin M$, and M is closed under multiplication. For example, if R is a domain, then the set $R - \{0\}$ is a multiplicative system in R, and more generally, for any ring R, if P is a prime ideal, then $R - P$ is a multiplicative system. Another important example is the set $M = \{a^n \mid n \geq 0\}$, where $a \in R$ is not nilpotent.

Given a multiplicative system $M \subseteq R$, we would like to embed R in a unitary overring S so that each element of M is invertible in S. In general, this is not possible; a necessary condition is that the ideal I of the next lemma is trivial.

(16.22) LEMMA.   *Let $M \subseteq R$ be a multiplicative system and let $I = \{r \in R \mid rm = 0$ for some $m \in M\}$. Then I is a proper ideal, and if every element of M is invertible in some unitary overring $S \supseteq R$, then $I = 0$.*

Proof.   Certainly $1 \notin I$ since $0 \notin M$, and so $I < R$. If $x, y \in I$, let $m, n \in M$ with $xm = 0$ and $yn = 0$. Then $mn \in M$ and $(x - y)(mn) = 0$. It follows that $x - y \in I$, which is therefore an additive subgroup of R. (Of course, $0 \in I$ and so $I \neq \varnothing$.) If $x \in I$ and $r \in R$, choose $m \in M$ with $xm = 0$. Then $(xr)m = 0$ and hence $xr \in I$, and this proves that I is an ideal.

Now suppose $S \supseteq R$ and that each element of M is invertible in S. Then if $x \in I$, choose $m \in M$ with $xm = 0$. Working in S, we find this yields $x = (xm)m^{-1} = 0$, as required.   ∎

If $M \subseteq R$ is a multiplicative system, we write $\mathcal{Z}(M)$ to denote the ideal I of the previous lemma. As the following theorem shows, this ideal is the only obstruction to the construction of an overring S in which the elements of M are invertible. We note that if R is a domain, then $\mathcal{Z}(M) = 0$ for every multiplicative system $M \subseteq R$. For domains, therefore, we can always construct the overring S.

(16.23) THEOREM.   *Let $M \subseteq R$ be a multiplicative system and assume that $\mathcal{Z}(M) = 0$. Then there exists a unitary overring $S \supseteq R$ such that every element of M is a unit in S and every element of S has the form rm^{-1} for some $r \in R$ and $m \in M$.*

Proof.   Let $\mathcal{F}$ be the set of ordered pairs (a, m), where $a \in R$ and $m \in M$. (To motivate the calculations that follow, think of (a, m) as the formal "fraction" a/m.) For elements (a, m) and $(b, n) \in \mathcal{F}$, write $(a, m) \sim (b, n)$ if $an = bm$. We claim that this is an equivalence relation on $\mathcal{F}$. The reflexive and symmetric properties are trivial and we need only to check transitivity. Suppose then that $(a, m) \sim (b, n)$ and $(b, n) \sim (c, l)$. Then $an = bm$ and $bl = cn$. Thus $anl = bml = cnm$, and so $n(al - cm) = 0$ and $al - cm \in \mathcal{Z}(M) = 0$. Therefore $(a, m) \sim (c, l)$, as required.

Now let S be the set of equivalence classes under $\sim$. We use the notation $[a/m]$ to denote the class containing (a, m), so that $[a/m] = [b/n]$ iff $an = bm$. Our next goal is to define addition and multiplication on S so that S becomes a ring.

If $[a/m], [b/n] \in S$, define $[a/m][b/n] = [ab/mn]$. (Note that $mn \in M$, so that the right side makes sense.) We need to prove that this multiplication is well defined. In other words, if $[a/m] = [a'/m']$ and $[b/n] = [b'/n']$, we must prove that $[ab/mn] = [a'b'/m'n']$. We have $am' = a'm$ and $bn' = b'n$. Multiplying these, we get $(ab)(m'n') = (a'b')(mn)$, as required.

Next we define addition in S by the formula

$$\left[\frac{a}{m}\right] + \left[\frac{b}{n}\right] = \left[\frac{an + bm}{mn}\right].$$

We must show that this is well defined and hence we need to deduce that

$(an + bm)(m'n') = (a'n' + b'm')(mn)$, assuming $am' = a'm$ and $bn' = b'n$. This follows readily, and we omit the details.

The commutativity and associativity of these operations and the distributive law are easy to check, as are the facts that $[0/1]$ is an additive identity and $[1/1]$ is a multiplicative identity. Also, $[a/m] + [-a/m] = [0/m]$ and $[0/m] = [0/1]$, since $0 \cdot 1 = 0 \cdot m$. At this point, we know that S is a ring.

Next we embed R into S. Let $\theta : R \to S$ be defined by $\theta(a) = [a/1]$. It is trivial to check that this is a ring homomorphism. If $a \in \ker(\theta)$, then $[a/1] = [0/1]$. From this we conclude that $a = 0$, and thus θ maps R isomorphically into S. We identify R with $\theta(R) \subseteq S$ via θ, and we note that this makes S into a unitary overring of R since $\theta(1) = [1/1]$ is the unity of S.

If $m \in M \subseteq R$, then m is identified with $[m/1] \in S$, and we note that $[m/1][1/m] = [m/m] = [1/1]$. Thus $m = [m/1]$ is a unit in S and $m^{-1} = [1/m]$. Finally, if $[a/m] \in S$, we can write $[a/m] = [a/1][1/m] = am^{-1}$, and the proof is complete. ∎

Proof of Theorem 16.17.   Apply Theorem 16.23 with $M = R - \{0\}$ in the domain R. To show that the resulting ring S is a field, let $s \in S$ with $s \neq 0$. Then $s = am^{-1}$ for some $a, m \in R$ with $m \neq 0$. Also $a \neq 0$ since $s \neq 0$, and so a is invertible in S. Writing $t = ma^{-1} \in S$, we have $st = 1$, and this shows that every nonzero element of S is invertible, as required. ∎

The ring S whose existence is guaranteed by Theorem 16.23 is called the *localization* of R at M. To justify the phrase "*the* localization," we should show that (up to isomorphism) only one ring can satisfy the conclusion of Theorem 16.23. Actually, we can prove a bit more than this.

(16.24) THEOREM.   *Let $\theta : R_1 \to R_2$ be a ring isomorphism and let $M_1 \subseteq R_1$ be a multiplicative system. Write $M_2 = \theta(M_1)$ and let S_1 and S_2 be rings that contain R_1 and R_2 and satisfy the conclusions of Theorem 16.23. Then θ has a unique extension $\varphi : S_1 \to S_2$ such that φ is an isomorphism.*

Proof.   Since φ is supposed to extend θ, we must have $\varphi(a) = \theta(a)$ for $a \in R$, and so $\varphi(am^{-1}) = \theta(a)\theta(m)^{-1}$ for $a \in R_1$ and $m \in M_1$. This proves that φ (if it exists) is unique, and it tells us how to define φ on S_1.

If $s \in S_1$, write $s = am^{-1}$ with $a \in R_1$ and $m \in M_1$, and define $\varphi(s) = \theta(a)\theta(m)^{-1}$. (We are using the fact that $\theta(m) \in M_2$ so that $\theta(m)$ is invertible in S_2.)

We need to show that φ is well defined. If also $s = bn^{-1}$ with $b \in R_1$ and $n \in M_1$, then we must establish that $\theta(a)\theta(m)^{-1} = \theta(b)\theta(n)^{-1}$. Since $am^{-1} = bn^{-1}$, we have $an = bm$ and so $\theta(a)\theta(n) = \theta(b)\theta(m)$, and the desired equality follows.

Next we must show that φ really is an isomorphism of S_1 onto S_2. For instance, if $s = am^{-1}$ and $t = bn^{-1}$, we must compute $\varphi(s + t)$. We have $s + t = (s + t)(mn)(mn)^{-1} = (an + bm)(mn)^{-1}$, and this enables us to compute that $\varphi(s + t) = \varphi(s) + \varphi(t)$, as required. The rest of the proof is even more trivial and we omit the details.   ■

(16.25) COROLLARY. *Let $M \subseteq R$ be as in Theorem 16.23, and let S_1 and S_2 be two rings that satisfy the conclusions of that theorem. Then $S_1 \cong S_2$ via an isomorphism that is the identity on R.*   ■

It is interesting to note that we have not referred to the proof of Theorem 16.23, only to its statement. In fact, the real meaning of Theorem 16.24 and Corollary 16.25 is that one never needs to refer back to the *construction* of the localization S. All of the information we might ever need (up to isomorphism) is contained in the *statement* of Theorem 16.23.

What do we do if we want to localize at some multiplicative system $M \subseteq R$, but we find that $\mathcal{Z}(M) > 0$? This is not really a serious problem, since we can just "mod out" the obstruction.

(16.26) LEMMA. *Let $M \subseteq R$ be a multiplicative system and let $I = \mathcal{Z}(M)$. Write $\overline{R} = R/I$, where we write $\overline{a}$ to denote the image of a under the canonical homomorphism. Then $\overline{M}$ is a multiplicative system in $\overline{R}$ and $\mathcal{Z}(\overline{M}) = 0$.*

Proof.   Certainly, $\overline{M}$ is closed under multiplication in $\overline{R}$ and $\overline{1} \in \overline{M}$ since $1 \in M$. If $\overline{0} \in \overline{M}$, then there exists $m \in M \cap I$, and so $mn = 0$ for some $n \in M$. Since $mn \in M$, this is a contradiction, and so $\overline{0} \notin \overline{M}$. It follows that $\overline{M}$ is a multiplicative system in $\overline{R}$.

Now let $\overline{r} \in \mathcal{Z}(\overline{M})$. Then $\overline{rm} = \overline{0}$ for some $m \in M$, and thus $rm \in I$. It follows that there exists $n \in M$ with $(rm)n = 0$. Then $r \in I$ since $mn \in M$, and we have $\overline{r} = \overline{0}$ as desired.   ■

If $M \subseteq R$ is a multiplicative system and $I = \mathcal{Z}(M)$, we can first construct $\overline{R} = R/I$ and then localize at $\overline{M}$ to get a ring S containing $\overline{R}$. This gives the following corollary.

(16.27) COROLLARY. *Let $M \subseteq R$ be a multiplicative system. Then there exists a ring S and a homomorphism $\theta : R \rightarrow S$ such that*

a. $\ker(\theta) = \mathcal{Z}(M)$,
b. $\theta(1) = 1$,
c. $\theta(M)$ *consists of invertible elements, and*
d. *every element $s \in S$ is of the form $s = \theta(a)\theta(m)^{-1}$ for some $a \in R$ and $m \in M$.*   ■

In fact, analogs of Theorem 16.24 and Corollary 16.25 hold in this situation, too, although we do not give the precise statements or proofs. Even when $\mathcal{Z}(M) \neq 0$, the ring S of Corollary 16.27 is usually called the "localization" of R at M. It is important to remember that in this generality, it is not true that R is embedded in S; there is only a homomorphism $R \rightarrow S$ with kernel equal to $\mathcal{Z}(M)$.

16E

We conclude this chapter with a result concerning modules over PIDs. Recall that the fundamental theorem of abelian groups (7.10) asserts that a finite abelian group is necessarily a direct sum of cyclic subgroups. This can be generalized considerably. First, the finiteness requirement can be weakened: the result holds for all finitely generated abelian groups. Next, we note that abelian groups are really just $\mathbb{Z}$-modules. An analog of the fundamental theorem holds not just for finitely generated $\mathbb{Z}$-modules, but also for finitely generated modules over any PID.

Although the structure theorem on finitely generated modules over PIDs is not especially deep or difficult, we have decided not to present it here. We prove an important special case in this section, and that enables us to extend Theorem 7.10 to all finitely generated abelian groups. This case is also used in Chapter 28.

An R-module M is said to be *torsion free* if, whenever we have $mr = 0$ with $m \in M$ and $r \in R$, either $m = 0$ or $r = 0$. Recall that M is free if it has a basis: a generating set that is linearly independent over R. Also, by Lemma 14.20 an R-module is free iff it is a direct sum of submodules isomorphic to the regular module $R^{\bullet}$. An abelian group is a free $\mathbb{Z}$-module, therefore, iff it is a direct sum of infinite cyclic subgroups.

(16.28) THEOREM.   *Let M be a finitely generated, torsion-free R-module, where R is a PID. Then M is free.*

(16.29) LEMMA.   *Suppose $a, b \in R$ are not both zero and have no common prime divisor, where R is a PID. Then there exist $u, v \in R$ such that $au + bv = 1$.*

Proof.   Let $I = \{au + bv \mid u, v \in R\}$. Then I is a nonzero ideal of R, and so we can write $I = (r)$ for some element $r \in R$. If r is a unit, then $I = R$ and $1 \in I$, as desired. Otherwise, there exists some prime of R that divides r, and this prime thus divides every member of I. Since $a, b \in I$, this contradicts the hypothesis. ∎

Proof of Theorem 16.28.   We work by induction on the number n of generators for M. If $n = 0$, then $M = 0$ and the result is trivially true. Assume that $n > 0$ and fix one of the given generators x. Note that if $x \in N \subseteq M$, where N is a submodule, then the R-module M/N is generated by $n - 1$ elements.

Since R is noetherian, the module M is noetherian by Theorem 12.19. We consider submodules of the form yR for elements $y \in M$, and we write $\mathcal{X} = \{yR \mid x \in yR\}$. Clearly $\mathcal{X}$ is nonempty, and thus we can choose a maximal element $N \in \mathcal{X}$.

We claim that M/N is torsion free. In other words, we must show that if $z \in M$ and $za \in N$ for some nonzero element $a \in R$, then $z \in N$. Choose a with as few prime factors as possible such that $za \in N$. Now $N = yR$ for some element $y \in M$, and thus we have $za = yb$, where $b \in R$. If a and b have some common prime divisor π, write $a = a_0 \pi$ and $b = b_0 \pi$, and note that $(za_0 - yb_0)\pi = 0$. Since M is torsion free, this implies that $za_0 = yb_0 \in N$, and this contradicts our choice of a. It follows that a and b have no common prime divisor, and by Lemma 16.29 we can write $1 = au + bv$, where $u, v \in R$.

We have $y = y1 = y(au + bv) = yau + zav = (yu + zv)a$. It follows that $N = yR$ is contained in $(yu + zv)R$, and this submodule clearly lies in $\mathcal{X}$. By the maximality of N, we deduce that $N = (yu + zv)R$ and thus $yu + zv \in N$. Since $y \in N$, this yields $zv \in N$, and it follows that $z = z1 = z(au + bv) = (za)u + (zv)b$ lies in N. This proves our assertion that M/N is torsion free.

Because M/N is generated by $n - 1$ elements, we can apply the inductive hypothesis and conclude that M/N is free, and thus it is projective by Corollary 14.25. By Lemma 14.28, therefore, we deduce that N is a direct summand of M, and we can write $M = N \dotplus F$, where F is a submodule isomorphic to M/N, and so it is free. Because M is torsion free, it follows that $N = yR$ is isomorphic to the regular R-module $R^\bullet$. Also, since F is free, it is isomorphic to a sum of copies of $R^\bullet$ by Lemma 14.20, and it follows by Lemma 14.20 that M is free, as required.   ∎

(16.30) COROLLARY. *Let A be a finitely generated abelian group. Then A is the direct sum of cyclic subgroups.*

Proof. Let $T \subseteq A$ be the set of all elements of finite order in A. It is routine to check that T is a subgroup (called the *torsion* subgroup) of A and that A/T is torsion free as a $\mathbb{Z}$-module. Since A is finitely generated, the factor group A/T is also finitely generated, and it follows by Theorem 16.28 that A/T is a free $\mathbb{Z}$-module. By Lemma 14.28 we conclude that A is the direct sum of T and a free $\mathbb{Z}$-submodule F. Since F is a direct sum of cyclic groups, it suffices to show that T is also a direct sum of cyclic groups, and this follows from Theorem 7.10 if we can show that T is finite.

Now $T \cong A/F$ and so T is finitely generated. Suppose that T has n generators and that e is the maximum of the orders of these generators. Every element of T is a $\mathbb{Z}$-linear combination of the generators with nonnegative coefficients less than e. It follows that there are at most e^n possible choices for coefficients, and thus $|T| \le e^n$ and T is finite.   ∎

Problems

16.1 Let $n \in \mathbb{Z}$ and write α to denote either of the square roots of n in $\mathbb{C}$. Show that $\mathbb{Z}[\alpha] = \{a + b\alpha \mid a, b \in \mathbb{Z}\}$.

16.2 Let $R \subseteq \mathbb{C}$ be defined by $R = \{a + bi\sqrt{5} \mid a, b \in \mathbb{Z}\}$, so that R is a subring of $\mathbb{C}$ by Problem 16.1. Prove that R is not a UFD.

HINT: If $r \in R$, then $|r|^2 \in \mathbb{Z}$. Note that $2 \cdot 3 = (1 + i\sqrt{5})(1 - i\sqrt{5})$.

16.3 Let R be a UFD with fraction field F and suppose that $\alpha \in F$. Show that it is possible to write $\alpha = a/b$, with $a, b \in R$ and such that a and b are coprime in R.

16.4 Let R be a UFD and let F be the field of fractions for R. If $f(\alpha) = 0$, where $f \in R[X]$ is monic and $\alpha \in F$, show that $\alpha \in R$.

NOTE: A corollary is the fact that if $m \in \mathbb{Z}$ and m is not an nth power in $\mathbb{Z}$, then $\sqrt[n]{m}$ is irrational.

16.5 Let F be a finite field. Show that $F[X]$ contains irreducible polynomials of arbitrarily large degree.

NOTE: In fact, there are irreducible polynomials of every degree greater than zero. This is harder to prove, however.

16.6 Let R be a noetherian UFD and suppose that whenever $a, b \in R$ are not both zero and have no common prime divisor, there exist elements $u, v \in R$ such that $au + bv = 1$. Show that R is a PID.

16.7 Let R be a PID with field of fractions F, and let S be a ring with $R \subseteq S \subseteq F$.
a. If $\alpha \in S$, show that $\alpha = a/b$, with $a, b \in R$ and $1/b \in S$.
b. Show that S is a PID.

HINT: Use Problem 16.6 for part (a). If $I \subseteq S$ is an ideal, consider $J = R \cap I$.

16.8 The ring R is said to be *local* if it has a unique maximal ideal. (Note that the maximal ideal of a local ring consists precisely of the nonunits of R.) Suppose that $a, b \in R$, where R is local, and that $a|b$ and $b|a$. Show that $b = ua$ for some unit $u \in R$.

NOTE: By Theorem 14.31 a commutative artinian ring R is a direct sum of a finite number of minimally potent ideals and each of these, viewed as a ring, is local by Lemma 14.34. It follows easily that if $a|b$ and $b|a$ in any artinian (commutative) ring, then $b = ua$ for some unit u.

16.9 Let R be the ring of all continuous real valued functions on the interval $-1 \le x \le 1$. (Addition and multiplication are defined pointwise.) Let

$$f(x) = \begin{cases} 2x + 1 & \text{for } -1 \le x \le -1/2 \\ 0 & \text{for } -1/2 < x < 1/2 \\ 2x - 1 & \text{for } 1/2 \le x \le 1, \end{cases}$$

and let $g(x) = |f(x)|$. Show that $f, g \in R$ and that $f|g$ and $g|f$ but that there is no unit $u \in R$ with $g = uf$.

16.10 Let F be any field and let $g, h \in F[X]$ be coprime polynomials of positive degree. Let Y be a new indeterminate and let E be a field of fractions for $F[Y]$, so that $F \subseteq F[Y] \subseteq E$ and we can view $g, h \in E[X]$. Define $f \in E[X]$ by $f(X) = g(X) - Yh(X)$. Show that f is irreducible in $E[X]$.

16.11 Let P be a prime ideal of R and write $M = R - P$. Let S be the localization of R at M with $\theta : R \to S$ the natural homomorphism (with $\ker(\theta) = \mathcal{Z}(M)$). Finally, let Q be the ideal of S generated by $\theta(P)$.

    a. Show that $Q = \{\theta(a)/\theta(b) \mid a \in P \text{ and } b \in R - P\}$.

    b. Show that Q is the unique maximal ideal of S (and so S is a local ring).

NOTE: This explains the word "localization".

16.12 Let R be a subring of a field E. (Note that this forces R to be a domain.)

    a. Show that R is a unitary subring of E.

    b. Show that there exists a unique subfield $F \subseteq E$ (with $F \supseteq R$) such that F is a field of fractions for R.

16.13 Let F be a field and suppose that E is a field of fractions for the domain $R = F[X_1, X_2, \ldots, X_n]$. Show that each permutation of the indeterminates X_i extends to an automorphism σ of E with the property that $\sigma(\alpha) = \alpha$ for all $\alpha \in F$.

HINT: Show first that R has such an automorphism.

CHAPTER SEVENTEEN

Field Extensions

17A

In group theory it is customary to ask questions about the subgroups of a given group. In field theory this point of view is usually reversed; we start with some particular field F and examine fields that contain it. (These are the *extension fields* of F.) For instance, we may have in mind a polynomial $f \in F[X]$ for which we seek *roots*, that is, elements $\alpha \in F$ such that $f(\alpha) = 0$. It may be that f has no roots or has insufficiently many roots in F, and so we can look in various extensions of F in the hope of finding more roots.

Note that if $F \subseteq E$ is a field extension, then the unity element of E is automatically the unity of F, and the characteristic of E is the same as that of F. Also $F[X] \subseteq E[X]$, and so any polynomial with coefficients in F can be viewed as lying in $E[X]$. We mention that in the field extension $F \subseteq E$, the field F is often called the *ground* field.

Let $F \subseteq E$ be fields. An element $\alpha \in E$ is said to be *algebraic* over F if there exists some nonzero polynomial $f \in F[X]$ such that $f(\alpha) = 0$. (For instance, the real number $\sqrt{2}$ is algebraic over $\mathbb{Q}$ since it is a root of $f(X) = X^2 - 2 \in \mathbb{Q}[X]$.) An element of E that is not algebraic over F is *transcendental* over F. (The real numbers e and π, for instance, are known to be transcendental over $\mathbb{Q}$, although it is far from a triviality to prove this.)

To produce examples of transcendental elements in extensions of an arbitrary field F, consider the polynomial ring $F[t]$ in the indeterminate t. The field of fractions of the domain $F[t]$ is denoted $F(t)$, and it is usually called the field of *rational functions* over F in the indeterminate t. Note that the elements of $F(t)$ are fractions of the form $f(t)/g(t)$, where $0 \neq g \in F[t]$. These "rational functions" are not exactly functions, however, since, for instance, $1/t$ and $1/(t - 1)$ have no values at $t = 0$ and $t = 1$, respectively.

Because of the natural embeddings $F \subseteq F[t]$ and $F[t] \subseteq F(t)$, we usually view $F \subseteq F(t)$ and we have a field extension. The element $t \in F(t)$ is transcendental over F since if $f(t) = 0$ for some $f \in F[X]$, then necessarily $f = 0$. (This is because the evaluation homomorphism $F[X] \to F[t]$ obtained by substituting t for X is an isomorphism.)

In any field extension $F \subseteq E$, the elements $\alpha \in F$ are certainly algebraic over F (since α is a root of $X - \alpha$). As we shall see, in the rational function field $F(t)$, the only elements that are algebraic over F are the elements of the ground field F.

We need to introduce the notion of the "degree" of a field extension in order to have a method for deciding whether or not an element is algebraic.

Suppose F is a field and R is an integral domain that contains F. (Note that this forces R to be a unitary overring of F.) In this situation, we may view R as a vector space over F. Vector addition is just the ordinary addition in R, and scalar multiplication is defined as multiplication in R by elements in F.

In this situation, if R has finite F-dimension, we write $|R : F|$ to denote this dimension; otherwise, we write $|R : F| = \infty$. (Some authors use the notation $[R : F]$ for this number.) In particular, if $E \supseteq F$ is a field extension, then $|E : F|$ is the *degree* of the field extension. Thus, for example, $\mathbb{R} \subseteq \mathbb{C}$ is an extension of degree $|\mathbb{C} : \mathbb{R}| = 2$, and $\mathbb{Q} \subseteq \mathbb{R}$ is an infinite degree extension. Of course, $F \subseteq F$ always has degree 1 and, conversely, if $F \subseteq E$ has degree 1, then $E = F$.

(17.1) LEMMA. *Let $F \subseteq R$, where F is a field and R is a domain. If $|R : F| < \infty$, then R is a field.*

Proof. Let $\alpha \in R$ with $\alpha \neq 0$. Our object is to find $\beta \in R$ with $\alpha\beta = 1$, proving that α is invertible.

Define $\theta : R \to R$ by $\theta(r) = \alpha r$ and observe that θ is an F-linear operator on R and that $\ker(\theta) = 0$ since R is a domain and $\alpha \neq 0$. By elementary linear algebra, therefore,

$$\dim_F(\theta(R)) = \dim_F(R) - \dim_F(\ker(\theta)) = \dim_F(R),$$

and since $\dim_F(R)$ is finite, we conclude that the subspace $\theta(R)$ cannot be proper. In particular, $1 \in \theta(R)$, as desired. ∎

If $F \subseteq E$ is a field extension and $\alpha \in E$, recall that $F[\alpha]$ denotes $\{f(\alpha) \mid f \in F[X]\}$. In particular, $F[\alpha]$ is a subring of E containing F, and since it is contained in a field, $F[\alpha]$ is certainly a domain. In general, $F[\alpha]$ may not be a field, and we define $F(\alpha)$ to be the set of all quotients u/v (computed in E), where $u, v \in F[\alpha]$ and $v \neq 0$. Then $F(\alpha)$ is a (the) field of fractions of $F[\alpha]$. It is easy to see that $F(\alpha)$ is the unique minimal subfield of E containing both F and α. It is called the subfield of E *generated* by α over F.

(17.2) LEMMA. *Let $F \subseteq E$ be a field extension and let $\alpha \in E$.*

   a. *If α is transcendental over F, then $F[\alpha] \cong F[X]$.*
   b. *If α is algebraic over F, then $|F[\alpha] : F| \leq \deg(f)$, where $f \in F[X]$ is any nonzero polynomial such that $f(\alpha) = 0$.*

Proof.   Let $\theta : F[X] \to E$ be the evaluation homomorphism at α. Then $F[\alpha] = \theta(F[X])$, and if α is transcendental, we see that $\ker(\theta) = 0$ and (a) is proved.

To prove (b), it suffices to show that the elements $1, \alpha, \alpha^2, \ldots, \alpha^{n-1}$ span $F[\alpha]$ over F, where $n = \deg(f)$. An arbitrary element of $F[\alpha]$ has the form $g(\alpha)$ for some polynomial $g \in F[X]$. By the division algorithm (16.3), we have $g = fq + r$, where $q, r \in F[X]$ and either $\deg(r) < n$ or $r = 0$. Since $f(\alpha) = 0$, we have $g(\alpha) = r(\alpha)$, and certainly $r(\alpha)$ is an F-linear combination of $1, \alpha, \ldots, \alpha^{n-1}$, as desired.   ∎

(17.3) THEOREM.   *Let $F \subseteq E$ be a field extension and let $\alpha \in E$. Then the following are equivalent:*

i.   *α is algebraic over F.*
ii.   *$|F[\alpha] : F| < \infty$.*
iii.   *$F[\alpha]$ is a field.*
iv.   *$F[\alpha] = F(\alpha)$.*

Proof.   First we note that (iii) and (iv) are clearly equivalent. That (i) implies (ii) is immediate from Lemma 17.2(b); that (ii) implies (iii) follows from Lemma 17.1. If (i) is false, then α is transcendental over F and Lemma 17.2(a) yields that $F[\alpha] \cong F[X]$, which is certainly not a field. This establishes that (iii) implies (i).   ∎

(17.4) COROLLARY.   *Let $F \subseteq E$ be a finite degree field extension. Then every element of E is algebraic over F.*

Proof.   If $\alpha \in E$, then $F[\alpha] \subseteq E$, and so $|F[\alpha] : F| \le |E : F| < \infty$. Therefore α is algebraic by Theorem 17.3.   ∎

A field extension in which every element is algebraic over the ground field is called an *algebraic extension*. By Corollary 17.4 finite degree extensions are necessarily algebraic. The converse is not true, however. For instance, if we let $\mathbb{A} = \{\alpha \in \mathbb{C} \mid \alpha \text{ is algebraic over } \mathbb{Q}\}$, then in fact $\mathbb{A}$ is a field and $|\mathbb{A} : \mathbb{Q}| = \infty$. (That $\mathbb{A}$ is a field is our next theorem, and that it has infinite degree over $\mathbb{Q}$ appears as Corollary 17.10.) Of course, $\mathbb{A}$ is algebraic over $\mathbb{Q}$. The field $\mathbb{A}$ is called the field of *algebraic numbers*, and the subfields of $\mathbb{C}$ that are of finite degree over $\mathbb{Q}$ are all contained in $\mathbb{A}$ by Corollary 17.4. These fields are called *algebraic number fields*. Note that $\mathbb{A}$ itself is not an algebraic number field.

(17.5) THEOREM.   *Let $F \subseteq E$ be any field extension. Then*

$$A = \{\alpha \in E \mid \alpha \text{ is algebraic over } F\}$$

is a subfield of E containing F.

We need a lemma.

(17.6) LEMMA.   *Let $F \subseteq E \subseteq L$ be fields. Then $|L : F|$ is finite iff both $|E : F|$ and $|L : E|$ are finite. In this case, $|L : F| = |L : E||E : F|$.*

Proof. If $|L : F| < \infty$, then certainly $|E : F| < \infty$ since E is an F-subspace of L. Also, any F-basis for L over F spans L over E, and so $|L : E| < \infty$.

Now assume that $|L : E| = m$ and $|E : F| = n$ are both finite, and choose bases $l_1, l_2, \ldots, l_m$ for L over E and $e_1, e_2, \ldots, e_n$ for E over F. It suffices to show that the elements $l_i e_j$ are distinct and form a basis for L over F, where $1 \le i \le m$ and $1 \le j \le n$.

First suppose $\sum a_{ij} l_i e_j = 0$ for some choice of coefficients $a_{ij} \in F$. Then

$$0 = \sum_i \left(\sum_j a_{ij} e_j \right) l_i ,$$

and since $\sum a_{ij} e_j \in E$ and the l_i are linearly independent over E, we conclude that $\sum a_{ij} e_j = 0$ for each i. It follows from the linear independence of the e_j over F that $a_{ij} = 0$ for all i, j. We now know that the elements $l_i e_j$ are distinct and linearly independent over F.

To see that these elements span L over F, let $l \in L$ and write $l = \sum x_i l_i$ with $x_i \in E$. Each coefficient x_i can be written in the form $x_i = \sum a_{ij} e_j$ with $a_{ij} \in F$, and thus

$$l = \sum_{i,j} a_{ij} e_j l_i ,$$

as required. ∎

Proof of Theorem 17.5. Of course, $F \subseteq A$, and so we need to show that A is a subfield of E. In particular, if $\alpha, \beta \in A$, we must show that $\alpha - \beta \in A$ and that $\alpha/\beta \in A$ if $\beta \ne 0$.

Now $F[\alpha]$ is a field by Lemma 17.2, and since β is algebraic over F, it is certainly algebraic over the field $F[\alpha] \supseteq F$. Thus $(F[\alpha])[\beta]$ is a field and we have $F \subseteq F[\alpha] \subseteq (F[\alpha])[\beta]$. Now $|F[\alpha] : F| < \infty$ and $|(F[\alpha])[\beta] : F[\alpha]| < \infty$, and hence $|(F[\alpha])[\beta] : F| < \infty$ by Lemma 17.6. By Corollary 17.4 the elements of $(F[\alpha])[\beta]$ are algebraic over F. Since this field contains both $\alpha - \beta$ and α/β (if $\beta \ne 0$), these elements are algebraic and thus lie in A. ∎

In particular, we have now proved that the algebraic numbers $\mathbb{A}$ form a field. Note that it is trivial to find polynomials over $\mathbb{Q}$ that have, for instance, the algebraic numbers $\sqrt[3]{2}$ and $\sqrt{3}$ as roots. We know that $\sqrt[3]{2} + \sqrt{3}$ must be algebraic over $\mathbb{Q}$, but it is certainly not completely obvious how to find an explicit polynomial $f \in \mathbb{Q}[X]$ such that $f(\sqrt[3]{2} + \sqrt{3}) = 0$.

(17.7) COROLLARY. Let $F \subseteq E$ and $E \subseteq L$ be field extensions and assume E is algebraic over F. If $\alpha \in L$ is algebraic over E, then α is algebraic over F.

Before we give the proof of Corollary 17.7, we introduce some simplifying notation. If $F \subseteq E$ and $\alpha, \beta \in E$, then $(F[\alpha])[\beta]$ and $(F[\beta])[\alpha]$ are easily seen to be the same thing, namely, the set of all $f(\alpha, \beta)$ as f runs over $F[X, Y]$. We write

$F[\alpha, \beta]$ to denote this object (which is a field if both α and β are algebraic over F). Similarly, we write

$$F[\alpha_1, \alpha_2, \ldots, \alpha_n] = (\cdots(((F[\alpha_1])[\alpha_2])[\alpha_3])\cdots)[\alpha_n]$$

for $\alpha_1, \alpha_2, \ldots, \alpha_n \in E$, and we observe that the result is independent of the order in which we take the α_i. If each α_i is algebraic over F, then $F[\alpha_1, \ldots, \alpha_n]$ is a field, and by repeated application of Lemma 17.6, we see that it has finite degree over F.

Proof of Corollary 17.7. Let $f(\alpha) = 0$ with $0 \neq f \in E[X]$, and write $f(X) = e_n X^n + \cdots + e_0$, where $e_i \in E$ for $0 \leq i \leq n$. Since each e_i is algebraic over F, we see that $K = F[e_0, e_1, \ldots, e_n]$ is a field and that it is of finite degree over F. Also, $f \in K[X]$, and so α is algebraic over K and $|K[\alpha] : K| < \infty$. It follows that $|K[\alpha] : F| < \infty$ and thus α is algebraic over F by Corollary 17.4.   ∎

In particular, if $\alpha \in \mathbb{C} - \mathbb{A}$, then α is transcendental over $\mathbb{Q}$, and thus by Corollary 17.7, α must be transcendental over $\mathbb{A}$. An extension such as $\mathbb{A} \subseteq \mathbb{C}$ where every element not in the ground field is transcendental is said to be *totally transcendental*. The following corollary gives another example of a totally transcendental extension.

(17.8) COROLLARY. *Let $F \subseteq E$, where $E = F(\alpha)$ and α is transcendental over F. If $\beta \in E - F$, then β is transcendental over F and $|E : F(\beta)| < \infty$.*

Proof. Since $F(\alpha)$ is the field of fractions of $F[\alpha]$, we can write $\beta = u/v$, where $u = f(\alpha)$ and $0 \neq v = g(\alpha)$ for polynomials $f, g \in F[X]$. Let $K = F(\beta)$ and define $p \in K[X]$ by $p(X) = f(X) - \beta g(X)$. Observe that $p(\alpha) = u - \beta v = 0$.

Since $\beta \notin F$ and $g \neq 0$, the polynomial $\beta g(X)$ does not lie in $F[X]$ and thus $\beta g(X) \neq f(X)$. It follows that p is nonzero, and thus α is algebraic over K and $|K[\alpha] : K| < \infty$.

Now $K[\alpha] = K(\alpha) \supseteq F(\alpha) = E$ and thus $|E : K| < \infty$. Since α is transcendental over F, we have $|E : F| = \infty$, and thus $|K : F| = \infty$ by Lemma 17.6. It follows that β is transcendental over F.   ∎

We now examine a little more closely the set of polynomials that have some particular algebraic element as a root.

(17.9) THEOREM. *Let $F \subseteq E$ be fields and suppose $\alpha \in E$ is algebraic over F. Let $I = \{f \in F[X] \mid f(\alpha) = 0\}$. Then I is an ideal of $F[X]$ and the following hold:*

a. *I contains a unique monic irreducible polynomial.*
b. *If $f \in I$ is irreducible, then $\deg(f) = |F[\alpha] : F|$.*
c. *If $f \in I$ is irreducible, then $I = (f)$.*
d. *$F[\alpha] \cong F[X]/I$.*

Proof. Since I is the kernel of the evaluation homomorphism from $F[X]$ onto $F[\alpha]$, it is certainly an ideal, and (d) is immediate.

Because $F[X]$ is a PID, we can write $I = (g)$ for some polynomial $g \in F[X]$, and we note that $g \neq 0$ since α is algebraic. Also, g is nonconstant since $g(\alpha) = 0$. Multiplying by a scalar if necessary, we can assume that g is monic.

If $f \in I$, then we can write $f = gh$ for some polynomial h. If f is irreducible, we deduce that $\deg(h) = 0$ and hence $I = (f)$, proving (c). This also shows that if $f \in I$ is irreducible, then $\deg(f) = \deg(g)$. If, in addition, f is monic, then $f = g$, and this proves the uniqueness in (a).

To see that g is irreducible, we suppose that $g = g_1 g_2$. Since $g(\alpha) = 0$, we have $g_1(\alpha) g_2(\alpha) = 0$, and so one of $g_i(\alpha) = 0$. Assuming that $g_1(\alpha) = 0$, we have $g_1 \in I$ and hence $g | g_1$. Since also $g_1 | g$, it follows that g_2 is a unit. Thus g is irreducible and the proof of (a) is complete.

What remains is to show that $\deg(g) = |F[\alpha] : F|$. By Lemma 17.2(b) we have $\deg(g) \geq |F[\alpha] : F|$, and so it suffices to find $n = \deg(g)$ elements of $F[\alpha]$ that are linearly independent over F. We claim, in fact, that the elements $1, \alpha, \alpha^2, \ldots, \alpha^{n-1}$ are linearly independent. To prove this, suppose that

$$\sum_{i=0}^{n-1} a_i \alpha^i = 0,$$

where $a_i \in F$. Write $f(X) = \sum a_i X^i$ and note that $f(\alpha) = 0$, so that $f \in I$ and we can write $f = gh$. From its definition, however, we see that the polynomial f cannot have a degree as large as $n = \deg(g)$. It follows that $f = 0$ and hence all of the coefficients $a_i = 0$. The result now follows. ∎

The unique monic irreducible polynomial $m \in F[X]$ for which $m(\alpha) = 0$ is called the *minimal polynomial* for α over F, and we write $m = \min_F(\alpha)$. Note that the name "minimal" polynomial is explained by the observation that $\deg(m) = |F[\alpha] : F|$, whereas by Lemma 17.2(b) we have $\deg(f) \geq |F[\alpha] : F|$ for every nonzero polynomial $f \in F[X]$ such that $f(\alpha) = 0$.

In addition to the fact that $\deg(m) = |F[\alpha] : F|$, where $m = \min_F(\alpha)$, probably the most useful consequence of Theorem 17.9 is that m divides every polynomial $f \in F[X]$ for which $f(\alpha) = 0$. Of course, the fact that m is defined to be monic is irrelevant to these conclusions; it is needed only to guarantee the uniqueness of $\min_F(\alpha)$.

As an example of how Theorem 17.9 can be applied, consider a prime number p and an integer $n \geq 1$. By the Eisenstein criterion (16.21) we know that the polynomial $X^n - p$ is irreducible in $\mathbb{Q}[X]$. Since the real number $\sqrt[n]{p}$ is a root of this polynomial (which is certainly monic), we see that $X^n - p$ is the minimal polynomial of $\sqrt[n]{p}$ over $\mathbb{Q}$, and we conclude that $|\mathbb{Q}[\sqrt[n]{p}] : \mathbb{Q}| = n$. In particular, since $\sqrt[n]{p}$ lies in the field $\mathbb{A}$ of algebraic numbers, we see that $|\mathbb{A} : \mathbb{Q}| \geq n$ for all integers $n \geq 1$.

(17.10) COROLLARY. *The field $\mathbb{A}$ of algebraic numbers is an infinite degree algebraic extension of $\mathbb{Q}$.* ∎

As an example of how additional information can be obtained by combining Theorem 17.9 with Lemma 17.6, we will show how to deduce that the polynomial $X^n - p$ is irreducible, not only over $\mathbb{Q}$ but even over certain larger fields. Suppose that $\mathbb{Q} \subseteq F \subseteq \mathbb{C}$ and $|F : \mathbb{Q}| = m$, where $\gcd(m, n) = 1$. We will show that $X^n - p$ is irreducible over F. (We cannot use the Eisenstein criterion to prove this, since we do not have F written as the field of the field of fractions of a UFD.)

Let $E = F[\sqrt[n]{p}]$ and write $f = \min_F(\sqrt[n]{p})$, so that $|E : F| = \deg(f)$ and $|E : \mathbb{Q}| = |E : F||F : \mathbb{Q}| = m \deg(f)$. By Lemma 17.6 we also see that $|E : \mathbb{Q}|$ is divisible by $|\mathbb{Q}[\sqrt[n]{p}] : \mathbb{Q}| = n$. Thus n divides $m \deg(f)$, and hence n divides $\deg(f)$ by the coprimeness assumption.

Now write $g(X) = X^n - p \in F[X]$. Then $g(\sqrt[n]{p}) = 0$, and so $f \mid g$ by Theorem 17.9 and this gives $\deg(f) \leq n$. Combining this with the result of the previous paragraph, we get $\deg(f) = n$ and thus $f = g$. Therefore g is irreducible in $F[X]$, as claimed.

17B

A very pretty application of our techniques so far is the following result of E. Artin.

(17.11) THEOREM (Artin). *Let $F \subseteq E$ be a finite degree field extension. Then $E = F[\alpha]$ for some element $\alpha \in E$ iff there are just finitely many fields K with $F \subseteq K \subseteq E$.*

A field extension of the form $F(\alpha) \supseteq F$, generated by a single element α, is often called a *simple* extension, and the generating element α is said to be a *primitive* element. (Of course, under the assumption that $|E : F| < \infty$, we have $F(\alpha) = F[\alpha]$, and so Artin's theorem says that a finite degree extension is simple iff there are only finitely many intermediate fields.)

It seems to be necessary to subdivide the proof of Theorem 17.11 into two cases, depending on whether or not F is a finite field. To handle the easier case where F is finite, we need the following lemma, whose proof we defer.

(17.12) LEMMA. *Let G be a finite subgroup of the multiplicative group $E^\times$ of a field E. Then G is cyclic.*

Proof of Theorem 17.11. Let $\mathcal{F} = \{K \mid F \subseteq K \subseteq E\}$, the set of intermediate fields, and suppose $E = F[\alpha]$. Let $f = \min_F(\alpha)$ and write $\mathcal{P} = \{g \in E[X] \mid g \text{ is monic and } g \mid f\}$. By unique factorization in $E[X]$, it is clear that $\mathcal{P}$ is a finite set. We will show that $\mathcal{F}$ is finite by constructing an injective map $\mathcal{F} \to \mathcal{P}$.

Let $K \in \mathcal{F}$ and write $g_K = \min_K(\alpha)$. Then g_K is monic, and since $f \in F[X] \subseteq K[X]$ and $f(\alpha) = 0$, we conclude that $g_K \mid f$ and so $g_K \in \mathcal{P}$. We must show that the map $K \mapsto g_K$ is injective, and so we want to be able to recover K from a knowledge of g_K. Let $a_0, a_1, \ldots, a_r$ be the coefficients of g_K and define $L = F[a_0, a_1, \ldots, a_r]$ so that $L \subseteq K$. We propose to prove that $L = K$, and this will show that g_K determines K.

Now $g_K \in L[X]$ and g_K is irreducible when viewed as a polynomial over $K \supseteq L$. It follows that g_K is irreducible in $L[X]$ and hence $g_K = \min_L(\alpha)$. Since also $g_K = \min_K(\alpha)$ and $K[\alpha] = E = L[\alpha]$, we have

$$|E : L| = \deg(g_K) = |E : K|.$$

However, $L \subseteq K \subseteq E$ and so $|E : L| = |E : K||K : L|$. It follows that $|K : L| = 1$ and $K = L$, as required.

Conversely, we now assume that $\mathcal{F}$ is finite and we show by induction on $|E : F|$ that the extension $E \subseteq F$ is simple. This certainly holds when $E = F$, and so we assume that $E > F$ and we choose $\alpha \in E - F$. Then $E \supseteq F[\alpha] > F$ and so $|E : F[\alpha]| < |E : F|$. Since there are only finitely many fields intermediate between $F[\alpha]$ and E (because these constitute a subset of the finite set $\mathcal{F}$), the inductive hypothesis yields that $E = (F[\alpha])[\beta] = F[\alpha, \beta]$ for some element $\beta \in E$.

Now assume $|F|$ is infinite and consider the fields $K_t = F[\alpha + t\beta]$ as t runs over the elements of F. Each K_t lies in $\mathcal{F}$, and so by the "pigeon-hole" principle, $K_t = K_s$ for some distinct elements $s, t \in F$. Thus

$$\alpha + s\beta \in K_s \quad \text{and} \quad \alpha + t\beta \in K_s,$$

and subtraction yields $(s - t)\beta \in K_s$. Since $0 \neq s - t \in F \subseteq K_s$, we deduce that $\beta \in K_s$. Since $s \in F \subseteq K_s$, we get $\alpha \in K_s$, and this yields

$$E = F[\alpha, \beta] \subseteq K_s = F[\alpha + s\beta] \subseteq E.$$

We have equality here, and hence $\alpha + s\beta$ is a primitive element for E, as desired.

Finally, suppose $|F| < \infty$. Since E has a finite F-basis and there are only finitely many possibilities for each coefficient in an F-linear combination of this basis, it follows that $|E|$ is finite. By Lemma 17.12 we deduce that $E^\times$ is a cyclic group, and we can write $E^\times = \langle \alpha \rangle$. It follows that $E = F[\alpha]$ and hence $E \supseteq F$ is a simple extension. ∎

As we shall see when we study Galois theory in Chapter 18, if F has characteristic zero, then every finite degree extension of F is simple. This is proved by an appeal to Theorem 17.11, using Galois theory to verify that there are only finitely many intermediate fields.

To prove Lemma 17.12, we need the following well-known fact.

(17.13) LEMMA. *Let $f \in F[X]$, where F is a field.*

a. *If $\alpha \in F$, then $f(\alpha) = 0$ iff $(X - \alpha)|f$ in $F[X]$.*
b. *If $\deg(f) = n$, then f has at most n roots in F.*

Proof. Assume that $\alpha \in F$. By the division algorithm (16.3) we can write $f(X) = (X - \alpha)q(X) + r(X)$, where either $r = 0$ or $\deg(r) < 1$. In either case, r is a constant and we can view $r \in F$. Substitution of α for X yields $f(\alpha) = r$, and thus $r = 0$ iff $f(\alpha) = 0$. We see that $r = 0$, however, precisely when $(X - \alpha)\big|f$, and this proves (a).

We prove (b) by induction on n, observing that if $n = 0$, then f is a nonzero constant and so it has no roots. Supposing that $n > 0$, we may certainly assume that f has at least one root α (or else there is nothing to prove). By (a) we can write $f(X) = (X-\alpha)q(X)$ for some polynomial $q \in F[X]$ with $\deg(q) = n-1$. If $\beta \in F$ is any root of f other than α, we have $0 = f(\beta) = (\beta - \alpha)q(\beta)$ and thus $q(\beta) = 0$. By the inductive hypothesis applied to q, there are at most $n - 1$ such elements β and these, together with α, give at most n roots for f.   ∎

Proof of Lemma 17.12.   By Corollary 7.14 (of the fundamental theorem of abelian groups) it suffices, in order to show that G is cyclic, to show for each prime $p\big||G|$ that G contains at most one subgroup of order p. If $\alpha \in G$ lies in such a subgroup, then $\alpha^p = 1$ and so α is a root of the polynomial $X^p - 1$ in E. By Lemma 17.13 there are at most p such roots, and so there are not enough elements of order dividing p in G to form more than one subgroup of order p.   ∎

The following is an easy consequence of Artin's theorem (17.11).

(17.14)   COROLLARY.   *Let $F \subseteq E$, where $E = F(\alpha)$ for some element $\alpha \in E$. Assume that α is algebraic over F. If $F \subseteq K \subseteq E$, then $K = F(\beta)$ for some element $\beta \in K$.*

Proof.   Since α is algebraic, we have $E = F[\alpha]$ and $|E : F| < \infty$. By Theorem 17.11 there are only finitely many fields intermediate between F and E, and thus there surely are only finitely many between F and K. Since $|K : F| \leq |E : F| < \infty$, Theorem 17.11 applies and guarantees that K is a simple extension of F.   ∎

It may be surprising that Corollary 17.14 remains true even without the hypothesis that α is algebraic over F. Of course, if α is transcendental, we cannot use Artin's theorem to prove this result, and an essentially different argument is needed. The version of Corollary 17.14 in which α is transcendental is known as Lüroth's theorem, and we prove it in Chapter 24 when we study transcendental extensions.

An example of a finite-degree field extension that is not simple is given in the problems at the end of this chapter.

17C

We study isomorphisms now. (Note that we are not concerned with "homomorphisms" of fields since a nonzero ring homomorphism defined on a field automatically has a trivial kernel because a field has no nonzero proper ideals.)

If $F_1 \subseteq E_1$ and $F_2 \subseteq E_2$ are field extensions, we say that they are *isomorphic* if there is an isomorphism θ of E_1 onto E_2 such that θ carries F_1 onto F_2. It should be obvious in this case that the extension $E_2 \supseteq F_2$ enjoys the "same properties" as $E_1 \supseteq F_1$. For instance, E_2 is algebraic or is simple or is of finite degree over F_2

iff E_1 is algebraic, simple, or of finite degree over F_1. In fact, if the extensions are isomorphic, then we have $|E_2 : F_2| = |E_1 : F_1|$.

Suppose φ is an isomorphism of F_1 onto F_2. Given extensions $E_i \supseteq F_i$, we often want to investigate the existence of a surjective isomorphism $\theta : E_1 \rightarrow E_2$ that not only carries F_1 onto F_2 but also agrees with the given isomorphism φ on F_1. In this case we say that θ *extends* φ and that the field extensions $E_i \supseteq F_i$ are isomorphic *with respect to* φ. Perhaps the most important case of this is where $F_1 = F_2$ and φ is the identity map. If $F \subseteq E_1$ and $F \subseteq E_2$ and $\theta : E_1 \rightarrow E_2$ extends the identity map on F, we say that θ is an *F-isomorphism*, and if θ maps onto E_2, we say that E_1 and E_2 are *F-isomorphic*.

Note that if $\varphi : F_1 \rightarrow F_2$ is an isomorphism, then φ induces a ring isomorphism $\widehat{\varphi} : F_1[X] \rightarrow F_2[X]$, where $\widehat{\varphi}(f)$ is computed by applying φ to each coefficient of $f \in F_1[X]$. (We usually suppress the parentheses and write $\widehat{\varphi}f$ in place of $\widehat{\varphi}(f)$.)

The following result enables us to construct isomorphisms between field extensions. It is crucial for our development of Galois theory in the next chapter.

(17.15) LEMMA. *Let $\varphi : F_1 \rightarrow F_2$ be an isomorphism of F_1 onto F_2 and suppose $F_i \subseteq E_i$ for $i = 1, 2$. Let $\alpha_i \in E_i$ and $f_i \in F_i[X]$ with $f_i(\alpha_i) = 0$. Suppose that $\widehat{\varphi}f_1 = f_2$ and that f_1 is irreducible in $F_1[X]$. Then φ extends to an isomorphism θ of $F_1[\alpha_1]$ onto $F_2[\alpha_2]$ such that $\theta(\alpha_1) = \alpha_2$.*

Proof.   Each element of $F_1[\alpha_1]$ has the form $g(\alpha_1)$ for some $g \in F_1[X]$. We can thus define $\theta : F_1[\alpha] \rightarrow F_2[\alpha]$ by setting

$$\theta(g(\alpha_1)) = (\widehat{\varphi}g)(\alpha_2),$$

although, of course, we must show that this is well defined. Suppose therefore that $g(\alpha_1)$ can also be written as $h(\alpha_1)$ for some other polynomial $h \in F_1[X]$. We must establish that $(\widehat{\varphi}g)(\alpha_2) = (\widehat{\varphi}h)(\alpha_2)$.

Now α_1 is a root of the polynomial $g - h \in F_1[X]$ and it is also a root of the irreducible polynomial $f_1 \in F_1[X]$. By Theorem 17.9(c) we conclude that $f_1|(g - h)$, and therefore $f_2 = \widehat{\varphi}f_1$ must divide $\widehat{\varphi}g - \widehat{\varphi}h$. Since $f_2(\alpha_2) = 0$, this yields $(\widehat{\varphi}g)(\alpha_2) = (\widehat{\varphi}h)(\alpha_2)$ as required, and θ is well defined.

It is now routine to show that θ respects addition and multiplication in $F_1[\alpha_1]$. (The key to this is to recall that $g(\alpha_1) + h(\alpha_1) = (g + h)(\alpha_1)$, and similarly for multiplication. We omit the details.) It follows that $\theta : F_1[\alpha_1] \rightarrow F_2[\alpha_2]$ is a ring homomorphism. It is surjective because $\widehat{\varphi}$ maps $F_1[X]$ onto $F_2[X]$, and we conclude that θ is an isomorphism of $F_1[\alpha_1]$ onto $F_2[\alpha_2]$.

We can compute $\theta(a)$ for $a \in F_1$ by writing $a = g(\alpha_1)$, where $g(X) = a$, the constant polynomial. Then $\theta(a) = (\widehat{\varphi}g)(\alpha_2) = \varphi(a)$, and so θ does extend φ. Finally, $\alpha_1 = g(\alpha_1)$ where $g(X) = X$. Thus $\theta(\alpha_1) = (\widehat{\varphi}g)(\alpha_2) = \alpha_2$, as desired. ∎

(17.16) COROLLARY. *Let $F \subseteq E_1$ and $F \subseteq E_2$ be two field extensions, and let $f \in F[X]$ be irreducible. Suppose $\alpha_i \in E_i$ with $f(\alpha_i) = 0$. Then $F[\alpha_1]$ and $F[\alpha_2]$ are F-isomorphic via an isomorphism that carries α_1 to α_2.* ∎

As an application of Corollary 17.16, we observe that the two subfields $\mathbb{Q}[\sqrt[4]{2}]$ and $\mathbb{Q}[i\sqrt[4]{2}]$ of $\mathbb{C}$ are $\mathbb{Q}$-isomorphic. (This may seem a bit surprising, since one of these fields is real and the other contains imaginary numbers.) The point here is that both $\sqrt[4]{2}$ and $i\sqrt[4]{2}$ are roots of the polynomial $X^4 - 2 \in \mathbb{Q}[X]$, and this polynomial is irreducible by the Eisenstein criterion.

17D

A nonzero polynomial $f \in F[X]$ is said to *split* if every irreducible divisor of f has degree 1. (In particular, nonzero constant polynomials split.) In other words, f splits if it has the form

$$f(X) = a(X - \alpha_1)(X - \alpha_2) \cdots (X - \alpha_n),$$

where a and each α_i lie in F. (Note that a must be the leading coefficient of f and $n = \deg(f)$.) For example, the polynomial $X^4 - 1$ does not split in $\mathbb{Q}[X]$ since it has the irreducible factor $X^2 + 1$ of degree 2. The same polynomial $X^4 - 1$ does split in $\mathbb{C}[X]$ (or, as is often said, it splits *over* $\mathbb{C}$) since it factors as

$$X^4 - 1 = (X + 1)(X - 1)(X + i)(X - i).$$

In fact, every polynomial in $\mathbb{C}[X]$ splits. This theorem, called the "fundamental theorem of algebra," has no purely algebraic proof, although proofs exist that use techniques from complex function theory, from topology, and from elementary analysis. It may seem to be a serious defect of the discipline of algebra that its own fundamental theorem cannot be proved without help from other branches of mathematics.

In defense of the subject of this book, we observe that the fundamental theorem of algebra is a theorem about the complex number field $\mathbb{C}$, and that $\mathbb{C}$ (unlike $\mathbb{Q}$, for instance) is *not* an object defined by algebra. Certainly, $\mathbb{C}$ is constructed from $\mathbb{R}$ by an algebraic technique, but $\mathbb{R}$ itself is constructed by the use of some sort of limit process to "fill in the holes" in $\mathbb{Q}$. Since the definition of $\mathbb{C}$ involves ideas outside of algebra, and since any theorem about $\mathbb{C}$ must ultimately refer to its definition, there could not possibly be a purely algebraic proof of such a theorem.

Nevertheless, there does exist a proof of the fundamental theorem of algebra that is *almost* purely algebraic. It uses Galois theory and group theory and only very elementary analysis facts. We give that proof in Chapter 22, after we develop the required Galois theory.

(17.17) LEMMA.   *Assume that $f \in F[X]$ splits, so that*

$$f(X) = a \prod_{i=1}^{n} (X - \alpha_i)$$

for some elements a and α_i in F. Then the α_i are the only roots of f in any extension field of F.

Proof. Let $E \supseteq F$ and suppose that $\alpha \in E$ is a root of f. We have $0 = f(\alpha) = a \prod(\alpha - \alpha_i)$, and thus one of the factors $\alpha - \alpha_i$ must be zero. ∎

The reader should note that we did not state Lemma 17.17 by saying that "the α_i are all of the roots of f." One should not speak of "roots of f" in the abstract; this makes sense only inside a field or some other algebraic structure.

By Lemma 17.17 we see that if $f \in F[X]$ splits, then it is futile to try to find additional roots for f by looking in extension fields of F. Note that this is true even if f has fewer than its theoretical maximum number of $\deg(f)$ roots. (It may be that not all of the α_i are distinct.)

If, on the other hand, $f \in F[X]$ does not split, then there exists an extension field $E \supseteq F$ in which f has at least one additional root. This is because by the following result, we can always choose $E \supseteq F$ such that f splits over E. Then $f(X) = a \prod(X - \alpha_i)$ with all $\alpha_i \in E$, and since we assumed that f does not split over F, it follows that there is some root α_i in E that does not lie in F.

(17.18) THEOREM. *Let $0 \neq f \in F[X]$. Then there exists $E \supseteq F$ such that f splits over E.*

The key to the proof of Theorem 17.18 is the following result. Its proof is easy but somewhat subtle, and it deserves careful thought.

(17.19) LEMMA. *Let $f \in F[X]$ be irreducible. Then there exists a field $E \supseteq F$, such that f has a root in E.*

Proof. We know by Theorem 17.9(d) that if we can find $E \supseteq F$ and $\alpha \in E$ with $f(\alpha) = 0$, then $F[\alpha] \cong F[X]/(f)$. This tells us where to look for a field in which f has a root.

Let I be the ideal (f) in $F[X]$. Since f is irreducible, Lemma 16.9 implies that I is a maximal ideal of $F[X]$, and thus $F[X]/I$ is a field, which we call E. We denote the canonical homomorphism $F[X] \to E$ by an overbar, so that if $g \in F[X]$, then $\overline{g}$ is the coset $I + g \in E$.

We have $F \subseteq F[X]$, as usual, and $F \cap I = 0$, so that $F \cong \overline{F} \subseteq E$. We identify each element $a \in F$ with its image $\overline{a} \in E$ and henceforth we view $F \subseteq E$.

Since an overbar denotes a homomorphism, it respects evaluation of polynomials (which involves only addition and multiplication). By our identification, $\overline{a} = a$ for $a \in F$, and from this it follows that

$$g(\overline{h}) = \overline{g(h)}$$

for all $g, h \in F[X]$. In particular, if we take $g = f$ and $h(X) = X$, this says that

$$f(\overline{X}) = \overline{f(X)} = \overline{f} = 0,$$

where the last equality holds since $f \in I$. Thus $\overline{X} \in E$ is a root of f. ∎

It is instructive to examine the proof of Lemma 17.19 in the case where $F = \mathbb{R}$ and $f(X) = X^2 + 1$. We want, therefore, to construct an extension field of $\mathbb{R}$ in which -1 has a square root. Let $E = \mathbb{R}/(X^2 + 1)$ and consider the coset $(X^2 + 1) + X$, which was denoted $\overline{X}$ in the proof. If we square this element of E, we get the coset $(X^2 + 1) + X^2$. This is the same coset, however, as $(X^2 + 1) + (-1)$, which is the "-1" element of E. We have succeeded, therefore, in constructing a field in which -1 has a square root.

Proof of Theorem 17.18.   Write

$$f(X) = g(X) \prod_{i=1}^{r} (X - \alpha_i),$$

where $g \in F[X]$ and all $\alpha_i \in F$, and where we allow $r = 0$. We use induction on $\deg(g)$, and we note that if $\deg(g) = 0$, then f splits over F and there is nothing to prove. We may assume, therefore, that $\deg(g) > 0$, and so g has some irreducible factor. By Lemma 17.19 that factor, and thus g itself, has a root β in some extension field $K \supseteq F$. Then $X - \beta$ divides $g(X)$ in $K[X]$ and we have

$$f(X) = g_0(X)(X - \beta) \prod_{i=1}^{r} (X - \alpha_i),$$

where $g_0(X) \in K[X]$. Since $\deg(g_0) < \deg(g)$, the inductive hypothesis applies and f splits over some field $E \supseteq K$.   ∎

Suppose $f \in F[X]$ and f splits over E, where $E \supseteq F$. We say that E is a *splitting field* for *f over F* if, in addition, E is generated over F by the roots of f in E.

(17.20) LEMMA.   *Let* $f \in F[X]$ *and assume that* $L \supseteq F$ *and that* f *splits over* L. *Then* L *contains a unique splitting field* E *for* f *over* F. *Also,* E *is contained in every other intermediate field* K *(between* F *and* L) *such that* f *splits over* K.

Proof.   Write

$$f(X) = a \prod_{i=1}^{n} (X - \alpha_i),$$

where $a \in F$ and all $\alpha_i \in L$. Set $E = F[\alpha_1, \alpha_2, \ldots, \alpha_n]$ and note that E is a field since the α_i are algebraic over F. It is clear that E is a splitting field for f over F.

Now suppose that $F \subseteq K \subseteq L$ and that f splits over K. By Lemma 17.17 all roots of f in L lie in K, and it follows that all $\alpha_i \in K$ and so $E \subseteq K$. If K is also a splitting field for f over F, then K is also generated by roots of f, and all of these lie among the α_i by Lemma 17.17 again. This forces $K \subseteq E$, and thus $K = E$.   ∎

(17.21) COROLLARY. *Let $0 \neq f \in F[X]$. Then there exists a splitting field E for f over F.* ∎

Note that a splitting field for a polynomial $f \in F[X]$ is always a finite-degree extension of F.

In addition to the uniqueness in Lemma 17.20, splitting fields enjoy a different kind of uniqueness: given any two splitting fields for $f \in F[X]$ over F, they must be F-isomorphic. We prove a more general result first.

(17.22) THEOREM. *Let $\varphi : F_1 \to F_2$ be a surjective field isomorphism and let $f_i \in F_i[X]$ with $\widehat{\varphi} f_1 = f_2$. Suppose E_i is a splitting field for f_i over F_i. Then φ extends to an isomorphism of E_1 onto E_2.*

Proof. We work by induction on $|E_1 : F_1|$. If this degree is 1, then f_1 splits in $F_1[X]$ and thus f_2 splits in $F_2[X]$, and it follows by Lemma 17.20 that $E_2 = F_2$, and there is nothing to prove.

Suppose that $|E_1 : F_1| > 1$. Then f_1 does not split over F_1 and we can choose an irreducible factor g_1 of f_1 in $F_1[X]$ with $\deg(g_1) > 1$. We let $g_2 = \widehat{\varphi} g_1$ and note that $g_2 \big| f_2$.

Since f_1 splits over E_1, we see that g_1 does too, and thus g_1 has some root $\alpha_1 \in E_1$. Similarly, we can choose a root α_2 for g_2 in E_2.

Now let $K_i = F_i[\alpha_i]$ for $i = 1, 2$. By Lemma 17.15 there is an isomorphism θ of K_1 onto K_2 that extends φ, and we note that $\widehat{\theta} f_1 = f_2$. We will show (by the inductive hypothesis) that θ extends to an isomorphism $E_1 \to E_2$.

We claim that E_i is a splitting field for f_i over K_i. Certainly, f_i splits in $E_i[X]$, and since E_i is generated by roots of f_i over F_i, those same roots generate E_i over K_i. Now $|K_1 : F_1| = \deg(g_1) > 1$ by Theorem 17.9(b), and therefore $|E_1 : K_1| < |E_1 : F_1|$. The inductive hypothesis now implies that θ extends to an isomorphism of E_1 onto E_2. ∎

(17.23) COROLLARY. *Let $f \in F[X]$ and suppose that $E_1 \supseteq F$ and $E_2 \supseteq F$ and that each of E_1 and E_2 is a splitting field for f over F. Then E_1 and E_2 are F-isomorphic.* ∎

17E

Given a field F, we seek an extension field $E \supseteq F$ such that every polynomial $f \in F[X]$ (with $f \neq 0$) splits over E. We say that a field E is *algebraically closed* if every nonzero polynomial $f \in E[X]$ splits. (Thus the fundamental theorem of algebra is the assertion that $\mathbb{C}$ is algebraically closed.) Also, we say that a field $E \supseteq F$ is an *algebraic closure* of F if E is algebraic over F and every nonzero polynomial $f \in F[X]$ splits over E. A field E, therefore, is algebraically closed iff it is an algebraic closure of itself. In fact, more is true.

(17.24) LEMMA. *Let $F \subseteq E$ be an algebraic field extension. The following are then equivalent:*

i.   *E is algebraically closed.*
ii. *E is an algebraic closure of F.*
iii. *There does not exist $L > E$ with L algebraic over F.*
iv. *There does not exist $L > E$ with L algebraic over E.*

Proof.   That (i) implies (ii) is a triviality, since if every nonzero $f \in E[X]$ splits over E, then certainly every nonzero $f \in F[X]$ splits over E. Now assume (ii) and suppose that $L \supseteq E$ with L algebraic over F. To prove (iii), we need to show that $L = E$, and so we consider an element $\alpha \in L$. By hypothesis, α is algebraic over F and so we can find $f \in F[X]$ with $f \neq 0$ and $f(\alpha) = 0$. We know from (ii) that f splits over E and thus $\alpha \in E$ by Lemma 17.17. Thus $L = E$ as desired.

   If $L \supseteq E$ and L is algebraic over E, then L is algebraic over F by Corollary 17.7, and it follows that (iii) implies (iv).

   Finally, assume (iv). If $0 \neq f \in E[X]$, then by Theorem 17.18 there exists a splitting field L for f over E. Since L is certainly algebraic over E, condition (iv) implies that $L = E$, and thus f splits over E and (i) holds.   ∎

We can obtain a few other elementary facts about algebraic closures and algebraically closed fields.

(17.25)   COROLLARY.   *Let $F \subseteq L$ and assume that L is algebraically closed. Let $E = \{\alpha \in L \mid \alpha$ is algebraic over $F\}$. Then E is the unique algebraic closure of F contained in L.*

Proof.   By Theorem 17.5 E is a field and E is clearly algebraic over F. If $0 \neq f \in F[X]$, then f splits over L and thus L contains a splitting field K for f over F. Then $K \subseteq E$ and so f splits over E and E is an algebraic closure for F.

   If $E_0 \subseteq L$ is also an algebraic closure of F, then certainly $E_0 \subseteq E$, and equality follows by Lemma 17.24 since E is algebraic over E_0.   ∎

Assuming the truth of the fundamental theorem of algebra, that $\mathbb{C}$ is algebraically closed, it follows from Corollary 17.25 that the field $\mathbb{A}$ of algebraic numbers is an algebraic closure of $\mathbb{Q}$, and thus $\mathbb{A}$ is algebraically closed by Lemma 17.24.

(17.26)   COROLLARY.   *Let E be a field with the property that every polynomial $f \in E[X]$ with $\deg(f) \geq 1$ has at least one root in E. Then E is algebraically closed.*

Proof.   Let $0 \neq f \in E[X]$. We need to show that f splits, and so we consider an irreducible factor g of f in $E[X]$, and we show that $\deg(g) = 1$. By hypothesis, g has a root α in E and thus $(X - \alpha)\big|g(X)$ in $E[X]$. The irreducibility of g now forces $g(X) = a(X - \alpha)$ for some $a \in E$.   ∎

In view of Corollary 17.26 and the close connection between algebraic closures and algebraically closed fields, it is tempting to conjecture that the analog of Corollary 17.26 for algebraic closures ought to be true. In other words, if $E \supseteq F$ is an algebraic extension and if every $f \in F[X]$ with $\deg(f) \geq 1$ has a root in E, we might expect this to guarantee that E is an algebraic closure of F, and thus that E is algebraically closed.

In fact, the conjecture of the previous paragraph is true, but it is surprisingly difficult to prove. We give a proof of this result after we have developed the necessary theory. (See Theorem 19.22.)

17F

Do algebraic closures always exist?

(17.27) THEOREM. *Let F be any field. Then there exists an algebraic closure for F.*

In principle, this is easy to prove. We "simply" list all of the nonzero polynomials $f \in F[X]$ and then, one by one, construct splitting fields. To be a little more precise, suppose that at some stage in the process we have obtained an algebraic extension $E \supseteq F$ and that $f \in F[X]$ is the next polynomial we wish to split. We can find a splitting field L for f over E and we observe that L is algebraic over F. All of the polynomials in $F[X]$ that we had already split (over E) still split over L, and in addition, the polynomial f splits over L. We continue like this until we exhaust our list of polynomials.

Obviously, the problems with this argument are "listing" the polynomials, finding a "next" polynomial on the list, and "exhausting" the list. In fact, these difficulties can be overcome using the "well ordering" theorem of set theory, which depends on the axiom of choice.

Another approach to a proof of Theorem 17.27 is suggested by the fact that an algebraic closure of F is precisely a maximal algebraic extension. (This follows by Lemma 17.24.) Why not prove Theorem 17.27 by using Zorn's lemma (11.17) to produce a maximal field among the algebraic extensions of F? This is the method we use, but we must be very careful.

If Zorn's lemma can be used to produce a maximal algebraic extension of F, one might think that the same argument could also be used to prove that a maximal field extension $E \supseteq F$ (with no restrictions) exists. That cannot be, however, since $E(X) > E \supseteq F$. There must be something wrong, therefore, with the argument that the partially ordered set $\mathcal{P}$ of all field extensions of F, ordered by containment, has a maximal element by Zorn's lemma. The problem is not that we have failed to verify the hypothesis of Zorn's lemma that every linearly ordered subset of $\mathcal{P}$ has an upper bound; that hypothesis does hold. The difficulty is more subtle: $\mathcal{P}$ is not a set.

In order to avoid this set theoretic difficulty in our proof of Theorem 17.27, we work inside some fixed large set. The purpose of the next lemma is to assure that the set we choose will be large enough. We assume some familiarity with elementary cardinal arithmetic.

(17.28) LEMMA. *Let* $F \subseteq E$ *be an algebraic extension of fields. Then the cardinality of* E *cannot exceed the cardinality of* $F[X]$.

We use the notation $|S|$ to denote the cardinal number of a set S, even if S is infinite. We mention that if F is infinite, then $|F[X]| = |F|$, and it follows in this case that $|E| = |F|$. If F is finite, on the other hand, then $F[X]$ is countable, and hence E is either finite or countably infinite. None of these facts is needed, however.

Proof of Lemma 17.28.   Let S be the set of all ordered pairs (f, α), where $f \in F[X]$ is nonzero and $\alpha \in E$ with $f(\alpha) = 0$. Since for each polynomial f, the number of α such that (f, α) lies in S is finite, we have $|S| \leq \aleph_0 |F[X]| = |F[X]|$. On the other hand, E maps injectively into S via $\alpha \mapsto (\min_F(\alpha), \alpha)$, and thus $|E| \leq |S|$. ∎

We digress briefly for the following corollary .

(17.29) COROLLARY.   *There exist real numbers transcendental over* $\mathbb{Q}$.

Proof.   There are only countably many polynomials in $\mathbb{Q}[X]$. Since $|\mathbb{R}|$ is uncountable, Lemma 17.28 guarantees that $\mathbb{R}$ is not algebraic over $\mathbb{Q}$. ∎

Proof of Theorem 17.27.   Fix any set S with cardinality exceeding that of $F[X]$. (For instance, we could take S to be the power set of $F[X]$.) Choose an arbitrary injective map of F into S and identify F with its image under this map so that we may assume $F \subseteq S$.

Many subsets $E \subseteq S$ can be made into fields by defining addition and multiplication on these sets. We denote a field constructed like that by a pair $(E, \mathcal{S})$, where $E \subseteq S$ is the underlying set of the field and $\mathcal{S}$ denotes the algebraic structure that makes E into a field. (In other words, $\mathcal{S}$ is a pair of functions $E \times E \rightarrow E$ that define addition and multiplication on E.)   In particular, our original field is the pair $(F, \mathcal{F})$ for some fixed structure $\mathcal{F}$.

We write $(E_1, \mathcal{S}_1) \leq (E_2, \mathcal{S}_2)$ if $E_1 \subseteq E_2$ as subsets of S, and the restriction of the algebraic structure $\mathcal{S}_2$ to E_1 is exactly the structure $\mathcal{S}_1$. In other words, $(E_1, \mathcal{S}_1) \leq (E_2, \mathcal{S}_2)$ means exactly that $(E_2, \mathcal{S}_2)$ is an extension field of $(E_1, \mathcal{S}_1)$.

Now let $\mathcal{P}$ be the set of all fields $(E, \mathcal{S})$ with $E \subseteq S$ and such that $(F, \mathcal{F}) \leq (E, \mathcal{S})$, and this is an algebraic field extension. Note that $\mathcal{P}$ is nonempty since $(F, \mathcal{F}) \in \mathcal{P}$.

We intend to apply Zorn's lemma to the poset $\mathcal{P}$, and so we must check that if $\mathcal{L} \subseteq \mathcal{P}$ is linearly ordered, then $\mathcal{L}$ has an upper bound in $\mathcal{P}$. To construct this upper bound $(B, \mathcal{B})$, we put $B = \bigcup E$ as $(E, \mathcal{S})$ runs over $\mathcal{L}$, and we define $\mathcal{B}$ as follows. If $x, y \in B$, then since $\mathcal{L}$ is linearly ordered, we can find $(E, \mathcal{S}) \in \mathcal{L}$ with $x, y \in E$. We define xy and $x + y$ to have the same values that they have in $(E, \mathcal{S})$. We should check that this is well defined and really does make $(B, \mathcal{B})$ into an algebraic field extension of $(F, \mathcal{F})$, but this is routine and we omit the details.

Now let $(M, \mathcal{M})$ be a maximal element of $\mathcal{P}$. To show that this field is an algebraic closure of $(F, \mathcal{F})$, we show that there does not exist a proper algebraic extension of the field $(M, \mathcal{M})$. Suppose L were such an extension. By Lemma 17.28, $|L| < |S|$ and thus $|L - M| < |S - M|$ and there exists an injection of the set $L - M$ into $S - M$. Identify $L - M$ with its image in $S - M$. Now $L \subseteq S$, and we transport the field structure on L via this identification. What results is $(L, S) \in \mathcal{P}$ with $(L, S) > (M, \mathcal{M})$, contradicting the maximality of $(M, \mathcal{M})$. $\blacksquare$

We close this chapter with the corresponding uniqueness result.

(17.30) THEOREM. *Let $\varphi : F_1 \to F_2$ be a surjective isomorphism and let $E_i \supseteq F_i$ be an algebraic closure. Then φ extends to an isomorphism of E_1 onto E_2.*

Proof. Let $\mathcal{P}$ be the set of all pairs (K, θ), where $F_1 \subseteq K \subseteq E_1$ and θ is an isomorphism of K into E_2 that extends φ. Partially order $\mathcal{P}$ by setting $(K_1, \theta_1) \geq (K_2, \theta_2)$ if $K_1 \supseteq K_2$ and θ_2 is the restriction of θ_1 to K_2. Check that this poset satisfies the hypotheses of Zorn's lemma and choose a maximal element (M, μ) of $\mathcal{P}$.

We claim that M is an algebraic closure of F_1. To see this, let $0 \neq f \in F_1[X]$ and note that since f splits over E_1, we can apply Lemma 17.20 and obtain a splitting field K for f over M with $K \subseteq E_1$. Similarly, there exists a splitting field $L \subseteq E_2$ for $\widehat{\mu} f$ over $\mu(M)$. By Theorem 17.22, μ extends to an isomorphism θ of K onto L. Then $(K, \theta) \in \mathcal{P}$ and we have $(K, \theta) \geq (M, \mu)$. The maximality of (M, μ) thus forces $K = M$, and so f splits over M. It follows that M is an algebraic closure of F_1.

Now $E_1 \supseteq M$ is an algebraic extension, and so Lemma 17.24 implies that $E_1 = M$, and thus μ is an isomorphism of E_1 into E_2 that extends φ. Since E_1 is an algebraic closure of F_1 and μ is an isomorphism, we see that $\mu(E_1)$ is an algebraic closure of $\mu(F_1) = \varphi(F_1) = F_2$. Since E_2 is an algebraic extension of $\mu(E_1)$, we can apply Lemma 17.24 again and conclude that $\mu(E_1) = E_2$. $\blacksquare$

(17.31) COROLLARY. *Let F be any field and let E_1 and E_2 be algebraic closures for F. Then E_1 and E_2 are F-isomorphic.* $\blacksquare$

Problems

17.1 A field F of prime characteristic p is said to be *perfect* if the map $\alpha \mapsto \alpha^p$ is a surjection on F.

   a. Show that every finite field is perfect.

   b. Let F be an arbitrary field of characteristic $p \neq 0$. Show that the field $F(X)$ is not perfect.

   HINT: Recall that $(\alpha + \beta)^p = \alpha^p + \beta^p$ if $\alpha, \beta \in F$ and F has prime characteristic p.

17.2 Let $f \in F[X]$ be irreducible and of prime degree p and let $E \supseteq F$ with $|E : F| < \infty$. If f is not irreducible in $E[X]$, show that $p \,|\, |E : F|$.

HINT: Consider a field $L \supseteq E$ in which f has some root α.

17.3 Let $F \subseteq E \subseteq L$ and suppose $\alpha \in L$ is algebraic over F. Let $f = \min_E(\alpha)$.

a. Show that all roots of f in any extension of L are algebraic over F.
b. Show that all of the coefficients of f are algebraic over F.

17.4 Let $F \subseteq E$ be a totally transcendental extension and let $f \in F[X]$ be irreducible. Show that f is irreducible in $E[X]$.

HINT: Apply Problem 17.3(b) when L is a splitting field for f over E.

17.5 Let $f, g \in F[X]$ and assume that $f \mid g$ in $E[X]$ for some field $E \supseteq F$. Show that $f \mid g$ in $F[X]$.

17.6 Let E be the splitting field for $f(X) = X^p - 2$ over $\mathbb{Q}$ in $\mathbb{C}$, where p is a prime number.

a. Show that $E = \mathbb{Q}[\alpha, \epsilon]$, where $\alpha = \sqrt[p]{2}$ and $\epsilon = e^{2\pi i/p}$.
b. Show that $X^{p-1} + X^{p-2} + \cdots + X + 1$ is the minimal polynomial of ϵ over $\mathbb{Q}$.
c. Show that $|E : \mathbb{Q}| = p(p-1)$.

HINT: For part (b), recall that this polynomial was proved to be irreducible in Chapter 16.

17.7 Let F have prime characteristic p and let $a \in F$. Show that the polynomial $f(X) = X^p - a$ either splits or is irreducible in $F[X]$.

HINT: What can you say about all of the roots of f in a splitting field?

17.8 Let $F \subseteq \mathbb{C}$ and assume $\epsilon = e^{2\pi i/p} \in F$, where p is a prime number. Let $a \in F$ be arbitrary. Show that the polynomial $f(X) = X^p - a$ either splits or is irreducible in $F[X]$.

17.9 Let $\alpha = \sqrt{3} + \sqrt[3]{2}$ in $\mathbb{C}$.

a. Show that α is a primitive element for the extension $\mathbb{Q} \subseteq \mathbb{Q}[\sqrt{3}, \sqrt[3]{2}]$.
b. Find a polynomial of degree 6 in $\mathbb{Q}[X]$ that has α as a root, and deduce that this polynomial must be irreducible.

HINT: For part (a), show first that $\sqrt{3} \in \mathbb{Q}[\alpha]$.

17.10 Let $\alpha = \sqrt{2 + \sqrt{2}}$ in $\mathbb{C}$ and let $f = \min_{\mathbb{Q}}(\alpha)$.

a. Compute f.
b. Let E be a splitting field for f over $\mathbb{Q}$. Compute $|E : \mathbb{Q}|$.

17.11 Let $F \subseteq E$ be a field extension and let $\alpha, \beta \in E$.

a. Show that $(F(\alpha))(\beta) = (F(\beta))(\alpha)$.
b. Show that β is algebraic over $F(\alpha)$ iff α is algebraic over $F(\beta)$.

NOTE: We write $F(\alpha, \beta)$ for $(F(\alpha))(\beta)$.

17.12 Let $F \subseteq E$ with $\alpha, \beta \in E$ and assume that α is transcendental over F and β is transcendental over $F(\alpha)$. Let m and n be positive integers.

   a. Show that $|F(\alpha) : F(\alpha^n)| = n$.
   b. Show that $|F(\alpha, \beta) : F(\alpha^n, \beta^m)| = mn$.

HINT: For part (b), write

$$F(\alpha^n, \beta^m) \subseteq F(\alpha, \beta^m) \subseteq F(\alpha, \beta)$$

and apply part (a) twice. This uses both parts of Problem 17.11.

17.13 Let F have prime characteristic p and let $E = F(Y, Z)$, where Y and Z are indeterminates. Let $L = F(Y^p, Z^p) \subseteq E$.

   a. Show that $\alpha^p \in L$ for all $\alpha \in E$.
   b. Show that $E \neq L[\alpha]$ for any $\alpha \in E$.

17.14 Let $f \in F[X]$ with $\deg(f) = n$ and let $E \supseteq F$ be a splitting field for f over F. Show that $|E : F| \leq n!$.

17.15 Let $F \subseteq L$ be an algebraic field extension and let $\theta : L \to L$ be an F-isomorphism. Show that $\theta(L) = L$.

HINT: A polynomial $f \in F[X]$ must have as many roots in $\theta(L)$ as it does in L.

CHAPTER EIGHTEEN

Galois Theory

18A

In Galois theory we associate a finite group, called the "Galois group," with each finite-degree field extension. The object of this is that in many cases, questions about the field extension can be answered by investigating the group. For instance, given a polynomial $f \in \mathbb{Q}[X]$, we might like to "solve" the equation $f(X) = 0$ in $\mathbb{C}$ and write down the roots of f explicitly. For example, if $f(X) = X^3 - 2$, the three complex roots are $\sqrt[3]{2}$ and $\sqrt[3]{2}(-1 \pm \sqrt{-3})/2$. (We have written $\sqrt{-3}$ rather than the more standard $i\sqrt{3}$ in order to emphasize that the roots of $X^3 - 2$ can all be expressed by using the rational numbers, addition, subtraction, multiplication, division, and radicals.) The obvious question here is whether or not this can be done for an arbitrary nonzero polynomial in $\mathbb{Q}[X]$. For any given polynomial f, this question can be interpreted in the field extension $E \supseteq \mathbb{Q}$, where E is the (unique) splitting field for f in $\mathbb{C}$. Galois proved that the roots of f cannot always be expressed by the use of standard arithmetic operations (including radicals). More precisely, Galois showed that f can be "solved by radicals" precisely when the group associated with the extension $\mathbb{Q} \subseteq E$ is a solvable group.

There exist polynomials of degree 5 where the associated Galois group is the symmetric group S_5, which is not solvable. It follows that there cannot exist any general "formula" for "solving" all polynomials of degree 5 (and involving nothing worse than radicals).

In Galois's time (about 1830), formulas for solving polynomials of degree ≤ 4 were known, and so it must have been quite a shock to discover that no such formula could exist for degree 5. (It is interesting that Galois's theorem on polynomials of degree 5 was proved independently and approximately simultaneously by N. Abel. It is an amazing and tragic coincidence that each man proved his result at a very young age and then died shortly thereafter; Galois was killed in a duel at age 21 and Abel died at age 27.)

We mention Galois's polynomial solvability theorem now because this is probably the most famous and spectacular application of Galois theory. We have decided, however, to explore other aspects of the theory of field extensions and other applications (including the theory of finite fields and Euclidean geometric constructions) before we present a precise statement and proof of Galois's theorem in Chapter 22. Impatient readers may skip all but Sections 19A and 20A and all of Chapter 21. The present chapter lays the foundations for the whole theory, however, and should be read in its entirety.

Let $F \subseteq E$ be an arbitrary field extension. (We define the Galois group in this great generality, although additional hypotheses have to be assumed in order to prove anything interesting.) We consider F-isomorphisms from E onto itself. These are the F-*automorphisms* of E, the automorphisms that fix all elements of F. It is easily seen that the set of F-automorphisms of E forms a subgroup of $\text{Aut}(E)$. This subgroup is the *Galois group* of the extension and it is denoted $\text{Gal}(E/F)$. (Note that the virgule "/" has no formal meaning; it is written as a shorthand for "with respect to" and it is pronounced "over." In the formal sense, $\text{Gal}(\cdot/\cdot)$ is a function of two "variables," each a field. The "/" serves merely to separate the two arguments of the function.)

As an example, we compute $\text{Gal}(\mathbb{C}/\mathbb{R})$. First we note that if $\sigma \in \text{Gal}(\mathbb{C}/\mathbb{R}))$, then $-1 = \sigma(-1) = \sigma(i^2) = \sigma(i)^2$. Thus $\sigma(i) = \pm i$. Also, if $a, b \in \mathbb{R}$, then $\sigma(a) = a$ and $\sigma(b) = b$. If $\sigma(i) = i$, we get $\sigma(a + bi) = a + bi$, and if $\sigma(i) = -i$, then $\sigma(a + bi) = a - bi$. It follows that $\text{Gal}(\mathbb{C}/\mathbb{R})$ contains exactly two elements: the identity and complex conjugation. In general, it will be proved that if $E \supseteq F$ with $|E : F| < \infty$, then $\text{Gal}(E/F)$ is finite. (Note that if $|E : F| = 1$, then $\text{Gal}(E/F) = 1$.) If $|E : F| = \infty$, then $\text{Gal}(E/F)$ can be infinite.

It should be mentioned that our definition of a Galois group, though quite standard today, was not the original one. Galois defined his groups only in the case that E is a splitting field over F for some polynomial $f \in F[X]$, and his construction depended on the particular polynomial f. Galois's Galois group, when defined, turns out to be isomorphic to "ours." His group, however, consists of certain permutations of the roots of f rather than of field automorphisms.

The function $\text{Gal}(\cdot/\cdot)$ is a way to get groups from fields. We also need to be able to go in the other direction. Given a subgroup $H \subseteq \text{Aut}(E)$ for some field E, we define $\text{Fix}(H) = \{\alpha \in E \mid \sigma(\alpha) = \alpha \text{ for all } \sigma \in H\}$. It is easily seen that $\text{Fix}(H)$ is a subfield of E; it is called the *fixed field* of H.

Observe that, in general, $\text{Fix}(\text{Gal}(E/F)) \supseteq F$ and $\text{Gal}(E/\text{Fix}(H)) \supseteq H$ for $H \subseteq \text{Aut}(E)$. Also, we have $\text{Fix}(\text{Gal}(\mathbb{C}/\mathbb{R})) = \mathbb{R}$.

Now let $F \subseteq E$ be an arbitrary field extension and let $G = \text{Gal}(E/F)$. We want to study connections between the set $\mathcal{F} = \{K \mid F \subseteq K \subseteq E\}$ of intermediate fields between F and E and the set $\mathcal{G} = \{H \mid H \subseteq G\}$ of subgroups of G. We clearly have maps $\text{Fix}(\cdot) : \mathcal{G} \to \mathcal{F}$ and $\text{Gal}(E/\cdot) : \mathcal{F} \to \mathcal{G}$. The following lemma is a triviality.

(18.1) LEMMA. *Define* $f : \mathcal{G} \to \mathcal{F}$ *and* $g : \mathcal{F} \to \mathcal{G}$ *by* $f = \text{Fix}(\cdot)$ *and* $g = \text{Gal}(E/\cdot)$, *where* $\mathcal{F}$ *and* $\mathcal{G}$ *are as above. Then:*

a. $g(f(H)) \supseteq H$ and $f(g(K)) \supseteq K$ for $H \in \mathcal{G}$ and $K \in \mathcal{F}$.
b. If $H_1 \subseteq H_2$, then $f(H_1) \supseteq f(H_2)$ for $H_i \in \mathcal{G}$.
c. If $K_1 \subseteq K_2$, then $g(K_1) \supseteq g(K_2)$ for $K_i \in \mathcal{F}$.   ∎

Whenever we have two partially ordered sets and two maps between them that satisfy the conclusions of Lemma 18.1, we say that the maps establish a *Galois connection* between the two sets. (Galois connections arise frequently in various areas of mathematics. In the final chapter of this book, for example, we study the Galois connection between sets of polynomials in n indeterminates and their sets of zeros in n-dimensional "graph paper.") The following result lists a few consequences of this situation. It is valid for any Galois connection, since its proof uses only Lemma 18.1 and does not refer to the definitions of the actual sets and maps.

(18.2) LEMMA.   *Assume the situation and notation of Lemma 18.1. Then f and g define inverse bijections between the subsets $\mathcal{F}_0 \subseteq \mathcal{F}$ and $\mathcal{G}_0 \subseteq \mathcal{G}$ defined by*

$$\mathcal{F}_0 = \{f(H) \mid H \in \mathcal{G}\} \quad \text{and} \quad \mathcal{G}_0 = \{g(K) \mid K \in \mathcal{F}\}.$$

Also,

$$\mathcal{F}_0 = \{K \in \mathcal{F} \mid f(g(K)) = K\} \quad \text{and} \quad \mathcal{G}_0 = \{H \in \mathcal{G} \mid g(f(H)) = H\}.$$

Proof.   Certainly, $g(\mathcal{F}_0) \subseteq g(\mathcal{F}) = \mathcal{G}_0$, and so g defines a map $\mathcal{F}_0 \to \mathcal{G}_0$. We let $K \in \mathcal{F}_0$ and compute $f(g(K))$ as follows. We have $f(g(K)) \supseteq K$ by Lemma 18.1(a). Also, we can write $K = f(H)$ for some $H \in \mathcal{G}$ since $K \in \mathcal{F}_0$, and so $g(K) = g(f(H)) \supseteq H$ by Lemma 18.1(a). Thus $f(g(K)) \subseteq f(H) = K$ by Lemma 18.1(b), and it follows that $f(g(K)) = K$. Similarly, f maps $\mathcal{G}_0$ into $\mathcal{F}_0$ and $g(f(H)) = H$ for all $H \in \mathcal{G}_0$. Thus f and g are inverse bijections, as claimed.

If $K \in \mathcal{F}_0$, we have proved that $f(g(K)) = K$, and it is trivial that, conversely, if $f(g(K)) = K$, then $K \in f(\mathcal{G}) = \mathcal{F}_0$. The characterization of $\mathcal{G}_0$ follows similarly.   ∎

Sometimes, in the situation of Lemma 18.2 for a general Galois connection, the elements of $\mathcal{F}_0$ and $\mathcal{G}_0$ are called the *closed* elements of $\mathcal{F}$ and $\mathcal{G}$. This is because $\mathcal{F}_0$ is the image of the map $f(g(\cdot))$, which is a "closure" operator on $\mathcal{F}$. (In general, a *closure* operator is a function $X \mapsto \overline{X}$ on sets such that $\overline{X} \supseteq X$ and $\overline{X} = \overline{\overline{X}}$.) Similarly $g(f(\cdot))$ is a closure operator on $\mathcal{G}$.

Since it is our object to use the structure of the group $G = \mathrm{Gal}(E/F)$ to get information about the field extension $E \supseteq F$ and, in particular, about the intermediate fields of this extension, it is obviously important to investigate which subgroups and intermediate fields are closed. The more closed objects there are, the better, since it is the closed members of $\mathcal{F}$ and of $\mathcal{G}$ that are most closely tied to one another via the maps $f = \mathrm{Fix}(\cdot)$ and $g = \mathrm{Gal}(E/\cdot)$.

As it turns out, the situation could hardly be better. As we shall see, if $|E : F| < \infty$, then $\mathcal{G}_0 = \mathcal{G}$; every subgroup is closed. It is not true in general that every intermediate field is closed, but if the ground field F happens to lie in $\mathcal{F}_0$, then this is true. Furthermore, relatively straightforward necessary and sufficient conditions can be given for a finite-degree field extension $E \supseteq F$ to have the property that $F \in \mathcal{F}_0$. (Of course, the results mentioned in this paragraph are not merely consequences of the trivial fact that we have a Galois connection; their proofs lie considerably deeper.)

The condition $F \in \mathcal{F}_0$ does not always hold for finite degree field extensions, and so we give a name to the "good" situation. We say that $E \supseteq F$ is a *Galois* field extension if $|E : F|$ is finite and $F = \text{Fix}(\text{Gal}(E/F))$; in other words, F is closed in $\mathcal{F}$.

Note that if $F \subseteq E$ is any finite degree field extension and we let $K = \text{Fix}(\text{Gal}(E/F))$, then E is Galois over K by Lemma 18.2, since K lies in the image of the map $f = \text{Fix}(\cdot)$. In particular, of course, the trivial extension $F \subseteq F$ is always Galois.

How can we hope to prove that some particular finite-degree extension $F \subseteq E$ is Galois? What we need to show is that every element $\alpha \in E - F$ is moved by some element $\sigma \in \text{Gal}(E/F)$. What is wanted, therefore, is something that can be used to produce sufficiently many Galois group elements. The tools we use to produce these F-automorphisms of E are Corollary 17.16 and Theorem 17.22. Part (c) of the following lemma shows how we can to use these results to prove that Galois groups are "large."

(18.3) LEMMA. *Let $F \subseteq E$ and write $G = \text{Gal}(E/F)$. Let $f \in F[X]$ with $f \neq 0$ and write $\Omega = \{\alpha \in E \mid f(\alpha) = 0\}$. Assuming that Ω is nonempty, the following hold.*

a. *The action of G on E permutes the elements of Ω.*
b. *If the elements of Ω generate E over F, then G is isomorphically embedded in $\text{Sym}(\Omega)$.*
c. *If we assume that f is irreducible and also that E is a splitting field over F for some polynomial in $F[X]$, then G acts transitively on Ω.*

Proof. Part (a) is a triviality since if $\alpha \in \Omega$ and $\sigma \in G$, then $0 = \sigma(0) = \sigma(f(\alpha)) = f(\sigma(\alpha))$, where the last equality follows because the coefficients of f are fixed by σ since they lie in F. This shows that $\sigma(\alpha) \in \Omega$, as required for (a).

Suppose now that E is generated over F by the elements of Ω. Since G acts on Ω by (a), we have a natural homomorphism $G \to \text{Sym}(\Omega)$. To prove (b), we must show that the kernel K of this map is trivial. We have $\Omega \subseteq \text{Fix}(K)$ and, of course, $F \subseteq \text{Fix}(K)$. It follows that the elements of K all act trivially on the whole field E. In other words, K contains just the identity automorphism.

Now assume that E is a splitting field for g over F, where $g \in F[X]$, and assume also that f is irreducible in $F[X]$. If $\alpha, \beta \in \Omega$, we need to find $\sigma \in G$ such that $\sigma(\alpha) = \beta$. Since f is irreducible over F and α and β are roots of f, we know by Corollary 17.16 that there exists an F-isomorphism φ of $F[\alpha]$ onto $F[\beta]$ that carries α to β.

Now E is a splitting field for g over each of $F[\alpha]$ and $F[\beta]$. Since φ is an F-isomorphism and $g \in F[X]$, we have $\widehat{\varphi} g = g$, and thus by Theorem 17.22, φ extends to an isomorphism σ of E onto E. Because σ extends the F-isomorphism φ, we see that σ acts trivially on F and hence $\sigma \in G$. Also, since $\varphi(\alpha) = \beta$, we deduce that σ carries α to β, as required.   ∎

We shall show (using Lemma 18.3) that a finite-degree field extension $E \supseteq F$ is Galois iff the extension is both "normal" and "separable." We discuss the first of these two properties next.

18B

An algebraic field extension $F \subseteq E$ is a *normal* extension if for every element $\alpha \in E$, the minimal polynomial $\min_F(\alpha)$ splits over E. Equivalently, E is normal over F if every irreducible polynomial $f \in F[X]$ that happens to have a root in E splits over E. (Note that an irreducible polynomial over F that has some element $\alpha \in E$ as a root is just a scalar multiple of $\min_F(\alpha)$.)

It is obvious, for example, that if E is an algebraic closure of F, then E is normal over F, and it is always true that the trivial extension $F \subseteq F$ is normal. Slightly more interesting is the fact that if $|E : F| = 2$, then E is normal over F. To see this, let $\alpha \in E$ and let $f = \min_F(\alpha)$. Then $\deg(f) = |F[\alpha] : F| \leq |E : F| = 2$, and so $\deg(f) = 1$ or 2. If $\deg(f) = 1$, then f certainly splits over E, and if $\deg(f) = 2$, then since $X - \alpha$ divides $f(X)$, we can write $f(X) = (X - \alpha)g(X)$ in $E[X]$, where $\deg(g) = 1$. In this case, too, we see that f splits.

An example of a nonnormal field extension is $\mathbb{Q}[\sqrt[3]{2}] \supseteq \mathbb{Q}$, working inside $\mathbb{C}$. Since the polynomial $X^3 - 2$ is irreducible by the Eisenstein criterion, we see that it is the minimal polynomial of $\sqrt[3]{2}$ over $\mathbb{Q}$. This polynomial has some nonreal roots in $\mathbb{C}$, and so it cannot split over $\mathbb{Q}[\sqrt[3]{2}] \subseteq \mathbb{R}$. It follows that the extension $\mathbb{Q}[\sqrt[3]{2}] \supseteq \mathbb{Q}$ is not normal.

We should mention that not all books on this subject agree on terminology. The word "normal" is sometimes defined to be what we have called "Galois."

The following theorem is a useful characterization of finite-degree normal extensions.

(18.4) THEOREM. *Let $F \subseteq E$ be a field extension with $|E : F| < \infty$. The following are then equivalent:*

i.   *E is normal over F.*
ii. *E is a splitting field over F for some polynomial $g \in F[X]$.*
iii. *For every field extension $L \supseteq E$, all F-isomorphisms from E into L map E onto E.*

We need the following easy lemma.

(18.5) LEMMA. *Let $E \supseteq F$ be a finite-degree field extension and let $\alpha \in E$. Then there exists a field $L \supseteq E$ and a polynomial $g \in F[X]$ such that*

  a. *L is a splitting field for g over F,*
  b. *every irreducible factor of g in F[X] has a root in E, and*
  c. $g(\alpha) = 0$.

Proof.   Let $\alpha_1, \alpha_2, \ldots, \alpha_n$ be a basis for E over F and let g be the product of the minimal polynomials of α and of all α_i over F. Assertions (b) and (c) are clear.

  Let L be a splitting field for g over E (which exists by Corollary 17.21). Now g splits over L, and so to prove (a) we need to show that L is generated over F by roots of g. Certainly, L is generated over E by roots of g, and by the construction of g, we also know that E is generated over F by roots of g, and (a) follows.   ∎

  We mention that in the preceding proof, we could have used Theorem 17.27, that algebraic closures exist. In general, we avoid the use of algebraic closures and instead use splitting fields whenever that is feasible.

Proof of Theorem 18.4.   Assume (i), so that E is normal over F. Use Lemma 18.5 to choose $g \in F[X]$ and $L \supseteq E$ such that L is a splitting field for g over F and every irreducible factor of g in $F[X]$ has a root in E. By the normality of E, each of these irreducible factors of g splits over E, and thus g splits over E. It follows that $E = L$ (by Lemma 17.20, for instance). This proves (ii).

  Now assume (ii). Thus E is a splitting field for some polynomial $g \in F[X]$ over F, and we suppose $L \supseteq E$. If $\theta : E \rightarrow L$ is an F-isomorphism, then $\theta(E)$ is a splitting field for $\widehat{\theta}g = g$ over $\theta(F) = F$. It follows that $\theta(E) = E$ by the uniqueness of splitting fields within larger fields (Lemma 17.20). This is assertion (iii).

  Finally, assume (iii). To prove that E is normal over F, let $\alpha \in E$ and write $f = \min_F(\alpha)$. We need to show that f splits over E. By Lemma 18.5 choose $L \supseteq E$ and $g \in F[X]$ such that L is a splitting field for g over F and $g(\alpha) = 0$. Then $f \mid g$, and hence f splits over L and Lemma 18.3(c) applies. It follows that for each root β of f in L, there exists $\sigma \in \text{Gal}(L/F)$ such that $\sigma(\alpha) = \beta$. By (iii), however, σ maps E to E and thus $\beta \in E$. This proves that all of the roots of f in L lie in E. Since f splits over L, it must also split over E, as required.   ∎

(18.6) COROLLARY.   *Let $F \subseteq E$ be a finite degree extension. Then there exists $L \supseteq E$ such that L is a finite degree normal extension of F.*

Proof.   By Lemma 18.5 there exists a field $L \supseteq E$ such that L is a splitting field for some polynomial $g \in F[X]$ over F. By Theorem 18.4 we know that E is normal over F.   ∎

18C

Next we discuss the notion of "separability" for polynomials, elements, and field extensions. We begin with the following definition: A polynomial $f \in F[X]$ of degree n is said to have *distinct roots* if f has n (different) roots in every field $E \supseteq F$ over which f splits.

(18.7) LEMMA.   *Let $0 \neq f \in F[X]$. The following are then equivalent:*

   i.   *f has distinct roots.*
   ii.   *If $K \supseteq F$ and $\alpha \in K$, then $(X - \alpha)^2$ does not divide f in $K[X]$.*
   iii.   *There exists $K \supseteq F$ such that f has $\deg(f)$ roots in K.*

Proof.   Assume (i) and let $K \supseteq F$ and $\alpha \in K$. If $(X - \alpha)^2 \big| f(X)$, we can write $f(X) = (X - \alpha)^2 g(X)$ for some polynomial $g \in K[X]$. Replacing K by a larger field if necessary, we can assume that g splits over K. Then f splits over K and thus has n distinct roots in K, where $n = \deg(f)$. One of these roots is α, and the remaining $n - 1$ roots must be roots of g. This is a contradiction, however, since $\deg(g) = n - 2$.

Now assume (ii) and choose $K \supseteq F$ to be a splitting field for f over F. Writing $f(X) = a \prod(X - \alpha_i)$ with $\alpha_i \in K$, we see by (ii) that no two of the α_i can be equal. It follows that there are $n = \deg(f)$ distinct α_i, and this proves (iii).

Finally, assume (iii). We need to show that f has $n = \deg(f)$ roots in every extension field $L \supseteq F$ over which f splits. This is certainly true if $L \supseteq K$, and so it suffices to show that f has equal numbers of roots in any two extensions L_1 and L_2 of F over which it splits. Every root of f in L_i, however, lies in the unique splitting field E_i for f over F contained in L_i. Since E_1 and E_2 are F-isomorphic and $f \in F[X]$, it follows that f has equal numbers of roots in these fields.   ∎

Whether or not f has distinct roots is really a property of the polynomial itself and is independent of the particular field in which we consider its coefficients to lie. More precisely, we have the following corollary.

(18.8) COROLLARY.   *Let $F \subseteq K$ and let $0 \neq f \in F[X]$. Then f has distinct roots as a member of $F[X]$ iff it has distinct roots as a member of $K[X]$.*

Proof.   Let $L \supseteq K$ be a field over which f splits. Then by Lemma 18.7 we see that f has distinct roots viewed in $F[X]$ iff f has $n = \deg(f)$ roots in L. By Lemma 18.7, again, this happens iff f has distinct roots when viewed in $K[X]$.   ∎

(18.9) COROLLARY.   *Let $f \in F[X]$ have distinct roots. If $g \big| f$ in $F[X]$, then g has distinct roots.*

Proof.   If $K \supseteq F$ and $\alpha \in K$, then $(X - \alpha)^2$ does not divide $f(X)$ in $K[X]$. It thus cannot divide $g(X)$, and the result follows by Lemma 18.7.   ∎

Finally, we are ready to discuss separability. A nonzero polynomial $f \in F[X]$ is *separable* over F if each irreducible factor of f in $F[X]$ has distinct roots. (Note that we do not require f itself to have distinct roots.) In contrast with Corollary 18.8, the separability of f does depend on the field F. For instance, even if f is inseparable over F, it may be separable when viewed as lying in $E[X]$, where E is some field extension of F. In particular, if f splits over E, then it is certainly separable over E because the irreducible factors of f in $E[X]$ have degree 1 and hence have distinct roots.

(18.10) LEMMA. *Let $F \subseteq E$ and let $f \in F[X]$ be separable over F. Then f is separable over E.*

Proof. Let h be an irreducible factor of f in $E[X]$. By unique factorization in $E[X]$, there must be some irreducible factor g of f in $F[X]$ such that $h \big| g$ in $E[X]$. Since g has distinct roots, Corollary 18.9 guarantees that h does, too. ∎

It seems appropriate at this point to give an example of an inseparable polynomial. Let K be any field of prime characteristic p and let $F = K(Y)$ be the field of rational functions over K in the indeterminate Y. Since F is the field of fractions of $K[Y]$ and since Y is a prime element in this UFD, we can apply the Eisenstein criterion and conclude that the polynomial $f(X) = X^p - Y \in F[X]$ is irreducible.

We claim that f is not separable over F. Since f is irreducible, we must show that f does not have distinct roots. Let α be a root of f in some extension field $E \supseteq F$. Thus $0 = f(\alpha) = \alpha^p - Y$ and $\alpha^p = Y$. Since F has characteristic p, we see that $f(X) = X^p - Y = X^p - \alpha^p = (X - \alpha)^p$ in $E[X]$. Thus $(X - \alpha)^2 \big| f(X)$ in $E[X]$, and f does not have distinct roots by Lemma 18.7. The following result tells us that there can be no easier example.

(18.11) THEOREM. *Let F be a field and suppose that $F[X]$ contains an inseparable polynomial. Then F must be infinite and have prime characteristic.*

Because of Theorem 18.11, we never need to worry about separability in the situations where we most often work: when the fields are of characteristic zero or are finite. In these cases, all polynomials are separable. Except for the psychological benefit of this reassurance, Theorem 18.11 is not relevant to our development of the basic facts of Galois theory, and so it seems best not to interrupt this development now. We therefore defer the proof of this result to the next chapter, where separability and inseparability are studied in depth.

Let $F \subseteq E$ be any field extension. An element $\alpha \in E$, algebraic over F, is said to be *separable over F* if $\min_F(\alpha)$ is separable or, equivalently, has distinct roots. (Note that if $\alpha \in F$, then certainly α is separable over F.) An extension $E \supseteq F$ is *separable* if every element $\alpha \in E$ is separable over F.

(18.12) COROLLARY. *Suppose that $F \subseteq K \subseteq E$ and that E is a separable extension of F. Then E is separable over K and K is separable over F.*

Proof.   Let $\alpha \in E$ and write $f = \min_F(\alpha)$, so that f has distinct roots. Let $g = \min_K(\alpha)$ and note that $g \mid f$ since $f \in K[X]$ and $f(\alpha) = 0$. It follows by Corollary 18.9 that g has distinct roots. The first assertion is now proved, and the second is a triviality.   ∎

18D

We are now ready to state and prove one of the key results of Galois theory: the characterization of Galois extensions.

(18.13) THEOREM.   *Let $F \subseteq E$ with $|E : F| < \infty$. The following are then equivalent:*

i.   *E is a Galois extension of F.*
ii.   *E is both separable and normal over F.*
iii.   *E is a splitting field over F for some separable polynomial over F.*

We extract most of the proof that (i) implies (ii) as a separate lemma that will be used again.

(18.14) LEMMA.   *Let $G \subseteq \mathrm{Aut}(E)$ and write $F = \mathrm{Fix}(G)$. Let $\alpha \in E$ and assume that $\Lambda = \{\sigma(\alpha) \mid \sigma \in G\}$ is a finite set. Then α is algebraic over F. Furthermore, if we write $f = \min_F(\alpha)$, then*

a.   *f has distinct roots,*
b.   *f splits over E,*
c.   *Λ is the set of all roots of f in E, and*
d.   *$\deg(f) = |\Lambda|$.*

Proof.   Define $p \in E[X]$ by

$$p(X) = \prod_{\beta \in \Lambda}(X - \beta).$$

Since each element $\sigma \in G$ permutes Λ, it follows that the corresponding automorphism $\hat{\sigma}$ of $E[X]$ permutes the factors of p, and thus $\hat{\sigma} p = p$. Therefore σ fixes the coefficients of $p(X)$. Because this holds for all $\sigma \in G$, these coefficients must lie in $\mathrm{Fix}(G) = F$, and thus $p \in F[X]$. Since $\alpha \in \Lambda$, we have $p(\alpha) = 0$ and hence α is algebraic over F and $f \mid p$.

By Lemma 18.3(a) every element of Λ is a root of f, and thus $\deg(f) \geq |\Lambda| = \deg(p)$ and we conclude that $f = p$. Also, f has $|\Lambda| = \deg(f)$ roots in E. Everything now follows.   ∎

Proof of Theorem 18.13.   Assume (i) and let $\alpha \in E$. To prove (ii), we must show that $f = \min_F(\alpha)$ has distinct roots and splits over E. Write $G = \mathrm{Gal}(E/F)$ and $\Lambda = \{\sigma(\alpha) \mid \sigma \in G\}$. By Lemma 18.3(a) every element of Λ is a root of f and so $|\Lambda| \leq \deg(f) < \infty$. Now $F = \mathrm{Fix}(G)$ by (i), and thus Lemma 18.14 applies and proves (ii).

Now assume (ii). Since E is normal over F, Theorem 18.4 tells us that E is a splitting field over F for some polynomial $g \in F[X]$. If f is any monic irreducible factor of g, then f has some root $\alpha \in E$, and since α is separable over F by (ii), the polynomial f has distinct roots. Therefore g is separable over F and (iii) is proved.

To show that (iii) implies (i), assume that E is a splitting field over F for some polynomial $g \in F[X]$ that is separable over F. Let $G = \mathrm{Gal}(E/F)$ and write $K = \mathrm{Fix}(G)$. We need to show that $K = F$.

There is nothing to prove if $E = F$, and so we can work by induction on $|E : F|$ and assume that $E > F$. Since E is generated over F by roots of g, there must be some root α of g with $\alpha \in E - F$. Then $|E : F[\alpha]| < |E : F|$ and E is a splitting field for g over $F[\alpha]$. Also, g is separable over $F[\alpha]$ by Lemma 18.10, and thus the inductive hypothesis guarantees that E is Galois over $F[\alpha]$. Since $\mathrm{Gal}(E/F[\alpha]) \subseteq G$, we have

$$F[\alpha] = \mathrm{Fix}(\mathrm{Gal}(E/F[\alpha])) \supseteq \mathrm{Fix}(G) = K.$$

Let $f = \min_F(\alpha)$. Since $g(\alpha) = 0$, we have $f \mid g$ and thus f splits over E. Write Ω to denote the set of roots of f in E, and note that since g is separable, we know that f has distinct roots and hence $|\Omega| = \deg(f)$. By Lemma 18.3 we deduce that G acts on Ω and that the whole set Ω is a single orbit.

Now let $h = \min_K(\alpha)$. Since $K = \mathrm{Fix}(G)$, we see that $G \subseteq \mathrm{Gal}(E/K)$, and thus by Lemma 18.3(a), G permutes the roots of h in E. Since α is one of these roots, so also is each element $\sigma(\alpha)$ for all $\sigma \in G$. It follows that every element of Ω is a root of h, and we have

$$|K[\alpha] : K| = \deg(h) \geq |\Omega| = \deg(f) = |F[\alpha] : F|.$$

Since $F \subseteq K \subseteq F[\alpha]$, we see that $K[\alpha] = F[\alpha]$, and our inequality yields

$$|F[\alpha] : K||K : F| = |F[\alpha] : F| \leq |K[\alpha] : K| = |F[\alpha] : K|.$$

We deduce that $|K : F| = 1$ and $K = F$, as required. ∎

The following easy corollary of Theorem 18.13 is an important step in our development of Galois theory.

(18.15) COROLLARY. *Let $F \subseteq K \subseteq E$ and assume that E is Galois over F. Then E is Galois over K.*

Proof. We know that E is a splitting field over F for some polynomial $g \in F[X]$ that is separable over F. Then g is separable over K by Lemma 18.10, and E is clearly a splitting field for g over K. The result follows by Theorem 18.13. ∎

What does Corollary 18.15 say in the language of Galois connections? Let $F \subseteq E$ and write $G = \mathrm{Gal}(E/F)$. In the notation of Lemma 18.2, the maps

$f = \text{Fix}(\cdot)$ and $g = \text{Gal}(E/\cdot)$ define inverse bijections between the subsets $\mathcal{F}_0$ and $\mathcal{G}_0$ of $\mathcal{F}$ and $\mathcal{G}$. Recall that $\mathcal{F}_0$ consists precisely of those intermediate fields K between F and E such that $K = f(g(K))$. In other words, $K \in \mathcal{F}_0$ iff E is Galois over K. By Corollary 18.15, therefore, if E is Galois over F, then $\mathcal{F}_0 = \mathcal{F}$ and the entire set $\mathcal{F}$ of intermediate fields is in bijective correspondence with $\mathcal{G}_0$.

We will see that for any finite-degree field extension $F \subseteq E$, we always have $\mathcal{G}_0 = \mathcal{G}$. It follows that for Galois extensions, the maps f and g define bijections between the sets of all intermediate fields and all subgroups. This is part of the "fundamental theorem" of Galois theory, toward which we are working.

18E

So far, we have not attempted to compute the order of $\text{Gal}(E/F)$.

(18.16) LEMMA. *Let $F \subseteq E$ with $|E : F| < \infty$. Then $|\text{Gal}(E/F)| < \infty$.*

Proof. Write $G = \text{Gal}(E/F)$ and choose elements $\alpha_i \in E$ such that $E = F[\alpha_1, \alpha_2, \ldots, \alpha_n]$. (For example, we could take the α_i to be a basis for E over F.) Now let f be the product of the minimal polynomials of the α_i over F, and write Ω to denote the set of all roots of f in E. Then Ω is finite and generates E over F. By Lemma 18.3(b) we conclude that G is isomorphically embedded in the finite group $\text{Sym}(\Omega)$. ∎

The following important (and perhaps surprising) theorem is now easy to prove.

(18.17) THEOREM (Primitive element). *Let $F \subseteq E$ be a finite degree separable extension. Then $E = F[\alpha]$ for some $\alpha \in E$.*

We give part of the proof as a separate lemma.

(18.18) LEMMA. *Let $F \subseteq E$ be a finite-degree separable extension. Then there exists $L \supseteq E$ such that L is Galois over F.*

Proof. By Lemma 18.5 there exists a field $L \supseteq E$ and a polynomial $g \in F[X]$ such that L is a splitting field for g over F and every monic irreducible factor of g in $F[X]$ has a root in E. Each such factor, therefore, has the form $\min_F(\alpha)$ for some $\alpha \in E$, and hence it has distinct roots since α is separable over F. It follows that the polynomial g is separable over F, and so L is Galois over F by Theorem 18.13. ∎

Proof of Theorem 18.17. Use Lemma 18.18 to choose $L \supseteq E$ with L Galois over F. By Corollary 18.15 and Lemma 18.2, the set of fields K with $F \subseteq K \subseteq L$ is in bijective correspondence with some set of subgroups of $G = \text{Gal}(L/F)$. (See the discussion following Corollary 18.15.) Since G is finite by Lemma 18.16, it has only finitely many subgroups and hence there are only finitely many fields intermediate between F and L. It follows that there are only finitely many fields intermediate between F and E, and therefore by Artin's theorem (17.11) we conclude that E is a simple extension of F. ∎

We mention that by Theorem 18.11 (which we have not yet proved) every finite-degree extension in characteristic zero is separable, and hence by Theorem 18.17 every such extension has a primitive element.

As a consequence of Theorem 18.17 we can sharpen Lemma 18.16 dramatically.

(18.19) COROLLARY. *Let* $F \subseteq E$ *with* $|E : F| < \infty$. *Then* $|\mathrm{Gal}(E/F)|$ *divides* $|E : F|$. *Furthermore,* $|\mathrm{Gal}(E/F)| = |E : F|$ *iff* E *is Galois over* F.

Proof. Let $G = \mathrm{Gal}(E/F)$ and consider first the case where E is Galois over F. Now E is separable over F by Theorem 18.13, and so by the primitive element theorem (18.17) we have $E = F[\alpha]$ for some element $\alpha \in E$.

Let $f = \min_F(\alpha)$ so that $|E : F| = \deg(f)$. Write Λ to denote the orbit of α under the action of the finite group G. Since $F = \mathrm{Fix}(G)$, we can apply Lemma 18.14 and deduce that $|\Lambda| = \deg(f)$. Thus $|\Lambda| = |E : F|$.

We can also compute the cardinality of the orbit Λ in a different way. By the fundamental counting principle we know that $|\Lambda| = |G : G_\alpha|$, where G_α is the stabilizer of α in G. Since $E = F[\alpha]$ is contained in $\mathrm{Fix}(G_\alpha)$, it follows that $G_\alpha = 1$, and this yields $|G| = |\Lambda| = |E : F|$, as desired.

Now we consider the general case, where E may not be Galois over F. Let $K = \mathrm{Fix}(G)$, and note that by Lemma 18.2 we have $G = \mathrm{Gal}(E/K)$, since G lies in the image of the map $\mathrm{Gal}(E/\cdot)$. It follows that E is Galois over K, and so $|E : K| = |G|$ by the first part of the proof. Thus $|E : F| = |G||K : F|$, and so $|G|$ divides $|E : F|$. Also, $|G| = |E : F|$ iff $K = F$, and the result follows. ∎

18F

For our next result, we begin with a group instead of a field extension.

(18.20) THEOREM. *Let* $G \subseteq \mathrm{Aut}(E)$, *where* E *is any field, and write* $F = \mathrm{Fix}(G)$. *If* G *is finite, then*

a. $|G| = |E : F|$,
b. $G = \mathrm{Gal}(E/F)$, *and*
c. E *is Galois over* F.

Proof. Let $\alpha \in E$ and let Λ denote the orbit of α under G. Since G is finite, Λ is certainly finite and Lemma 18.14 applies. We conclude that α is separably algebraic over F. Also, $|F[\alpha] : F| = \deg(\min_F(\alpha)) = |\Lambda| \leq |G|$.

Since the degrees $|F[\alpha] : F|$ are bounded as α runs over E, we can choose α to maximize this degree, and we claim that $F[\alpha] = E$. If this were not the case, we could choose $\beta \in E$ with $F[\alpha, \beta] > F[\alpha]$. Each element of $F[\alpha, \beta]$ is separable over F, and so by the primitive element theorem (18.17) we can write $F[\alpha, \beta] = F[\gamma]$ for some element $\gamma \in E$. This gives $|F[\gamma] : F| > |F[\alpha] : F|$, which contradicts the choice of α. This proves that $E = F[\alpha]$, as claimed, and we conclude that $|E : F| \leq |G|$.

Now $G \subseteq \mathrm{Gal}(E/F)$, and by Corollary 18.19, $|\mathrm{Gal}(E/F)| \leq |E : F|$ with equality iff E is Galois over F. We have

$$|G| \leq |\mathrm{Gal}(E/F)| \leq |E : F| \leq |G|,$$

and so equality holds and $G = \mathrm{Gal}(E/F)$. This proves (a), (b), and (c). (Note that (c) also follows directly from (a) and (b) since $F = \mathrm{Fix}(G) = \mathrm{Fix}(\mathrm{Gal}(E/F))$ and $|E : F| < \infty$.) ∎

Given a finite-degree field extension $F \subseteq E$, write $G = \mathrm{Gal}(E/F)$. We know that G is finite by Lemma 18.16, and thus if $H \subseteq G$ is an arbitrary subgroup, then Theorem 18.20 can be applied to the finite group H, and we conclude that $H = \mathrm{Gal}(E/\mathrm{Fix}(H)) = g(f(H))$ in the notation of Lemma 18.2. Thus every subgroup of G is in the set $\mathcal{G}_0$ of that lemma.

(18.21) THEOREM (Fundamental, of Galois theory).   *Let $E \supseteq F$ be a Galois extension with $G = \mathrm{Gal}(E/F)$. Write*

$$\mathcal{F} = \{K \mid F \subseteq K \subseteq E\} \quad and \quad \mathcal{G} = \{H \mid H \subseteq G\}.$$

a. *The maps $f = \mathrm{Fix}(\cdot)$ and $g = \mathrm{Gal}(E/\cdot)$ are inverse bijections between $\mathcal{F}$ and $\mathcal{G}$. These maps reverse containments.*
b. *If $g(K) = H$, then $|E : K| = |H|$ and $|K : F| = |G : H|$. In particular, $|E : F| = |G|$.*
c. *If $g(K) = H$ and $\sigma \in G$, then $g(\sigma(K)) = H^\sigma$, the conjugate of H by σ in G. Also, $H \triangleleft G$ iff K is Galois over F, and in this case, $\mathrm{Gal}(K/F) \cong G/H$.*

Proof.   We know by Lemma 18.1 that f and g are containment reversing, and by Lemma 18.2 they define inverse bijections between the subsets $\mathcal{F}_0 \subseteq \mathcal{F}$ and $\mathcal{G}_0 \subseteq \mathcal{G}$ of "closed" fields and groups. Since E is Galois over F, Corollary 18.15 implies that $\mathcal{F}_0 = \mathcal{F}$. That $\mathcal{G}_0 = \mathcal{G}$ follows by Theorem 18.20, and this completes the proof of (a).

Now assume that $g(K) = H$ (or, equivalently, that $f(H) = K$), where $K \in \mathcal{F}$ and $H \in \mathcal{G}$. Since E is Galois over K by Corollary 18.15, we have $|H| = |\mathrm{Gal}(E/K)| = |E : K|$ by Corollary 18.19, and in particular, taking $K = F$ and $H = G$, we have $|G| = |E : F|$. Dividing these equations yields $|G : H| = |E : F|/|E : K| = |K : F|$, and the proof of (b) is complete.

We continue to assume that $g(K) = H$. If $\sigma \in G$, then $\sigma(K)$ is a field, and so $\sigma(K) \in \mathcal{F}$. We want to show that $g(\sigma(K)) = H^\sigma$. Since function composition is relevant here, we write our operators on the right. If $\tau \in H$ and $\alpha \in K$, then $\alpha\tau = \alpha$ since $H = \mathrm{Gal}(E/K)$. This yields $(\alpha\sigma)(\sigma^{-1}\tau\sigma) = (\alpha\tau)\sigma = \alpha\sigma$, and we conclude that $H^\sigma \subseteq \mathrm{Gal}(E/K\sigma) = g(K\sigma)$. Since σ is a field automorphism, we have $|E : K| = |E\sigma : K\sigma| = |E : K\sigma|$, and so part (b) yields

$$|H^\sigma| = |H| = |E : K| = |E : K\sigma| = |g(K\sigma)|.$$

Thus $H^\sigma = g(K\sigma)$, as desired.

If K is Galois over F, then since K is normal over F, we have $\sigma(K) = K$ for all $\sigma \in G$ by Theorem 18.4. It follows that $H^\sigma = H$ for all $\sigma \in G$, and thus $H \triangleleft G$.

Conversely, if $H \triangleleft G$, then $H^\sigma = H$ and hence $\sigma(K) = K$ for all $\sigma \in G$. For all such σ, therefore, the restriction $\rho(\sigma)$ of σ to K is an automorphism of K fixing all elements of F. In other words, ρ defines a map of G into $\text{Gal}(K/F)$, and this map is clearly a group homomorphism. Since $\rho(G) \subseteq \text{Gal}(K/F)$, we have
$$F \subseteq \text{Fix}(\text{Gal}(K/F)) \subseteq \text{Fix}(\rho(G)) \subseteq \text{Fix}(G) = F ,$$
and thus $F = \text{Fix}(\text{Gal}(K/F))$ and K is Galois over F. The map $\text{Fix}(\cdot)$ is thus injective on subgroups of $\text{Gal}(K/F)$, and we deduce from the above chain of containments that $\text{Gal}(K/F) = \rho(G)$.

An element $\sigma \in G$ lies in $\ker(\rho)$ precisely when σ fixes all elements of K. Thus $\ker(\rho) = \text{Gal}(E/K) = H$ and hence
$$G/H \cong \rho(G) = \text{Gal}(K/F) ,$$
completing the proof.   ■

A common error in applying the fundamental theorem of Galois theory is to forget that the maps f and g reverse containment: that big subgroups correspond to small subfields, and *vice versa*. Let the reader beware.

18G

Let's work out an example. Let $f(X) = X^4 - 2 \in \mathbb{Q}[X]$, and note that f is irreducible by the Eisenstein criterion. Write $\alpha = \sqrt[4]{2}$, so that α is the unique positive root of f in $\mathbb{R}$. We can list four roots for f in $\mathbb{C}$: α, $-\alpha$, $i\alpha$ and $-i\alpha$, and this shows that f has distinct roots and, in particular, that f is separable. (The separability of f also follows from Theorem 18.11 since we are working in characteristic zero.)

Let E be the splitting field for f over $\mathbb{Q}$ in $\mathbb{C}$. By Theorem 18.13, therefore, E is Galois over $\mathbb{Q}$. We determine the Galois group $G = \text{Gal}(E/\mathbb{Q})$, and then, using the fundamental theorem of Galois theory, we find some intermediate fields between $\mathbb{Q}$ and E.

First we note that $\mathbb{Q}[\alpha] \subseteq E$ and that this containment must be proper since $\mathbb{Q}[\alpha] \subseteq \mathbb{R}$, and so $i\alpha \notin \mathbb{Q}[\alpha]$. Also, $i = (i\alpha)/\alpha \in E$, and so $\mathbb{Q}[\alpha, i] \subseteq E$. In fact, $\mathbb{Q}[\alpha, i] = E$ since it is clear that all four roots of f in $\mathbb{C}$ lie in this field. Now $|\mathbb{Q}[\alpha] : \mathbb{Q}| = 4$ since $f = \min_{\mathbb{Q}}(\alpha)$. Also, $|\mathbb{Q}[\alpha, i] : \mathbb{Q}[\alpha]| \le 2$ since i is a root of the polynomial $X^2 + 1$ over $\mathbb{Q}[\alpha]$. But $\mathbb{Q}[\alpha, i] = E > \mathbb{Q}[\alpha]$, and we conclude that $|E : \mathbb{Q}[\alpha]| = 2$ and thus $|E : \mathbb{Q}| = 8$.

Now write $G = \text{Gal}(E/\mathbb{Q})$, so that $|G| = 8$. The field $\mathbb{Q}[\alpha]$ is an intermediate field that is not Galois over $\mathbb{Q}$ since it is not normal over $\mathbb{Q}$. (The $\mathbb{Q}$-irreducible polynomial f has a root in this field but does not split.) Therefore the subgroup $H \subseteq G$ corresponding to the field $\mathbb{Q}[\alpha]$ satisfies $|H| = |E : \mathbb{Q}[\alpha]| = 2$ and H is not normal in G.

Up to isomorphism, there is just one group of order 8 that has a nonnormal subgroup of order 2, and so we conclude that $G \cong D_8$. Another way we could see this is to appeal to Lemma 18.3(b), which tells us that G is isomorphically embedded in the symmetric group S_4 on four symbols. Observe that D_8 is also embedded in S_4 because D_8 acts faithfully on the four corners of a square. The images of G and of D_8 are Sylow 2-subgroups of S_4, and so they must be conjugate by Sylow's theorem, and hence they are isomorphic. It follows that $G \cong D_8$.

We can go further and actually determine which eight permutations of $\Omega = \{\alpha, -\alpha, i\alpha, -i\alpha\}$ are induced by elements of G. Complex conjugation is an automorphism of $\mathbb{C}$ and thus determines a $\mathbb{Q}$-isomorphism of E into $\mathbb{C}$. Since E is normal over $\mathbb{Q}$, Theorem 18.4 implies that complex conjugation defines a $\mathbb{Q}$-isomorphism τ of E onto itself. Thus $\tau \in G$ and τ induces the transposition $(i\alpha, -i\alpha)$ on Ω.

Since G is transitive on Ω, there exists $\sigma \in G$ such that $(\alpha)\sigma = i\alpha$. Now $(i)\sigma$ must be a root of $X^2 + 1$, and so $(i)\sigma = \pm i$. Replacing σ by $\tau\sigma$ if necessary, we can assume that $(i)\sigma = i$, and we then calculate that $(i\alpha)\sigma = (i)\sigma \cdot (\alpha)\sigma = i(i\alpha) = -\alpha$. It follows that the permutation induced on Ω by σ is the 4-cycle $(\alpha, i\alpha, -\alpha, -i\alpha)$. Now $\langle\sigma\rangle$ is cyclic of order 4 and $\tau \notin \langle\sigma\rangle$. Thus $G = \langle\sigma, \tau\rangle$ and we could, if we wished, list all eight permutations of Ω induced by the elements of G. (We mention that according to Galois's original definition, this group of permutations of Ω is the Galois group of the polynomial f.)

We now consider intermediate fields. We see that $\sqrt{2}$, i, and $i\sqrt{2}$ all lie in E, and that the fields $\mathbb{Q}[\sqrt{2}]$, $\mathbb{Q}[i]$, and $\mathbb{Q}[i\sqrt{2}]$ are intermediate fields of degree 2 over $\mathbb{Q}$. The first of these, being real, is clearly unequal to either of the others. Also, $\mathbb{Q}[i] \neq \mathbb{Q}[i\sqrt{2}]$, or else $(i\sqrt{2})/i = \sqrt{2}$ would lie in this field and this would force $\mathbb{Q}[\sqrt{2}] = \mathbb{Q}[i]$, which is not the case. We have therefore found three different intermediate fields of degree 2 over $\mathbb{Q}$. Since $G = D_8$ has exactly three subgroups of index 2, these are all of the degree 2 intermediate fields.

We could now go on to find all of the intermediate fields K with $|K : \mathbb{Q}| = 4$. There must be five of these because D_8 has exactly five subgroups of order 2. Although we do not present the relevant calculations, we list the fields:

$$\mathbb{Q}[\alpha], \quad \mathbb{Q}[i\alpha], \quad \mathbb{Q}[\sqrt{2}, i], \quad \mathbb{Q}[\alpha + i\alpha], \quad \mathbb{Q}[\alpha - i\alpha].$$

The reader might wish to prove that no two of these are identical. (Note, however, that there do exist isomorphisms among some of these fields.)

18H

We close this chapter with a result that is useful in applications of Galois theory.

(18.22) THEOREM (Natural irrationalities). *Let $F \subseteq E \subseteq K$ be fields and suppose that E is Galois over F. Let $F \subseteq L \subseteq K$ and write $M = L \cap E$. Assume that no proper subfield of K contains both L and E. Then K is Galois over L, and restriction of automorphisms to E defines an isomorphism of $\mathrm{Gal}(K/L)$ onto $\mathrm{Gal}(E/M)$. In particular, $|K : L| = |E : M|$.*

We make a few observations before we proceed with the proof. Note that there is no assumption made that K or L is of finite degree or is even algebraic over F. In fact, the case where K is transcendental over F occurs in some applications of the theorem.

The necessity of assuming that no proper subfield of K contains both L and E does not really limit the usefulness of this result. Usually, in applications of Theorem 18.22, we start with some large field extension $C \supseteq F$ and two intermediate fields E and L (where E is Galois over F). We can take K to be the subfield of C generated by E and L. This field, $\langle L, E \rangle$, is called the *compositum* of L and E in C. It is constructed by intersecting all subfields of C that contain both L and E, and it should be obvious that it satisfies the hypothesis in Theorem 18.22 that no proper subfield contains both L and E.

Proof of Theorem 18.22.   Since E is Galois over F, there is some polynomial $f \in F[X]$, separable over F, such that E is a splitting field for f over F. Now f splits over K, and we claim that, in fact, K is a splitting field for f over L. Certainly K contains such a splitting field K_0, and since E is generated over F by roots of f, and all of these lie in K_0, we conclude that $E \subseteq K_0$. Since also $L \subseteq K_0$ and $K_0 \subseteq K$, we conclude that $K_0 = K$, as claimed. Since f is separable over L by Lemma 18.10, it follows that K is Galois over L.

If $\sigma \in \mathrm{Gal}(K/L)$, then $\sigma(E) = E$ by Theorem 18.4, since E is normal over F. It follows that the restriction of σ to E is an automorphism of E that fixes all elements of $L \cap E = M$. Therefore restriction defines a homomorphism $\rho : \mathrm{Gal}(K/L) \to \mathrm{Gal}(E/M)$.

If $N = \ker(\rho)$, then each element of N acts trivially on E, and so $E \subseteq \mathrm{Fix}(N)$. Since, clearly, $L \subseteq \mathrm{Fix}(N)$, our hypothesis on K yields that $\mathrm{Fix}(N) = K$. Thus $N \subseteq \mathrm{Gal}(K/K) = 1$, and so ρ is injective.

Let $M_1 = \mathrm{Fix}(\rho(\mathrm{Gal}(K/L)))$. Then $M_1 \supseteq M$ and every element of M_1 is fixed by every automorphism $\sigma \in \mathrm{Gal}(K/L)$. It follows that $M_1 \subseteq \mathrm{Fix}(\mathrm{Gal}(K/L)) = L$. Since also $M_1 \subseteq E$, we have $M_1 \subseteq E \cap L = M$ and thus $M_1 = M$, and we have

$$\mathrm{Fix}(\rho(\mathrm{Gal}(K/L))) = \mathrm{Fix}(\mathrm{Gal}(E/M)) \,.$$

It follows by the fundamental theorem that $\rho(\mathrm{Gal}(K/L)) = \mathrm{Gal}(E/M)$, and ρ is surjective, as required.

Finally,

$$|E : M| = |\mathrm{Gal}(E/M)| = |\mathrm{Gal}(K/L)| = |K : L| \,,$$

and the proof is complete.   ∎

(18.23) COROLLARY.   *In the situation of Theorem 18.22, assume that $|K : F|$ is finite. Then $|K : E| = |L : M|$.*

Proof.   We have

$$|K : E| = |K : M|/|E : M| = |K : M|/|K : L| = |L : M| \,,$$

as required.   ∎

Problems

18.1 Let $F \subseteq E$ be an arbitrary field extension. Show that E is normal over F iff E is the union of all of those intermediate fields K (with $F \subseteq K \subseteq E$) such that K is a splitting field over F of some polynomial in $F[X]$.

18.2 Let $F \subseteq L$ be a normal (but not necessarily finite degree) field extension. Let $F \subseteq E \subseteq L$. Show that E is normal over F iff every F-isomorphism σ of E into L maps E into E.

18.3 Let $F \subseteq E \subseteq L$ and suppose that L is normal over F but that for no field K with $E \subseteq K < L$ is K normal over F. In this situation, we say that L is a *normal closure* for E over F. If $F \subseteq E$ is an arbitrary algebraic extension, show that there exists $L \supseteq E$ such that L is a normal closure for E over F. Show also that L is unique up to E-isomorphism.

HINT: For existence, work inside on algebraic closure $\overline{E}$ for E. Use algebraic closures for uniqueness, too.

18.4 Let $F \subseteq E$ be a finite-degree extension and let $L \supseteq E$ be a splitting field over F for some polynomial $g \in F[X]$ with the property that every irreducible factor of g in $F[X]$ has a root in E. (Note that L exists by Lemma 18.5.) Prove that L is a normal closure for E over F. (See Problem 18.3 for the definition.)

18.5 In the situation of Problem 18.4, assume that E is separable over F. Prove that L is Galois over F and that $|L : F|$ divides $|E : F|!$.

18.6 Suppose $E \supseteq F$ is Galois and $\mathrm{Gal}(E/F)$ is isomorphic to a transitive subgroup of the symmetric group S_n. Show that E is a splitting field over F for some irreducible polynomial $f \in F[X]$ with $\deg(f) = n$.

18.7 Let $f, g \in F[X]$ and suppose $E \supseteq F$ is a splitting field over F both for f and for g. Show that f is separable over F iff g is separable over F.

18.8 Let $F \subseteq E$ be a finite degree normal extension and let K and L be intermediate fields. Assume that E is separable over K and over L. Show that E is separable over $K \cap L$.

18.9 Let E be the splitting field for $X^4 - 2$ over $\mathbb{Q}$ in $\mathbb{C}$. Show that $E = \mathbb{Q}[i + \sqrt[4]{2}]$.

HINT: Find at least five different elements in the orbit of $i + \sqrt[4]{2}$ under $\mathrm{Gal}(E/\mathbb{Q})$.

18.10 Let E be as in Problem 18.9. The group $G = \mathrm{Gal}(E/\mathbb{Q})$ has a unique cyclic subgroup of order 4. Find the intermediate field corresponding to this subgroup.

18.11 Let $F \subseteq E$ be a Galois extension and let K and L be intermediate fields. Show that K and L are F-isomorphic iff the subgroups of $G = \mathrm{Gal}(E/F)$ corresponding to K and L are conjugate in G.

18.12 Let $p_1, p_2, \ldots, p_n$ be different prime numbers and let

$$E = \mathbb{Q}[\sqrt{p_1}, \sqrt{p_2}, \ldots, \sqrt{p_n}]$$

in $\mathbb{R}$.

a. Show that E is Galois over $\mathbb{Q}$.

b. Show that $\mathrm{Gal}(E/\mathbb{Q})$ is elementary abelian of order 2^n.

HINT: For part (b), show that the fields $\mathbb{Q}[\sqrt{k}]$ are all different as k runs over the $2^n - 1$ different nontrivial products of distinct members of the set $\{p_1, p_2, \ldots, p_n\}$.

18.13 In the situation of Problem 18.12, show that for each subscript i, there exists $\sigma_i \in \mathrm{Gal}(E/\mathbb{Q})$ such that $\sigma_i(\sqrt{p_i}) = -\sqrt{p_i}$ and $\sigma_i(\sqrt{p_j}) = \sqrt{p_j}$ for $j \neq i$. Use this to show that $\sqrt{p_1}, \sqrt{p_2}, \ldots, \sqrt{p_n}$ are linearly independent over $\mathbb{Q}$.

18.14 In the situation of Problems 18.12 and 18.13, show that $E = \mathbb{Q}[\sqrt{p_1} + \sqrt{p_2} + \cdots + \sqrt{p_n}]$.

HINT: Show that the orbit of $\sqrt{p_1} + \cdots + \sqrt{p_n}$ under $\mathrm{Gal}(E/\mathbb{Q})$ contains 2^n different elements.

18.15 Let E be an algebraic closure of F and let $F \subseteq K \subseteq E$ with K Galois over F. Show that restriction of automorphisms to K defines homomorphism of $\mathrm{Gal}(E/F)$ onto $\mathrm{Gal}(K/F)$.

18.16 Let E be an algebraic closure of F and assume that E is separable over F. Show that $\mathrm{Fix}(\mathrm{Gal}(E/F)) = F$.

18.17 Let $F \subseteq E$ be an algebraic extension (of possibly infinite degree). Let $G = \mathrm{Gal}(E/F)$. Show that G is "residually finite." In other words, show that the intersection of all normal subgroups of G having finite index is trivial.

18.18 Show that $\mathrm{Aut}(\mathbb{R}) = 1$.

HINT: If $\sigma \in \mathrm{Aut}(\mathbb{R})$ and $a \in \mathbb{R}$ with $a > 0$, show that $\sigma(a) > 0$.

18.19 Let $F \subseteq L$ and suppose K and E are intermediate fields such that E is Galois over F and $E \cap K = F$. If $\alpha \in K$ is algebraic over F and $\min_F(\alpha) = f$, show that f is irreducible in $E[X]$.

18.20 Give an example that shows that Problem 18.19 becomes false if the hypothesis that E is Galois over F is dropped.

18.21 Let $\alpha = \sqrt{2 + \sqrt{2}}$ in $\mathbb{C}$ and let $f = \min_{\mathbb{Q}}(\alpha)$ as in Problem 17.10. Let E be a splitting field for f over $\mathbb{Q}$. Show that $\mathrm{Gal}(E/\mathbb{Q})$ is cyclic of order 4.

18.22 Find an example of fields $F \subseteq K \subseteq E$ such that K is Galois over F and E is Galois over K, but E is not Galois over F.

18.23 Let $E = F(X_1, X_2, \ldots, X_n)$, where the X_i are distinct indeterminates, and write $\Omega = \{X_1, \ldots, X_n\} \subseteq E$. Let

$$G = \{\sigma \in \text{Gal}(E/F) \mid \sigma(\Omega) = \Omega\}.$$

Show that G is a subgroup of $\text{Gal}(E/F)$ isomorphic to $\text{Sym}(\Omega)$.

18.24 Let G be any finite group and F any field. Show that there exist fields L and E with $F \subseteq L \subseteq E$ and where E is Galois over L with $\text{Gal}(E/L) \cong G$.

HINT: Use Problem 18.23.

18.25 Assume the situation and notation of Problem 18.23 and define $f \in E[X]$ by

$$f(X) = (X - X_1)(X - X_2) \cdots (X - X_n).$$

Write

$$f(X) = X^n + c_{n-1}X^{n-1} + \cdots + c_1X + c_0$$

and set

$$K = F[c_0, c_1, \ldots, c_{n-1}] \subseteq E.$$

Show that $K = \text{Fix}(G)$.

HINT: Observe that $|E : K| \le n!$.

NOTE: The coefficients c_i are polynomials in $X_1, X_2, \ldots, X_n$. If we define

$$s_i = (-1)^i c_{n-i}$$

for $i = 1, 2, \ldots, n$, we get the *elementary symmetric functions*. We have more to say about these later.

CHAPTER NINETEEN

Separability and Inseparability

19A

As we saw in Chapter 18, the notion of separability is crucial in Galois theory. Since Theorem 18.11 guarantees that everything is separable in characteristic zero and for finite fields, however, one rarely needs to worry about separability in most applications. The purpose of this chapter is to study how, and to what extent, separability can fail. In particular, we prove Theorem 18.11.

Suppose f is a polynomial with real coefficients. We can study the (real) roots of f by considering the graph of $y = f(x)$. These roots, of course, are the points where the graph touches or crosses the x-axis, and a real root α occurs with multiplicity exceeding 1 precisely when the x-axis is actually tangent to the graph at $x = \alpha$. In other words, f fails to have distinct roots if there is some real number α that is simultaneously a root of both f and its derivative f'. We can generalize the notion of the derivative to obtain a necessary and sufficient condition for a polynomial to have distinct roots over an arbitrary field. The key idea here is that of a "derivation."

Suppose R is any ring. (We need not assume commutativity for the definition, and in fact, even the associativity of the multiplication in R is irrelevant.) An additive homomorphism $\delta : R \to R$ is a *derivation* of R if

$$\delta(xy) = x\delta(y) + \delta(x)y$$

for all $x, y \in R$. For example, if $R = \mathbb{R}[X]$, then the ordinary derivative d/dX of elementary calculus is a derivation.

Our next example of a derivation is interesting only when R is noncommutative. (We are, as always, assuming associativity in our ring.) Fix an element $a \in R$ and define the map $\delta = \delta_a : R \to R$ by $\delta(x) = xa - ax$. (The reader should check that δ really is a derivation. It is the *inner* derivation induced by a.)

The set $\mathrm{Der}(R)$ of all derivations of R is closed under addition (where, of course, $\alpha + \beta$ is defined by the formula $(\alpha + \beta)(x) = \alpha(x) + \beta(x)$). In general, however, Der(R) is not closed under multiplication (function composition). As if to compensate for this, however, there is another way to get a new derivation from two old ones. It turns out that if $\alpha, \beta \in \text{Der}(R)$, then the map $\alpha\beta - \beta\alpha$ is also a derivation. (As usual, we write $\alpha\beta$ to denote the composite function "α followed by β.")

We do not intend to pursue our exploration of the set Der(R) much further, but having said this much, we cannot resist mentioning that Der(R), together with its addition operation and the operation that assigns to the pair α, β the element $\alpha\beta - \beta\alpha$, is an example of an algebraic structure called a "Lie ring."

A *Lie ring L* is an additive abelian group that has a (usually nonassociative) "multiplication" in which the "product" of a and b (commonly denoted $[a, b]$) satisfies the following conditions:

1. $[a + b, c] = [a, c] + [b, c]$,
2. $[a, b + c] = [a, b] + [a, c]$,
3. $[a, a] = 0$, and
4. $[[a, b], c] + [[b, c], a] + [[c, a], b] = 0$ for all $a, b, c \in L$.

In Der(R), if we define the bracket operation by $[\alpha, \beta] = \alpha\beta - \beta\alpha$ (where $\alpha\beta$ and $\beta\alpha$ are defined by function composition), then Der(R) becomes a Lie ring.

Suppose F is a field and R is a ring that happens to be an F-algebra. (The two important examples of this for our purposes are $R = F[X]$ and $R = E$, an extension field of F.) A derivation $\delta \in \text{Der}(R)$ is said to be an F-*derivation* if δ is an F-linear operator on R. For example, the calculus derivative d/dX is an $\mathbb{R}$-derivation on $\mathbb{R}[X]$.

(19.1) LEMMA.   *Let $\delta \in \text{Der}(R)$, where R is a commutative ring. Then*

$$\delta(1) = 0 \quad and \quad \delta(a^n) = na^{n-1}\delta(a)$$

for all $a \in R$ and all integers $n \geq 1$.

We remind the reader that if $b \in R$ and $n \geq 0$, then by definition, nb is the sum of n copies of b.

Proof of Lemma 19.1.   First we compute

$$\delta(1) = \delta(1 \cdot 1) = 1\delta(1) + \delta(1)1 = 2\delta(1),$$

and so $\delta(1) = 0$. The case $n = 1$ is trivial, and so we assume $n > 1$ and we complete the proof by induction on n. We have

$$\begin{aligned}
\delta(a^n) = \delta(a \cdot a^{n-1}) &= a\delta(a^{n-1}) + \delta(a)a^{n-1} \\
&= a \cdot (n - 1)a^{n-2}\delta(a) + \delta(a)a^{n-1} \\
&= na^{n-1}\delta(a),
\end{aligned}$$

as required.   ■

(19.2) LEMMA.   *Let R be an F-algebra spanned over F by some subset $B \subseteq R$. Let $\delta : R \to R$ be an F-linear map and suppose*

$$\delta(uv) = u\delta(v) + \delta(u)v$$

for all $u, v \in B$. Then δ is a derivation.

Proof.   We define two functions of two variables by the formulas

$$\alpha(u, v) = \delta(uv) \quad \text{and} \quad \beta(u, v) = u\delta(v) + \delta(u)v,$$

where u and v run over R. Note that each of these functions is linear in each of its variables (when the other variable is held fixed). This follows from the linearity of δ and the distributive law in R.

By hypothesis, $\alpha(u, v) = \beta(u, v)$ if $u, v \in B$, and thus if we hold $u \in B$ fixed, we see from the linearity of α and β in the second variable that $\alpha(u, v) = \beta(u, v)$ for all $v \in R$.

Now hold $v \in R$ fixed. Since $\alpha(u, v) = \beta(u, v)$ for all $u \in B$, linearity in the first variable of both α and β yields that $\alpha(u, v) = \beta(u, v)$ for all $u \in R$. We have now shown that $\alpha = \beta$ on all of $R \times R$. In other words, δ is a derivation. ∎

(19.3) LEMMA.   *Let $R = F[X]$, where F is any field. Then there exists a unique F-derivation δ of R with $\delta(X) = 1$. We have*

$$\delta\left(\sum_{i=0}^{n} a_i X^i\right) = \sum_{i=1}^{n} i a_i X^{i-1}.$$

Proof.   If δ exists, then by Lemma 19.1 we have $\delta(X^i) = iX^{i-1}$ for $i > 0$, and $\delta(X^0) = 0$. The formula for δ given in the statement of the theorem follows by F-linearity, and in particular, δ is unique.

For existence, we define an F-linear map $\delta : R \to R$ by setting $\delta(X^0) = 0$ and $\delta(X^i) = iX^{i-1}$ for $i > 0$. (This is valid since $B = \{X^i \mid i \geq 0\}$ is an F-basis for R.) To prove that δ is a derivation, it suffices by Lemma 19.2 to show that

$$\delta(X^i X^j) = X^i \delta(X^j) + \delta(X^i)X^j$$

for all $i, j \geq 0$. We omit the routine verification of this formula. ∎

The derivation of $F[X]$ defined in Theorem 19.3 is called *formal differentiation*. If $f(X) \in F[X]$, the formal derivative of this polynomial is denoted $f'(X)$. It is important to note that if $F \subseteq E$ (so that f may be viewed either in $F[X]$ or in $E[X]$), then the formal derivative f' is the same in both cases. This follows from the explicit formula in Theorem 19.3.

(19.4) THEOREM.   *Let $0 \neq f \in F[X]$. Then f has distinct roots iff f and f' have no common root in any extension field of F.*

Proof.   Suppose $E \supseteq F$ and $\alpha \in E$ with $f(\alpha) = 0 = f'(\alpha)$. Write $f(X) = (X - \alpha)g(X)$ with $g \in E[X]$. Differentiating (formally), we have

$$f'(X) = (X - \alpha)g'(X) + g(X).$$

This yields

$$0 = f'(\alpha) = g(\alpha),$$

and so $g(X) = (X - \alpha)h(X)$ for some polynomial $h \in E[X]$. It follows that $(X - \alpha)^2 | f(X)$ in $E[X]$, and f does not have distinct roots by Lemma 18.7.

Conversely, suppose f does not have distinct roots. By Lemma 18.7 there exists a field $E \supseteq F$ and an element $\alpha \in E$ such that $(X - \alpha)^2 | f(X)$ in $E[X]$. Writing $f(X) = (X - \alpha)^2 h(X)$ with $h \in E[X]$ and differentiating (using Lemma 19.1), we obtain

$$f'(X) = 2(X - \alpha)h(X) + (X - \alpha)^2 h'(X).$$

Thus $f'(\alpha) = 0$, and since also $f(\alpha) = 0$, the proof is complete. ∎

Note that the analogy between the ordinary calculus derivative and the formal derivative is not perfect. For instance, it is not always true that $\deg(f') = \deg(f) - 1$ when $\deg(f) > 0$. If $f(X) = a_n X^n + \cdots + a_1 X + a_0$ with $a_n \neq 0$, then $f'(X) = na_n X^{n-1} + \cdots + a_1$, and so f' cannot have degree exceeding $n - 1$. If, however, we are working in prime characteristic p and it happens that p divides $n = \deg(f)$, then $na_n = 0$. In this situation, f' does not have degree $n - 1$. An extreme example of this happens for the polynomial $f(X) = X^p - 1$ in characteristic p. In this case $f'(X) = 0$, and so the degree of f' is undefined. In characteristic zero, of course, we do have $\deg(f') = \deg(f) - 1$ if $\deg(f) > 0$.

(19.5) COROLLARY.   *Let $f \in F[X]$ be irreducible. Then f has distinct roots iff f' is a nonzero polynomial.*

Proof.   If f does not have distinct roots, then Theorem 19.4 yields $f(\alpha) = 0 = f'(\alpha)$ for some element α in some field $E \supseteq F$. Since f is irreducible, it divides every polynomial in $F[X]$ that has α as a root (by Theorem 17.9(c)). Thus $f | f'$, and since f' cannot have degree as large as $\deg(f)$, the only possibility is that $f' = 0$.

Conversely, assume $f' = 0$. Let $E \supseteq F$ be a field in which f has some root α. Then $f(\alpha) = 0 = f'(\alpha)$, and hence f does not have distinct roots by Theorem 19.4. ∎

(19.6) COROLLARY.   *Let $f \in F[X]$ and assume that f is irreducible but that it does not have distinct roots. Then the characteristic $\mathrm{char}(F) = p \neq 0$ and $f(X) = g(X^p)$ for some irreducible polynomial $g \in F[X]$.*

Proof.   By Corollary 19.5 we know that $f' = 0$. If we write

$$f(X) = \sum_{i=0}^{n} a_i X^i,$$

we have

$$0 = f'(X) = \sum_{i=1}^{n} i a_i X^{i-1}$$

and thus $i a_i = 0$ for $0 \leq i \leq n$. Since $a_i \neq 0$ for some $i > 0$, we conclude that we are in prime characteristic p and that $a_i = 0$ whenever $p \nmid i$.

We can now rewrite

$$f(X) = \sum_{j=0}^{n/p} a_{pj} X^{pj} = g(X^p),$$

where

$$g(X) = \sum_{j=0}^{n/p} a_{pj} X^j .$$

To see that g must be irreducible, we observe that a proper factorization $g = hk$ would yield the proper factorization $f(X) = h(X^p)k(X^p)$.   ∎

Let F be a field of prime characteristic p. We know that $(a + b)^p = a^p + b^p$ and, of course, $(ab)^p = a^p b^p$. The map $a \mapsto a^p$ is therefore a field isomorphism from F into F. Its image $\{a^p \mid a \in F\}$ is a subfield, usually denoted F^p. The field F is *perfect* if $F^p = F$. In other words, F is perfect if it has prime characteristic p and every element of F is a pth power in F. For example, if F is finite, then since $|F^p| = |F|$, we see that $F^p = F$. Finite fields, therefore, are always perfect.

For an example of a field that is not perfect, let $F = K(t)$, where $K = \mathbb{Z}/p\mathbb{Z}$ (or K is any other field of characteristic $p \neq 0$) and t is an indeterminate. (Thus F is the field of fractions of the polynomial ring $K[t]$.) We claim that t does not lie in F^p, and so F is not perfect. To see this, suppose $t = a^p$ for some element $a \in F$. Then $a = f(t)/g(t)$, where $f, g \in K[t]$ and $f, g \neq 0$. Thus $tg(t)^p = f(t)^p$. However, $\deg(f^p) = p \deg(f)$ and similarly $\deg(g^p) = p \deg(g)$. This gives $1 + p \deg(g) = p \deg(f)$, and this is impossible since the right side is divisible by p and the left side is not.

We are now ready to prove Theorem 18.11, which we generalize slightly as follows.

(19.7) THEOREM.   *Suppose $f \in F[X]$ is an inseparable polynomial. Then F has prime characteristic p and F is not perfect. In particular, F is infinite.*

Proof.   We may assume that f is irreducible, and so it does not have distinct roots. By Corollary 19.6 we have $\operatorname{char}(F) = p \neq 0$ and $f(X) = g(X^p)$ for some polynomial $g \in F[X]$.

If F were perfect, each coefficient of $g(X)$ would be a pth power and we could write $g(X) = \sum a_i^p X^i$, where $a_i \in F$. Then

$$f(X) = g(X^p) = \sum a_i^p X^{pi} = \left(\sum a_i X^i \right)^p .$$

Since f is irreducible, however, it cannot be the pth power of some other polynomial. This is a contradiction, and we conclude that F is not perfect.   ∎

(19.8) COROLLARY.   *Suppose either that F has characteristic zero or that it is perfect of prime characteristic. Then every algebraic field extension of F is separable.*   ∎

We mention one more result along these lines. It follows from Corollary 19.6.

(19.9) COROLLARY.   *Let F have prime characteristic p and let $f \in F[X]$ be irreducible. Then $f(X) = g(X^{p^n})$ for some integer $n \geq 0$ and some separable irreducible polynomial $g \in F[X]$.*

Proof.   If f has distinct roots, take $g = f$ and $n = 0$. Otherwise, by Corollary 19.6 we can write $f(X) = h(X^p)$ for some irreducible polynomial $h \in F[X]$. Since $\deg(h) < \deg(f)$, we can work by induction on the degree and write $h(X) = g(X^{p^n})$ for some integer n and some separable irreducible polynomial g. Then $f(X) = g(X^{p^{n+1}})$, and the result follows.   ∎

19B

Let $F \subseteq E$ be a finite degree field extension. Recall that by definition, this extension is separable iff every element $\alpha \in E$ is separable over F or, equivalently, $\min_F(\alpha)$ has distinct roots. The extreme opposite case is where no element of E is separable over F except, of course, for the elements of F itself. In this situation, we say that E is *purely inseparable* over F.

A trivial example of a purely inseparable extension is $F \subseteq F$; a more interesting example will be forthcoming shortly. Note that since a nontrivial purely inseparable extension $F < E$ is not separable, such an extension can exist only if F has prime characteristic and is not perfect.

(19.10) THEOREM.   *Suppose $F \subseteq E$ is an algebraic extension with $\mathrm{char}(F) = p \neq 0$. The following are then equivalent:*

i.   *E is purely inseparable over F.*
ii.   *For each element $\alpha \in E$, there exists $n \geq 0$ such that $\alpha^{p^n} \in F$.*
iii.   *Each element of E has minimal polynomial over F of the form $X^{p^n} - a$ for some integer $n \geq 0$ and some element $a \in F$.*

Before proving Theorem 19.10, we give a corollary that allows us to construct examples of purely inseparable extensions.

(19.11) COROLLARY.   *Suppose that $E = F[\alpha]$ and that $\alpha^{p^n} \in F$ for some integer $n \geq 0$. Then E is purely inseparable over F.*

Proof.   Let $\beta \in E$. By Theorem 19.10 it suffices to find some p-power power of β that lies in F. We can write $\beta = f(\alpha)$ for some polynomial $f \in F[X]$, and so

$$\beta = a_m \alpha^m + \cdots + a_1 \alpha + a_0,$$

where the coefficients $a_i \in F$. We have

$$\beta^{p^n} = a_m^{p^n} \alpha^{mp^n} + \cdots + a_1^{p^n} \alpha^{p^n} + a_0^{p^n},$$

and this lies in F since $\alpha^{p^n} \in F$. ∎

If F is any field of prime characteristic p and F is not perfect, choose $a \in F$ with $a \notin F^p$ and let $f(X) = X^p - a$. Then f has no root in F, and so if E is a splitting field for f over F, we can choose $\alpha \in E$ with $f(\alpha) = 0$, and we have $F[\alpha] > F$. Since $\alpha^p = a \in F$, Corollary 19.11 guarantees that $F[\alpha]$ is a purely inseparable extension. This shows that every imperfect field of prime characteristic has a nontrivial purely inseparable extension.

Actually, $F[\alpha] = E$ in the above situation. To see this, note that $f(X) = X^p - \alpha^p = (X - \alpha)^p$, and so α is the only root of f in E. Since E is generated over F by roots of f, we conclude that $F[\alpha] = E$, as claimed.

Proof of Theorem 19.10. Suppose that E is purely inseparable over F and let $\alpha \in E$. Write $f = \min_F(\alpha)$ and note that by Corollary 19.9 we can write $f(X) = g(X^{p^n})$ for some integer $n \geq 0$, where $g \in F[X]$ is irreducible and separable over F. We have $g(\alpha^{p^n}) = f(\alpha) = 0$, and it follows that $g = \min_F(\alpha^{p^n})$. Since g is separable, we see that the element α^{p^n} is separable over F, and by the pure inseparability of E over F, we conclude that $\alpha^{p^n} \in F$. This shows that (i) implies (ii).

Now assume (ii) and let $\alpha \in E$. Then $\alpha^{p^n} \in F$ for some integer $n \geq 0$, and thus α is a root of $g(X) = X^{p^n} - \alpha^{p^n} \in F[X]$. Since $g(X) = (X - \alpha)^{p^n}$, we see that any monic irreducible factor f of g in $F[X]$ has the form $(X - \alpha)^r$, where $r > 0$. In particular, $f(\alpha) = 0$, and it follows that $f = \min_F(\alpha)$ and f is uniquely determined. It follows that g is a power of f, and so r is a divisor of p^n and we can write $r = p^m$ for some integer $m \geq 0$. Thus $f(X) = X^{p^m} - a$, where $a = \alpha^{p^m}$. Since $f = \min_F(\alpha)$, this proves (iii).

Finally, we assume (iii) and let $\alpha \in E$ be separable over F. To show that $\alpha \in F$, we let $f = \min_F(\alpha)$ and note that by (iii), f has the form $X^{p^n} - a$. Necessarily then, $a = \alpha^{p^n}$ and we have $f(X) = (X - \alpha)^{p^n}$. Since $f \in F[X]$ is irreducible and separable over F, it has distinct roots and cannot be divisible by $(X - \alpha)^2$. It follows that $n = 0$ and $\alpha = a \in F$, as required. ∎

(19.12) COROLLARY. *Let* $\text{char}(F) = p \neq 0$ *and suppose that* $E \supseteq F$ *is a purely inseparable extension. The following then hold:*

a. *If* $F \subseteq K \subseteq E$, *then* K *is purely inseparable over* F *and* E *is purely inseparable over* K.

b. *If* $|E : F| < \infty$, *then* $|E : F|$ *is a power of* p.

Proof. For (a), note that each element $\alpha \in E$ satisfies $\alpha^{p^n} \in F$ for some $n \geq 0$, depending on α. It is a triviality, therefore, that for each $\alpha \in K$, we have $\alpha^{p^n} \in F$ for some n, and thus K is purely inseparable over F by Theorem 19.10. It is equally trivial that for each $\alpha \in E$, we have $\alpha^{p^n} \in K$ for some n, and so E is purely inseparable over K.

We use induction on $|E : F|$ to prove (b). We may assume $E > F$, and we choose $\alpha \in E - F$. Then $|E : F[\alpha]| < |E : F|$, and $|E : F[\alpha]|$ is a power of p by the inductive hypothesis and part (a). Also, $|F[\alpha] : F| = \deg(\min_F(\alpha))$, which is a power of p by Theorem 19.10. The result follows.   ∎

The converse of Corollary 19.12(a) is also valid.

(19.13) COROLLARY.   *Suppose that $F \subseteq K \subseteq E$ and that $F \subseteq K$ and $K \subseteq E$ are purely inseparable extensions. Then E is purely inseparable over F.*

Proof.   We may assume that $E > F$, and thus either $E > K$ or $K > F$ and hence $\mathrm{char}(F) = p \neq 0$. Let $\alpha \in E$. By Theorem 19.10, $\alpha^{p^n} \in K$ for some $n \geq 0$ and thus, by Theorem 19.10 again, $(\alpha^{p^n})^{p^m} \in F$ for some $m \geq 0$. Therefore $\alpha^{p^{n+m}} \in F$, and so E is purely inseparable over F.   ∎

Our next result shows that purely inseparable extensions can arise "naturally."

(19.14) THEOREM.   *Let $F \subseteq E$ be an algebraic field extension and let*

$$S = \{\alpha \in E \mid \alpha \text{ is separable over } F\}.$$

Then S is a field. It is the unique intermediate field that is separable over F and over which E is purely inseparable.

We begin with a preliminary result.

(19.15) LEMMA.   *Let $E = F[\alpha, \beta]$, where α and β are separable over F. Then E is a separable extension of F.*

Proof.   Let f be the product of the minimal polynomials of α and β over F, and observe that f is separable over F. Let L be a splitting field for f over E and note that L is then a splitting field for f over F. It follows by Theorem 18.13 that L is separable over F, and so E is also separable over F.   ∎

Proof of Theorem 19.14.   Let $\alpha, \beta \in S$. By Lemma 19.15 we see that the whole field $F[\alpha, \beta]$ is contained in S. In particular, $\alpha - \beta \in S$ and also (if $\beta \neq 0$) $\alpha/\beta \in S$. It follows that S is a field. By its definition, it is obvious that S is separable over F.

To see that E is purely inseparable over S, we may assume that F has prime characteristic p, or else $S = E$ and there is nothing to prove. Let $\alpha \in E$ and write $f = \min_F(\alpha)$. By Corollary 19.9 we can write $f(X) = g(X^{p^n})$ for some integer $n \geq 0$ and separable irreducible polynomial $g \in F[X]$. Then $g(\alpha^{p^n}) = f(\alpha) = 0$, and it follows that $g = \min_F(\alpha^{p^n})$. Because g is separable, we see that $\alpha^{p^n} \in S$. Since some p-power power of every element of E lies in S, we conclude that E is purely inseparable over S by Theorem 19.10.

Finally, suppose that $F \subseteq T \subseteq E$, where T is a field separable over F and over which E is purely inseparable. By the definition of S, we clearly have

$T \subseteq S$ and hence, since E is purely inseparable over T, it follows by Corollary 19.12(a) that S is purely inseparable over T. On the other hand, since S is separable over F, it is also separable over T. (This is Corollary 18.12.) At this point, we know that S is both separable and purely inseparable over T, and we deduce that $S = T$. ∎

(19.16) COROLLARY. *Let* $F \subseteq E$ *be a finite degree inseparable extension. Then* $|E : F|$ *is divisible by the characteristic of* F.

Proof. First observe that $\mathrm{char}(F) = p$, a prime. If S is as in Theorem 19.14, then $S < E$ since E is not separable over F. Also, since E is purely inseparable over S, Corollary 19.12(b) tells us that $|E : S|$ is a power of p. The result now follows. ∎

The following result is the separable version of Corollary 19.13. It can also be viewed as a converse of Corollary 18.12.

(19.17) COROLLARY. *Let* $F \subseteq L \subseteq E$, *where* E *is separable over* L *and* L *is separable over* F. *Then* E *is separable over* F.

Proof. As in Theorem 19.14, let S be the field consisting of those elements of E that are separable over F, and note that since L is separable over F, we have $L \subseteq S \subseteq E$. We are given that E is separable over L, and so by Corollary 18.12 we see that E is separable over S. By Theorem 19.14, however, E is purely inseparable over S and it follows that $S = E$ and E is separable over F, as required. ∎

By Theorem 19.14 an arbitrary algebraic field extension $F \subseteq E$ can be viewed as two successive extensions: first we have the separable extension $F \subseteq S$ and then the purely inseparable extension $S \subseteq E$. This suggests the question of whether or not it is possible to reverse the order and to find an intermediate field K such that $F \subseteq K$ is purely inseparable and $K \subseteq E$ is separable. In general, this cannot be done, but using Galois theory, we can easily handle the case of finite-degree normal extensions.

(19.18) THEOREM. *Let* $F \subseteq E$, *where* $|E : F| < \infty$ *and* E *is normal over* F. *Let* $K = \mathrm{Fix}(\mathrm{Gal}(E/F))$. *Then* K *is purely inseparable over* F *and* E *is separable over* K.

Proof. That E is separable over K is immediate since E is Galois over K by Theorem 18.20(c). To see that K is purely inseparable over F, let $\alpha \in K$ and assume α is separable over F. Write $f = \min_F(\alpha)$ and note that f splits over E by the normality assumption.

If $\alpha \notin F$, then $\deg(f) \geq 2$, and since f splits and has distinct roots, there exists $\beta \in E$ with $\beta \neq \alpha$ and $f(\beta) = 0$. Now Lemma 18.3(c) applies since E is normal over F, and we conclude that $\sigma(\alpha) = \beta$ for some $\sigma \in \mathrm{Gal}(E/F)$.

This is a contradiction since $\alpha \in \text{Fix}(\text{Gal}(E/F))$ and $\beta \neq \alpha$. We conclude that $\alpha \in F$, and thus K is purely inseparable over F. ∎

Suppose $F \subseteq E$ is a finite-degree extension and that S is as in Theorem 19.14, the field of separable elements of E (over F). The degree $|S : F|$ is called the *separable part* of the degree of the extension, and we write $|E : F|_{\text{sep}} = |S : F|$. The extension $F \subseteq E$ is separable, therefore, precisely when $|E : F| = |E : F|_{\text{sep}}$. If E is inseparable over F, then $|E : F|_{\text{sep}}$ is a divisor of $|E : F|$, and by Corollary 19.12 the quotient is a power of the characteristic.

What happens if, as in Theorem 19.18, the finite-degree extension $F \subseteq E$ has an intermediate field K such that K is purely inseparable over F and E is separable over K? In fact, if K exists, then as one might guess, it is true that $|E : K| = |E : F|_{\text{sep}}$. This is the content of the next theorem.

(19.19) THEOREM. *Let $F \subseteq E$ be a finite degree extension and let K and L be intermediate fields such that $L \supseteq F$ and $E \supseteq K$ are separable extensions and $E \supseteq L$ and $K \supseteq F$ are purely inseparable extensions. Then $|L : F| = |E : K|$.*

We begin with a lemma.

(19.20) LEMMA. *Suppose that $F \subseteq K$ is a purely inseparable extension. If $f \in F[X]$ is a separable irreducible polynomial, then f remains irreducible in $K[X]$.*

Proof. Suppose that g is a monic irreducible divisor of f in $K[X]$. Our goal is to show that $\deg(g) = \deg(f)$.

Working in some splitting field E for f over K, we can write $g = \prod(X - \alpha_i)$, where each root α_i of g is also a root of f and so is separable over F. If S is the field of all elements of E separable over F, then each factor $X - \alpha_i$ of g lies in $S[X]$, and it follows that $g \in S[X]$. The coefficients of g, therefore, lie in $S \cap K$, which is equal to F since K is purely inseparable over F. Now $g \in F[X]$ and $g \mid f$. Since f is irreducible in $F[X]$, the result follows. ∎

Proof of Theorem 19.19. If $\alpha \in L$, then by Lemma 19.20 we know that $\min_F(\alpha)$ is irreducible in $K[X]$, and thus it is also the minimal polynomial of α over K. It follows that $|F[\alpha] : F| = |K[\alpha] : K|$.

Because L is separable over F, we can use the primitive element theorem (18.17) and apply the result of the preceding paragraph to an element α such that $L = F[\alpha]$. This yields $|L : F| = |K[\alpha] : K| \leq |E : K|$.

To obtain the reverse inequality, choose a primitive element β for the extension $K \subseteq E$. Since E is purely inseparable over L, we know from Theorem 19.10 that $\beta^{p^n} \in L$ for some integer $n \geq 0$, where $p = \text{char}(F)$. (Note that if F has characteristic zero, then $E = L$ and $K = F$ and there is nothing to prove. We assume, therefore, that p is a prime.) Since $E = K[\beta]$, we deduce from Corollary 19.11 that E is purely inseparable over $K[\beta^{p^n}]$. However, E must be separable over this field by Corollary 18.12, since E is separable over K. It follows that $E = K[\beta^{p^n}]$.

Since $\beta^{p^n} \in L$, the result of the first paragraph applies, and we have

$$|E : K| = |K[\beta^{p^n}] : K| = |F[\beta^{p^n}] : F| \leq |L : F|,$$

as required.   ∎

(19.21) COROLLARY. *Let* $F \subseteq U \subseteq E$ *be fields with* $|E : F| < \infty$. *Then* $|E : U|_{\text{sep}}|U : F|_{\text{sep}} = |E : F|_{\text{sep}}$.

Proof. By Theorem 19.14 there exist fields S and T with $F \subseteq S \subseteq U \subseteq T \subseteq E$ such that $S \supseteq F$ and $T \supseteq U$ are separable extensions and $U \supseteq S$ and $E \supseteq T$ are purely inseparable. Another application of Theorem 19.14 gives us a field V with $S \subseteq V \subseteq T$ such that $V \supseteq S$ is separable and $T \supseteq V$ is purely inseparable.

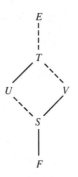

Figure 19.1

In Figure 19.1 we have denoted separable extensions by solid lines and purely inseparable extensions by dashed lines. Now V is separable over F by Corollary 19.17, and E is purely inseparable over V by Corollary 19.13. By the uniqueness in Theorem 19.14, therefore, V is the field consisting of those elements of E that are separable over F. By definition, therefore, $|E : F|_{\text{sep}} = |V : F|$.

We have $|V : F| = |V : S||S : F|$. The second factor equals $|U : F|_{\text{sep}}$ by definition, and the first factor, $|V : S|$, is equal to $|T : U|$ by Theorem 19.19. Finally, $|T : U| = |E : U|_{\text{sep}}$ by definition, and the proof is complete.   ∎

19C

As our final item in this chapter, we prove the following result, which was promised in Chapter 17.

(19.22) THEOREM. *Let* $F \subseteq E$ *be an algebraic field extension and suppose that every nonconstant polynomial* $f \in F[X]$ *has at least one root in* E. *Then* E *is algebraically closed.*

Proof.   First suppose E is separable over F. It suffices to show that E is an algebraic closure of F, and so we let $0 \neq f \in F[X]$ and show that f splits in $E[X]$. We may assume that f is irreducible and monic. Since f has some root $\alpha \in E$, we know that $f = \min_F(\alpha)$, and thus f has distinct roots by the separability assumption.

Now let L be a splitting field for f over F. Since f is a separable polynomial (over F), the field extension $L \supseteq F$ is Galois and hence is separable. Thus $L = F[\beta]$ for some element $\beta \in L$ by the primitive element theorem, and we write $g = \min_F(\beta)$.

By hypothesis, g has some root $\gamma \in E$, and so $L = F[\beta]$ is F-isomorphic to $F[\gamma] \subseteq E$ by Corollary 17.16. It follows that $F[\gamma]$ is a splitting field for f over F, and thus f splits in $E[X]$. This completes the proof in the separable case.

We now assume that $\mathrm{char}(F) = p \neq 0$. Let

$$K = \{\alpha \in E \mid \alpha^{p^n} \in F \text{ for some integer } n \geq 0\}.$$

Note that if $\alpha, \beta \in K$, we can find a single integer n such that $\alpha^{p^n} \in F$ and $\beta^{p^n} \in F$. It follows that $(\alpha - \beta)^{p^n} = \alpha^{p^n} - \beta^{p^n} \in F$, and also (if $\beta \neq 0$) we have $(\alpha/\beta)^{p^n} \in F$. We conclude that K is a field.

Next we show that K is perfect. Given $\alpha \in K$, we have $\alpha^{p^n} \in F$, and so the polynomial $X^{p^{n+1}} - \alpha^{p^n}$ lies in $F[X]$ and thus has a root $\beta \in E$, by hypothesis. Then

$$0 = \beta^{p^{n+1}} - \alpha^{p^n} = (\beta^p - \alpha)^{p^n}$$

and thus $\beta^p = \alpha$. Also, $\beta^{p^{n+1}} = \alpha^{p^n} \in F$, and so $\beta \in K$. We have thus shown that α is a pth power in K and hence K is perfect, as claimed.

Since K is perfect, we know that E is separable over K, and so by the first part of the proof, we will be done if we can show that every nonconstant polynomial $g \in K[X]$ has a root in E. Write $g(X) = a_n X^n + \cdots + a_1 X + a_0$, with $a_i \in K$. By the definition of K, we can choose an integer r large enough so that $(a_i)^{p^r} \in F$ for $0 \leq i \leq n$. Then

$$g(X)^{p^r} = a_n^{p^r} X^{np^r} + \cdots + a_1^{p^r} X^{p^r} + a_0^{p^r} \in F[X].$$

By hypothesis, then, there exists $\alpha \in E$ with $g(\alpha)^{p^r} = 0$, and thus $g(\alpha) = 0$. ∎

Problems

19.1   Let R be an integral domain with field of fractions E. If δ is any derivation of R, show that there exists a derivation Δ of E such that $\Delta(r) = \delta(r)$ for $r \in R$.

19.2   Let F be any field. Show that there exists a field extension $E \supseteq F$ such that E has a nonzero F-derivation.

19.3   Let $F \subseteq E$ be a field extension and let δ be an F-derivation of E.

    a. If $f \in F[X]$ and $\alpha \in E$, show that $\delta(f(\alpha)) = f'(\alpha)\delta(\alpha)$.

    b. Let $K = \{\alpha \in E \mid \delta(\alpha) = 0\}$. Show that K is a subfield of E, that $K \supseteq F$, and that δ is K-linear.

    c. Assuming that E is algebraic over F, show that E is purely inseparable over K, where K is as in part (b).

19.4 Let $E = F[\alpha]$, where $E > F$ and $\alpha^{p^n} \in F$ for some integer $n \geq 0$. Show that there exists an F-derivation δ on E for which $\delta(\alpha) = 1$.

    HINT: Let $m = |E : F|$. Define δ on $1, \alpha, \alpha^2, \dots, \alpha^{m-1}$ and show that you have defined a derivation.

19.5 Let $F \subseteq E$ be a finite-degree extension. Show that this extension is separable iff the only F-derivation of E is the zero map.

19.6 Let R be a (not necessarily commutative) ring.

    a. If $\sigma, \tau \in \text{Der}(R)$, show that $\sigma\tau - \tau\sigma$ is also a derivation.

    b. For $a \in R$, define $\delta_a : R \to R$ by $(x)\delta_a = xa - ax$. Show that $\delta_a \in \text{Der}(R)$. (The map δ_a is the *inner derivation* induced by a.)

    c. In the situation of part (a), if σ and τ are both inner, show that $\sigma\tau - \tau\sigma$ is also inner.

19.7 Let R be a (possibly noncommutative) ring. For $a, b \in R$, write $[a, b] = ab - ba$. Show that R is a Lie ring with respect to the "bracket" operation.

    NOTE: Part of proving that R is a Lie ring involves showing that $[a, b, c] + [b, c, a] + [c, a, b] = 0$ for $a, b, c \in R$, where we write $[x, y, z]$ to mean $[[x, y], z]$. This fact is called the *Jacobi identity*.

19.8 Let L be a Lie ring with respect to a "multiplication" denoted $[a, b]$. A derivation of L is thus an additive endomorphism δ such that $\delta([a, b]) = [\delta(a), b] + [a, \delta(b)]$. If $a \in L$, show that the map $x \mapsto [x, a]$ is a derivation of L.

$\sqrt{}$**19.9** Let $F \subseteq E$ be a finite degree normal extension.

    a. Let $S = \{\alpha \in E \mid \alpha \text{ is separable over } F\}$. Show that S is Galois over F and that restriction defines an isomorphism of $\text{Gal}(E/F)$ onto $\text{Gal}(K/F)$.

    b. Show that $|E : F|_{\text{sep}} = |\text{Gal}(E/F)|$.

    NOTE: Although part (b) can be done using Theorems 19.18 and 19.19, the point of the problem is to use part (a) instead.

$\sqrt{}$**19.10** Let $F \subseteq E$ be a finite degree normal extension and let L be an intermediate field. Show that $|L : F|_{\text{sep}}$ is the number of different F-isomorphisms of L into E.

    HINT: When are the restrictions of two elements of $\text{Gal}(E/F)$ to L equal?

$\sqrt{}$**19.11** Let $E = F[\alpha]$ and let $f = \min_F(\alpha)$. Show that $|E : F|_{\text{sep}}$ is the number of different roots f has in some (and hence any) splitting field for f over F.

19.12 Suppose char$(F) = p$, a prime, and let $f, g \in F[X]$ have distinct roots. If $f(X^{p^a}) = g(X^{p^b})$, show that $f = g$ and $a = b$.

19.13 Let $F \subseteq E$ be purely inseparable with $|E : F| = n < \infty$. If $\alpha \in E$, show that $\alpha^n \in F$.

   HINT: If $\alpha \notin F$, show that $|F[\alpha] : F[\alpha^p]| = p$, where $p = \text{char}(F)$.

19.14 Let $F \subseteq S \subseteq E$ where S is separable over F and E is purely inseparable over S. Let $0 \neq p = \text{char}(F)$. If $\alpha \in E$, choose $n \geq 0$ minimal such that $\alpha^{p^n} \in S$ and write $\alpha^{p^n} = a$.

   a. Show that $X^{p^n} - a$ is irreducible in $S[X]$.
   b. Show that $|F[\alpha] : F[a]| = p^n$.
   c. Show that $f(X) = g(X^{p^n})$, where $f = \min_F(\alpha)$ and $g = \min_F(a)$.
   d. Show that $F[a] = F[\alpha] \cap S$.

19.15 Let $F \subseteq E$ be purely inseparable and let $f \in E[X]$ be monic and irreducible. Show that there exists an integer $m \geq 1$ such that f^m is an irreducible polynomial in $F[X]$.

CHAPTER TWENTY

Cyclotomy and Geometric Constructions

20A

An element $\epsilon \in F$ is said to be a *root of unity* in the field F if it has finite order in the multiplicative group $F^\times$ of F. In particular, ϵ is an nth root of unity if $\epsilon^n = 1$ for some positive integer n, and it is a *primitive* nth root of unity if it actually has order n in $F^\times$ (in other words, $\epsilon^m \neq 1$ for $1 \leq m < n$).

If ϵ is a root of unity in the complex number field $\mathbb{C}$, we have $|\epsilon|^n = |\epsilon^n| = 1$ for some integer n, and thus $|\epsilon| = 1$. The complex roots of unity, therefore, all lie on the unit circle in the complex plane, and hence they are of the form $e^{i\theta} = \cos\theta + i\sin\theta$. In particular, $\epsilon = e^{i\theta}$ is an nth root of unity iff $e^{in\theta} = 1$. The complex number $e^{i\theta}$ is thus an nth root of unity precisely when $\theta = 2\pi k/n$ for some integer k. It follows that there are exactly n nth roots of unity in $\mathbb{C}$ and that they divide the unit circle into n equal arcs. (The word "cyclotomy," meaning "circle dividing," refers to this situation in $\mathbb{C}$ and, by extension, to the theory of roots of unity in arbitrary fields.)

(20.1) LEMMA. *Let F be a field and let $n \geq 1$. The set of nth roots of unity in F forms a cyclic subgroup of $F^\times$ having order dividing n. The number of nth roots of unity in F is exactly equal to n iff F contains a primitive nth root of unity.*

Proof. The set C of nth roots of unity in F is just the set of roots of the polynomial $X^n - 1$, and so it is finite. It is clear that $C \subseteq F^\times$ is a subgroup, which must be cyclic by Lemma 17.12. Write $|C| = m$ and $C = \langle \epsilon \rangle$. Then $m = o(\epsilon)$, and since $\epsilon^n = 1$, we have $m \mid n$.

If $m = n$, then ϵ is a primitive nth root of unity. Conversely, if F contains a primitive nth root of unity δ, then $\delta \in C$ and so $\delta^m = 1$. Thus $n = o(\delta) \leq m$, and we have $n = m$. ∎

The following corollary is immediate from properties of cyclic groups.

(20.2) COROLLARY.   *Suppose $\epsilon \in F$ is a primitive nth root of unity.*

   a. *The elements ϵ^k for $0 \leq k < n$ are distinct and are all of the nth roots of unity in F.*
   b. *The elements ϵ^k with $0 \leq k < n$ and $\gcd(k, n) = 1$ are all of the primitive nth roots of unity in F.*
   c. *F contains precisely $\varphi(n)$ primitive nth roots of unity, where φ is Euler's function.*   ∎

Suppose a field F fails to contain a primitive nth root of unity. Can we simply adjoin one? How many different fields can we get this way?

(20.3) LEMMA.   *A field F has an extension that contains a primitive nth root of unity iff $\mathrm{char}(F)$ does not divide n. If $E \supseteq F$ and $\epsilon \in E$ is a primitive nth root of unity, then $F[\epsilon]$ is a splitting field for $X^n - 1$ over F. In particular, $F[\epsilon]$ is uniquely determined (up to F-isomorphism) by F and n.*

Proof.   Suppose $\mathrm{char}(F)\,|\,n$. Then $\mathrm{char}(F) = p$, a prime, and we can write $n = pm$. Then $X^n - 1 = (X^m - 1)^p$, and this polynomial can have at most m roots in any extension field of F. Thus no extension can contain n nth roots of unity, and so by Lemma 20.1 none contains a primitive nth root.

   If $\mathrm{char}(F) \nmid n$, then zero is the only root of the formal derivative f' of $f(X) = X^n - 1$. Thus f and f' have no common root, and hence f has distinct roots. If E is a splitting field for f over F, we see that f has n roots in E, and thus E contains a primitive nth root by Lemma 20.1. The first assertion of the lemma is now proved.

   If $F \subseteq E$ and $\epsilon \in E$ is a primitive nth root, then $F[\epsilon]$ contains n nth roots, and so $f(X)$ has n roots in $F[\epsilon]$ and hence it splits. Since ϵ is one of the roots of f, we see that $F[\epsilon]$ is actually a splitting field for f over F.   ∎

We now return to the complex number field $\mathbb{C}$. Of course, $\mathbb{C}$ contains the primitive nth root of unity $e^{2\pi i/n}$ for every integer $n \geq 1$, and so $\mathbb{C}$ contains exactly $\varphi(n)$ primitive nth roots. The nth *cyclotomic polynomial*, denoted $\Phi_n(X)$, is the monic polynomial in $\mathbb{C}[X]$ whose roots are precisely the primitive nth roots of unity. Thus

$$\Phi_n(X) = \prod_\epsilon (X - \epsilon) = \prod_k (X - e^{2\pi i k/n}),$$

where in the first product, ϵ runs over the primitive complex nth roots of unity and in the second, k runs over the integers coprime to n such that $0 \leq k < n$. We see that $\deg(\Phi_n) = \varphi(n)$.

   Let's look at a few examples. We see that $\Phi_1(X) = X - 1$ and $\Phi_2(X) = X + 1$ since 1 and -1 are respectively the unique primitive nth roots of unity in $\mathbb{C}$ for $n = 1, 2$. The primitive cube roots of unity in $\mathbb{C}$ are ω and ω^2, where $\omega = e^{2\pi i/3} = (-1 + i\sqrt{3})/2$. Thus $\Phi_3(X) = (X - \omega)(X - \omega^2) = X^2 + X + 1$. To compute Φ_4, we observe that $\pm i$ are the primitive 4th roots of unity, and so $\Phi_4(X) = (X - i)(X + i) = X^2 + 1$.

To compute other examples, we need a better method.

(20.4) LEMMA.
$$X^n - 1 = \prod_{d\mid n} \Phi_d(X),$$

where the product runs over all positive divisors d of n.

Note that a comparison of the degrees of the polynomials on both sides of the equation in Lemma 20.4 yields the standard formula $n = \sum_{d\mid n} \varphi(d)$.

Proof of Lemma 20.4.   The roots of $X^n - 1$ in $\mathbb{C}$ are all of the nth roots of unity in $\mathbb{C}$, and thus $X^n - 1 = \prod(X - \epsilon)$, where ϵ runs over all of these roots. Now each nth root of unity in $\mathbb{C}$ has multiplicative order d for some positive divisor d of n, and so it is a primitive dth root of unity. Conversely, each primitive dth root of unity for each divisor d of n is an nth root of unity. Thus, by grouping the factors $(X - \epsilon)$ in the factorization of $X^n - 1$, we obtain

$$X^n - 1 = \prod_{d\mid n} \prod_{\epsilon} (X - \epsilon),$$

where ϵ runs over primitive dth roots of unity in the inner product. The result now follows.   ∎

(20.5) COROLLARY.   *All coefficients of Φ_n lie in $\mathbb{Z}$.*

Proof.   By Lemma 20.4 we can write

$$\Phi_n(X) = (X^n - 1)/ \prod_{\substack{d\mid n \\ d<n}} \Phi_d(X).$$

Working by induction on n, we may assume that the denominator on the right above lies in $\mathbb{Z}[X]$. Since each cyclotomic polynomial is monic, this denominator is monic, and so Φ_n can be computed in $\mathbb{Z}[X]$ by polynomial long division.   ∎

Lemma 20.4 can be used for explicit computation of cyclotomic polynomials. For instance, if p is prime, then

$$X^p - 1 = \Phi_p(X)\Phi_1(X),$$

and since $\Phi_1(X) = X - 1$, this yields

$$\Phi_p(X) = \frac{X^p - 1}{X - 1} = X^{p-1} + X^{p-2} + \cdots + X + 1.$$

We work out a few more examples. We have

$$\Phi_6(X) = \frac{X^6 - 1}{\Phi_1(X)\Phi_2(X)\Phi_3(X)} = \frac{X^6 - 1}{(X^3 - 1)\Phi_2(X)}$$

$$= \frac{X^3 + 1}{X + 1} = X^2 - X + 1.$$

Similarly,

$$\Phi_{15}(X) = \frac{X^{15} - 1}{(X^5 - 1)\Phi_3(X)} = \frac{X^{10} + X^5 + 1}{X^2 + X + 1}$$

$$= X^8 - X^7 + X^5 - X^4 + X^3 - X + 1,$$

where the last equality is obtained by long division.

As we will see in the problems at the end of the chapter, $\Phi_{pn}(X) = \Phi_n(X^p)$ if p is a prime divisor of n. This yields, for instance, $\Phi_8(X) = \Phi_4(X^2) = X^4 + 1$ and $\Phi_{12}(X) = \Phi_6(X^2) = X^4 - X^2 + 1$.

The reader will undoubtedly have noticed that in all the examples of cyclotomic polynomials Φ_n that we have computed, all of the nonvanishing coefficients are ± 1, and it is tempting to conjecture that this is always so. In fact, further computations show that this "conjecture" holds for all values of n up to 104, but that it fails when $n = 105$. (The author is unaware of any other "simple" wrong conjecture that works for so many cases before failing.) A theorem of Migotti illuminates this phenomenon. It asserts that in order for Φ_n to have a coefficient other than 0, 1, or -1, it is necessary that n be divisible by at least three different odd prime numbers. The smallest positive integer that satisfies this condition is $3 \cdot 5 \cdot 7 = 105$. A proof of Migotti's result can be found in the problems.

20B

Let ϵ_n denote the complex number $e^{2\pi i/n}$, so that ϵ_n is a primitive nth root of unity in $\mathbb{C}$. We write $\mathbb{Q}_n = \mathbb{Q}[\epsilon_n]$, the nth *cyclotomic field*. It should be clear that $\mathbb{Q}_n = \mathbb{Q}[\epsilon]$ for any complex primitive nth root of unity ϵ; our choice of ϵ_n was made only for definiteness of notation. Note that $\mathbb{Q}_n$ could also have been defined as the splitting field of $X^n - 1$ over $\mathbb{Q}$ in $\mathbb{C}$ (by Lemma 20.3), and so $\mathbb{Q}_n$ is Galois over $\mathbb{Q}$. We propose to compute the Galois group $\mathrm{Gal}(\mathbb{Q}_n/\mathbb{Q})$.

As we shall see, $\mathrm{Gal}(\mathbb{Q}_n/\mathbb{Q})$ can be described (up to isomorphism) in two different ways: as $\mathrm{Aut}(C)$ where C is a cyclic group of order n, or as the group U_n of units in the ring $\mathbb{Z}/n\mathbb{Z}$.

(20.6) LEMMA.   *Let C be a cyclic group of order n. Then $\mathrm{Aut}(C) \cong U_n$ and, in particular, $\mathrm{Aut}(C)$ is abelian. If n is prime, $\mathrm{Aut}(C)$ is cyclic of order $n - 1$.*

Proof.   It is no loss to assume that C is the additive group of the ring $\mathbb{Z}/n\mathbb{Z}$. If u is any unit of this ring, then multiplication by u defines an automorphism of C, and this defines a group homomorphism from U_n to $\mathrm{Aut}(C)$.

If $u \in U_n$ determines the identity automorphism of C, then $1 = 1 \cdot u = u$, and so the homomorphism $U_n \to \mathrm{Aut}(C)$ is injective. It is surjective since $|U_n| = \varphi(n) = |\mathrm{Aut}(C)|$ by Theorem 2.11.

Of course, U_n is abelian since $\mathbb{Z}/n\mathbb{Z}$ is a commutative ring. If n is prime, then $\mathbb{Z}/n\mathbb{Z}$ is a field, and so U_n is cyclic (of order $n - 1$) by Lemma 17.12. ∎

(20.7) LEMMA. *Let $F \subseteq E$ and suppose $E = F[\epsilon]$, where ϵ is a root of unity. Then E is Galois over F and $\mathrm{Gal}(E/F)$ is isomorphic to a subgroup of $\mathrm{Aut}(C)$, where $C = \langle \epsilon \rangle \subseteq E^\times$. In particular, $\mathrm{Gal}(E/F)$ is abelian.*

Proof.   Let n be the multiplicative order of ϵ, so that ϵ is a primitive nth root of unity and E is a splitting field for $f(X) = X^n - 1$ over F. Since E contains n nth roots of unity, the polynomial f has n roots in E, and so it has distinct roots and is separable. It follows that E is Galois over F.

If $\sigma \in \mathrm{Gal}(E/F)$, then $\sigma(\epsilon)$ is an nth root of unity and hence $\sigma(\epsilon) \in C$ by Corollary 20.2. It follows that σ maps C into itself, and since $|C| < \infty$, we conclude that the restriction of σ to C is an automorphism of C. Restriction thus defines a homomorphism $\mathrm{Gal}(E/F) \to \mathrm{Aut}(C)$, and this map has a trivial kernel since if $\sigma(\epsilon) = \epsilon$ for $\sigma \in \mathrm{Gal}(E/F)$, then $\sigma = 1$ because $E = F[\epsilon]$. ∎

By Lemma 20.7 we know that $\mathrm{Gal}(\mathbb{Q}_n/\mathbb{Q})$ is isomorphic to a subgroup of $\mathrm{Aut}(C)$, where C is a cyclic group of order n. Our next task is to compute the order of this Galois group. Since $\mathbb{Q}_n$ is Galois over $\mathbb{Q}$, we have

$$|\mathrm{Gal}(\mathbb{Q}_n/\mathbb{Q})| = |\mathbb{Q}_n : \mathbb{Q}| = \deg(\min_{\mathbb{Q}}(\epsilon_n)).$$

We shall see that $\min_{\mathbb{Q}}(\epsilon_n) = \Phi_n$. It follows that $|\mathrm{Gal}(\mathbb{Q}_n/\mathbb{Q})| = \varphi(n)$, and thus the Galois group is isomorphic to the full automorphism group of a cyclic group of order n and to the full group U_n of units in $\mathbb{Z}/n\mathbb{Z}$.

We know that $\Phi_n \in \mathbb{Q}[X]$ and it is monic and has ϵ_n as a root. The additional information needed to show that Φ_n is the minimal polynomial of ϵ_n is irreducibility.

(20.8) THEOREM (Gauss).   *The cyclotomic polynomial $\Phi_n(X)$ is irreducible in $\mathbb{Q}[X]$ for all integers $n \geq 1$.*

Proof.   Let $f \in \mathbb{Q}[X]$ be a monic irreducible factor of Φ_n and write $\Phi_n = fg$ with $g \in \mathbb{Q}[X]$. We assume that $\deg(g) > 0$ and derive a contradiction.

Since $\Phi_n \in \mathbb{Z}[X]$, Gauss's lemma (16.19) applies, and there exist $\alpha, \beta \in \mathbb{Q}$ with $\alpha f, \beta g \in \mathbb{Z}[X]$ and $\Phi_n = (\alpha f)(\beta g)$. Since f and g are monic, this forces $\alpha, \beta \in \mathbb{Z}$, and because $\alpha\beta = 1$, we deduce that $\alpha, \beta = \pm 1$ and $f, g \in \mathbb{Z}[X]$.

Since $fg = \Phi_n$, the complex primitive nth roots of unity are partitioned into two disjoint subsets: the roots of f and the roots of g. If ϵ and δ are roots of f and g, respectively, it follows by Corollary 20.2(b) that we can write $\delta = \epsilon^k$ with $\gcd(k, n) = 1$. We choose roots ϵ and δ of f and g so that k is as small as possible. If p is a prime divisor of k, then $(\epsilon^p)^{k/p} = \delta$, and since $k/p < k$, it follows from the minimality of k that the primitive nth root of unity ϵ^p is not one of the roots of f. It is a root of g, therefore, and the minimality of k now forces $p = k$. Thus $\epsilon^p = \delta$, and so ϵ is a root of the polynomial $g(X^p)$. Since $f = \min_{\mathbb{Q}}(\epsilon)$ by the irreducibility of f in $\mathbb{Q}[X]$, we deduce that $f(X)$ divides $g(X^p)$ in $\mathbb{Q}[X]$. Writing $g(X^p) = f(X)h(X)$, we see that $h \in \mathbb{Z}[X]$. (This follows via polynomial long division since f and g have coefficients in $\mathbb{Z}$ and f is monic.)

Now we come to the "trick": we work mod p. Let $F = \mathbb{Z}/p\mathbb{Z}$, so that F is a field. For each polynomial $k \in \mathbb{Z}[X]$, we can read the coefficients mod p and obtain the polynomial $\overline{k} \in F[X]$. Also, the map $f \mapsto \overline{f}$ is a ring homomorphism from $\mathbb{Z}[X]$ to $F[X]$, and so

$$\overline{g(X^p)} = \overline{f}(X)\,\overline{h}(X).$$

Now, if $\alpha \in F$ and $\alpha \neq 0$, then $\alpha^{p-1} = 1$ since $F^\times$ is a group of order $p - 1$. It follows that $\alpha^p = \alpha$, and this holds for all elements of F. In particular, each coefficient of $\overline{g}$ is equal to its own pth power, and since char$(F) = p$, we conclude that

$$\overline{g}(X)^p = \overline{g}(X^p) = \overline{g(X^p)}.$$

Thus $\overline{f}$ divides $\overline{g}^p$ in $F[X]$, and so $\overline{f}$ and $\overline{g}$ have a common irreducible factor. (Note that $\overline{f}$ has positive degree since f is monic.)

Since $fg = \Phi_n$ divides $X^n - 1$ in $\mathbb{Z}[X]$, we see that

$$\overline{f}\,\overline{g} \text{ divides } \overline{X^n - 1},$$

and it follows that in $F[X]$, the polynomial $X^n - 1$ is divisible by the square of the common irreducible factor of $\overline{f}$ and $\overline{g}$. We conclude that $X^n - 1 \in F[X]$ does not have distinct roots. This is a contradiction, however, since $p \nmid n$, and so the derivative of $X^n - 1$ in $F[X]$ has only the root zero, which is not a root of $X^n - 1$. ∎

(20.9) COROLLARY. $|\mathbb{Q}_n : \mathbb{Q}| = \varphi(n)$.

Proof. Since $\mathbb{Q}_n = \mathbb{Q}[\epsilon_n]$, we have $|\mathbb{Q}_n : \mathbb{Q}| = \deg(\min_\mathbb{Q}(\epsilon_n))$. By Theorem 20.8 we have $\min_\mathbb{Q}(\epsilon_n) = \Phi_n$, and the result follows since $\deg(\Phi_n) = \varphi(n)$. ∎

(20.10) COROLLARY. $\mathrm{Gal}(\mathbb{Q}_n/\mathbb{Q}) \cong \mathrm{Aut}(C) \cong U_n$, *where C is cyclic of order of n.*

Proof. By Lemma 20.7, $\mathrm{Gal}(\mathbb{Q}_n/\mathbb{Q})$ is isomorphically embedded in $\mathrm{Aut}(C)$ and $\mathrm{Aut}(C) \cong U_n$, which has order $\varphi(n)$. Since $\mathbb{Q}_n$ is Galois over $\mathbb{Q}$, we have $|\mathrm{Gal}(\mathbb{Q}_n/\mathbb{Q})| = |\mathbb{Q}_n : \mathbb{Q}| = \varphi(n)$, and the result follows. ∎

Given positive integers m and n, it is not hard to see that the compositum $\langle \mathbb{Q}_m, \mathbb{Q}_n \rangle$ in $\mathbb{C}$ is just the cyclotomic field $\mathbb{Q}_l$, where $l = \mathrm{lcm}(m, n)$, the least common multiple. We shall prove this and also the less trivial "dual" result, that $\mathbb{Q}_m \cap \mathbb{Q}_n = \mathbb{Q}_d$, where $d = \gcd(m, n)$. We first need some facts about Euler's function.

(20.11) LEMMA. *Let $m, n > 1$.*

    a. *If $\gcd(m, n) = 1$, then $U_{mn} \cong U_m \times U_n$.*
    b. *If $\gcd(m, n) = 1$, then $\varphi(mn) = \varphi(m)\varphi(n)$.*

c. *In general,* $\varphi(m)\varphi(n) = \varphi(l)\varphi(d)$, *where* $l = \mathrm{lcm}(m, n)$ *and* $d = \gcd(m, n)$.

Proof. For (a), note that we have a ring homomorphism $\mathbb{Z} \to (\mathbb{Z}/m\mathbb{Z}) \oplus (\mathbb{Z}/n\mathbb{Z})$ defined by $a \mapsto (a + (m\mathbb{Z}), a + (n\mathbb{Z}))$. The kernel of this homomorphism is the set of all integers that are multiples both of m and of n. Since m and n are coprime, this kernel is exactly $mn\mathbb{Z}$, and thus $\mathbb{Z}/mn\mathbb{Z}$ is isomorphic to a subring of $(\mathbb{Z}/m\mathbb{Z}) \oplus (\mathbb{Z}/n\mathbb{Z})$. Since $|\mathbb{Z}/mn\mathbb{Z}| = mn = |(\mathbb{Z}/m\mathbb{Z}) \oplus (\mathbb{Z}/n\mathbb{Z})|$, we conclude that $\mathbb{Z}/mn\mathbb{Z} \cong (\mathbb{Z}/m\mathbb{Z}) \oplus (\mathbb{Z}/n\mathbb{Z})$. Statement (a) now follows, since the units in an external direct sum of two rings are precisely those elements for which each coordinate is a unit. Conclusion (b) is then immediate.

In view of (b), it suffices to prove (c) in the case that m and n are powers of the same prime number. In that case, however, l is one of m or n and d is the other, and so there is nothing to prove. ∎

(20.12) THEOREM. *Let* $l = \mathrm{lcm}(m, n)$ *and* $d = \gcd(m, n)$. *Then*

a. $\langle \mathbb{Q}_m, \mathbb{Q}_n \rangle = \mathbb{Q}_l$ *and*
b. $\mathbb{Q}_m \cap \mathbb{Q}_n = \mathbb{Q}_d$,

where in (a), *the compositum is computed in* $\mathbb{C}$.

Proof. Since $m|l$, we see that ϵ_m is an lth root of unity and so lies in $\mathbb{Q}_l$ (which contains all complex lth roots of unity). Thus $\mathbb{Q}_m = \mathbb{Q}[\epsilon_m] \subseteq \mathbb{Q}_l$ and similarly $\mathbb{Q}_n \subseteq \mathbb{Q}_l$. It follows that $\langle \mathbb{Q}_m, \mathbb{Q}_n \rangle \subseteq \mathbb{Q}_l$. Also, since $d|m$ and $d|n$, similar reasoning gives $\mathbb{Q}_d \subseteq \mathbb{Q}_m \cap \mathbb{Q}_n$.

To complete the proof of (a), we observe that in the cyclic group $\langle \epsilon_l \rangle$, the subgroup $\langle \epsilon_m, \epsilon_n \rangle$ has order divisible by both m and n, and thus its order is divisible by l. It follows that $\langle \epsilon_m, \epsilon_n \rangle$ is the whole group $\langle \epsilon_l \rangle$, and so $\epsilon_l \in \langle \mathbb{Q}_m, \mathbb{Q}_n \rangle$. This gives $\mathbb{Q}_l \subseteq \langle \mathbb{Q}_m, \mathbb{Q}_n \rangle$, and (a) is proved.

To prove (b), we appeal to the natural irrationalities theorem (18.22). Since $\mathbb{Q}_n$ is Galois over $\mathbb{Q}$, we conclude that

$$|\mathbb{Q}_l : \mathbb{Q}_m| = |\langle \mathbb{Q}_m, \mathbb{Q}_n \rangle : \mathbb{Q}_m| = |\mathbb{Q}_n : \mathbb{Q}_n \cap \mathbb{Q}_m|.$$

By Corollary 20.9, however, we have

$$|\mathbb{Q}_l : \mathbb{Q}_m| = \frac{|\mathbb{Q}_l : \mathbb{Q}|}{|\mathbb{Q}_m : \mathbb{Q}|} = \frac{\varphi(l)}{\varphi(m)}$$

and similarly,

$$|\mathbb{Q}_n : \mathbb{Q}_d| = \frac{\varphi(n)}{\varphi(d)} = \frac{\varphi(l)}{\varphi(m)},$$

from Lemma 20.11(c). Thus

$$|\mathbb{Q}_l : \mathbb{Q}_m| = |\mathbb{Q}_n : \mathbb{Q}_m \cap \mathbb{Q}_n| \leq |\mathbb{Q}_n : \mathbb{Q}_d| = \frac{\varphi(l)}{\varphi(m)} = |\mathbb{Q}_l : \mathbb{Q}_m|$$

and we have equality. Statement (b) now follows. ∎

20C

We can use cyclotomic polynomials to prove an important special case of a deep theorem of number theory due to Dirichlet. Combining this with Corollary 20.10, we obtain the following result.

(20.13) THEOREM. *Let G be any finite abelian group. Then there exists a field $E \subseteq \mathbb{C}$, Galois over $\mathbb{Q}$, and such that*

$$\mathrm{Gal}(E/\mathbb{Q}) \cong G\,.$$

An extremely deep theorem of Schafarevitch asserts that Theorem 20.13 remains true if the word "abelian" is replaced by "solvable." It is not known whether or not the word can be deleted altogether.

Dirichlet's theorem says that given coprime positive integers n and m, there exist infinitely many prime numbers $p \equiv m \bmod n$. In other words, there are infinitely many primes of the form $m + kn$ as k runs over positive integers. We will prove the case where $m = 1$.

(20.14) THEOREM. *For each positive integer n, there are infinitely many primes p such that $p \equiv 1 \bmod n$.*

Nearly all the work is in the following lemma.

(20.15) LEMMA. *Let p be a prime divisor of $\Phi_n(m)$ where $n, m \geq 1$ are integers. Then $p \nmid m$. If also $p \nmid n$, then $p \equiv 1 \bmod n$.*

Proof. By Lemma 20.4 we know that $\Phi_n(X)$ divides $X^n - 1$ in $\mathbb{Z}[X]$, and it follows that $\Phi_n(m)$ divides $m^n - 1$ in $\mathbb{Z}$. We conclude that $m^n \equiv 1 \bmod p$ and, in particular, $p \nmid m$. Thus m defines some element $\overline{m}$ in the multiplicative group $(\mathbb{Z}/p\mathbb{Z})^\times$, and we write d to denote the order of $\overline{m}$ in this group. By Lagrange's theorem, $d \mid (p - 1)$ (since the group $(\mathbb{Z}/p\mathbb{Z})^\times$ has order $p - 1$) and it suffices to show that $d = n$. Since $m^n \equiv 1 \bmod p$, we certainly have $d \mid n$, and we can write $n = de$ for some positive integer e.

It suffices to show that if $e > 1$, then $p \mid n$. Since we are assuming that d is a proper divisor of n, we have

$$(X^d)^e - 1 = X^n - 1 = \Phi_n(X)(X^d - 1)g(X)\,,$$

where either $g = 1$ or g is a product of some cyclotomic polynomials. Dividing by $X^d - 1$ yields

$$1 + X^d + (X^d)^2 + \cdots + (X^d)^{e-1} = \Phi_n(X)g(X)\,,$$

and so $\Phi_n(m)$ divides $1 + m^d + \cdots + (m^d)^{e-1}$, and thus $1 + m^d + \cdots + (m^d)^{e-1}$ is divisible by p. Since $m^d \equiv 1 \bmod p$, however, this yields $e \equiv 0 \bmod p$. Recalling that $e \mid n$, we conclude that $p \mid n$, as required. ∎

(20.16) COROLLARY. *Let* $n \geq 1$ *and suppose* p *is a prime divisor of* $\Phi_n(n)$. *Then* $p \equiv 1 \bmod n$. ∎

To complete the proof of Theorem 20.14, we need the following easy result.

(20.17) LEMMA. *Let* $n > 1$. *Then* $|\Phi_n(x)| > x - 1$ *for all real* $x \geq 2$.

Proof. Since $x > 1$, we see that the point on the unit circle closest to x in the complex plane is 1. It follows that $|x - \epsilon| > x - 1$ for every nontrivial complex root of unity ϵ. Since $\Phi_n(x)$ is a product of $\varphi(n)$ factors of the form $x - \epsilon$, it follows that $|\Phi_n(x)| > (x - 1)^{\varphi(n)} \geq x - 1$, where the final inequality holds because $x - 1 \geq 1$. ∎

Proof of Theorem 20.14. We may assume that $n > 1$. For each integer $k \geq 1$, we let $N_k = \Phi_{kn}(kn)$. Then N_k is an integer, and by Lemma 20.17 we see that $|N_k| > 1$, and so N_k has some prime divisor p_k. By Corollary 20.16 we have $p_k \equiv 1 \bmod (nk)$, and hence each prime $p_k \equiv 1 \bmod n$. To see that we have found infinitely many different primes, note that since $1 < p_k \equiv 1 \bmod (nk)$, we have $p_k > nk$, and thus there are arbitrarily large primes among the p_k. ∎

Proof of Theorem 20.13. By the fundamental theorem of abelian groups, write $G \cong C_1 \times C_2 \times \cdots \times C_r$, where C_i is a cyclic group of order n_i. Now, by Theorem 20.14 choose distinct prime numbers p_i with $p_i \equiv 1 \bmod n_i$ for $1 \leq i \leq r$. Let $n = p_1 p_2 \cdots p_r$ and note that

$$\text{Gal}(\mathbb{Q}_n/\mathbb{Q}) \cong U_n \cong U_{p_1} \times U_{p_2} \times \cdots \times U_{p_r},$$

where we use Lemma 20.11(a) repeatedly to get the second isomorphism.

Now U_{p_i} is cyclic of order $p_i - 1$, and since n_i divides $p_i - 1$, we can choose $V_i \subseteq U_{p_i}$ with index n_i. Then

$$\frac{U_{p_1} \times \cdots \times U_{p_r}}{V_1 \times \cdots \times V_r} \cong G,$$

and so $\text{Gal}(\mathbb{Q}_n/\mathbb{Q})$ has a subgroup H such that $\text{Gal}(\mathbb{Q}_n/\mathbb{Q})/H \cong G$.

Let $E = \text{Fix}(H)$. Then E is Galois over $\mathbb{Q}$ since $H \triangleleft \text{Gal}(\mathbb{Q}_n/\mathbb{Q})$, and we have $\text{Gal}(E/\mathbb{Q}) \cong \text{Gal}(\mathbb{Q}_n/\mathbb{Q})/H \cong G$. ∎

We mention that if $\mathbb{Q} \subseteq E \subseteq \mathbb{C}$, where E is arbitrarily chosen so that E is Galois over $\mathbb{Q}$ with abelian Galois group, then E is contained in some cyclotomic field $\mathbb{Q}_n$. We do not prove this deep result.

20D

Next we show how to use field theory to settle certain questions about geometric constructibility. We use the Euclidean rules that specify that only a compass and a straightedge are allowed as tools, and we consider the three classical "hard" problems: squaring a circle, duplicating a cube, and trisecting an angle. In addition, we prove Gauss's theorem concerning which regular n-gons are constructible.

We begin by mentioning that the classical Euclidean rules for the use of a compass do not allow one to open the compass to the length of a certain line segment AB, and then move the compass and draw a circle of radius AB that is not centered at A or B. It is a fact (which we do not prove) that this operation can be simulated by a sequence of "legal" uses of the compass and straightedge. For this reason, we can ignore the question of whether or not compass use is legal in the Euclidean sense.

Given a geometric figure, say a circle or a triangle, printed on a piece of paper, to *square* the figure means to construct a square having area equal to that of the given figure. It turns out that it is possible to square a triangle, but it is not possible to square a circle. As we shall see, the proof of this impossibility depends on the theorem of Lindemann that π is transcendental, which we cannot prove here.

Given a line segment equal to the edge of some cube, to *duplicate* the cube means to construct a line segment equal to the edge of a cube with volume exactly twice that of the given cube. We will show that this is impossible. It is also impossible to construct an angle equal to one-third of an arbitrary given angle. It must be remembered, however, that the impossibility of these constructions depends on our assumed restriction to only compass and straightedge. It is easy to manufacture other tools that allow, for instance, the trisection of an arbitrary given angle.

We begin with a very general construction problem. Imagine that the plane of our constructions is the complex plane $\mathbb{C}$ (so that points are complex numbers) and that the points $0, 1 \in \mathbb{C}$ have been marked in advance. We say that a point $\alpha \in \mathbb{C}$ is *constructible* if there is a finite sequence of compass and straightedge operations that construct α. For instance, the number $2 \in \mathbb{C}$ can be constructed as follows: with the straightedge, draw a line through the given points 0 and 1 and then draw the circle centered at 1 and passing through 0. The other point where this circle crosses the line through 0 and 1 is 2. After 2 has been constructed, a similar procedure yields 3, and it should be easy to see that every element of $\mathbb{Z}$ is constructible. The problem is to characterize the subset $K \subseteq \mathbb{C}$ of constructible complex numbers. Our main result is the following theorem.

(20.18) THEOREM.   *Let $\alpha \in \mathbb{C}$. Then the following are equivalent:*

i.   *α is constructible.*
ii. *α lies in a subfield of $\mathbb{C}$ that is normal and is of 2-power degree over $\mathbb{Q}$.*
iii. *There exist fields $E_i \subseteq \mathbb{C}$ with*

$$\mathbb{Q} = E_0 \subseteq E_1 \subseteq E_2 \subseteq \cdots \subseteq E_r$$

and such that $\alpha \in E_r$ and $|E_i : E_{i-1}| = 2$ for $1 \leq i \leq r$.

Note that we can paraphrase the equivalence of (i) and (ii) in Theorem 20.18 by saying that a complex number is constructible iff it is algebraic and a splitting field over $\mathbb{Q}$ for its minimal polynomial over $\mathbb{Q}$ has 2-power degree over $\mathbb{Q}$. Note also that by Theorem 20.18, if $f \in \mathbb{Q}[X]$ is irreducible and one root of f is constructible, then every root is constructible.

(20.19) COROLLARY.   *If $\alpha \in \mathbb{C}$ is constructible, then α is algebraic over $\mathbb{Q}$ and $|\mathbb{Q}[\alpha] : \mathbb{Q}|$ is a power of 2.*   ∎

We mention that the converse of Corollary 20.19 is false: there exist noncon-structible complex numbers α for which $|\mathbb{Q}[\alpha] : \mathbb{Q}|$ is a power of 2. Any root of an irreducible polynomial over $\mathbb{Q}$ of 2-power degree whose splitting field is not of 2-power degree over $\mathbb{Q}$ is an example. A polynomial with this property, for instance, is $X^4 + 3X^2 + 3X + 6$. This is irreducible by the Eisenstein criterion, and when viewed over $\mathbb{Z}/2\mathbb{Z}$, it factors as a product of a linear and an irreducible cubic poly-nomial. By Theorem 28.23 it follows that the Galois group of a splitting field for this polynomial over $\mathbb{Q}$ contains an element of order 3. The splitting field does not, therefore, have 2-power degree over $\mathbb{Q}$.

We derive some consequences before proceeding with the proof of Theorem 20.18. First, note that we can certainly construct a circle of radius 1 (and area π) in our complex plane. If there were a general procedure that squared circles, we could apply it to this unit circle and construct a square with edge length equal to $\sqrt{\pi}$, and this would imply that $\sqrt{\pi}$ is in the constructible set K. Since π is not algebraic, neither is $\sqrt{\pi}$, and this contradicts Corollary 20.19. This shows (assuming that π is transcendental) that it is impossible to square an arbitrary circle with compass and straightedge.

The impossibility of the other two classical construction tasks is much more elementary.

(20.20) COROLLARY.   *No compass and straightedge construction can trisect a 60° angle.*

Proof.   The complex number $\epsilon_6 = e^{2\pi i/6}$ is certainly constructible; it is an inter-section point of the two unit circles centered at 0 and 1. The angle from ϵ_6 to 0 to 1 is 60°. If a compass and straightedge construction of a trisector of this angle were possible, then the point $\epsilon_{18} = e^{2\pi i/18}$ would be constructible. This contradicts Corollary 20.19, however, since $|\mathbb{Q}[\epsilon_{18}] : \mathbb{Q}| = \varphi(18) = 6$ by Corollary 20.9, and 6 is not a power of 2.   ∎

(20.21) COROLLARY.   *No compass and straightedge construction can dupli-cate a unit cube.*

Proof.   If the corollary were not true, we could construct a line segment of length $\sqrt[3]{2}$, given a segment of length 1. It would follow that $\sqrt[3]{2}$ is constructible. This contradicts Corollary 20.19, however, since the minimal polynomial of this element is $X^3 - 2$ by the Eisenstein criterion.   ∎

We begin work now toward a proof of Theorem 20.18.

(20.22) LEMMA.   *Given line segments of lengths $1, x,$ and y, it is possible to construct a segment of length x/y.*

Proof.   Draw two lines through some point P, as shown in Figure 20.1, and mark off points Y and E on these lines such that $PE = 1$ and $PY = y$. Next extend PY in the direction of Y and mark off the point U with $YU = x$. Construct UV parallel to YE with V on line PE, as shown. Using similar triangles, we can easily see that $EV = x/y$.   ∎

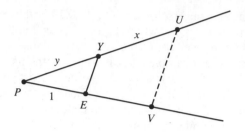

Figure 20.1

(20.23) LEMMA. *Given line segments of length* 1 *and* x, *it is possible to construct a segment of length* $\sqrt{x}$.

Proof. On a line through some point P, mark off points E and X on opposite sides of P such that $PE = 1$ and $PX = x$. Construct the midpoint of EX and draw a semicircle centered at this point with diameter EX. Next erect a perpendicular PS to EX, as shown in Figure 20.2. Since angle ESX is necessarily 90°, an argument that uses similar triangle gives $PS = \sqrt{x}$. ∎

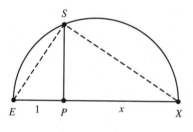

Figure 20.2

(20.24) THEOREM. *The constructible subset* $K \subseteq \mathbb{C}$ *is a subfield of* $\mathbb{C}$.

Proof. To prove that K is a field, it suffices (since $K > \{0\}$) to show for α, $\beta \in K$, with $\beta \neq 0$, that $\alpha - \beta$ and α/β lie in K.

Certainly, $-\beta \in K$. Since $0, \alpha, -\beta \in K$, if these points are not collinear, then $\alpha - \beta$ can be constructed as the fourth vertex of an appropriate parallelogram with vertices 0, α, and $-\beta$. (We leave to the reader the easier case where the three points are collinear.)

To show that α/β is constructible, construct a ray out from 0 whose angle with the positive real axis is $\theta - \varphi$, where θ and φ, respectively, are the angles of the rays from 0 through α and through β with the positive real axis. The point α/β lies on the ray just constructed, at distance $|\alpha|/|\beta|$ from 0. By Lemma 20.22 we can construct a segment of length $|\alpha|/|\beta|$, and this shows that α/β is constructible. ∎

(20.25) LEMMA. *Let $\alpha \in \mathbb{C}$ with $\alpha^2 \in K$. Then $\alpha \in K$.*

Proof.   Construct the line through 0 that bisects the angle between the ray from 0 through α^2 and the positive real axis. The point α is one of the two points on this line at distance $\sqrt{|\alpha^2|}$ from 0, and so it is constructible by Lemma 20.23.   ∎

By the previous two results, K is a subfield of $\mathbb{C}$ that is closed under taking square roots. In fact, K is the unique smallest subfield of $\mathbb{C}$ that is square-root closed.

Our next result contains the essence of the proof that (i) implies (ii) in Theorem 20.18.

(20.26) LEMMA. *Let $E \subseteq \mathbb{C}$ be closed under complex conjugation and let $\alpha \in \mathbb{C}$ be constructible in one step from the points of E. Then*

$$|E[\alpha] : E| \leq 2.$$

Proof.   There are three possibilities for the "one step" that constructs α: (1) α is the intersection of two lines, each determined by two points of E, (2) α is an intersection of a line determined by two points of E with a circle centered at a point of E and with radius equal to the distance between two points of E, or (3) α is an intersection of two circles, each centered at a point of E and each having radius equal to the distance between two points of E. (Note that since we are assuming that E is closed under complex conjugation, the square of the distance between any two points of E is itself an element of E.) The three possibilities are illustrated in Figure 20.3.

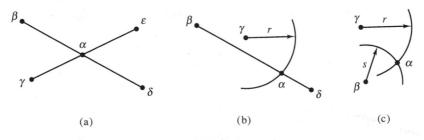

(a)	(b)	(c)

Figure 20.3

In the diagrams, β, γ, δ, and ϵ are in E, and the radii r and s satisfy r^2, $s^2 \in E$. Our object is to find a polynomial equation of degree ≤ 2 satisfied by α with coefficients in E. In each case, this is done by finding two expressions for the complex conjugate $\bar{\alpha}$ and equating them.

The assumption that α is on the line determined by β and δ (in cases (a) and (b)) implies that $(\alpha - \beta)/(\delta - \beta)$ is real. This yields

$$\bar{\alpha} = \bar{\beta} + \frac{\alpha - \beta}{\delta - \beta}(\bar{\delta} - \bar{\beta}).$$

In case (a), we get a similar expression for $\bar{\alpha}$ involving α, ϵ, and γ. Equating these, we obtain

$$\bar{\beta} + \frac{\alpha - \beta}{\delta - \beta}(\bar{\delta} - \bar{\beta}) = \bar{\epsilon} + \frac{\alpha - \epsilon}{\gamma - \epsilon}(\bar{\gamma} - \bar{\epsilon}),$$

a linear relation satisfied by α with coefficients in E. Now if this is not an identity that holds for all values of α, we can solve for α and deduce that $\alpha \in E$. If the above relation is an identity, substitute β for α to obtain

$$\bar{\beta} = \bar{\epsilon} + \frac{\beta - \epsilon}{\gamma - \epsilon}(\bar{\gamma} - \bar{\epsilon}).$$

This forces β to be on the line determined by γ and ϵ. Then $\alpha = \beta \in E$. In case (a), therefore, $|E[\alpha] : E| = 1$.

The assumption that α is at distance r from γ (in cases (b) and (c)) can be written as $(\alpha - \gamma)(\bar{\alpha} - \bar{\gamma}) = r^2$, and this yields

$$\bar{\alpha} = \bar{\gamma} + \frac{r^2}{\alpha - \gamma}.$$

In case (b), we can equate this with our earlier expression for $\bar{\alpha}$. Multiplying by $\alpha - \gamma$, we obtain

$$(\alpha - \gamma)\left[\bar{\beta} + \frac{\alpha - \beta}{\delta - \beta}(\bar{\delta} - \bar{\beta})\right] = \bar{\gamma}(\alpha - \gamma) + r^2,$$

which is a quadratic relation satisfied by α with coefficients in E. This relation is definitely not an identity since the substitution of γ for α would yield $r^2 = 0$, a contradiction. We conclude that $|E[\alpha] : E| \leq 2$ in case (b).

Finally, in case (c), α is at distance s from β. This yields an expression for $\bar{\alpha}$ that can be equated with the previous formula. Multiplication by $(\alpha - \gamma)(\alpha - \beta)$ then yields the relation

$$\bar{\gamma}(\alpha - \gamma)(\alpha - \beta) + r^2(\alpha - \beta) = \bar{\beta}(\alpha - \gamma)(\alpha - \beta) + s^2(\alpha - \gamma),$$

which is quadratic in α and has coefficients in E. This is not an identity since substitution of γ for α would yield $\gamma = \beta$ (since $r^2 \neq 0$), and this is false. We deduce that $|E[\alpha] : E| \leq 2$ in case (c). ∎

At last, we are ready to prove our main result on constructibility.

Proof of Theorem 20.18.   Let $\alpha \in K$. Then α can be constructed in a finite number m of "steps," starting with the given members $0, 1 \in K$, which require no steps. We prove by induction on m that there exists a normal extension $L \supseteq \mathbb{Q}$ in $\mathbb{C}$ with 2-power degree over $\mathbb{Q}$, such that α and all of the intermediate points used in the construction of α lie in L. This shows that (i) implies (ii).

If $m = 0$, then $\alpha \in \mathbb{Q}$ and there is nothing to prove. If we assume that $m > 0$, then the inductive hypothesis yields a field E normal with 2-power degree over $\mathbb{Q}$, and such that E contains all of the points used in the construction of α. Since E is normal over $\mathbb{Q}$, it is invariant under complex conjugation and Lemma 20.26 applies. We deduce that $|E[\alpha] : E| \leq 2$, and we may assume that in fact $|E[\alpha] : E| = 2$, or else we can simply take $L = E$.

Let $f = \min_E(\alpha)$ and write g to denote the product of the distinct images of f under $\mathrm{Gal}(E/\mathbb{Q})$. We take L to be the splitting field for g over E in $\mathbb{C}$. It suffices to show that L has 2-power degree over E and that L is normal over $\mathbb{Q}$.

First we compute degrees. We observe that if β is any root of g in L, then $\deg(\min_E(\beta)) = 2$, and thus adjoining β to any field that contains E results in an extension of degree 1 or 2. Since we obtain L from E by successively adjoining such roots β, we see that $|L : E|$ is a power of 2, as required.

To see that L is normal over $\mathbb{Q}$, we observe first that since E is Galois over $\mathbb{Q}$ and g is invariant under $\mathrm{Gal}(E/\mathbb{Q})$, we have $g \in \mathbb{Q}[X]$. Also, E is a splitting field over $\mathbb{Q}$ for some polynomial $h \in \mathbb{Q}[X]$. It follows that L is a splitting field over $\mathbb{Q}$ for the polynomial gh, and thus L is normal over $\mathbb{Q}$, as required.

Now assume (ii). We thus have $\alpha \in E$ for some field $E \subseteq \mathbb{C}$ with E normal and of 2-power degree over $\mathbb{Q}$. Because E is Galois over $\mathbb{Q}$ and the Galois group G is a 2-group, it follows that there exists a chain of subgroups of G, each of index 2 in the preceding, running from G to the identity. By the fundamental theorem of Galois theory, the fixed fields of these subgroups form the tower of fields whose existence is asserted by (iii).

Finally, we prove that (iii) implies (i). We are given the tower of fields E_i and we show by induction on i that $E_i \subseteq K$. Since K is a field by Theorem 20.24, we have $E_0 = \mathbb{Q} \subseteq K$, and our induction starts. It suffices now to show that if $U \subseteq V \subseteq \mathbb{C}$ are subfields with $U \subseteq K$ and $|V : U| = 2$, then $V \subseteq K$. We have $V = U[\gamma]$, where γ is a root of some quadratic polynomial with coefficients in U. It follows from the quadratic formula that we can also write $V = U[\delta]$, where $\delta^2 \in U \subseteq K$. By Lemma 20.25 we conclude that $\delta \in K$, and since K is a field, we have $V \subseteq K$, as required.   ∎

We close this chapter with a discussion of Gauss's theorem, which determines the integers $n \geq 3$ for which it is possible to construct a regular n-gon with a compass and straightedge. This question is interesting only when n is odd, and the only odd values of n for which the ancients knew an affirmative answer are 3, 5, and 15. As we shall see, it follows from Gauss's result that it is not possible to construct regular n-gons for $n = 7, 9, 11$, or 13, but it is possible for $n = 17$. (In fact, Gauss described an explicit construction for a regular 17-gon.)

A prime number p is called a *Fermat* prime if $p = 2^a + 1$ for some integer $a \geq 1$. For example, the values $a = 1, 2, 4, 8$, and 16 yield the Fermat primes 3, 5, 17, 257, and 65,537. (These five were known to Fermat himself, and no other Fermat primes have been discovered since.) Although we do not actually use the following result, it does help to put Fermat primes into perspective.

(20.27) LEMMA. *If $2^a + 1$ is prime with $a \geq 1$, then $a = 2^n$ for some integer n.*

Proof.   If a is not a power of 2, then we can factor $a = bc$ with b odd and $c < a$. In this case, we have $1 < 2^c + 1 < 2^a + 1$, and we will show that $2^c + 1$ divides $2^a + 1$, which cannot therefore be prime. We have

$$\frac{X^b + 1}{X + 1} = X^{b-1} - X^{b-2} + X^{b-3} - \cdots - X + 1.$$

Substitution of $X = 2^c$ shows that $(2^a + 1)/(2^c + 1)$ is an integer, as required.   ∎

It follows that the next possible Fermat prime after $2^{16} + 1 = 65{,}537$ is $2^{32} + 1$. This number is not prime, however, since one can check that 641 is a factor.

(20.28) THEOREM (Gauss).   *Let $n \geq 3$. Then it is possible to construct a regular n-gon with a compass and straightedge precisely when $\varphi(n)$ is a power of 2. This happens iff n is of the form*

$$n = 2^e p_1 p_2 \cdots p_r,$$

where $e \geq 0$ and the p_i are distinct Fermat primes. (We allow $r = 0$.)

Proof.   We observe first that the nth roots of unity in the complex plane form a regular n-gon inscribed in the unit circle (centered at the origin). If the primitive nth root ϵ_n lies in the constructible field K, therefore, it is possible to construct a regular n-gon. Conversely, if any regular n-gon is constructible, then it is easy to inscribe such a polygon in the unit circle, with one vertex of the polygon at the point 1, and this implies that $\epsilon_n \in K$.

Since the cyclotomic field $\mathbb{Q}_n$ is the splitting field in $\mathbb{C}$ for the minimal polynomial over $\mathbb{Q}$ of ϵ_n, Theorem 20.18 tells us that ϵ_n is constructible iff $|\mathbb{Q}_n : \mathbb{Q}|$ is a power of 2. We know, however, that $|\mathbb{Q}_n : \mathbb{Q}| = \varphi(n)$, and this completes the proof of the first assertion.

Now write $n = q_1^{e_1} q_2^{e_2} \cdots q_k^{e_k}$, where the q_i are distinct primes and the exponents $e_i > 0$. By Lemma 20.11 we know that $\varphi(n)$ is the product of the numbers $\varphi(q_i^{e_i})$, and thus $\varphi(n)$ is a power of 2 iff each of these factors is a power of 2. The question we must resolve, therefore, is: When is it true that $\varphi(p^e)$ is a power of 2 for primes p and positive exponents e? It is easy to see that $\varphi(p^e) = p^e - p^{e-1} = p^{e-1}(p - 1)$, and it follows that $\varphi(2^e)$ is always a power of 2. For odd primes p (and positive exponents e), we see that $\varphi(p^e)$ is a power of 2 iff $e = 1$ and p is a Fermat prime. The result now follows.   ∎

Since there are just five known Fermat primes, it may be that there are no more than 31 odd numbers $n \geq 3$ for which a regular n-gon is constructible. The largest of these is the product of the five known Fermat primes.

Problems

20.1 Suppose p is a prime and $n \geq 1$. Show that

$$\Phi_{pn}(X) = \begin{cases} \Phi_n(X^p) & \text{if } p \mid n \\ \Phi_n(X^p)/\Phi_n(X) & \text{if } p \nmid n. \end{cases}$$

20.2 Suppose $2 \nmid n$, where $n \geq 1$. Show that $\Phi_{2n}(X) = \Phi_n(-X)$.

20.3 Show that $\Phi_n(0) = \pm 1$ for all $n \geq 1$. If n is odd and $n \geq 3$, show that $\Phi_n(0) = 1$.

20.4 If $\Phi_n(X) = \sum a_i X^i$, show that $a_i = \pm a_{\varphi(n)-i}$ for all i with $0 \leq i \leq \varphi(n)$. Show that the sign depends on n but not on i.

20.5 Let q be prime and assume that $q \nmid m$, where $m \geq 1$.

a. Show that $\Phi_q(X^m) = \sum_{i=0}^{q-1} (X^q)^{e_i} X^i$ for suitable exponents $e_i \geq 0$.

b. Show that $\Phi_q(X^m)(X-1) = \sum_{i=0}^{q-1} [(X^q)^{a_i} - (X^q)^{b_i}] X^i$ for suitable exponents a_i, $b_i \geq 0$.

c. Show that $\Phi_q(X^m)/\Phi_q(X)$ is a polynomial with all coefficients in the set $\{0, 1, -1\}$.

d. If m is prime, show that all coefficients of $\Phi_{qm}(X)$ lie in $\{0, 1, -1\}$.

HINT: Use part (b) for (c). For part (d), use (c) together with Problem 20.1.

20.6 (**Migotti**) If $n = 2^a p^b q^c$, where p and q are primes, show that all coefficients of $\Phi_n(X)$ lie in $\{0, 1, -1\}$.

HINT: Use Problems 20.1 and 20.2 to reduce to the situation of Problem 20.5(d).

20.7 If n is divisible by p^2 for some prime p, show that the sum of all primitive nth roots of unity in $\mathbb{C}$ is zero.

20.8 If m and n are coprime, show that Φ_n is irreducible in $\mathbb{Q}_m[X]$.

20.9 If m and n are coprime, show that some automorphism of $\mathbb{Q}_n$ carries ϵ_n to $(\epsilon_n)^m$.

20.10 Find explicitly a polynomial $f \in \mathbb{Q}[X]$ such that the associated Galois group over $\mathbb{Q}$ is of order 3.

HINT: Work in $\mathbb{Q}_7$.

20.11 Let p be a prime and suppose that $\delta_1, \delta_2, \ldots, \delta_r$ are (not necessarily distinct) primitive pth roots of unity in $\mathbb{C}$. If $\sum \delta_i \in \mathbb{Q}$, show that $r \geq p - 1$ and that all of the primitive pth roots of unity occur among the δ_i.

20.12 Write $\epsilon = \epsilon_p$, where $p > 2$ is prime, and let

$$\alpha = \sum_{i=0}^{p-1} \epsilon^{i^2} .$$

Show that $\mathbb{Q}[\alpha]$ is the unique subfield of $\mathbb{Q}_p$ that has degree 2 over $\mathbb{Q}$.

NOTE: The quantity α is called a *Gauss sum*.

20.13 If α is the Gauss sum in Problem 20.12, show that $\alpha\bar{\alpha} = p$.

HINT: $\sum_{i,j} \epsilon^{i^2 - j^2} = \sum_{u,v} \epsilon^{uv}$, where i, j, u, and v run over $\{0, 1, \dots, p-1\}$.

20.14 Let α be the Gauss sum of Problems 20.12 and 20.13.

a. If $p \equiv 1 \bmod 4$, show that $\bar{\alpha} = \alpha$ and so $\alpha = \pm\sqrt{p}$.
b. If $p \equiv 3 \bmod 4$, show that $\bar{\alpha} = -\alpha$ and so $\alpha = \pm i \sqrt{p}$.

HINT: Use Problem 20.12 for part (a). For part (b), observe that when $p \equiv 3 \bmod 4$, we have $i^2 \not\equiv -j^2 \bmod p$ unless $i \equiv 0 \equiv j \bmod p$. Use this to compute $\alpha + \bar{\alpha}$.

NOTE: A remarkable result of Gauss asserts that always $\alpha = \sqrt{p}$ or $i\sqrt{p}$.

20.15 Suppose $\beta \in \mathbb{Q}_n \cap \mathbb{R}$ and $\beta^m \in \mathbb{Q}$ for some integer $m \geq 1$. Show that $\beta^2 \in \mathbb{Q}$.

HINT: Prove that $\min_{\mathbb{Q}}(\beta)$ splits over $\mathbb{Q}_n \cap \mathbb{R}$. If γ is any root of this polynomial, note that $|\gamma| = |\beta|$.

20.16 Suppose that $\beta \in \mathbb{Q}_n$ is purely imaginary and that $\beta^m \in \mathbb{Q}$ for some integer $m \geq 1$. Show that $\beta^2 \in \mathbb{Q}$.

20.17 Let $F = \mathbb{Z}/p\mathbb{Z}$ for some prime p, and let $\overline{\Phi}_n$ denote the image of Φ_n in $F[X]$. Assume that $\overline{\Phi}_n$ is irreducible in $F[X]$. Let E be a splitting field for $X^n - 1$ over F and let m be the order of the group of nth roots of unity in E.

a. Show that $\mathrm{Gal}(E/F) \cong U_m \cong U_n$.
b. Show that if $m \neq n$, then $p = 2$, $n = 2m$, and m is odd.
c. Show that either $p \nmid n$ or $p = 2$ and $4 \nmid n$.
d. Show that $\gcd(n, p-1) \leq 2$.

HINT: For part (a), note that U_m is a homomorphic image of U_n and that $\varphi(m) \geq |\mathrm{Gal}(E/F)| \geq \varphi(n)$. For part (d), show that $|C \cap F| \leq 2$, where C is the group of nth roots of unity in E. Use the fact that $\mathrm{Gal}(E/F) \cong \mathrm{Aut}(C)$.

20.18 Let E be the splitting field of $X^n - p$ over $\mathbb{Q}$ in $\mathbb{C}$, where p is prime. Show that $|E : \mathbb{Q}|$ is equal to either $n\varphi(n)$ or $n\varphi(n)/2$.

HINT: Write $F = \mathbb{Q}_n$ and show that $F \subseteq E$. Let $m = |E : F|$ and show that $F \subseteq E$. Let $m = |E : F|$ and show that $m \mid n$ by considering irreducible factors of $X^n - p$ in $F[X]$. Show that $p^{m/n} \in F$ and use Problem 20.15.

20.19 Let $K \subseteq \mathbb{C}$ be the field of constructible complex numbers. Show that K is the unique smallest subfield of $\mathbb{C}$ with the property that every element has a square root in the field.

20.20 Show that it is possible to square any given triangle using only a compass and a straightedge.

CHAPTER TWENTY-ONE

Finite Fields

21A

We begin this chapter on finite fields with a very general observation. Every field E has a unique minimal subfield, a subfield contained in every subfield of E. This *prime subfield* of E is constructed simply as the intersection of all subfields. It turns out that the prime subfield of E is determined up to isomorphism by the characteristic of E.

(21.1) LEMMA. *Let E be any field and suppose F is its prime subfield. If* $\mathrm{char}(E) = p > 0$, *then* $F \cong \mathbb{Z}/p\mathbb{Z}$, *and if* $\mathrm{char}(E) = 0$, *then* $F \cong \mathbb{Q}$.

Proof. Since $1 \in F$, we see that $n \cdot 1 \in F$ for all integers $n \in \mathbb{Z}$. In the case that $\mathrm{char}(E) = p$, a prime, the elements $n \cdot 1$ for $0 \le n < p$ form a subfield of F isomorphic to $\mathbb{Z}/p\mathbb{Z}$, and this subfield must be all of F by the minimality of F.

If $\mathrm{char}(E) = 0$, then the elements $n \cdot 1$ are all nonzero for $n \ne 0$. It follows that the elements $(m \cdot 1)/(n \cdot 1)$ of F, where $m, n \in \mathbb{Z}$ and $n \ne 0$ form a subfield of F isomorphic to $\mathbb{Q}$. By minimality again, this must be the whole field F. ∎

We can now begin our study of finite fields. If E is finite, then clearly its prime subfield is finite and so $\mathrm{char}(E) \ne 0$. The characteristic is therefore some prime number.

(21.2) LEMMA. *Let F be a finite field and suppose $E \supseteq F$ with $|E : F| = n < \infty$. Then E is finite and $|E| = |F|^n$.*

Proof. Let $\alpha_1, \alpha_2, \ldots, \alpha_n$ be an F-basis for E. Then every element of E is uniquely of the form $\sum a_i \alpha_i$, where the coefficients a_i lie in F. It follows that $|E|$ is equal to the number of n-tuples $(a_1, a_2, \ldots, a_n)$ of elements of F, and this number is equal to $|F|^n$. ∎

(21.3) COROLLARY. *Let E be any finite field and let $p = \mathrm{char}(E)$. Then p is prime and $|E| = p^n$, where $n = |E : F|$ and F is the prime subfield of E. In particular, $|E|$ is a prime power.*

Proof. We have seen that p must be prime. Now $|F| = p$ by Lemma 21.1, and since E is finite, we have $n = |E : F| < \infty$ and thus $|E| = p^n$ by Lemma 21.2. ∎

As we shall see, for every prime number p and positive integer n, there actually does exist a field of order p^n, and this field is unique is up to isomorphism.

(21.4) LEMMA. *Let L be a field of prime characteristic p and let q be a power of p. Then L contains a subfield of order q iff the polynomial $X^q - X$ splits in $L[X]$. In this case, $E = \{\alpha \in L \mid \alpha^q = \alpha\}$ is the unique subfield of L with order q.*

Proof. Since $\mathrm{char}(L) = p$ and q is a power of p, we see that $(\alpha - \beta)^q = \alpha^q - \beta^q$ for $\alpha, \beta \in E$. It follows that $\{\alpha \in L \mid \alpha^q = \alpha\}$ is an additive subgroup of L, which we call E. Since also $(\alpha/\beta)^q = \alpha^q/\beta^q$ for $\beta \neq 0$, we conclude that E is a subfield.

Writing $f(X) = X^q - X$, we see that E is precisely the set of roots of f in L, and thus $|E| \leq \deg(f) = q$. Note that the formal derivative $f'(X) = -1$, and so f has distinct roots by Theorem 19.4. Thus $|E| = q$ iff f splits over L.

To complete the proof, we must show that if $K \subseteq L$ with $|K| = q$, then f splits over L and $K = E$. Both of these conclusions will follow when we show that $K \subseteq E$, since we know that $|E| \leq q$ and that f splits iff equality holds. Suppose, then, that $\alpha \in K$. Certainly, if $\alpha = 0$, then $\alpha \in E$. If $\alpha \neq 0$, then $\alpha^{q-1} = 1$ since α is in the group $K^\times$ of order $q - 1$. Thus $\alpha^q = \alpha$ and $\alpha \in E$ in this case, too. ∎

(21.5) COROLLARY. *Let q be a power of the prime number p. Then there exists a field of order q, and every such field is isomorphic to a splitting field for $f(X) = X^q - X$ over $F = \mathbb{Z}/p\mathbb{Z}$. In particular, all fields of order q are isomorphic.*

Proof. Let L be a splitting field for f over F, and note that by Lemma 21.4 there is a subfield $E \subseteq L$ with $|E| = q$. This proves existence.

Now let E_0 be any field of order q. Let F_0 be the prime subfield of E_0, and note that $F_0 \cong F$ since $\mathrm{char}(E_0) = p$ by Corollary 21.3. By Lemma 21.4 the polynomial f splits over E_0, and so E_0 contains a splitting field L_0 for f over F_0. By Corollary 17.23 (uniqueness of splitting fields) we have $L_0 \cong L$. Thus

$$q = |E_0| \geq |L_0| = |L| \geq |E| = q,$$

and we conclude that $E_0 = L_0 \cong L$, as required. ∎

Finite fields are often called *Galois fields*, and we write $GF(q)$ to denote "the" Galois field of order q, where q is any prime power. (Of course, $GF(q)$ is defined only up to isomorphism.) We stress that if q is not prime, then $GF(q)$ is definitely not isomorphic to $\mathbb{Z}/q\mathbb{Z}$ since the latter ring is not a field in this case.

(21.6) COROLLARY. *Let $F = \mathbb{Z}/p\mathbb{Z}$, where p is prime, and let n be a positive integer. Then there exists an irreducible polynomial $f \in F[X]$ with $\deg(f) = n$.*

Proof. Let $E = GF(p^n)$ and identify F with the prime subfield of E. By Lemma 17.12 we know that $E^\times$ is a cyclic group, and we can write $E^\times = \langle \alpha \rangle$ for some $\alpha \in E$. Then $E = F[\alpha]$ and $\deg(\min_F(\alpha)) = |E : F| = n$, where the last equality holds by Lemma 21.2.   ∎

By studying the subfield structure of finite fields, we can sharpen and generalize Corollary 21.6 and actually count the number of irreducible polynomials of degree n in $F[X]$, where F is any finite field. The key result on subfields is the following theorem.

(21.7) THEOREM. *Let $E = GF(q^n)$, where q is a prime power. If $m \geq 1$ is an integer, then the following are equivalent:*

  i.   *E has a subfield of order q^m.*
  ii.   *m divides n.*
  iii. *$q^m - 1$ divides $q^n - 1$.*

Also, if q is prime, then every subfield of E has order q^m for some integer $m \mid n$.

Proof. Assume (i) and let $F \subseteq E$ with $|F| = q^m$. Since $|E : F| < \infty$, we can write $d = |E : F|$, and we have $|E| = |F|^d$ by Lemma 21.2. Then $q^n = |E| = (q^m)^d$, and so $md = n$, proving (ii).

Now, assuming (ii), we write $n = md$. Substitution of q^m for X in the identity $X^d - 1 = (X - 1)(1 + X + X^2 + \cdots + X^{d-1})$ shows that $q^m - 1$ divides $q^n - 1$, and (iii) is proved.

The cyclic group $E^\times$ of order $q^n - 1$ has a subgroup with order equal to any given divisor of $q^n - 1$. Assuming (iii), therefore, we can find a subgroup $H \subseteq E^\times$ of order $q^m - 1$. Clearly, each of the q^m elements of $H \cup \{0\}$ is a root of $X^{q^m} - X$, and so we see that this polynomial splits over E. By Lemma 21.4, E contains a subfield of order q^m, proving (i).

Finally, if q is prime, then $q = \operatorname{char}(E)$, and so if $F \subseteq E$, then $|F| = q^m$ for some integer $m \geq 1$. Then $m \mid n$ by the fact that (i) implies (ii).   ∎

The equivalence of (ii) and (iii) in Theorem 21.7 is a purely arithmetic result, independent of field theory. Our proof that (ii) implies (iii) did not rely on fields and would work just as well for any integer $q > 1$, even if q is not a prime power. There is also a direct proof that (iii) implies (ii), which works in this generality. Simply write $n = am + r$ with $0 \leq r < m$. Then $q^n = (q^m)^a q^r$, and working mod $(q^m - 1)$, we have $q^m \equiv 1$ and so $q^n \equiv q^r$. By (iii), we have $q^n \equiv 1$ and thus $q^r \equiv 1$. It follows that $q^m - 1$ divides $q^r - 1$. Since $q^r - 1 < q^m - 1$, however, this forces $q^r - 1 = 0$, and so $r = 0$ and (ii) holds.

(21.8) COROLLARY. *Let $E, F \subseteq L$, where $|E| = q^n$ and $|F| = q^m$ for some prime power q. Then $|E \cap F| = q^d$, where $d = \gcd(m, n)$. In particular, $F \subseteq E$ if $m \mid n$.*

Proof. By Theorem 21.7 each of E and F has a subfield of order q^d. By Lemma 21.4, however, L has at most one subfield of this order, and this forces $|E \cap F| \geq q^d$. Now write $q = p^a$, where $p = \text{char}(F)$. Then $|E \cap F| = p^e$ for some integer e, and we have $e \mid (ma)$ and $e \mid (na)$. (This is by Theorem 21.7 since $|F| = p^{ma}$ and $|E| = p^{na}$.) Thus $e \leq \gcd(ma, na) = da$, and so $|E \cap F| = p^e \leq p^{da} = q^d$. Thus $|E \cap F| = q^d$, as required.

If $m \mid n$, then $d = m$ and $|F| = |F \cap E|$. By Lemma 21.4 we conclude that $F = F \cap E \subseteq E$.   ∎

(21.9) COROLLARY. *Let $F \subseteq E$ with $|F| = q < \infty$ and $|E : F| = n < \infty$. Then for each divisor m of n, there exists a unique subfield K with $F \subseteq K \subseteq E$ and such that $|K| = q^m$. There are no other intermediate fields between F and E.*

Proof. If $F \subseteq K \subseteq E$, write $|K : F| = m$. Then $m \mid n$ and $|K| = q^m$. Also, K is unique of its order by Lemma 21.4.

Now let $m \mid n$. By Theorem 21.7 there exists $K \subseteq E$ with $|K| = q^m$. Since $|F| = q^1$ and $1 \mid m$, we have $F \subseteq K$ by Corollary 21.8.   ∎

In particular, if $E = GF(p^n)$, where p is prime, we can apply Corollary 21.9 with F being the prime subfield of E of order p. It follows that E has exactly as many subfields as n has divisors: there is one subfield isomorphic to $GF(p^m)$ for each divisor m of n, and there are no other subfields.

An alternative method for studying intermediate fields in extensions of finite fields is via Galois theory, and we digress briefly to discuss this approach. If $F \subseteq E$ is a Galois extension of fields with $|E : F| = n$ and with cyclic Galois group G, then $|G| = n$ and G has a unique subgroup of index m for each divisor m of n, and there are no other subgroups. By the fundamental theorem of Galois theory, there is a unique field K with $F \subseteq K \subseteq E$ and such that $|K : F| = m$, and there are no other intermediate fields. To prove Corollary 21.9 by Galois theory, therefore, the following result is sufficient.

(21.10) THEOREM. *Let $F \subseteq E$ be finite fields with $|F| = q$. Then E is Galois over F and $G = \text{Gal}(E/F)$ is cyclic. In fact, $G = \langle \sigma \rangle$, where $\sigma : E \to E$ is defined by $\sigma(\alpha) = \alpha^q$ for $\alpha \in E$.*

Proof. Let $p = \text{char}(F)$. The map $\alpha \mapsto \alpha^p$ is an isomorphism of E into itself, and since E is finite, this map is surjective and is an automorphism of E. Since σ is a power of this map, it follows that $\sigma \in \text{Aut}(E)$. By Lemma 21.4 we know that $F = \{\alpha \in E \mid \alpha^q = \alpha\} = \text{Fix}(\langle \sigma \rangle)$. It follows from Theorem 18.20 that E is Galois over F and $\text{Gal}(E/F) = \langle \sigma \rangle$.   ∎

21B

We now turn to the problem of counting the irreducible polynomials of degree n in $E[X]$, where $E = GF(q)$. We can restrict our attention to monic polynomials; the full count can then be obtained through multiplication by $q - 1$.

There is an inductive approach to the problem. If we already knew the total numbers of monic irreducible polynomials of degree m in $E[X]$ for all integers $m < n$, then using unique factorization, we could, in principle, compute the number of monic reducible polynomials of degree n. Since we know that the total number of monic polynomials of degree n in $E[X]$ is q^n, we could calculate the number of these that are irreducible by subtraction.

There is a much better way to do this. The following result gives both an estimate and a precise count.

(21.11) THEOREM. *Let N denote the number of monic irreducible polynomials of degree n over $GF(q)$. Then*

a. $N \geq \dfrac{1}{n} \left(q^n - \sum_r q^{n/r} \right)$, *where the sum is over all primes $r \,|\, n$, and*

b. $N = \dfrac{1}{n} \sum_s \mu(s) q^{n/s}$, *where s runs over all divisors of n that are products of distinct primes (allowing $s = 1$) and where $\mu(s) = (-1)^k$ when s is a product of k primes.*

We mention that the above function μ is the *Möbius function* of number theory. Its definition is extended to all positive integers s by setting $\mu(s) = 0$ if s is not a product of distinct primes.

(21.12) LEMMA. *Let $E \subseteq L$, where $|E| = q$ and $|L| = q^n$, and let*

$$T = \{\alpha \in L \mid E[\alpha] = L\}.$$

Then every monic irreducible polynomial of degree n over E has exactly n roots in T, and each element of T is a root of exactly one such polynomial.

Proof. Let $f \in E[X]$ be monic and irreducible of degree n, and let K be a splitting field for f over L. If $\alpha \in K$ is a root of f, then $f = \min_E(\alpha)$ and hence $|E[\alpha] : E| = \deg(f) = n$. By Lemma 21.2 we deduce that $|E[\alpha]| = q^n = |L|$, and so $E[\alpha] = L$ by Lemma 21.4. It follows that $\alpha \in T$.

Since E is finite, Theorem 19.7 guarantees that f is separable over E. Because f is irreducible, it has distinct roots, and thus f has exactly n roots in K. All of these are in T, and the first assertion is proved.

Now let $\alpha \in T$ and write $f = \min_F(\alpha)$. Then $f \in E[X]$ is monic and irreducible; it has α as a root and has degree equal to $|E[\alpha] : E| = |L : E| = n$. If $g \in E[X]$ is any other monic polynomial that has α as a root, then $f \,|\, g$, and so if $\deg(g) = n$, we have $g = f$.   ∎

(21.13) COROLLARY.   *Assume the notation of Lemma 21.12 and let N denote the number of monic irreducible polynomials of degree n in E[X]. Then N = |T|/n.*   ∎

Proof of Theorem 21.11(a).   Let $L = GF(q^n)$ and let E be the subfield of L of order q. (This exists by Theorem 21.7.) Writing $T = \{\alpha \in L \mid E[\alpha] = L\}$, we have $N = |T|/n$ by Corollary 21.13.

If $\beta \in L - T$, then $E \subseteq E[\beta] < L$ and so $|E[\beta]| = q^m$, where m is some proper divisor of n. Let r be any prime divisor of n/m and let K_r be the subfield of order $q^{n/r}$ in L. Then $m \mid (n/r)$, and so $\beta \in E[\beta] \subseteq K_r$ by Corollary 21.8.

Conversely, if r is any prime divisor of n, then $E \subseteq K_r < L$, and so no element of K_r lies in T. It follows that $L - T = \bigcup K_r$, where r runs over all prime divisors of n. In particular,

$$|L - T| \leq \sum_r |K_r| = \sum_r q^{n/r}.$$

Thus $|T| \geq q^n - \sum q^{n/r}$, and the inequality in Theorem 21.11(a) follows.   ∎

To complete the proof of Theorem 21.11, we need the "inclusion-exclusion" principle from elementary combinatorics.

(21.14) LEMMA (Inclusion-exclusion).   *Let $\mathcal{X}$ be a collection of subsets of some finite set U. Then*

$$\left|\bigcup \mathcal{X}\right| = -\sum_{\mathcal{P}} (-1)^{|\mathcal{P}|} \left|\bigcap \mathcal{P}\right|,$$

where $\mathcal{P}$ runs over all nonempty subsets of $\mathcal{X}$.

For instance, if $|\mathcal{X}| = 3$, we can write $\mathcal{X} = \{A, B, C\}$. Then Lemma 21.14 says that

$$|A \cup B \cup C| = |A| + |B| + |C| - |A \cap B| - |A \cap C| - |B \cap C| + |A \cap B \cap C|.$$

Proof of Lemma 21.14.   We may assume that $\bigcup \mathcal{X} = U$, and we adopt the usual convention that if $\mathcal{P} = \varnothing$, then $\bigcap \mathcal{P} = U$. Then the formula of Lemma 21.14 is equivalent to showing that the quantity

$$s = \sum_{\mathcal{P} \subseteq \mathcal{X}} (-1)^{|\mathcal{P}|} \left|\bigcap \mathcal{P}\right|$$

is equal to zero. (Here, we are summing over all subsets of $\mathcal{X}$, including $\varnothing$.)
For $u \in U$, write $\mathcal{X}_u = \{A \in \mathcal{X} \mid u \in A\}$ and note that $\mathcal{X}_u \neq \varnothing$. We have

$$s = \sum_{\mathcal{P} \subseteq \mathcal{X}} \sum_{u \in \bigcap \mathcal{P}} (-1)^{|\mathcal{P}|} = \sum_{u \in U} \sum_{\mathcal{P} \subseteq \mathcal{X}_u} (-1)^{|\mathcal{P}|}$$

by interchanging the order of summation. We claim that for each element $u \in U$, the inner sum on the right vanishes. To see this, write $|\mathcal{X}_u| = n$ and note that

$$\sum_{\mathcal{P} \subseteq \mathcal{X}_u} (-1)^{|\mathcal{P}|} = \sum_{i=0}^{n} \binom{n}{i}(-1)^i = (1-1)^n = 0.$$

It follows that $s = 0$, as required.   ∎

Proof of Theorem 21.11(b).   We know that $N = |T|/n$ and that $T = L - \bigcup K_r$, where $|L| = q^n$ and $|K_r| = q^{n/r}$ and r runs over all prime divisors of n. Let $\mathcal{P}$ be any nonempty subset of $\mathcal{X} = \{K_r\}$. Then $\bigcap \mathcal{P}$ is the subfield of L of order q^c, where c is the greatest common divisor of all those numbers n/r for which $K_r \in \mathcal{P}$. (This follows by repeated applications of Corollary 21.8.) Therefore $|\bigcap \mathcal{P}| = q^{n/s}$, where s is the product of the relevant primes. We write $k(s)$ to denote the number of prime factors of s (so that $k(s) = |\mathcal{P}|$), and then the inclusion-exclusion principle tells us that

$$\left| \bigcup_r K_r \right| = -\sum_{s>1}(-1)^{k(s)}q^{n/s} = -\sum_{s>1}\mu(s)q^{n/s},$$

where s runs over all square-free nontrivial divisors of n. Because $|L| = q^n = \mu(1)q^{n/1}$, we can include the term corresponding to $s = 1$. We have

$$|T| = |L| - \left| \bigcup_r K_r \right| = \sum_{s \geq 1}\mu(s)q^{n/s},$$

and the result follows.   ∎

Suppose we choose a polynomial of degree n at random from $E[X]$, where $E = GF(q)$. What is the probability that the chosen polynomial is irreducible? The answer is very nearly $1/n$. To see this, observe that it is no loss to assume that our polynomial is monic. Since there are exactly q^n monic polynomials of degree n, the probability we wish to compute is N/q^n, where N is as in Theorem 21.11. By Theorem 21.11(a), the probability is $1/n - \epsilon$, where $\epsilon = \left(\sum q^{n/r}\right) / nq^n$. Now

$$\sum_r q^{n/r} \leq \sum_{i=0}^{n/2} q^i < q^{1+n/2}.$$

It follows that $\epsilon < q/nq^{n/2}$, and of course this is tiny for even moderately large n.

Suppose we want to find explicitly an irreducible polynomial of fairly large degree over some finite field. For instance, suppose we seek an irreducible polynomial of degree 100 over $GF(2)$. Since the probability that a random polynomial of degree 100 is irreducible is nearly $1/100$, all we need is a fast procedure for deciding whether or not a particular polynomial is irreducible. Given such a decision algorithm, we can simply pick polynomials at random and test them. Unless we are extraordinarily unlucky, this procedure will soon produce the desired irreducible polynomial.

A fast irreducibility check is available, and using it, one can find an irreducible polynomial of degree 100 over $GF(2)$ in just a few seconds on even a small computer. The algorithm to which we refer is from E. Berlekamp, and we present it next.

21C

The Berlekamp algorithm is a useful computational procedure for factoring polynomials over finite fields into irreducible factors. In particular, it enables one to decide whether or not a given polynomial is itself irreducible. In fact, a considerably simplified version of the algorithm suffices for this latter task.

We need to discuss the polynomial analog of the greatest common divisor of two integers. If E is a field and $f, g \in E[X]$ are not both zero, we say that a polynomial $d \in E[X]$ is a *greatest common divisor* for f and g if $d|f$ and $d|g$ and no polynomial that divides both f and g has degree exceeding that of d. It is obvious that a greatest common divisor for f and g always exists, and it is easy to see by using unique factorization in $E[X]$ that "the" greatest common divisor of f and g is unique up to multiplication by nonzero scalars. In particular, f and g have a unique monic greatest common divisor, and by analogy with the standard notation from number theory, we write $\gcd(f, g)$ to denote this polynomial. (It should be noted that $\gcd(f, g)$ depends on f and g but not on the field E. We obtain the same monic greatest common divisor if we work over any field extension of E.)

Now let E be a finite field and assume that we know how to do arithmetic in E. (This is especially easy, for instance, if $E = \mathbb{Z}/p\mathbb{Z}$ for some prime p.) One of the things we need to be able to do is to construct a greatest common divisor for two given polynomials $f, g \in E[X]$ that are not both zero. The polynomial version of the familiar "Euclidean algorithm" is a fast way to do this, and we remind the reader how this works.

Given $f, g \in E[X]$ with $f \neq 0$, we can use polynomial "long division" to find $q, r \in E[X]$ such that $g = qf + r$ and either $r = 0$ or $\deg(r) < \deg(f)$. We observe that the greatest common divisors of f and g are exactly the greatest common divisors of r and f. In particular, if $r = 0$, then f is a greatest common divisor for f and g and there is nothing more to compute. Otherwise, we simply replace f by r and g by f and repeat the process. Since the polynomial r (which replaces f) has degree smaller than that of f, we see that this algorithm terminates after at most $1 + \deg(f)$ steps.

We are now ready to discuss the Berlekamp algorithm. Suppose we are given a polynomial $f \in E[X]$ with $\deg(f) = n \geq 1$, and we wish to factor f. We may certainly assume that f is monic, and so we know that $f = g_1 g_2 \cdots g_m$ for some monic irreducible polynomials $g_i \in E[X]$. Our problem is to determine the g_i explicitly.

The first step is to detect whether or not any two of the factors g_i are equal. For this purpose, we compute a greatest common divisor h for f and f', where f' is the formal derivative of f, and we consider three possibilities.

If $\deg(h) = 0$, then the g_i are distinct. To see this, note that if $g^2 | f$, then $g | f'$ and g is a common divisor for f and f'. If $\deg(g) > 0$, this contradicts the assumption that h has degree 0.

Suppose next that $0 < \deg(h) < \deg(f)$. In this case, h is a nontrivial proper divisor of f, and so by long division, we can find a polynomial $k \in E[X]$ with $f = hk$. In this case, the problem of factoring f has been reduced to factoring two polynomials of smaller degree.

The final possibility is that $\deg(h) = \deg(f)$. In this case, we necessarily have $f' = 0$, because otherwise $\deg(f') < \deg(f) = \deg(h)$ and h could not divide f'. Since $f' = 0$, we see that $f(X) = f_0(X^p)$ for some polynomial $f_0 \in E[X]$, where $p = \text{char}(E)$. Because E is finite, it is perfect, and so by taking pth roots of the coefficients of f_0, we can find $f_1 \in E[X]$ such that $f(X) = f_1(X)^p$. In this case, it suffices to factor f_1 in order to get the full factorization of f.

Note that in the second and third cases above, f is necessarily reducible. If our goal, therefore, is only to determine whether or not the given polynomial is irreducible, we can stop if $\deg(h) > 0$.

We can now assume that the g_i are distinct. The next step is to work in the ring $R = E[X]/(f)$. Before we do this, we make some general observations about factor rings of polynomial rings.

Let E be any field; let $f \in E[X]$ with $f \neq 0$ and, as usual, write (f) to denote the principal ideal of $E[X]$ generated by f. We consider $R = E[X]/(f)$ and note that since the ideal (f) is an E-vector subspace of $E[X]$, we can view R as an E-space in a natural way. We use overbars to denote the canonical homomorphism $E[X] \to R$, so that if $h \in E[X]$, we write $\overline{h}$ to denote the element $(f) + h$ of R.

(21.15) LEMMA. *Let $f \in E[X]$ with $\deg(f) = n \geq 1$ and write $R = F[X]/(f)$. As above, we use overbars for elements of R.*

a. *The elements $\overline{1}, \overline{X}, \overline{X}^2, \ldots, \overline{X}^{n-1}$ form an E-basis for R. In particular, $\dim_E(R) = n$.*

b. *If $f = g_1 g_2 \cdots g_m$, where $\deg(g_j) > 1$ and the g_j are pairwise coprime, then*

$$R \cong \frac{F[X]}{(g_1)} \oplus \cdots \oplus \frac{F[X]}{(g_m)},$$

an external direct sum of rings. This isomorphism is given by

$$\overline{h} \mapsto ((g_1) + h, \ldots, (g_m) + h).$$

We recall that in an external direct sum of rings, both addition and multiplication are defined componentwise. We mention also that (b) holds if the polynomial ring $E[X]$ is replaced by any PID. (The version of (b) for the ring $\mathbb{Z}$ is essentially the "Chinese remainder theorem." See the problems at the end of this chapter for more about this.)

Proof of Lemma 21.15. If $h \in E[X]$, we can write $h = qf + r$ with $q, r \in E[X]$ and either $r = 0$ or $\deg(r) < n$. Since $\overline{h} = \overline{r}$ in R, we see that $\overline{h}$ is a linear combination of $\overline{1}, \overline{X}, \ldots, \overline{X}^{n-1}$. That these vectors are linearly independent follows from the fact that the ideal (f) contains no polynomial with degree smaller than n. This proves (a).

For (b), we define the ring homomorphism

$$\theta : E[X] \rightarrow \frac{E[X]}{(g_1)} \oplus \cdots \oplus \frac{E[X]}{(g_m)}$$

by setting

$$\theta(h) = ((g_1) + h, (g_2) + h, \ldots, (g_m) + h).$$

Note that θ is E-linear as well as being a ring homomorphism. To prove (b), we need to establish that $\ker(\theta) = (f)$ and that θ is surjective.

We note that $\theta(h) = 0$ iff $h \in (g_j)$ for all subscripts j. In other words, $h \in \ker(\theta)$ iff $g_j | h$ for all j, and since $f = g_1 g_2 \cdots g_m$ and the factors are pairwise coprime, we see (using unique factorization in $F[X]$) that $h \in \ker(\theta)$ iff $f | h$. In other words, $\ker(\theta) = (f)$.

It follows that $R = E[X]/\ker(\theta) \cong \theta(F[X])$, and so R is isomorphic to a subring of the direct sum. Since the isomorphism is also an E-vector space isomorphism, it suffices to show that $\dim_E(R)$ is equal to the dimension of the whole direct sum.

By part (a), we have $\dim_E(R) = \deg(f)$. Similarly, $\dim_E(F[X]/(g_j)) = \deg(g_j)$, and so the dimension of the direct sum is equal to $\sum \deg(g_j) = \deg(f)$. This completes the proof. ∎

We return now to the Berlekamp algorithm. We have $f \in E[X]$, where $|E| = q$, and we have reduced the problem to the case where $f = g_1 g_2 \cdots g_m$, where the $g_j \in E[X]$ are distinct monic irreducible polynomials. We hold this notation fixed, remembering that the polynomial f is assumed to be known explicitly, whereas the integer m and the polynomials g_j are to be determined.

We say that a polynomial $h \in E[X]$ is a *factorization polynomial* for f if f divides the polynomial

$$\prod_{a \in E} (h(X) - a).$$

For example, any constant polynomial in $E[X]$ is a factorization polynomial for f since if h is constant, then the above product is zero. Another example of a factorization polynomial for f is f itself.

Much more interesting is the case where f has a factorization polynomial h such that $0 < \deg(h) < \deg(f)$. In this situation, it is easy to see that for some element $a \in E$, the polynomial $\gcd(f, h - a)$ is a proper nontrivial divisor of f. In this case, therefore, the factorization polynomial actually does yield a factorization of f. As we shall see, the Berlekamp algorithm succeeds in finding the complete factorization of f by finding sufficiently many factorization polynomials with degrees smaller than $n = \deg(f)$.

(21.16) LEMMA.   *Assume the previous notation and suppose $h \in E[X]$ is a factorization polynomial for f. Then there exists a unique m-tuple $(a_1, a_2, \ldots, a_m)$ with $a_i \in E$ such that $g_j \mid (h - a_j)$ for each subscript j, with $1 \leq j \leq m$.*

Proof.   Since $f \mid \prod(h - a)$, where a runs over E, each irreducible factor g_j of f must divide $h - a_j$ for some $a_j \in E$. Furthermore, a_j is uniquely determined since the polynomials $h - a$ are pairwise relatively prime as a runs over E.   ∎

We refer to the row vector $(a_1, a_2, \ldots, a_m)$ of Lemma 21.16 as the *factorization row* associated with the factorization polynomial h, and we write $\rho(h)$ to denote this vector. We view $\rho(h)$ as an element of E^m, the m-dimensional space of row vectors over E. Since we are holding fixed the numbering of the (not yet known) irreducible factors g_j of f, the factorization row $\rho(h)$ is well defined, although we do not yet have a procedure for computing it.

(21.17) THEOREM.   *Assume the previous notation and let $R = \overline{E[X]} = E[X]/(f)$. Let $\psi : R \to R$ be the map defined by $\psi(v) = v^q$, where $q = |E|$. Finally, let $U = \{v \in R \mid \psi(v) = v\}$. Then:*

  a. *ψ is an E-linear operator.*
  b. *$U \subseteq R$ is an E-subspace.*
  c. *If $h \in E[X]$ with $\overline{h} \in U$, then h is a factorization polynomial for f and the map $\varphi : U \to E^m$ given by $\varphi(\overline{h}) = \rho(h)$ is a well defined E-vector space isomorphism of U onto E^m.*
  d. *$\dim_E(U) = m$.*

Proof.   If $v, w \in R$, then $(v+w)^q = v^q + w^q$ since q is a power of the characteristic of E, and thus ψ is an additive homomorphism. Because $|E| = q$, we have $a^q = a$ for all $a \in E$, and so $\psi(av) = a^q v^q = av^q = a\psi(v)$, and (a) is proved. Part (b) follows immediately from (a). Note that (d) is a consequence of (c), and so it suffices to prove (c).

By Lemma 21.15 there is an isomorphism

$$\theta : R \to L_1 \oplus L_2 \oplus \cdots \oplus L_m ,$$

where $L_j = E[X]/(g_j)$ and the jth component of $\theta(\overline{h})$ is $(g_j) + h$ for $h \in E[X]$. Note that (g_j) is a maximal ideal of $E[X]$ since g_j is irreducible, and thus L_j is a field. The image of $E \subseteq E[X]$ under the canonical homomorphism $E[X] \to L_j$ is a subfield of L_j, which we denote E_j. Clearly, $E_j \cong E$ and we have $E_j = ((g_j) + E)/(g_j)$.

By Theorem 21.10 the extension $L_j \supseteq E_j$ is Galois, and since $|E_j| = q$, Theorem 21.10 tells us that the Galois group is generated by the map $\alpha \mapsto \alpha^q$. If $h \in E[X]$ with $\overline{h} \in U$, then $(\overline{h})^q = \overline{h}$, and so the projection of $\theta(\overline{h})$ into L_j equals its own qth power. It follows that this projection, $(g_j) + h$, lies in $\mathrm{Fix}(\mathrm{Gal}(L_j/E_j)) = E_j$, and so $(g_j) + h = (g_j) + a_j$ for some element $a_j \in E$ (depending on h). Thus g_j divides $h - a_j$, and it follows that h is a factorization polynomial for f and that $\rho(h) = (a_1, a_2, \ldots, a_m)$.

If we identify E^m with $E_1 \oplus \cdots \oplus E_m$ in the natural way, we have just seen that if $\overline{h} \in U$, then h is a factorization polynomial for f and we can write $\rho(h) = \theta(\overline{h})$. The restriction of θ to U is thus the map $\varphi : U \to E^m$ defined in the statement of (c). Since θ is an isomorphism on R, the map φ is certainly injective. To see that it is surjective, observe that every element of $E^m = E_1 \oplus \cdots \oplus E_m$ is of the form $\theta(v)$ for some $v \in R$, and it suffices to show that $v \in U$. This follows because each component of $\theta(v)$ lies in a field of order q and so equals its own qth power. It follows that $\theta(v^q) = \theta(v)$ and hence $v^q = v$ and $v \in U$. ∎

We can actually carry out computations with the linear transformation $\psi : R \to R$ of Theorem 21.17 by fixing an E-basis for R and working with matrices. By Lemma 21.15(a) we know that $\overline{1}, \overline{X}, \ldots, \overline{X}^{n-1}$ form a basis for R, where $n = \deg(f)$. To compute the matrix M for the linear operator ψ on R with respect to this basis, we need to express $\psi(\overline{X}^i) = \overline{X^{iq}}$ in terms of our basis. When this is done, the coefficient of $\overline{X}^j$ is the (i, j)-entry of M. (We must number the rows and columns of M from 0 to $n - 1$ to make this work.)

To express X^{iq} in terms of our basis, simply divide $f(X)$ into X^{iq} and compute the remainder $r_i(X)$. The (i, j)-entry of M is therefore the coefficient of X^j in $r_i(X)$ for $0 \le i, j \le n - 1$.

With the $n \times n$ matrix M in hand, we can easily compute the number m of irreducible factors of f by observing that

$$m = \dim_E(U) = \text{nullity}(M - I),$$

where I is the $n \times n$ identity matrix. In particular, f is irreducible iff nullity$(M-I) = 1$. (Note that the nullity of a matrix can be computed by the use of standard linear algebra algorithms such as Gaussian elimination.)

If we wish to find the irreducible factors g_j of f (and not merely to count them), we must do a little more work. Using Gaussian elimination, we compute a basis for the nullspace of $M - I$ (acting on rows). Viewing row vectors as polynomials, we obtain polynomials $h_1, h_2, \ldots, h_m$, each of degree $\le n - 1$ and such that $\overline{h}_1, \overline{h}_2, \ldots, \overline{h}_m$ form a basis for U. In particular, these polynomials h_i are factorization polynomials for f.

(21.18) COROLLARY. *Assume the previous notation and fix a monic divisor g of f (allowing $g = f$). Suppose $h_1, h_2, \ldots, h_m \in E[X]$ are polynomials in $E[X]$ whose images in R form a basis for U. For $1 \le i \le m$ and $a \in E$, write $k_{i,a} = \gcd(g, h_i - a)$. Then*

a. $g = \prod\limits_{a \in E} k_{i,a}$ *for each subscript i, and*

b. *g is irreducible iff for each i, one of the polynomials $k_{i,a}$ is equal to g.*

Of course, the polynomials $k_{i,a}$ depend on the particular polynomial g with which we start. We write $k_{i,a} = k_{i,a}^{(g)}$ when we wish to emphasize this dependence, and we refer to the polynomials $k_{i,a}^{(g)}$ as the *associated* factors of g. If g is known (for instance, $g = f$) and particular polynomials $h_1, h_2, \ldots, h_m$ have been found (as described earlier), then the associated factors of g can be explicitly computed.

Note that the significance of Corollary 21.18(b) is that if g has a proper factorization, then at least one of the m factorizations of g given in (a) is proper.

Proof of Corollary 21.18.   Since $\overline{h}_i \in U$, Theorem 21.17(c) tells us that h_i is a factorization polynomial for f. We write $a_{i,j} \in E$ to denote the jth entry of the factorization row $\rho(h_i)$ for $1 \le i, j \le m$. (Recall that this tells us that g_j divides $h_i - a_{i,j}$ for all subscripts i, j.)

Now g is the product of some subset of $\{g_j \mid 1 \le j \le m\}$, and the associated factor $k_{i,a}$ is the product of those factors g_j of g that happen to divide $h_i - a$. Furthermore, g_j divides $h_i - a$ iff $a = a_{i,j}$, and so each g_j that divides g occurs as a factor of $k_{i,a}$ for exactly one element $a \in E$. Part (a) follows.

To prove (b), note that if g is irreducible, then each of the factorizations of g given in (a) is trivial. Thus for each i, one associated factor $k_{i,a}$ is equal to g, and all of the others equal the constant polynomial 1.

Conversely, assume that for each i, there is an element $a(i) \in E$ such that $k_{i,a(i)} = g$. If g_j divides g, therefore, we see that g_j divides $k_{i,a(i)}$ and hence g_j divides $h_i - a(i)$. It follows that $a_{i,j} = a(i)$ whenever g_j divides g. If g were reducible, there would be at least two different subscripts j such that g_j divides g, and hence the $m \times m$ matrix $[a_{i,j}]$ would have two identical columns and so it would have dependent rows. The rows of this matrix, however, are the factorization rows $\rho(h_i)$, and these are linearly independent by Theorem 21.17(c) since the vectors $\overline{h}_i \in U$ are linearly independent. This contradiction completes the proof of (b).   ∎

We are now ready to describe an algorithm for computing the g_j. We also give three "shortcuts" for its implementation.

First compute the integer m and polynomials h_i for $1 \le i \le m$ as described earlier. Let i and S be "variables," where i is a positive integer and S is a list of positive degree monic divisors of f. Initialize $S = \{f\}$. When our process terminates, the entries in S will be the irreducible factors g_j of f in some order.

Loop through the values of i from 1 to m and do the following: let g run over S and a run over E and find all of those associated factors $k_{i,a}^{(g)}$ of g that have positive degree. Then replace S by the list of all the associated factors thus found.

(21.19) COROLLARY.   *The algorithm described above terminates with S being the list of all g_j.*

Proof.   Let S_i denote the state of the variable list S at the end of step i for $1 \le i \le m$, and write $S_0 = \{f\}$, the initial state. We claim that for each i, the entries in S_i are pairwise coprime and that their product is equal to f. This is certainly true for $i = 0$. Working by induction, we have for $i > 0$,

$$f = \prod_{g \in S_{i-1}} g = \prod_g \prod_{a \in E} k_{i,a}^{(g)} .$$

Since distinct members of S_{i-1} are coprime, their associated factors are certainly coprime. Also, for any fixed $g \in S_{i-1}$, the factors $k_{i,a}^{(g)}$ are pairwise coprime as a runs over E. Since S_i is precisely the set of all $k_{i,a}^{(g)}$ not equal to 1, where g runs over S_{i-1} and a runs over E, our claim is established.

To complete the proof, it suffices to show that every element $g \in S_m$ is irreducible. Given $g \in S_m$, we know that g divides some entry of S_i for each i with $1 \le i \le m$. Thus g divides some polynomial of the form $k_{i,a}^{(u)}$ for each i, where u is some suitable polynomial and $a \in E$. It follows that g divides $h_i - a$ for some element $a \in E$ (depending on i) for $1 \le i \le m$. We conclude that $k_{i,a}^{(g)} = g$, and thus g is irreducible by Corollary 21.18(b).   ■

Now we can give the promised shortcuts. First, keep a running count of the number of polynomials in S, and note that each time S is changed, its cardinality increases. Since the final S has exactly m elements, we can stop as soon as we reach a list S with m members. (Recall that we do know m at this point.) No further changes can occur.

Next, for a fixed subscript i and a fixed member $g \in S$, while running through the elements $a \in E$ seeking nontrivial associated factors $k_{i,a}$ for g, keep track of the total degree of the nontrivial associated factors already found. When this total reaches $\deg(g)$, there is no need to run through the rest of E.

Finally, and of lesser significance, note that the polynomial $\bar{1}$ is in U. If we arrange (as we can) that $h_m = 1$, then the associated polynomials $k_{m,a}$ cannot yield a proper factorization, and so we need only loop from $i = 1$ to $i = m - 1$.

21D

Perhaps our next result does not belong in the "commutative" part of this book since we do not assume commutativity. We prove it instead.

(21.20) THEOREM (Wedderburn).   *Let D be a finite division ring. Then D is commutative (and thus is a finite field).*

(21.21) LEMMA.   *Let D be a division ring and suppose $a \in D$. Then the centralizer $\mathbf{C}_D(a) = \{x \in D \mid xa = ax\}$ is a subdivision ring of D. Also, the center $\mathbf{Z}(D) = \{x \in D \mid xa = ax \text{ for all } a \in D\}$ is a subfield.*

Proof.   Certainly, $\mathbf{C}(a)$ is closed under addition, subtraction, and multiplication, and since $1 \in \mathbf{C}(a)$, it is a (unitary) subring of D. We need to show that if $0 \ne x \in \mathbf{C}(a)$, then the inverse x^{-1} of x in D lies in $\mathbf{C}(a)$. In other words, we need $x^{-1}a = ax^{-1}$. We have

$$x^{-1}a = x^{-1}(ax)x^{-1} = x^{-1}(xa)x^{-1} = ax^{-1},$$

as required.

The second statement follows since $\mathbf{Z}(D) = \bigcap \mathbf{C}(a)$ for $a \in D$.   ■

Proof of Theorem 21.20.   Let $Z = \mathbf{Z}(D)$, so that Z is a field by Lemma 21.21, and we can write $|Z| = q$, where q is some prime power. For each element $a \in D$, we have $Z \subseteq \mathbf{C}(a)$, and this makes the division ring $\mathbf{C}(a)$ into a Z-vector space. Writing $d(a) = \dim_Z(\mathbf{C}(a))$, we deduce that $|\mathbf{C}(a)| = q^{d(a)}$. In particular, $|D| = q^n$, where $n = d(1)$.

Now $D^{\times}$ is a finite group, and we choose a set S of representatives for the noncentral conjugacy classes of $D^{\times}$. (Of course, once we show that D is commutative, we will have $S = \varnothing$.)

If $a \in S$, then the class K_a of a in $D^{\times}$ contains exactly $|D^{\times} : \mathbf{C}_{D^{\times}}(a)| = (q^n - 1)/(q^{d(a)} - 1)$ elements, and we have

$$q^n - 1 = |D^{\times}| = |Z^{\times}| + \sum_{a \in S} |K_a| = (q - 1) + \sum_{a \in S} \frac{q^n - 1}{q^{d(a)} - 1}.$$

Since $(q^n - 1)/(q^{d(a)} - 1)$ is an integer, it follows by Theorem 21.7 that $d(a)$ divides n. Because $d(a) < n$, we deduce from Lemma 20.4 that the cyclotomic polynomial $\Phi_n(X)$ divides $(X^n - 1)/(X^{d(a)} - 1)$, and so each summand $(q^n - 1)/(q^{d(a)} - 1)$ is a multiple of $\Phi_n(q)$. Also, $q^n - 1$ is a multiple of $\Phi_n(q)$, and it follows from the above equation that $\Phi_n(q)$ divides $q - 1$. We conclude that $|\Phi_n(q)| \leq q - 1$. By Lemma 20.17, however, $\Phi_n(q) > q - 1$ when $n > 1$. It follows that $n = 1$ and thus $D = Z$ is commutative.   ∎

Problems

21.1   Show that there exists $\epsilon \in GF(8)$ such that $GF(8)^{\times} = \langle \epsilon \rangle$ and $\epsilon^3 + \epsilon + 1 = 0$. Write the addition table for the additive group of this field using the names $0, 1, \epsilon, \epsilon^2, \dots, \epsilon^6$ for the elements. In particular, what is $1 + \epsilon^2$?

21.2   Let $E = GF(25)$ and let $F = \mathbb{Z}/5\mathbb{Z}$ be the prime subfield. Show that there exists an element $\epsilon \in E$ with $\epsilon^2 = 3$. Show that 1 and ϵ form a basis for E over F and that $1 + \epsilon$ is a generator for the cyclic group $E^{\times}$.

HINT: Compute $(1 + \epsilon)^{12}$ and $(1 + \epsilon)^8$.

21.3   Let q be a prime power and let $n \geq 1$ be an integer. Show that the group $GL(n, q)$ of nonsingular $n \times n$ matrices over $GF(q)$ contains an element of order $q^n - 1$.

HINT: Let $E = GF(q^n)$ and consider the F-linear transformation $E \to E$ induced by multiplication by a generator of $E^{\times}$, where $F \subseteq E$ and $|F| = q$.

21.4   Suppose that the cyclotomic polynomial Φ_n is irreducible mod p for some prime p. Show that U_n is cyclic and deduce that either n divides 4 or n divides $2q$, where q is an odd prime power.

HINT: Use Problem 20.17. Show that U_8 is not cyclic.

21.5   Show that $X^4 + 1$ is never irreducible over any finite field.

21.6   Compute explicitly all irreducible polynomials of degree 4 over $\mathbb{Z}/2\mathbb{Z}$.

21.7 Let R be any commutative ring and let $I_1, I_2, \ldots, I_m$ be pairwise comaximal ideals of R. (This means that $I_i + I_j = R$ if $i \neq j$.) Let $D = \bigcap I_i$ and show that
$$R/D \cong (R/I_1) \oplus \cdots \oplus (R/I_m).$$

HINT: Show that $I_i + \bigcap_{j \neq i} I_j = R$, and deduce that there exists $a_i \in R$ with $a_i - 1 \in I_i$ and $a_i \in I_j$ for $j \neq i$.

NOTE: This is a generalization of the Chinese remainder theorem of number theory. Note that in a PID, the principal ideals generated by relatively prime elements are comaximal. Compare this with Lemma 21.15(b).

21.8 Let $F \subseteq E$, where $|F| = q < \infty$. Suppose $\alpha \in E$ is algebraic over F. Show that $|F[\alpha] : F|$ is the smallest positive integer n such that $\alpha^{q^n} = \alpha$, and that it divides every other such positive integer.

21.9 Let $F = GF(q)$ and write $f(X) = X^q - X - 1 \in F[X]$. Show that every irreducible factor of f in $F[X]$ has degree $p = \text{char}(F)$.

HINT: If α is a root of f in some extension field of F, show that $\alpha^{q^p} = \alpha$.

21.10 Let R be a not necessarily commutative finite ring containing no nonzero nilpotent elements. Show that R is commutative.

HINT: Use the Wedderburn-Artin theorem (14.15).

21.11 Let A be a finite abelian group and suppose $S \subseteq \text{End}(A)$ has the property that if $B\sigma \subseteq B$ for all $\sigma \in S$, where B is a subgroup of A, then either $B = 0$ or $B = A$. Show that there exists $\theta \in \text{End}(A)$ such that every nonzero member of $\mathbf{C}_{\text{End}(A)}(S)$ is a power of θ.

CHAPTER TWENTY-TWO

Roots, Radicals, and Real Numbers

22A

Chapter 20 dealt with roots of polynomials of the form $X^n - 1$, and Chapter 21 on finite fields was concerned essentially with the roots of polynomials of the form $X^q - X$, where q is a power of the characteristic. Much of the content of this chapter is also connected with the roots of certain special polynomials: those of the form $X^n - a$.

Since the radical symbol $\sqrt{}$ is often used to denote roots of such polynomials, we say that a field extension $F \subseteq E$ is a *radical extension* if $E = F[\alpha]$, where α is a root of some polynomial of the form $X^n - a$ with $a \in F$. In other words, $F \subseteq E$ is radical if E is generated over F by some element $\alpha \in E$ with $\alpha^n \in F$ for some positive integer n. Note that the cyclotomic extension $\mathbb{Q} \subseteq \mathbb{Q}_n$ is radical, and so also is every extension of finite fields.

We wish to prove Galois's theorem relating the "solvability" of polynomials with the solvability (in the group theoretic sense) of their associated Galois groups. Let $f \in \mathbb{Q}[X]$ with $\deg(f) > 0$. We are interested in the roots of f (in $\mathbb{C}$, say). Since $\mathbb{C}$ is algebraically closed, the existence of roots is not in question, although we may ask how the roots can be found and how they can be described. (We stress that we are concerned here with only *exact* roots; we are not interested in approximations.)

Ideally, we would like to have some sort of algorithm – a formula, for instance – into which we can plug the coefficients of $f \in \mathbb{Q}[X]$ and that would then output at least one root of f. Failing that, it might seem that at least we could expect to check computationally whether or not a given complex number is a root. Having such a check, we might, if we are extraordinarily lucky, be able to guess a root for f and then confirm our guess and thereby know a root.

Imagine a computer program designed either to find roots of rational polynomials and output them, or to accept complex numbers as input and decide whether or not they are roots of the given polynomial. How is the computer (or we) to describe a particular complex number? The most obvious difficulty is that $\mathbb{C}$ is uncountable, and yet the set of potential inputs and outputs is only countable. This problem is easy to resolve since we really do not need to work with the full field $\mathbb{C}$ of complex numbers, but only with the countable subfield $\mathbb{A}$ of algebraic numbers. (Note that since we are assuming that the coefficients of our polynomial are rational, there is no difficulty in describing them.)

There does exist a standard and accepted language for describing algebraic numbers: start with integers and use the four arithmetic operations together with the extraction of roots. Thus, for instance, $(3 + \sqrt{-5})/2$ and $\sqrt[5]{2 + \sqrt{-1}}$ are algebraic numbers. (Actually, each of these symbol strings refers to more than one algebraic number because the radical notation is ambiguous.) This language is not sufficient, however.

The point of Galois's theorem is that not every algebraic number can be described by the use of only addition, subtraction, multiplication, division, and radicals. One can write rational polynomials (of degree 5, for example) whose roots cannot be expressed in this form. If we limit ourselves to the standard radical notation, therefore, it would be impossible to have a computer program that would solve such polynomials, and we could not even find a root by guessing and confirming, no matter how lucky we are.

To be precise about which numbers can be described by radicals and arithmetic operations, we say that a field extension $F \subseteq E$ is a *repeated radical extension* if there exist fields F_i with

$$F = F_0 \subseteq F_1 \subseteq \cdots \subseteq F_r = E$$

so that each F_i is a radical extension of F_{i-1} for $1 \le i \le r$. In other words, there exist elements $\alpha_i \in F_i$ and integers $n_i \ge 1$ such that $F_i = F_{i-1}[\alpha_i]$ and $\alpha_i^{n_i} \in F_{i-1}$ for $1 \le i \le r$.

It should be clear that those complex numbers that lie in repeated radical extensions of $\mathbb{Q}$ are precisely the algebraic numbers that we can describe using elements of $\mathbb{Z}$, the arithmetic operations and radicals.

(22.1) LEMMA.   *Let $F \subseteq K$, $L \subseteq E$, where L is a repeated radical extension of F. Writing $\langle K, L \rangle$ to denote the compositum of K and L in E, we have*

    a. *$\langle K, L \rangle$ is a repeated radical extension of K, and*
    b. *if K is also a repeated radical extension of F, then so is $\langle K, L \rangle$.*

Proof.   Statement (b) is immediate from (a), and so we prove (a). We have $F = F_0 \subseteq F_1 \subseteq \cdots \subseteq F_r = L$, where $F_i = F_{i-1}[\alpha_i]$ and $\alpha_i^{n_i} \in F_{i-1}$ for $1 \le i \le r$. We define subfields $K_i \subseteq E$ inductively by setting $K_0 = K$ and $K_i = K_{i-1}[\alpha_i]$ for $1 \le i \le r$.

Since $F_0 = F \subseteq K = K_0$, it is trivial to show by induction on i that $F_i \subseteq K_i$ for $1 \le i \le r$. Thus $\alpha_i^{n_i} \in F_i \subseteq K_i$ for these values of i, and it follows that K_r is a repeated radical extension of K. Also $L = F_r \subseteq K_r$, and so

$$\langle K, L \rangle \subseteq K_r = K[\alpha_1, \alpha_2, \ldots, \alpha_r] \subseteq \langle K, L \rangle.$$

Thus $\langle K, L \rangle = K_r$, a repeated radical extension of F.   ∎

(22.2) COROLLARY. *Let $F \subseteq E$ and let R be the union of all of those intermediate fields K between F and E that are repeated radical extensions of F. Then R is a field. Furthermore, if $|E : F| < \infty$, then R is a repeated radical extension of F.*

Proof.   Let S be the set of all intermediate fields that are repeated radical extensions of F. If $\alpha, \beta \in R$, then $\alpha \in K$ and $\beta \in L$ for some fields $K, L \in S$. The elements $\alpha - \beta$ and α/β (if $\beta \neq 0$) lie in $\langle K, L \rangle$, and by Lemma 22.1 we have $\langle K, L \rangle \in S$ and thus $\langle K, L \rangle \subseteq R$. Since R contains $\alpha - \beta$ and α/β, we conclude that R is a field, as desired.

Now assume that $|E : F| < \infty$ and select $K \in S$ with $|K : F|$ as large as possible. If also $L \in S$, then $\langle K, L \rangle \in S$ and $|\langle K, L \rangle : F| \geq |K : F|$. By the choice of K, it follows that $\langle K, L \rangle = K$ and hence $L \subseteq K$. In other words, $R = K \in S$.   ∎

If $f \in F[X]$, we say that f is *solvable by radicals* over F if f splits over some field E that is a repeated radical extension of F. Note that we do not require that a splitting field for f over F be a repeated radical extension of F, only that it be contained in one.

(22.3) LEMMA.   *Let $f \in F[X]$ be irreducible. Then f is solvable by radicals over F iff f has a root in some repeated radical extension of F.*

Proof.   The "only if" direction is trivial, and so we assume that f has a root $\alpha \in L$, where $L \supseteq F$ is a repeated radical extension. We show that f is solvable by radicals.

By Lemma 18.5 there exists a field $E \supseteq L$ such that E is a splitting field for some polynomial over F. Since f is irreducible over F, we know from Lemma 18.3 that $\mathrm{Gal}(E/F)$ transitively permutes the roots of f in L.

Let R be, as in Corollary 22.2, the union of the repeated radical extensions of F in E. Since R is uniquely determined by the extension $F \subseteq E$, it is clear that the elements of $\mathrm{Gal}(L/F)$ map R to itself. Since $\alpha \in L \subseteq R$, we deduce that all of the roots of f in E lie in R. Because E is normal over F by Theorem 18.4, we see that f splits over E, and we conclude that f splits over R. But $|E : F| < \infty$, and so R is a repeated radical extension of F.   ∎

Finally, we state our main result.

(22.4) THEOREM (Galois).   *Assume $\mathrm{char}(F) = 0$ and let $f \in F[X]$. Then f is solvable by radicals over F iff $\mathrm{Gal}(E/F)$ is a solvable group, where E is a splitting field for f over F.*

Note that the symmetric group S_n is nonsolvable for $n \geq 5$. (This is because for these values of n, Theorem 6.17 tells us that the alternating group A_n is simple.) We shall show that the polynomial $f(X) = 2X^5 - 10X + 5 \in \mathbb{Q}[X]$ has a Galois group (over $\mathbb{Q}$) isomorphic to S_5, and it follows that this polynomial is not solvable by radicals over $\mathbb{Q}$. Since f is irreducible by the Eisenstein criterion, Lemma 22.3 tells us that none of its roots (in $\mathbb{C}$) lies in a repeated radical extension of $\mathbb{Q}$. It is thus impossible even to write down any root of this particular polynomial (if we limit ourselves to integers, arithmetic, and radicals). There certainly cannot exist, therefore, any general formula (involving nothing more exotic than radicals) that solves all quintic polynomial equations.

In order to compute the Galois group for the splitting field in $\mathbb{C}$ of $2X^5 - 10X + 5$ over $\mathbb{Q}$, we state a general lemma.

(22.5) LEMMA. *Let $f \in \mathbb{Q}[X]$ be irreducible and have prime degree p. Assume that f has $p-2$ real roots and two nonreal complex roots. Then* $\text{Gal}(E/\mathbb{Q}) \cong S_p$, *where E is a splitting field for f over $\mathbb{Q}$.*

Proof.   We may assume $E \subseteq \mathbb{C}$, and we write Ω to denote the set of roots of f in E. Then $G = \text{Gal}(E/\mathbb{Q})$ acts faithfully on Ω, and so G is isomorphically embedded in $\text{Sym}(\Omega) \cong S_p$.

Since f is irreducible, G acts transitively on Ω and thus $p \| G$. It follows that the image of G in S_p contains an element a of order p, a p-cycle. Also, since E is normal over $\mathbb{Q}$, complex conjugation maps E to itself, and so the restriction of complex conjugation to E defines an element of G that moves only two of the elements of Ω (the two nonreal roots). The image b of this element in S_p is thus a 2-cycle, a transposition.

We can assume that $b = (1, 2)$ in cycle notation. Also, the p-cycle a has some power that carries the symbol 1 to 2. We may replace a by this power and assume that $a = (1, 2, 3, \ldots, p)$. We complete the proof by showing that a and b generate the full symmetric group.

It suffices to show that $H = \langle a, b \rangle \subseteq S_p$ contains all of the transpositions. Note that $ab = (2, 3, \ldots, p - 1, p)$, and so the conjugates of b by the various powers of ab are $(1, 2), (1, 3), \ldots, (1, p)$. Conjugation of these by the powers of a yields all transpositions.   ∎

We have observed that the polynomial $f(X) = 2X^5 - 10X + 5$ is irreducible over $\mathbb{Q}$. To see that it satisfies the hypothesis of Lemma 22.5 and so is not solvable by radicals over $\mathbb{Q}$, we must show that it has exactly three real roots. Now $f'(X) = 10(X^4 - 1)$, and so the graph of $y = f(x)$ rises for $-\infty < x < -1$ and for $1 < x < \infty$, and it falls for $-1 < x < 1$. Since $f(-1) = 13 > 0$ and $f(1) = -3 < 0$, we see that f has exactly one root in each of the intervals $(-\infty, -1)$, $(-1, 1)$ and $(1, \infty)$. There are therefore exactly three real roots.

We mentioned previously that it is unknown which finite groups occur as Galois groups over $\mathbb{Q}$. We have just shown that S_5 occurs, and if we use Lemma 22.5, it is not terribly difficult to show that S_p occurs for all prime numbers p. (See the problems at the end of the chapter.) In fact, S_n occurs as a Galois group over $\mathbb{Q}$ for all integers $n \geq 1$, but a more subtle argument seems to be required when n is not prime.

22B

We begin now working toward the proof of Galois's theorem (22.4). We need the following useful lemma of Dedekind, but we digress briefly to discuss notation before we state it.

If E is a field and $\sigma \in \text{Aut}(E)$, we have been writing $\sigma(\alpha)$ to denote the element of E that results from the application of σ to $\alpha \in E$. Another common notation for this element is α^σ. In some formulas, this exponential notation appears to enhance clarity, and so we use it whenever it seems appropriate. Note that if σ and τ are automorphisms, then we have $(\alpha^\sigma)^\tau = \alpha^{\sigma\tau}$. This is *not* merely a consequence of the exponential notation; it holds because of our standing convention that if σ and τ are functions, then $\sigma\tau$ is the function obtained by doing σ first and then τ.

(22.6) LEMMA (Dedekind).   *Let E be any field and let S be a finite set of automorphisms of E. Suppose $f : S \to E$ is a function such that*

$$(*) \qquad \sum_{\sigma \in S} f(\sigma)\alpha^\sigma = 0$$

for all $\alpha \in E$. Then $f(\sigma) = 0$ for all $\sigma \in S$.

Lemma 22.6 can be paraphrased in an interesting way. If X is any nonempty set, then the set of all functions from X into E is an E-vector space with respect to pointwise addition and scalar multiplication. To say that a subset A of this vector space is linearly independent means that for every finite subset $S \subseteq A$, if we have any equation of the form $\sum f_\sigma \sigma = 0$, where σ runs over S and the coefficients f_σ lie in E, then all $f_\sigma = 0$. If we write $f(\sigma)$ for f_σ and use exponential notation for the application of σ to elements of X, then this says that if $\sum f(\sigma)x^\sigma = 0$ for all $x \in X$, then $f(\sigma) = 0$ for all $\sigma \in S$.

Now take $X = E$ in the above discussion. Then $\text{Aut}(E)$ is a subset of the set of all functions from E to E. We see that the content of Dedekind's lemma (22.6) is that $\text{Aut}(E)$ is a linearly independent subset of the E-space of all functions from E to E.

Proof of Lemma 22.6.   If $|S| = 1$, we have $S = \{\sigma\}$ and $f(\sigma)1^\sigma = 0$ by $(*)$. Since $1^\sigma = 1$, we have $f(\sigma) = 0$, as required. We may therefore assume that $|S| > 1$, and we work by induction on $|S|$.

Now fix $\tau \in S$ and $\beta \in E$. Multiplication of $(*)$ by β^τ yields

$$(**) \qquad \sum_{\sigma \in S} f(\sigma)\alpha^\sigma \beta^\tau = 0 \quad \text{for all } \alpha \in E.$$

We can also modify $(*)$ by substituting $\alpha\beta$ for α and using the fact that $(\alpha\beta)^\sigma = \alpha^\sigma \beta^\sigma$. This gives

$$(* * *) \qquad \sum_{\sigma \in S} f(\sigma)\alpha^\sigma \beta^\sigma = 0 \quad \text{for all } \alpha \in E.$$

Since the terms corresponding to $\sigma = \tau$ in ($**$) and ($*$ $*$ $*$) are identical, we can subtract these equations to obtain

$$\sum_{\sigma \in S - \{\tau\}} [f(\sigma)(\beta^\tau - \beta^\sigma)](\alpha^\sigma) = 0 \quad \text{for all } \alpha \in E.$$

This is an equation in the same form as ($*$), except that we are summing over the smaller set $S' = S - \{\tau\}$ and f has been replaced by the function $\sigma \mapsto f(\sigma)(\beta^\tau - \beta^\sigma)$. It follows by the inductive hypothesis that $f(\sigma)(\beta^\tau - \beta^\sigma) = 0$ for all $\sigma \in S'$. If $f(\sigma) \neq 0$ for some member $\sigma \in S'$, we conclude that $\beta^\sigma = \beta^\tau$, and this holds for all $\beta \in E$, so that $\sigma = \tau$, a contradiction.

We have now shown that $f(\sigma) = 0$ for all $\sigma \in S$, except possibly when $\sigma = \tau$. Since $|S| > 1$ and $\tau \in S$ is arbitrary, the result follows. ∎

(22.7) LEMMA. *Let $G \subseteq \text{Aut}(E)$ with $|G| < \infty$, and let $F = \text{Fix}(G)$. Suppose $\theta : G \to F^\times$ is a group homomorphism. Then there exists an element $\alpha \in E^\times$ such that for all $\tau \in G$, we have $\theta(\tau) = \alpha^\tau / \alpha$.*

Proof. View θ as a map $G \to E$. Certainly θ is not identically zero (since it is never zero), and so by Lemma 22.6 there exists an element $\gamma \in E$ with $\sum \theta(\sigma)\gamma^\sigma \neq 0$. Fix such an element γ and let α be the reciprocal of this sum. Thus

$$\frac{1}{\alpha} = \sum_{\sigma \in G} \theta(\sigma)\gamma^\sigma .$$

Fix an element $\tau \in G$. Since $\sigma\tau$ runs over G as σ does, we have

$$\frac{1}{\alpha} = \sum_\sigma \gamma^{\sigma\tau}\theta(\sigma\tau) = \sum_\sigma \gamma^{\sigma\tau}\theta(\sigma)\theta(\tau) = \left(\sum_\sigma \gamma^\sigma\theta(\sigma)\right)^\tau \theta(\tau),$$

where we have used the fact that $\theta(\sigma) \in F = \text{Fix}(G)$ to conclude that $\theta(\sigma)^\tau = \theta(\sigma)$. Since the sum on the right equals $1/\alpha$, we obtain $\alpha^\tau/\alpha = \theta(\tau)$, as required. ∎

In Chapter 23 we reconsider Lemma 22.7 and interpret it in a more general context.

(22.8) THEOREM (Kummer). *Let $F \subseteq E$ and assume that F contains a primitive nth root of unity. Then the following are equivalent:*

i. *E is Galois over F and $\text{Gal}(E/F)$ is cyclic of order dividing n.*
ii. *$E = F[\alpha]$ for some element $\alpha \in E$ with $\alpha^n \in F$.*

Since, in particular, (ii) says that E is a radical extension of F, Kummer's theorem provides a method for proving that a field extension is radical.

Proof of Theorem 22.8. Let $\epsilon \in F$ be a primitive nth root of unity. Assume (i) and write $G = \text{Gal}(E/F) = \langle \sigma \rangle$, where $o(\sigma) = m$ and $m | n$. Now $\langle \epsilon \rangle$ is cyclic of order n and so it contains an isomorphic copy of G. Let $\theta : G \to \langle \epsilon \rangle \subseteq F^{\times}$ be an isomorphism (into) and use Lemma 22.7 to get $\alpha \in E$ such that $\theta(\tau) = \alpha^{\tau}/\alpha$ for all $\tau \in G$.

Write $\delta = \theta(\sigma)$. Then δ is an nth root of unity in F and $\delta = \alpha^{\sigma}/\alpha$. Thus $\sigma(\alpha^{n}) = \sigma(\alpha)^{n} = (\alpha\delta)^{n} = \alpha^{n}$. We conclude that $\alpha^{n} \in \mathrm{Fix}(\langle\sigma\rangle) = F$, as required.

To show that $F[\alpha] = E$, let $\tau \in \mathrm{Gal}(E/F[\alpha]) \subseteq G$. Then $\theta(\tau) = \alpha^{\tau}/\alpha = 1$ and hence $\tau = 1$ since θ is injective. It follows the $\mathrm{Gal}(E/F[\alpha]) = 1$, and so $F[\alpha] = E$ by the fundamental theorem of Galois theory.

Conversely now, assume (ii). Then $E = F[\alpha]$ and $\alpha^{n} \in F$, and we let $f(X) = X^{n} - \alpha^{n}$ so that $f \in F[X]$. If δ is any nth root of unity in E, we have $f(\alpha\delta) = (\alpha\delta)^{n} - \alpha^{n} = 0$, and thus $\alpha\delta$ is a root of f in E. Since E contains n nth roots of unity, we see that f has $n = \deg(f)$ roots in E. It follows that f splits over E and has distinct roots. Since α is a root of f, we see that $E = F[\alpha]$ is a splitting field for f over F. Also, f is separable over F, and hence E is Galois over F, as required.

Now let $G = \mathrm{Gal}(E/F)$. We must show that G is cyclic and that $|G|$ divides n. We shall accomplish this by constructing an isomorphism from G into $\langle \epsilon \rangle$.

Since G permutes the roots of f, we see that for each automorphism $\tau \in G$, we have $\tau(\alpha) = \alpha\delta$ for some element $\delta \in \langle\epsilon\rangle$. Since we can certainly assume $\alpha \neq 0$ (or else $G=1$), we can define the map $\theta : G \to \langle\epsilon\rangle$ by $\theta(\tau)=\alpha^{\tau}/\alpha$ for $\tau \in G$. If $\sigma, \tau \in G$, we have

$$\theta(\sigma\tau) = \frac{\alpha^{\sigma\tau}}{\alpha} = \left(\frac{\alpha^{\sigma}}{\alpha}\right)^{\tau}\left(\frac{\alpha^{\tau}}{\alpha}\right) = \theta(\sigma)^{\tau}\theta(\tau).$$

However, $\epsilon \in F$ and thus $\theta(\sigma) \in F$ and $\theta(\sigma)^{\tau} = \theta(\sigma)$, and it follows that θ is a group homomorphism. Finally, if $\theta(\tau) = 1$, then $\alpha^{\tau} = \alpha$ and thus $\tau = 1$ since $E = F[\alpha]$. ∎

Kummer's theorem (22.8) provides a connection between radical extensions and Galois extensions with cyclic Galois group. For Galois's theorem, we need to extend this to an appropriate connection between repeated radical extensions and solvable Galois groups. We also need to be able to drop the assumption (essential in Theorem 22.8) that the ground field contains an appropriate root of unity.

The following technical result is used to obtain Galois's theorem (22.4) from Kummer's theorem (22.8).

(22.9) LEMMA. *Suppose $F = F_0 \subseteq F_1 \subseteq \cdots \subseteq F_r = L$, where for $1 \leq i \leq r$, each of the extensions $F_{i-1} \subseteq F_i$ is Galois with an abelian Galois group. Suppose $F \subseteq E \subseteq L$ and E is Galois over F. Then $\mathrm{Gal}(E/F)$ is solvable.*

Proof. If $r = 0$, then $L = F$ and so $E = F$ and $\mathrm{Gal}(E/F)$ is trivial. We may therefore assume that $r > 0$, and we work by induction on r.

Let $E_1 = \langle F_1, E \rangle$, the compositum, and write $K = E \cap F_1$, as illustrated in Figure 22.1. Note that by the natural irrationalities theorem (18.22) we know that E_1 is Galois over F_1. We have a tower of length $r - 1$ of abelian Galois extensions from F_1 to L, and so the inductive hypothesis applies and we conclude that $\text{Gal}(E_1/F_1)$ is solvable. This Galois group is isomorphic to $\text{Gal}(E/K)$, and so we deduce that $\text{Gal}(E/K)$ is solvable.

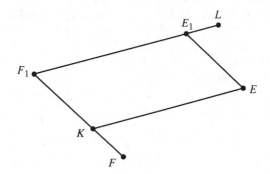

Figure 22.1

Now $F \subseteq K \subseteq F_1$ and $\text{Gal}(F_1/K) \triangleleft \text{Gal}(F_1/F)$, since the latter group is abelian by hypothesis. It follows by the fundamental theorem of Galois theory (applied to the Galois extension $F \subseteq F_1$) that K is Galois over F and $\text{Gal}(K/F) \cong \text{Gal}(F_1/F)/\text{Gal}(F_1/K)$. In particular, $\text{Gal}(K/F)$ is abelian.

Since K is Galois over F and $F \subseteq K \subseteq E$, the fundamental theorem of Galois theory (applied this time to the Galois extension $F \subseteq E$) tells us that $\text{Gal}(E/K) \triangleleft \text{Gal}(E/F)$ and $\text{Gal}(K/F) \cong \text{Gal}(E/F)/\text{Gal}(E/K)$. At this point, we know that $\text{Gal}(E/F)$ has the solvable normal subgroup $\text{Gal}(E/K)$ and that the corresponding factor group is abelian. It follows by Corollary 8.4 that $\text{Gal}(E/F)$ is solvable.   ∎

Finally, we are ready to prove Galois's theorem.

Proof of Theorem 22.4.   Suppose that $f \in F[X]$ and that $\text{Gal}(E/F)$ is solvable, where E is a splitting field for f over F. Let $n = |E : F|$. Since $\text{char}(F) = 0$, Lemma 20.3 guarantees the existence of an extension $E^* \supseteq E$ such that E^* contains a primitive nth root of unity ϵ, and we may assume that $E^* = E[\epsilon]$.

Write $F^* = F[\epsilon] \subseteq E^*$ and observe that $\langle F^*, E \rangle = E^*$ and E^* is Galois over F^* by the natural irrationalities theorem (18.22). Also, $\text{Gal}(E^*/F^*)$ is isomorphic to a subgroup of $\text{Gal}(E/F)$, and so it is solvable.

Write $G = \text{Gal}(E^*/F^*)$ and let

$$1 = G_0 \triangleleft G_1 \triangleleft \cdots \triangleleft G_r = G$$

be a composition series for G. Then G_i/G_{i-1} is simple for $1 \leq i \leq r$, and since G is solvable, we know that these composition factors are cyclic of prime order. Writing $|G_i/G_{i-1}| = p_i$ and putting $K_i = \text{Fix}(G_i)$, we have

$$F^* = K_r \subseteq K_{r-1} \subseteq \cdots \subseteq K_0 = E^* ,$$

and K_{i-1} is Galois over K_i with cyclic Galois group of order p_i. (All of this follows by repeated applications of the fundamental theorem.) Also, p_i divides $|G|$, which divides $|\text{Gal}(E/F)| = n$.

It follows by Kummer's theorem (22.8) that the extension $K_i \subseteq K_{i-1}$ is radical for $1 \le i \le r$. Since the extension $F \subseteq F^* = F[\epsilon]$ is also radical (because $\epsilon^n = 1 \in F$), we see that E^* is a repeated radical extension of F over which f splits. In other words, f is solvable by radicals over F.

Conversely, assume $f \in F[X]$ is solvable by radicals over F, and let L be a repeated radical extension of F over which f splits. Let E be the splitting field for f over F in L. Our task is to show that $\text{Gal}(E/F)$ is solvable.

We have $F = F_0 \subseteq F_1 \subseteq \cdots \subseteq F_r = L$ with $F_i = F_{i-1}[\alpha_i]$ and $\alpha_i^{n_i} \in F_{i-1}$ for $1 \le i \le r$. Let n be a common multiple of the integers n_i, and adjoin a primitive nth root of unity ϵ to L to get the field $L^* = L[\epsilon]$. We now define fields K_i as follows: $K_{-1} = F$, $K_0 = F[\epsilon]$, and for $1 \le i \le r$, put $K_i = K_{i-1}[\alpha_i]$. Note that $K_r = F[\epsilon, \alpha_1, \ldots, \alpha_r] = L[\epsilon] = L^*$ and also $F_i \subseteq K_i$ for $0 \le i \le r$. We now have

$$F = K_{-1} \subseteq K_0 \subseteq K_1 \subseteq \cdots \subseteq K_r = L^*$$

and $F \subseteq E \subseteq L^*$.

We will prove that $\text{Gal}(E/F)$ is solvable by appealing to Lemma 22.9, and so it suffices to show for $0 \le i \le r$ that each of the extensions $K_{i-1} \subseteq K_i$ is Galois with an abelian Galois group.

For $i = 0$, the extension $K_{i-1} \subseteq K_i$ is the extension $F \subseteq F[\epsilon]$, and this is Galois with an abelian group by Lemma 20.7. Now suppose $i > 0$. We have $K_i = K_{i-1}[\alpha_i]$, where $\alpha_i^{n_i} \in F_{i-1} \subseteq K_{i-1}$. Since K_{i-1} contains the primitive nth root of unity ϵ, and since n_i divides n, Kummer's theorem (22.8) implies that the extension $K_{i-1} \subseteq K_i$ is Galois with a cyclic Galois group. This completes the proof.   ∎

(22.10) COROLLARY.   Let $\text{char}(F) = 0$ and suppose $f \in F[X]$ with $\deg(f) \le 4$. Then f is solvable by radicals over F.

Proof.   The Galois group over F for a splitting field for f over F acts faithfully on the set of roots of f and so is isomorphic to a subgroup of the solvable symmetric group S_4. It is thus solvable.   ∎

As is well known, even more is true: not only are all the roots of f in Corollary 22.10 expressible in terms of radicals, but also there exist procedures for actually computing these roots.

22C

By definition, a nonzero polynomial $f \in F[X]$ is solvable by radicals over F if some splitting field for f over F is contained in some repeated radical extension of F. Are we being too general? Could it be that if f is solvable by radicals over F, then some (and hence every) splitting field for f over F actually is itself a repeated radical extension of F? The answer is no.

Take $F = \mathbb{Q}$ and $f(X) = X^3 - 4X + 2$. Since $\deg(f) = 3$, it is automatically true that f is solvable by radicals over $\mathbb{Q}$. We see that $f(0) > 0$ and $f(1) < 0$, and so it follows that f has three real roots, and we let E be the splitting field for f over $\mathbb{Q}$ in $\mathbb{R}$. We shall see that E is not a repeated radical extension of $\mathbb{Q}$. In fact, since f is irreducible, the following result implies that none of the roots of f can be contained in a real repeated radical extension of $\mathbb{Q}$.

(22.11) THEOREM. *Suppose $f \in \mathbb{Q}[X]$ is irreducible and splits over $\mathbb{R}$. If any root of f lies in a real repeated radical extension of $\mathbb{Q}$, then $\deg(f)$ is a power of 2.*

If $f \in \mathbb{Q}[X]$ has degree 3 and splits over $\mathbb{R}$, for example, then despite the fact that f is solvable by radicals and all of its (complex) roots are real, it is impossible to write any of its roots in terms of real nth roots of real numbers. The only possible radical expressions for the roots of f involve nonreal numbers. This phenomenon has long been known for polynomials of degree 3 (and is referred to as the *casus irreducibilis*), but by Theorem 22.11 it applies whenever the degree is not a power of 2.

The following result is the key step.

(22.12) THEOREM. *Let $F \subseteq E, L \subseteq \mathbb{R}$, where E is Galois over F and L is a repeated radical extension of F. Then $|E \cap L : F|$ is a power of 2.*

Proof of Theorem 22.11. Let E be the splitting field for f over $\mathbb{Q}$ in $\mathbb{R}$ and let $L \subseteq \mathbb{R}$ be a repeated radical extension of $\mathbb{Q}$ that contains some root α for f. By Theorem 22.12 we know that $|E \cap L : \mathbb{Q}|$ is a power of 2. Since $\alpha \in E \cap L$, we deduce that $\deg(f) = |\mathbb{Q}[\alpha] : \mathbb{Q}|$ is a power of 2, as desired. ∎

We begin work toward Theorem 22.12 with a lemma.

(22.13) LEMMA. *Suppose $F \subseteq K \subseteq E$, with $E = F[\alpha]$, and write $m = |E : K|$.*

    a. *If $\alpha^m \in K$, then $K = F[\alpha^m]$.*
    b. *If $\alpha^n \in F$ and F contains a primitive nth root of unity for some integer n, then $\alpha^m \in K$.*

Proof. For (a), assume $\alpha^m \in K$ and let $L = F[\alpha^m] \subseteq K$. Then α is a root of $f(X) = X^m - \alpha^m \in L[X]$ and $E = L[\alpha]$. Thus

$$|E : L| = \deg(\min_L(\alpha)) \le \deg(f) = m = |E : K| \le |E : L|,$$

and we have equality. We deduce that $K = L$, as required.

In the situation of (b), we can apply Kummer's theorem (22.8). We deduce that E is Galois over F and $\mathrm{Gal}(E/F)$ is cyclic. It follows that E is Galois over K and that we can write $\mathrm{Gal}(E/K) = \langle \sigma \rangle$ with $o(\sigma) = m$.

Now $\alpha^n \in F$ and so $\alpha^n = \sigma(\alpha^n) = \sigma(\alpha)^n$. We may certainly assume that $\alpha \neq 0$ and thus $(\sigma(\alpha)/\alpha)^n = 1$, and we have $\sigma(\alpha) = \alpha\delta$, where δ is some nth root of unity. We have $\delta \in F$ and thus $\sigma^k(\alpha) = \alpha\delta^k$ for integers $k \geq 1$. In particular, $\alpha = \sigma^m(\alpha) = \alpha\delta^m$ and so $\delta^m = 1$. Therefore $\sigma(\alpha^m) = \sigma(\alpha)^m = (\alpha\delta)^m = \alpha^m$ and $\alpha^m \in \text{Fix}(\langle\sigma\rangle) = K$.   ∎

Our next result proves the special case of Theorem 22.12 where $E \subseteq L$ and L is a (nonrepeated) radical extension of F. The proof of this case contains most of the work for Theorem 22.12.

(22.14) THEOREM. *Let $F \subseteq E \subseteq L \subseteq \mathbb{R}$, where E is Galois over F and L is a radical extension of F. Then $|E : F| \leq 2$.*

Proof.   Write $L = F[\alpha]$ with $\alpha^n \in F$, and let ϵ be a complex primitive nth root of unity. Write F^*, E^*, and L^* to denote $F[\epsilon]$, $E[\epsilon]$, and $L[\epsilon]$, respectively, and let $U = E^* \cap L$ and $V = F^* \cap L$. (See Figure 22.2.)

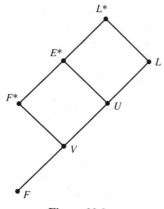

Figure 22.2

Let $m = |L^* : E^*|$. Since $L^* = F^*[\alpha]$ and $\alpha^n \in F^*$, Lemma 22.13(b) yields that $\alpha^m \in E^*$ and thus $\alpha^m \in E^* \cap L = U$. We claim that $|L : U| = m$, and thus by Lemma 22.13(a) we have $U = F[\alpha^m]$.

To compute $|L : U|$, we observe that E^* is Galois over E by Lemma 20.7. Since $L^* = \langle E^*, L \rangle$, the natural irrationalities theorem (18.22) gives $|L^* : L| = |E^* : U|$, and it follows that $|L : U| = |L^* : E^*| = m$. Thus $U = F[\alpha^m]$, as claimed.

Since V and E are both contained in U, we have $\langle V, E \rangle \subseteq U$, and we work to prove equality here. Observe first that $E \cap V = E \cap F^* \cap L = E \cap F^*$. Since E is Galois over F, the natural irrationalities theorem yields

$$|\langle V, E \rangle : V| = |E : E \cap V| = |E : E \cap F^*|.$$

Another application of the natural irrationalities theorem gives

$$|E^* : F^*| = |\langle E, F^* \rangle : F^*| = |E : E \cap F^*|,$$

and we deduce that $|\langle V, E \rangle : V| = |E^* : F^*|$.

Next we observe that

$$\langle F^*, U \rangle \subseteq E^* = \langle F^*, E \rangle \subseteq \langle F^*, U \rangle$$

and so $\langle F^*, U \rangle = E^*$. Also,

$$F^* \cap U = F^* \cap E^* \cap L = F^* \cap L = V ,$$

and since F^* is Galois over F by Lemma 20.7, the natural irrationalities theorem yields that $|E^* : U| = |F^* : V|$. It follows that $|E^* : F^*| = |U : V|$, and we now have

$$|\langle V, E \rangle : V| = |E^* : F^*| = |U : V|.$$

It follows that $U = \langle V, E \rangle$, as desired.

Now $\mathrm{Gal}(F^*/F)$ is abelian by Lemma 20.7, and thus V is Galois over F by the fundamental theorem. Each of the fields V and E, being Galois over F, is a splitting field over F of some polynomial, and it follows that the compositum U of these two fields is a splitting field for the product of these polynomials. We conclude that U is Galois over F.

Write $\beta = \alpha^m$ and recall that $U = F[\beta]$. Since some power of β lies in F, it follows that $\sigma(\beta)/\beta$ is a root of unity for all $\sigma \in \mathrm{Gal}(U/F)$. (We can certainly assume that $\beta \neq 0$.) Since the only roots of unity in the field $U \subseteq \mathbb{R}$ are ± 1, it follows that $\sigma(\beta) = \pm\beta$ for all $\sigma \in \mathrm{Gal}(U/F)$. We conclude that $|\mathrm{Gal}(U/F)| \leq 2$ since $U = F[\beta]$, and so $\sigma \in \mathrm{Gal}(U/F)$ is determined by its effect on β. We now have

$$|E : F| \leq |U : F| = |\mathrm{Gal}(U/F)| \leq 2$$

since U is Galois over F.   ∎

Proof of Theorem 22.12.   We work by induction on $|E : F|$. Since the case where $|E : F| = 1$ is trivial, we can assume $F < E$. Writing $D = E \cap L$, we must show that $|D : F|$ is a 2-power.

Suppose there exists an intermediate field K with $F < K < E$ and such that K is Galois over F. The inductive hypothesis yields that $|D \cap K : F| = |L \cap K : F|$ is a power of 2. Also, $\langle K, L \rangle$ is a repeated radical extension of K by Lemma 22.1(a), and E is Galois over K. The inductive hypothesis thus gives that $|E \cap \langle K, L \rangle : K|$ is a power of 2. Now $\langle D, K \rangle \subseteq E \cap \langle K, L \rangle$, and so $|\langle D, K \rangle : K|$ is a power of 2.

Since K is Galois over F, the natural irrationalities theorem yields that $|\langle D, K \rangle : D| = |K : K \cap D|$, and it follows that $|D : D \cap K| = |\langle D, K \rangle : K|$, which we know is a power of 2. Since $|D \cap K : F|$ is also known to be a power of 2, we see that in this case $|D : F|$ is a power of 2, as required.

We are left with the case that no intermediate field K with $F < K < E$ is Galois over F. By the fundamental theorem, it follows that $G = \mathrm{Gal}(E/F)$ is a simple group.

We may certainly assume that $D > F$, and we choose $\alpha \in D$ with $\alpha \notin F$. If $f = \min_F(\alpha)$, then f splits over E, and the splitting field for f over F in E is a Galois extension of F. We conclude that E is a splitting field for f over

F. Since α is a root of f and $\alpha \in L$, Lemma 22.3 tells us that f is solvable by radicals over F, and so G is a solvable group by Galois's theorem. Since G is simple, we conclude that it has prime order and hence $|E : F|$ is prime. But $F < D \subseteq E$, and so we have $D = E$ and $E \subseteq L$.

Now L is a repeated radical extension of F, and it is no loss to assume that no smaller repeated radical extension of F contains E. There exists a field M such that $F \subseteq M < L$, where L is a radical extension of M and M is a repeated radical extension of F. Now E is not contained in M, and since $|E : F|$ is prime, we see that $E \cap M = F$.

Since E is Galois over F, the natural irrationalities theorem shows that $\langle E, M \rangle$ is Galois over M, and thus by Theorem 22.14 we have $|\langle E, M \rangle : M| \leq 2$. It follows that $|E : F| = |E : E \cap M| = |\langle E, M \rangle : M| \leq 2$, and the proof is complete.   ∎

Finally, we can prove Theorem 22.11. In fact, we present a strengthened version.

(22.15) THEOREM.   *Let $F \subseteq \mathbb{R}$ and suppose $f \in F[X]$ is irreducible and splits over $\mathbb{R}$. If f has a root in a real repeated radical extension of F, then $|E : F|$ is a power of 2, where E is the splitting field for f over F in E.*

Proof.   Let $F \subseteq L \subseteq \mathbb{R}$, where L is a repeated radical extension of F that contains a root α of f. Write $D = E \cap L$, so that $\alpha \in D$. By Theorem 22.12 we know that $|D : F|$ is a power of 2. More generally, if K is any field intermediate between F and E, we will show that $|\langle D, K \rangle : K|$ is a power of 2. To see this, note that $\langle K, L \rangle$ is a repeated radical extension of K. Because E is Galois over K, we conclude that $|E \cap \langle K, L \rangle : K|$ is a power of 2 by Theorem 22.12. But $K \subseteq \langle D, K \rangle \subseteq E \cap \langle K, L \rangle$, and it follows that $|\langle D, K \rangle : K|$ is a power of 2, as claimed.

Now write $G = \mathrm{Gal}(E/F)$ and $H = \mathrm{Gal}(E/D)$. If $U \subseteq G$ is an arbitrary subgroup, then we apply the result of the previous paragraph to $K = \mathrm{Fix}(U)$. We see by the fundamental theorem that

$$\mathrm{Fix}(H \cap U) = \langle \mathrm{Fix}(H), \mathrm{Fix}(U) \rangle = \langle D, K \rangle,$$

and so $|U : H \cap U| = |\langle D, K \rangle : K|$, and this is a power of 2.

We must show that G is a 2-group, and so we let $P \in \mathrm{Syl}_p(G)$ for some odd prime p and we work to show that $P = 1$. Applying the result of the previous paragraph with $U = P$, we see that $|P : P \cap H|$ is a power of 2. Since it is also a p-power and $p \neq 2$, we conclude that $P \cap H = P$ and so $P \subseteq H$. Since P was an arbitrary Sylow p-subgroup of G, it follows that every conjugate of P in G is also contained in H, and thus H contains the subgroup N generated by all of these conjugates.

Let $K = \mathrm{Fix}(N)$. Since $N \subseteq H$, we have $D \subseteq K$, and thus K contains the root α of f. Also, $N \triangleleft G$ and so K is Galois and hence normal over F. It follows that f splits over K and hence $K = E$. We deduce that $N = 1$, and since $P \subseteq N$, the proof is complete.   ∎

22D

As we explained in Chapter 17, there cannot exist a purely algebraic proof of the so-called fundamental theorem of algebra, which asserts that $\mathbb{C}$ is algebraically closed. This is because the definition of $\mathbb{C}$ is based on that of $\mathbb{R}$, and in turn, $\mathbb{R}$ is constructed with the use of analysis ideas such as Cauchy sequences or Dedekind cuts. Nevertheless, we can use Galois theory and finite group theory to give a proof of the fundamental theorem of algebra that uses only a minuscule amount of analysis. Specifically, we assume the following lemma about $\mathbb{R}$ and $\mathbb{C}$.

(22.16) LEMMA.

a. *If $f \in \mathbb{R}[X]$ has odd degree, then f has a root in $\mathbb{R}$.*
b. *If $\alpha \in \mathbb{C}$, then there exists $\beta \in \mathbb{C}$ with $\beta^2 = \alpha$.* ∎

The purely algebraic part of the proof is contained in the following result.

(22.17) THEOREM. *Let $R \subseteq C$ be fields of characteristic zero with $|C : R| = 2$. Assume the following:*

i.   *Every polynomial $f \in R[X]$ with odd degree has a root in R.*
ii. *For each element $\alpha \in C$, there exists $\beta \in C$ with $\beta^2 = \alpha$.*

Then C is algebraically closed.

Proof. Suppose $L \supseteq C$ is an algebraic extension. By Lemma 17.24 it suffices to show that $L = C$, and so we assume there exists some element $\alpha \in L$ with $\alpha \notin C$, and we derive a contradiction. We have $|C[\alpha] : C| < \infty$ and thus $|C[\alpha] : R| < \infty$. By Lemma 18.5, there exists some field $E \supseteq C[\alpha]$ such that E is a splitting field for some polynomial over R. Since $\mathrm{char}(R) = 0$, we know that E is Galois over R.

Let $G = \mathrm{Gal}(E/R)$ and let $S \in \mathrm{Syl}_2(G)$. If $K = \mathrm{Fix}(S)$, then $|K : R| = |G : S|$, which is odd. For $\beta \in K$, let $f = \min_R(\beta)$. Then $\deg(f) = |R[\beta] : R|$, which divides $|K : R|$ and so is odd. Now f has a root in R by assumption (i), and since f is irreducible over R, we conclude that $\deg(f) = 1$ and $\beta \in R$. Thus $K = R$ and $S = G$, and we see that G is a 2-group.

Now let $H = \mathrm{Gal}(E/C) \subseteq G$. Then H is a 2-group and $|H| = |E : C| > 1$ since $\alpha \in E$ and $\alpha \notin C$. By Corollary 5.23 we can find a subgroup $M \subseteq H$ with $|H : M| = 2$. Let $F = \mathrm{Fix}(M)$. Then $|F : C| = |H : M| = 2$, and we can choose $\gamma \in F$ with $\gamma \notin C$.

The minimal polynomial $q(X)$ of γ over C is quadratic. Since C is closed under extraction of square roots by assumption (ii), we can use the quadratic formula to find a root for q in C. This is a contradiction because q is irreducible over C. ∎

(22.18) COROLLARY (Fundamental theorem of algebra). *The field $\mathbb{C}$ is algebraically closed.*

Proof.   Apply Theorem 22.17 with $R = \mathbb{R}$ and $C = \mathbb{C}$, using Lemma 22.16 to verify the hypotheses.   ∎

22E

We include one more topic that fits (more or less) into this chapter. Suppose that $F \subseteq E$ and that for *every* element $\alpha \in E$, there is some power of α that lies in F. This certainly holds if $F = E$. If $F < E$, it can hold if $\mathrm{char}(F) = p \neq 0$. (It holds, for instance, if E is finite or if E is purely inseparable over F.)

(22.19) THEOREM.   *Let $F \subseteq E$ and suppose that for each element $\alpha \in E$, there exists a positive integer n (depending on α) such that $\alpha^n \in F$. If $\mathrm{char}(F) = 0$, then $E = F$.*

We need an easy lemma about Euler's function.

(22.20) LEMMA.   *Let S be an infinite set of positive integers. Then $\{\varphi(m)|\, m \in S\}$ is infinite.*

Proof.   It would suffice to show that $\varphi(n) \geq \sqrt{n/2}$ for all integers $n \geq 2$. We use induction on n to prove slightly more: that $\varphi(n) \geq \sqrt{n/2}$ if n is even, and $\varphi(n) \geq \sqrt{n}$ if n is odd.

First consider the case where $n = p^e$ for some prime p and exponent $e > 0$. We have $\varphi(n) = p^e - p^{e-1} \geq n/2$. Since $n/2 \geq \sqrt{n}$ for $n \geq 4$, it suffices to check the cases $n = 2, 3$. For $n = 2$, we have $\varphi(n) = 1 = \sqrt{n/2}$, and for $n = 3$, $\varphi(n) = 2 > \sqrt{n}$.

If n is not a prime power, then we can write $n = rs$, where r and s are coprime and less than n. Then $\varphi(n) = \varphi(r)\varphi(s)$ by Lemma 20.11. We can assume that s is odd, and so $\varphi(s) \geq \sqrt{s}$ by the inductive hypothesis. Also, $\varphi(r) \geq \sqrt{r/2}$ and so $\varphi(n) \geq \sqrt{r/2}\sqrt{s} = \sqrt{n/2}$. Finally, if n is odd, then so is r, and we have $\varphi(r) \geq \sqrt{r}$. This yields $\varphi(n) \geq \sqrt{r}\sqrt{s} = \sqrt{n}$.   ∎

Proof of Theorem 22.19.   Assume that $\mathrm{char}(F) = 0$ and $E > F$, and identify the prime subfield of F with $\mathbb{Q}$. Let $\alpha \in E - F$ and define $K = \mathbb{Q}(\alpha)$ and $D = K \cap F$. Then $K > D$ and every element of K has a power in D. In particular, some power of α lies in D, and thus α is algebraic over D. We thus have $K = D(\alpha) = D[\alpha]$ and $|K : D| < \infty$.

By Lemma 18.5 there exists $L \supseteq K$ such that L is a splitting field over D, and since we are in characteristic zero, we know that L is Galois over D. Since $\alpha \in L$ and $\alpha \notin D$, we can fix an element $\sigma \in \mathrm{Gal}(L/F)$ with $\sigma(\alpha) \neq \alpha$. Now $\alpha^n \in D$ and so $\alpha^n = \sigma(\alpha^n) = \sigma(\alpha)^n$, and hence $(\sigma(\alpha)/\alpha)^n = 1$ and we can write $\sigma(\alpha) = \alpha\delta$, where δ is an nth root of unity in L.

For each integer $z \in \mathbb{Z} \subseteq \mathbb{Q} \subseteq D$, we can apply the above argument with $\alpha + z$ in place of α (since $\alpha + z \in K$). We conclude that $\sigma(\alpha + z) = (\alpha + z)\delta_z$, where $\delta_z \in L$ is a root of unity uniquely determined by z.

If $z, w \in \mathbb{Z}$, then z and w are fixed by σ, and we have

$$\sigma(\alpha + z) = \sigma((\alpha + w) + (z - w)) = \sigma(\alpha + w) + (z - w).$$

Since $\sigma(\alpha + z) = (\alpha + z)\delta_z$ and $\sigma(\alpha + w) = (\alpha + w)\delta_w$, this yields

$$(*) \qquad \alpha(\delta_z - \delta_w) = z(1 - \delta_z) - w(1 - \delta_w)$$

for all $z, w \in \mathbb{Z}$.

We show next that the map $z \mapsto \delta_z$ is injective. If we suppose that $\delta_z = \delta_w$, equation $(*)$ yields $0 = (z - w)(1 - \delta_z)$. If $z \neq w$, we conclude that $\delta_z = 1$, and thus

$$\alpha + z = (\alpha + z)\delta_z = \sigma(\alpha + z) = \sigma(\alpha) + z$$

and so $\alpha = \sigma(\alpha)$. This is not the case, and we conclude that the map $z \mapsto \delta_z$ is injective, as claimed.

When $z \neq w$, equation $(*)$ yields that $\alpha \in \mathbb{Q}[\delta_z, \delta_w]$, and in particular, α is algebraic over $\mathbb{Q}$. Thus $K = \mathbb{Q}(\alpha) = \mathbb{Q}[\alpha]$ is a finite-degree extension of $\mathbb{Q}$.

It follows that L has finite degree over $\mathbb{Q}$, and we obtain our final contradiction from the fact that L contains infinitely many roots of unity. Since L can contain at most m mth roots of unity for any integer $m \geq 1$, it follows that the set S of multiplicative orders of roots of unity in L is infinite. If $m \in S$, let $\delta \in L$ be a primitive mth root of unity. Then the cyclotomic polynomial Φ_m is the minimal polynomial for δ over $\mathbb{Q}$ by Gauss's theorem (20.8), and so

$$|L : \mathbb{Q}| \geq |\mathbb{Q}[\delta] : \mathbb{Q}| = \deg(\Phi_m) = \varphi(m).$$

Since this holds for all elements m in the infinite set S, Lemma 22.20 contradicts the fact that $|L : \mathbb{Q}| < \infty$. ∎

Problems

22.1 We define $F \subseteq E$ to be a *square-root extension* if $E = F[\alpha]$ with $\alpha^2 \in F$. If

$$F = F_0 \subseteq F_1 \subseteq \cdots \subseteq F_r = E,$$

where each F_i is a square-root extension, we say that $F \subseteq E$ is a *repeated square-root extension*. Assume $\text{char}(F) \neq 2$ and let $F \subseteq E$ be a Galois extension. Show that the following are equivalent:

i. E is a repeated square-root extension of F.
ii. E is contained in some repeated square-root extension of F.
iii. $|E : F|$ is a power of 2.

22.2 Suppose $F \subseteq E$ is a Galois extension with $\text{Gal}(E/F)$ abelian of order n, and assume that F contains a primitive nth root of unity. Show that E is a splitting field over F of some polynomial of the form

$$(X^{n_1} - a_1)(X^{n_2} - a_2) \cdots (X^{n_r} - a_r),$$

where $a_i \in F$.

HINT: Find intermediate fields K_i such that $E = \langle K_1, K_2, \ldots, K_r \rangle$ and $\mathrm{Gal}(K_i/F)$ is cyclic.

22.3 a. Given integers $m > 0$ and $n \geq 0$, let

$$f(X) = (X^{2n} + 2)(X - 2)(X - 4) \cdots (X - 2m).$$

For odd integers $k > 0$, write $g_k(X) = f(X) - (2/k)$. Show that g_k is irreducible in $\mathbb{Q}[X]$ and that if k is large enough, g_k has precisely m real roots.

b. Let $p \geq 3$ be prime. Show that there exists a polynomial of degree p in $\mathbb{Q}[X]$ such that the Galois group of its splitting field over $\mathbb{Q}$ is isomorphic to S_p.

HINT: It suffices to choose k so that $(2/k) < f(c)$ for all real numbers c such that $f'(c) = 0$ and $f(c) > 0$.

22.4 Let $F \subseteq E$ be a separable extension and suppose $\alpha \in E$ with $\alpha^n \in F$ for some integer $n > 0$. Show that $\alpha^m \in F$ for some integer m that is not divisible by $\mathrm{char}(F)$.

22.5 Let $F \subseteq E$ be a Galois extension and assume that $E = F[\alpha]$, where $\alpha^n \in F$. Let $G = \mathrm{Gal}(E/F)$.

a. Show that the commutator subgroup G' is cyclic.

b. If n is prime, show that G is cyclic.

22.6 Let $E = \mathbb{Q}[i, \sqrt[8]{2}] \subseteq \mathbb{C}$.

a. Show that E is Galois over $\mathbb{Q}$ and $|E : \mathbb{Q}| = 16$.

b. Let $F = \mathbb{Q}[i]$. Show that $\mathrm{Gal}(E/F)$ is cyclic.

c. Let $K = \mathbb{Q}[i\sqrt{2}]$. Show that $\mathrm{Gal}(E/K)$ is nonabelian.

NOTE: Since $E = K[\sqrt[8]{2}]$, part (c) shows that in Problem 22.5, the group G' can be nontrivial.

HINT: Let $\epsilon = e^{2\pi i/8}$ and note that $\epsilon \in E$. Write $\alpha = \sqrt[8]{2}$. Show that there exist $\sigma, \tau \in \mathrm{Gal}(E/\mathbb{Q})$ with $\sigma(\alpha) = \alpha\epsilon = \tau(\alpha)$, $\sigma(i) = i$, and $\tau(i) = -i$. Show that $\mathrm{Gal}(E/F) = \langle \sigma \rangle$ and $\mathrm{Gal}(E/K) = \langle \sigma^2, \tau \rangle$.

22.7 Let $f(X) = X^4 - 6X^2 + 7 \in \mathbb{Q}[X]$.

a. Show that f is irreducible.

b. Let E be the splitting field of f over $\mathbb{Q}$ in $\mathbb{C}$. Show that $E \subseteq \mathbb{R}$ and that $\mathrm{Gal}(E/\mathbb{Q})$ is a nonabelian 2-group.

HINT: Compute the roots of f and show that $|\mathbb{Q}[\alpha] : \mathbb{Q}| = 4$, where α is some root.

22.8 Let $F \subseteq E$ with $E = F[\alpha]$ and $\alpha^n \in F$. Assume that $\mathrm{char}(F)$ does not divide n and that $\gcd(n, |E : F|) = 1$. Show that E is Galois over F and that $\mathrm{Gal}(E/F)$ is abelian.

HINT: Adjoin a primitive nth root of unity.

CHAPTER TWENTY-THREE

Norms, Traces, and Discriminants

23A

In this chapter we show how to apply certain ideas from linear algebra in order to get information about finite degree field extensions. If $F \subseteq E$ is any field extension, then E can be viewed as a vector space over F. For each element $\alpha \in E$, we let $R_\alpha : E \to E$ be the map defined by multiplication by α. (Thus $(\beta)R_\alpha = \beta\alpha$ for all $\beta \in E$.) It is easy to see that R_α is an F-linear operator on E, in other words, an F-linear transformation from E to itself.

For any F-space V, the set $\mathrm{End}_F(V)$ of F-linear operators on V is a ring (which is necessarily noncommutative if $\dim_F(V) > 1$). Furthermore, the ring $\mathrm{End}_F(V)$ is an F-vector space (and, in fact, the ring operations are compatible with scalar multiplication, making $\mathrm{End}_F(V)$ into an F-algebra).

In the field extension situation, where $F \subseteq E$, it is routine to check that the map $\alpha \mapsto R_\alpha$ is an F-linear ring isomorphism from E into the ring $\mathrm{End}_F(E)$. We refer to this map $E \to \mathrm{End}_F(E)$ as the *natural F-representation* of E.

Returning to the general situation of an F-space V, we assume that $\dim_F(V) = n < \infty$. For each choice of an ordered basis $\mathcal{B} = \{v_1, v_2, \ldots, v_n\}$ for V, we get a map $R \mapsto [R]$ from the ring $\mathrm{End}_F(V)$ of F-linear operators on V to the ring $M_n(F)$ of $n \times n$ matrices over F. (The (i, j)-entry of $[R]$ is the coefficient of v_j in the expansion of $(v_i)R$ in terms of $\mathcal{B}$.) We say that the matrix $[R]$ *corresponds* to the linear operator R with respect to the basis $\mathcal{B}$. The map $R \mapsto [R]$ is an F-linear ring isomorphism from $\mathrm{End}_F(V)$ onto $M_n(F)$.

Although the particular matrix $[R]$ that corresponds to the operator R depends on $\mathcal{B}$, the similarity class of $[R]$ does not. (In other words, matrices that correspond to the same linear operator R with respect to different bases for V are similar to one another.) It follows that any object associated with square matrices and that is constant on similarity classes can be assigned to a linear operator. Examples are the determinant, the trace, and the characteristic polynomial. If $R \in \mathrm{End}_F(V)$, therefore, we can define $\det(R) = \det([R])$ and $\operatorname{tr}(R) = \operatorname{tr}([R])$, and we say that $\det(XI - [R])$ is the *characteristic polynomial* of R.

We are now ready to give the key definitions of this chapter. Let $F \subseteq E$ be a finite-degree field extension. If $\alpha \in E$, we write $T_{E/F}(\alpha) = \operatorname{tr}(R_\alpha)$ and $N_{E/F}(\alpha) = \det(R_\alpha)$. The maps $T_{E/F}$ and $N_{E/F}$ are, respectively, the *trace* and *norm* maps from E to F.

It is instructive to compute these maps for the extension $\mathbb{R} \subseteq \mathbb{C}$. First we choose a basis $\mathcal{B}$ (although the final result is independent of this choice). Taking $\mathcal{B} = \{1, i\}$, we compute R_z for $z \in \mathbb{C}$ by writing $z = a + bi$ with $a, b \in \mathbb{R}$. Then

$$[R_z] = \begin{bmatrix} a & b \\ -b & a \end{bmatrix}$$

and so $N_{\mathbb{C}/\mathbb{R}}(z) = a^2 + b^2 = |z|^2$ and $T_{\mathbb{C}/\mathbb{R}}(z) = 2a = 2\operatorname{Re}(z)$.

(23.1) LEMMA. *Let $F \subseteq E$ with $|E : F| = n < \infty$. Writing $N = N_{E/F}$ and $T = T_{E/F}$, we have the following:*

  a. *The trace map $T : E \to F$ is an F-linear transformation, and in particular, $T(\alpha + \beta) = T(\alpha) + T(\beta)$ for $\alpha, \beta \in E$.*
  b. *The norm map satisfies $N(\alpha\beta) = N(\alpha)N(\beta)$ for $\alpha, \beta \in E$. Also, N defines a homomorphism of multiplicative groups $E^\times \to F^\times$.*
  c. *If $\alpha \in F$, then $N(\alpha) = \alpha^n$ and $T(\alpha) = n\alpha$.*

Proof.   Note that the map $\operatorname{tr} : M_n(F) \to F$ is F-linear. Since T is the composite map

$$\alpha \mapsto R_\alpha \mapsto [R_\alpha] \mapsto \operatorname{tr}([R_\alpha])$$

and each step is F-linear, it follows that T is F-linear, proving (a).

Now N is the composite map

$$\alpha \mapsto R_\alpha \mapsto [R_\alpha] \mapsto \det([R_\alpha]),$$

where the first two steps are ring isomorphisms. Also, since $\det(AB) = \det(A)\det(B)$ for $A, B \in M_n(F)$, it follows that $N(\alpha\beta) = N(\alpha)N(\beta)$ for $\alpha, \beta \in E$.

If $\gamma \in F$, then $[R_\gamma]$ is the matrix γI, where I is the identity matrix, and thus (c) follows. In particular, $N(1) = 1$, and so if $0 \neq \alpha \in E$, we have $N(\alpha)N(\alpha^{-1}) = N(1) \neq 0$ and $N(\alpha) \neq 0$. Thus N maps $E^\times$ to $F^\times$, completing the proof of (b). ∎

It is often impossibly difficult to compute $N_{E/F}(\alpha)$ or $T_{E/F}(\alpha)$ directly from their definitions. Fortunately, there are easier methods, which we now develop.

Recall that if A is an $n \times n$ matrix over F, then its characteristic polynomial $f(X) = \det(XI - A)$ lies in $F[X]$. It is monic and has degree n.

We need the following familiar general fact.

(23.2) LEMMA.   *Let $f(X) = \prod(X - \alpha_i)$, where the roots α_i lie in some field. Writing*

$$f(X) = X^n + a_{n-1}X^{n-1} + \cdots + a_1 X + a_0,$$

we have $a_{n-u} = (-1)^u s_u$ for $1 \le u \le n$, where s_u is the sum of the products of the α_i taken u at a time. In other words,

$$s_u = \sum_I \prod_{i \in I} \alpha_i,$$

where I runs over all subsets of $\{1, 2, \ldots, n\}$ with cardinality u. In particular, $a_{n-1} = -\sum \alpha_i$ and $a_0 = (-1)^n \prod \alpha_i$.

Proof.   Expand the product by the distributive law, and note that each term of the result that has exactly $n - u$ of its factors equal to X has exactly u factors of the form $-\alpha_i$, and these have u different subscripts. ∎

The quantities s_u in Lemma 23.2 are called the *elementary symmetric functions* of the quantities $\alpha_1, \alpha_2, \ldots, \alpha_n$. When $n = 3$, for instance, the elementary symmetric functions of x, y, and z are $s_1 = x + y + z$, $s_2 = xy + xz + yz$, and $s_3 = xyz$.

(23.3) LEMMA.   *Let $A \in M_n(F)$ and let*

$$f(X) = X^n + a_{n-1}X^{n-1} + \cdots + a_1 X + a_0$$

be its characteristic polynomial. Then $\det(A) = (-1)^n a_0$ and $\mathrm{tr}(A) = -a_{n-1}$.

Proof.   Since $f(X) = \det(XI - A)$, we have $a_0 = f(0) = \det(-A) = (-1)^n \det(A)$, and the first statement follows.

To complete the proof, we note that if every off-diagonal entry of A were changed to zero, this would not affect the coefficient a_{n-1} of the characteristic polynomial. To see why this is so, observe that if $i \ne j$, then the determinantal minor corresponding to position (i, j) is the determinant of an $(n - 1) \times (n - 1)$ matrix with exactly $n - 2$ entries involving X. This contributes nothing to the X^{n-1}-term of the characteristic polynomial.

Since changing the off-diagonal entries of A to zero also has no effect on $\mathrm{tr}(A)$, we may assume that A is diagonal with (i, i)-entry equal to α_i, say. Then $f(X) = \det(XI - A) = \prod(X - \alpha_i)$, and so

$$\mathrm{tr}(A) = \sum \alpha_i = -a_{n-1}$$

by Lemma 23.2. ∎

Combining Lemmas 23.2 and 23.3, we obtain the following corollary .

(23.4) COROLLARY.   *Let $A \in M_n(F)$ have characteristic polynomial $f(X)$, and suppose $E \supseteq F$, where f splits over E. Writing $f(X) = \prod(X - \alpha_i)$ with $\alpha_i \in E$, we have $\det(A) = \prod \alpha_i$ and $\mathrm{tr}(A) = \sum \alpha_i$.* ∎

If $f(X) = \prod(X - \alpha_i)$, we say that $p = \prod \alpha_i$ and $s = \sum \alpha_i$ are, respectively, the product and sum of the roots of f, *counting multiplicities*. By Lemma 23.2, of course, p and s can be determined by inspection from f; it is not necessary to construct a splitting field and find the roots.

We are now ready to state our norm and trace evaluation theorem.

(23.5) THEOREM.   *Let $F \subseteq E$ with $|E : F| < \infty$, and suppose $\alpha \in E$. Let p and s denote the product and sum, respectively, of the roots of $\min_F(\alpha)$, counting multiplicities (in some splitting field). Then*

$$N_{E/F}(\alpha) = p^m \quad \text{and} \quad T_{E/F}(\alpha) = ms ,$$

where $m = |E : F[\alpha]|$.

Note that Theorem 23.5 includes Lemma 23.1(c) since if $\alpha \in F$, then $X - \alpha$ is its minimal polynomial and so $p = \alpha = s$. We need two preliminary results.

(23.6) LEMMA.   *Let $f, g \in F[X]$ be monic and assume that g is irreducible. If every root of f (in some splitting field) is also a root of g, then $f = g^m$ for some integer m.*

Proof.   If h is a monic irreducible factor of f in $F[X]$, choose a root α for h in our splitting field. Then $g(\alpha) = 0$ and so $g = \min_F(\alpha) = h$. The result follows since f is the product of its monic irreducible factors.   ∎

(23.7) LEMMA.   *Suppose $F \subseteq E$ with $|E : F| < \infty$. For $\alpha \in E$, let $g = \min_F(\alpha)$ and let f be the characteristic polynomial of the linear operator $R_\alpha \in \operatorname{End}_F(E)$, defined as before. Then $f = g^m$, where $m = |E : F[\alpha]|$.*

Proof.   Note that $\deg(f) = |E : F|$ and $\deg(g) = |F[\alpha] : F|$. Thus $\deg(f) = m \deg(g)$, and it suffices to prove that f is a power of g. By Lemma 23.6 it suffices to show that $g(\lambda) = 0$ whenever $f(\lambda) = 0$ for $\lambda \in L$, some splitting field for f over E.

Write $[R_\alpha] = A \in M_n(F)$ (with respect to some basis for E over F). Then f is the characteristic polynomial of the matrix A. We can view $A \in M_n(L)$ (and this does not affect the characteristic polynomial). If $f(\lambda) = 0$, we have $\det(\lambda I - A) = 0$, and it follows that there is some nonzero row vector $v \in L^n$ with $v(\lambda I - A) = 0$. Thus $vA = \lambda v$ and hence $vA^r = \lambda^r v$ for all integers $r \geq 0$. By taking linear combinations, we deduce that $vg(A) = g(\lambda)v$.

The standard F-representation $\beta \mapsto R_\beta$ and the map $R \mapsto [R]$ are F-linear isomorphisms, and so we have

$$g(A) = g([R_\alpha]) = [g(R_\alpha)] = [R_{g(\alpha)}] = [R_0] = 0 .$$

It follows that $g(\lambda)v = 0$, and thus since $v \neq 0$, we have $g(\lambda) = 0$, as required.   ∎

Proof of Theorem 23.5.   In the usual notation, let f be the characteristic polynomial of R_α and write $g = \min_F(\alpha)$. Then $f = g^m$ by Lemma 23.7, and thus p^m

is the product of the roots of f and ms is the sum of the roots of f, counting multiplicities. By Corollary 23.4

$$N_{E/F}(\alpha) = \det(R_\alpha) = p^m$$

and

$$T_{E/F}(\alpha) = \operatorname{tr}(R_\alpha) = ms,$$

as required. ∎

(23.8) COROLLARY. *Let $F \subseteq K \subseteq E$ with $|E : F| < \infty$. For $\alpha \in K$, we have*

$$T_{E/F}(\alpha) = |E : K|\, T_{K/F}(\alpha) \quad and \quad N_{E/F}(\alpha) = N_{K/F}(\alpha)^{|E:K|}.$$

Proof. In the notation of Theorem 23.5, we have $m = |E : K|m_0$, where $m_0 = |K : F[\alpha]|$. The result follows from Theorem 23.5. ∎

The following lemma enables us to use Galois groups to compute norms and traces. We use the exponential notation α^σ to denote the result of applying $\sigma \in \operatorname{Aut}(E)$ to $\alpha \in E$.

(23.9) LEMMA. *Let $F \subseteq K \subseteq E$ with E Galois over F, and write $G = \operatorname{Gal}(E/F)$ and $H = \operatorname{Gal}(E/K) \subseteq G$. Let U be any set of representatives for the right cosets of H in G. Then for $\alpha \in K$, we have*

$$\prod_{\sigma \in U}(X - \alpha^\sigma) = g^{|K:F[\alpha]|},$$

where $g = \min_F(\alpha)$.

Proof. Let $h(X) = \prod(X - \alpha^\sigma)$ as in the statement of the lemma. Note that h is independent of the choice of the set U of coset representatives (as it must be if the result is true). This is so because if $\tau \in H\sigma$, then since H fixes α, we have $\alpha^\tau = \alpha^\sigma$.

If $\tau \in G$, then right multiplication by τ permutes the right cosets of H in G, and so $U\tau$ is another set of representatives for the right cosets of H. It follows that the polynomial h is invariant under τ for all elements $\tau \in G$, and we conclude that $h \in F[X]$ since $F = \operatorname{Fix}(G)$.

Each root α^σ of h in E is a root of g since $g \in F[X]$, and thus by Lemma 23.6 we know that h is a power of g, and it suffices to show that $\deg(h)/\deg(g) = |K : F[\alpha]|$. We have $\deg(h) = |G : H| = |K : F|$, and since $\deg(g) = |F[\alpha] : F|$, the result follows. ∎

(23.10) COROLLARY. *Let $F \subseteq K \subseteq E$, where E is Galois over F, and let U be a set of representatives for the right cosets of $\operatorname{Gal}(E/K)$ in $\operatorname{Gal}(E/F)$. Then*

$$T_{K/F}(\alpha) = \sum_{\sigma \in U} \alpha^\sigma \quad and \quad N_{K/F}(\alpha) = \prod_{\sigma \in U} \alpha^\sigma$$

for $\alpha \in K$.

Proof. Let $g = \min_F(\alpha)$, and let p and s, respectively, denote the sum and product of the roots of g, counting multiplicities. By Lemma 23.9 we have $\sum \alpha^\sigma = |K : F[\alpha]|s$ and $\prod \alpha^\sigma = p^{|K:F[\alpha]|}$, where the sum and product are as in the statement of the corollary. The result now follows by Theorem 23.5.   ∎

The previous result allows us, in principle, to use Galois groups to compute norms and traces for any finite-degree separable extension $F \subseteq K$. In this situation, we can always find a field $E \supseteq K$ with E Galois over F. If we know $\mathrm{Gal}(E/F)$ and its subgroup $\mathrm{Gal}(E/K)$, then Corollary 23.10 gives us simple formulas for the norms and traces of elements of K.

We state separately the important special case of Corollary 23.10 where K is Galois over F, so that we can take $E = K$.

(23.11) COROLLARY.   *Let $F \subseteq K$ be Galois with $G = \mathrm{Gal}(K/F)$. Then*

$$T_{K/F}(\alpha) = \sum_{\sigma \in G} \alpha^\sigma \quad and \quad N_{E/F}(\alpha) = \prod_{\sigma \in G} \alpha^\sigma$$

for all elements $\alpha \in E$.   ∎

In the case of the extension $\mathbb{R} \subseteq \mathbb{C}$, Corollary 23.11 tells us that $T(z) = z + \bar{z}$ and $N(z) = z\bar{z}$. Of course, this agrees with our earlier calculation. We mention one more easy general result.

(23.12) COROLLARY.   *Let $F \subseteq E$ with $|E : F| < \infty$, and let $\sigma \in \mathrm{Gal}(E/F)$. Then*

$$T_{E/F}(\alpha) = T_{E/F}(\alpha^\sigma) \quad and \quad N_{E/F}(\alpha) = N_{E/F}(\alpha^\sigma)$$

for $\sigma \in E$.

Proof. By Theorem 23.5 it suffices to observe that $\min_F(\alpha) = \min_F(\alpha^\sigma)$ and that $|E : F[\alpha]| = |E : F[\alpha^\sigma]|$.   ∎

23B

Let $F \subseteq E$ be a finite degree extension, so that $N_{E/F}$ defines a homomorphism $E^\times \to F^\times$. There are situations when we would like to know the kernel of this homomorphism.

For example, suppose a is a positive integer. The equation $x^2 - ay^2 = 1$ is sometimes called "Pell's equation," and the object is to find positive integer solutions x and y. If a is square, so that $a = b^2$ for some positive integer b, Pell's equation becomes $(x + by)(x - by) = 1$, and this clearly has no positive integer solutions. Suppose a is nonsquare and let $E = \mathbb{Q}[\sqrt{a}]$. Then $\mathrm{Gal}(E/\mathbb{Q}) = \{1, \sigma\}$, where $\sigma(\sqrt{a}) = -\sqrt{a}$. If $N : E^\times \to \mathbb{Q}^\times$ is the norm map, we have $N(x + y\sqrt{a}) = (x + y\sqrt{a})(x - y\sqrt{a}) = x^2 - ay^2$ by Corollary 23.11. What we seek, therefore, are elements of $\ker(N)$ of the form $x + y\sqrt{a}$, where x and y are positive integers.

In general, how can we find elements of $\ker(N)$, where $N : E^\times \to F^\times$ is the norm map? By Corollary 23.12 we have $N(\alpha) = N(\alpha^\sigma)$ for $\sigma \in \mathrm{Gal}(E/F)$, and thus $\alpha^\sigma / \alpha \in \ker(N)$. In the situation of Pell's equation, let $\alpha = u - v\sqrt{a}$ with $u, v \in \mathbb{Q}$. Then

$$\frac{\alpha^\sigma}{\alpha} = \frac{u + v\sqrt{a}}{u - v\sqrt{a}} = \frac{(u + v\sqrt{a})^2}{u^2 - av^2} = \frac{(u^2 + av^2) + (2uv)\sqrt{a}}{u^2 - av^2},$$

and thus, if we take

$$x = \frac{u^2 + av^2}{u^2 - av^2} \quad \text{and} \quad y = \frac{2uv}{u^2 - av^2},$$

then $x + y\sqrt{a}$ lies in $\ker(N)$. We can therefore solve Pell's equation if we can find $u, v \in \mathbb{Q}$ such that the corresponding quantities x and y are positive integers. In fact, if u and v are positive integers that themselves satisfy Pell's equation, then $u^2 - av^2 = 1$, and thus

$$x = u^2 + av^2 \quad \text{and} \quad y = 2uv$$

are another (larger) solution.

We have proved that if, for some value of a, Pell's equation has a (positive-integer) solution, then it has infinitely many solutions. For example, if $a = 2$, then $(x, y) = (3, 2)$ is a solution of Pell's equation. Taking $(u, v) = (3, 2)$ yields the solution $(x, y) = (17, 12)$, and after that we get $(577, 408)$, and so on.

We should stress that we do not need any deep theory to see that if (u, v) solves Pell's equation, then so does $(u^2 + av^2, 2uv)$; this follows by simple calculation. The theory of norms, however, enabled us to discover this fact.

Returning to the general situation where $N : E^\times \to F^\times$ is the norm homomorphism, we can ask whether or not every element of $\ker(N)$ has the form α^σ / α for some element $\alpha \in E$ and some automorphism $\sigma \in \mathrm{Gal}(E/F)$. If E is Galois over F and $\mathrm{Gal}(E/F)$ is cyclic, the answer is yes, and, in fact, we need to use only one element of the Galois group. This is the famous "Theorem 90" of Hilbert.

(23.13) THEOREM (Hilbert's Theorem 90). *Let $F \subseteq E$ be Galois and assume that $\mathrm{Gal}(E/F) = \langle \sigma \rangle$. Then $N_{E/F}(\beta) = 1$ iff $\beta = \alpha^\sigma / \alpha$ for some $\alpha \in E^\times$.*

To prove Theorem 23.13, we need a refinement of Lemma 22.7. Suppose that G is a finite subgroup of $\mathrm{Aut}(E)$, and suppose we have a map $\theta : G \to E^\times$. By Lemma 22.7, if θ is a group homomorphism and its image lies in $F = \mathrm{Fix}(G)$, then θ is given by the formula $\theta(\tau) = \alpha^\tau / \alpha$, where α is some element of $E^\times$, independent of τ.

We seek a result like Lemma 22.7 that enables us to recognize every function $\theta : G \to E^\times$ that has the form $\theta(\tau) = \alpha^\tau / \alpha$. Are all of these functions group homomorphisms; do all of them map into F, or do we need a result more general than Lemma 22.7? If we choose $\alpha \in E^\times$ arbitrarily and define $\theta : G \to E^\times$ by the above formula, let us see how close we come to getting a homomorphism.

Fixing α, we have

$$\theta(\sigma\tau) = \frac{\alpha^{\sigma\tau}}{\alpha} = \left(\frac{\alpha^{\sigma}}{\alpha}\right)^{\tau}\left(\frac{\alpha^{\tau}}{\alpha}\right) = \theta(\sigma)^{\tau}\theta(\tau),$$

where $\sigma, \tau \in G$. We see that θ is not, in general, a homomorphism. In fact, θ is a homomorphism iff it maps into $\text{Fix}(G)$, and this is unlikely to hold for a random choice of $\alpha \in E^{\times}$.

The equations $\theta(\sigma\tau) = \theta(\sigma)^{\tau}\theta(\tau)$ for $\sigma, \tau \in G$ are often called *Noether's equations*. If $G \subseteq \text{Aut}(E)$ and $\theta : G \rightarrow E^{\times}$ satisfies Noether's equations, we say that θ is a *crossed homomorphism*. We see that the notion of a crossed homomorphism $\theta : G \rightarrow E^{\times}$ generalizes that of a homomorphism $\theta : G \rightarrow F^{\times}$, where $F = \text{Fix}(G)$. As we have seen, if $\alpha \in E^{\times}$ is arbitrary, then the map $\sigma \mapsto \alpha^{\sigma}/\alpha$ is always a crossed homomorphism, and so, with Lemma 22.7 in mind, it is reasonable to guess that every crossed homomorphism $G \rightarrow E^{\times}$ arises this way. This is true, and it can be proved using the argument that proved Lemma 22.7.

(23.14) LEMMA. *Let $G \subseteq \text{Aut}(E)$ with $|G| < \infty$, and let $\theta : G \rightarrow E^{\times}$ be a crossed homomorphism. Then there exists $\alpha \in E^{\times}$ such that $\theta(\sigma) = \alpha^{\sigma}/\alpha$ for all $\sigma \in G$.*

Since a homomorphism mapping G into $F = \text{Fix}(G)$ is a crossed homomorphism, we see that Lemma 23.14 includes Lemma 22.7.

Proof of Lemma 23.14.   Since θ is not identically zero, Dedekind's lemma (22.6) guarantees the existence of $\gamma \in E$ with $\sum \theta(\sigma)\gamma^{\sigma} \neq 0$. Define α so that

$$\alpha^{-1} = \sum_{\sigma} \theta(\sigma)\gamma^{\sigma}.$$

Fixing $\tau \in G$, we have

$$\alpha^{-1} = \sum_{\sigma} \theta(\sigma\tau)\gamma^{\sigma\tau} = \sum_{\sigma} \theta(\sigma)^{\tau}\theta(\tau)\gamma^{\sigma\tau} = \left(\sum_{\sigma} \theta(\sigma)\gamma^{\sigma}\right)^{\tau}\theta(\tau),$$

and we observe that the sum on the right equals $1/\alpha$. Thus $\theta(\tau) = \alpha^{\tau}/\alpha$, as required. ∎

Proof of Theorem 23.13.   We have seen that if $\beta = \alpha^{\sigma}/\alpha$ for some $\alpha \in E^{\times}$, then $N_{E/F}(\beta) = 1$, and so we prove the converse. Assume, then, that $N_{E/F}(\beta) = 1$. Since $\text{Gal}(E/F) = \{1, \sigma, \sigma^2, \ldots, \sigma^{n-1}\}$, where $\sigma^n = 1$, Corollary 23.11 yields

$$\prod_{i=0}^{n-1} \beta^{\sigma^i} = N_{E/F}(\beta) = 1.$$

For integers $k \geq 1$, define

$$\psi(k) = \prod_{i=0}^{k-1} \beta^{\sigma^i} \in E$$

and note that $\psi(1) = \beta$ and $\psi(n) = 1$. For $k, l \geq 1$, we have

$$\psi(k+l) = \psi(k) \prod_{i=0}^{l-1} \beta^{\sigma^{i+k}} = \psi(k) \left(\prod_{i=0}^{l-1} \beta^{\sigma^i} \right)^{\sigma^k} = \psi(k)\psi(l)^{\sigma^k}.$$

In particular, $\psi(k+n) = \psi(k)\psi(n)^{\sigma^k} = \psi(k)$ since $\psi(n) = 1$, and it follows that $\psi(k) = \psi(l)$ if $k \equiv l \bmod n$. We can therefore write $\theta(\sigma^k) = \psi(k)$ for $k \geq 1$, and this is a well defined function from G into $E^\times$. Also,

$$\theta(\sigma^l \sigma^k) = \psi(k+l) = \psi(k)\psi(l)^{\sigma^k} = \theta(\sigma^l)^{\sigma^k}\theta(\sigma^k),$$

and so θ is a crossed homomorphism.

By Lemma 23.14 there exists $\alpha \in E^\times$ such that $\theta(\tau) = \alpha^\tau/\alpha$ for all $\tau \in G$. In particular,

$$\frac{\alpha^\sigma}{\alpha} = \theta(\sigma) = \psi(1) = \beta,$$

as required.   ∎

23C

We digress to discuss crossed homomorphisms in a more general context. Suppose G is any group that acts on some abelian group A via automorphisms. (Recall that this means that G acts on A as a set and that, in addition, the map $a \mapsto a \cdot g$ is an automorphism of A for all $g \in G$. We usually write a^g for $a \cdot g$.) For example, we might have $G \subseteq \text{Aut}(E)$, where E is a field, and $A = E^\times$. (We could also have taken A to be the additive group of E.) Another example where G and A are arbitrary is the *trivial* action $a^g = a$ for all elements $a \in A$ and $g \in G$.

Let G act on A via automorphisms. A map $\theta : G \to A$ is a *crossed homomorphism* with respect to the given action if $\theta(xy) = \theta(x)^y\theta(y)$ for all $x, y \in G$. (If the action is trivial, then crossed homomorphisms are just homomorphisms.) In general, if $a \in A$ is arbitrary, it easy to check that the map $\theta : G \to A$ defined by $\theta(g) = a^{-1}a^g$ is a crossed homomorphism, and maps of this type, defined using particular elements $a \in A$, are called *principal* crossed homomorphisms. Lemma 23.14 could be restated in this language as the assertion that all crossed homomorphisms are principal if we take G to be a finite subgroup of $\text{Aut}(E)$ and $A = E^\times$.

Note that it cannot be true in general that all crossed homomorphisms are principal, since in the case of trivial action, there is just one principal crossed homomorphism: the constant map $g \mapsto 1$. In each situation, we would like to obtain some measure of the extent by which the assertion that "all crossed homomorphisms are principal" fails. As we shall see, the desired measure is the abelian group $H^1(G, A)$, called the "first cohomology group." This group is trivial precisely when all crossed homomorphisms are principal, and more generally, the probability that a random crossed homomorphism is principal is exactly the reciprocal of the order of the first cohomology group. In particular, Lemma 23.14 says that $H^1(G, E^\times) = 1$ if $G \subseteq \text{Aut}(E)$ and $|G| < \infty$.

If G acts on the abelian group A by automorphisms, then the nth cohomology group $H^n(G, A)$ for $n \geq 1$ is defined by considering certain abelian groups $C^n(G, A)$ and homomorphisms $\delta^n : C^n(G, A) \to C^{n+1}(G, A)$ for $n \geq 0$. These are constructed to have the property that for each $n \geq 1$, the composite map $\delta^{n-1}\delta^n$ is trivial, mapping $C^{n-1}(G, A)$ to the identity of $C^{n+1}(G, A)$. Writing $Z^n(G, A)$ and $B^n(G, A)$ to denote the kernel of δ^n and the image of δ^{n-1}, respectively, we have $B^n(G, A) \subseteq Z^n(G, A) \subseteq C^n(G, A)$ for $n \geq 1$. By definition, the factor group $Z^n(G, A)/B^n(G, A)$ is $H^n(G, A)$, the nth *cohomology group*. (The names of the other objects we have mentioned are as follows: the maps δ^n are the *coboundary* maps and the groups C^n, B^n, and Z^n are, respectively, the groups of *cochains*, *coboundaries*, and *cocycles*.)

There are a number of different ways to define the cochain groups C^n and the coboundary maps $\delta^n : C^n \to C^{n+1}$ that yield the same (up to isomorphism) cohomology groups. For our purposes, the most useful definition is to let $C^n(G, A)$ denote the group of all functions of n variables from G to A with pointwise operations. (The elements of $C^0(G, A)$, being functions of no variables, can be identified with the elements of A.) The first two coboundary maps δ^0 and δ^1 are as follows: If $a \in C^0(G, A)$, then $\delta^0(a) \in C^1(G, A)$ is defined by $(\delta^0(a))(x) = a(a^x)^{-1}$ for $x \in G$. If $\theta \in C^1(G, A)$, then $\delta^1(\theta) \in C^2(G, A)$ is the function $(\delta^1(\theta))(x, y) = \theta(x)^y \theta(y)\theta(xy)^{-1}$.

With these definitions, we see that $\theta \in Z^1(G, A) = \ker(\delta^1)$ iff θ is a crossed homomorphism, and $\theta \in B^1(G, A) = \mathrm{im}(\delta^0)$ iff θ is a principal crossed homomorphism. Thus

$$H^1(G, A) = \frac{\{\text{crossed homomorphisms}\}}{\{\text{principal crossed homomorphisms}\}},$$

and so $H^1(G, A)$ is trivial exactly when every crossed homomorphism is principal.

We should stress that the power of cohomology theory comes from the fact that there are different constructions for the cochain groups and coboundary maps, and these yield isomorphic cohomology groups. Thus, for instance, it might be possible to prove that $H^1(G, A)$ is trivial using some construction different from the one described here, and then to deduce from this that all crossed homomorphisms from G to A are principal.

For applications, H^1 and H^2 are the most important of the cohomology groups. Before ending this digression, therefore, we should define δ^2; that enables us to interpret $Z^2(G, A)$ and $H^2(G, A)$. Using the definitions of $C^2(G, A)$ and $C^3(G, A)$ as functions of two and three variables from G to A, we let $\varphi \in C^2(G, A)$. Then $\delta^2(\varphi) \in C^3(G, A)$ is defined by

$$\delta^2(\varphi)(x, y, z) = [\varphi(x, yz)\varphi(y, z)][\varphi(xy, z)\varphi(x, y)^z]^{-1}.$$

Thus φ is a 2-cocycle, an element of $Z^2(G, A)$, iff

$$\varphi(x, yz)\varphi(y, z) = \varphi(xy, z)\varphi(x, y)^z$$

for all $x, y, z \in G$. Functions with this property are sometimes called *factor sets*.

We show in the problems at the end of the chapter how this is relevant to the theory of extensions of groups.

23D

We now consider the trace map $T_{E/F} : E \to F$. It turns out that this map is interesting only when the extension $F \subseteq E$ is separable.

(23.15) THEOREM.   *Let $F \subseteq E$ with $|E : F| < \infty$.*

    a. *If E is inseparable over F, then $T_{E/F}$ is identically zero.*
    b. *If E is separable over F, then $T_{E/F}$ is surjective.*

Proof.   Suppose E is inseparable over F, and let S be the set of elements of E that are separable over F. Then $F \subseteq S < E$, and Theorem 19.14 implies that S is a field and E is purely inseparable over S. By Corollary 19.12(b), $|E : S|$ is a power of $p = \mathrm{char}(F)$ and, in particular, $|E : S|$ is divisible by p.

    If $\alpha \in S$, then Corollary 23.8 tells us that $T_{E/F}(\alpha) = |E : S| \, T_{S/F}(\alpha)$, and this vanishes since $p \,|\, |E : S|$. If $\alpha \in E - S$, write $g = \min_F(\alpha)$. Then g is inseparable, and so we can write $g(X) = h(X^p)$ for some $h \in F[X]$. (We are using Corollary 19.6 here.)   If $\deg(g) = n$, therefore, we see that $p\,|\,n$, and it follows that the coefficient of X^{n-1} in $g(X)$ is zero. The sum of the roots of g counting multiplicities is therefore zero, and it follows that $T_{E/F}(\alpha) = 0$ by Theorem 23.5. This completes the proof of (a).

    Assuming now that E is separable over F, we can choose $L \supseteq E$ such that L is Galois over F. Let $G = \mathrm{Gal}(L/F)$ and $H = \mathrm{Gal}(L/E)$, and let U be a set of representatives for the right cosets of H in G. Then for $\alpha \in E$, we have

$$T_{E/F}(\alpha) = \sum_{\sigma \in U} \alpha^\sigma$$

by Corollary 23.10. We seek some element α such that $T_{E/F}(\alpha) \neq 0$. This is sufficient because $T_{E/F}$ is an F-linear transformation, and any nonzero linear transformation to a one-dimensional space is necessarily surjective.

    By Dedekind's lemma (22.6) we know that there exists $\beta \in L$ such that $\sum \beta^\rho \neq 0$, where we are summing over all $\rho \in G$. Let $\alpha = \sum \beta^\tau$, summing over $\tau \in H$. It is clear that $\alpha \in \mathrm{Fix}(H) = E$, and we compute

$$\sum_{\sigma \in U} \alpha^\sigma = \sum_{\sigma \in U} \left(\sum_{\tau \in H} \beta^\tau \right)^\sigma = \sum_{\rho \in G} \beta^\rho \neq 0,$$

as desired. Note that the last equality holds since every element of G is uniquely of the form $\rho = \tau\sigma$ with $\tau \in H$ and $\sigma \in U$.   ∎

We can now prove the additive version of Lemma 23.14.

(23.16) COROLLARY.   *Let $F \subseteq E$ be Galois with $G = \mathrm{Gal}(E/F)$. Then $H^1(G, E^+)$ is trivial, where E^+ is the additive group of E.*

In other words, if $\theta : G \to E$ satisfies the additive form of Noether's equations: $\theta(\sigma\tau) = \theta(\sigma)^\tau + \theta(\tau)$, then there exists $\beta \in E$ with $\theta(\sigma) = \beta^\sigma - \beta$ for all $\sigma \in G$. In still other words: all crossed homomorphisms from G to E^+ are principal.

Proof of Corollary 23.16.   Given $\theta : G \to E$ satisfying $\theta(\sigma\tau) = \theta(\sigma)^\tau + \theta(\tau)$, let $\alpha \in E$ and define

$$\beta = \sum_{\sigma \in G} \theta(\sigma)\alpha^\sigma .$$

Then for fixed $\tau \in G$, we have

$$\beta = \sum_{\sigma \in G} \theta(\sigma\tau)\alpha^{\sigma\tau} = \sum_{\sigma}(\theta(\sigma)^\tau + \theta(\tau))\alpha^{\sigma\tau}$$

$$= \left(\sum_\sigma \theta(\sigma)\alpha^\sigma\right)^\tau + \theta(\tau)\left(\sum_\sigma \alpha^{\sigma\tau}\right),$$

and thus $\beta = \beta^\tau + \theta(\tau)\,T_{E/F}(\alpha)$ by Corollary 23.11.

Since E is Galois over F, it is separable, and by Theorem 23.15(b) we can choose $\alpha \in E$ so that $T_{E/F}(\alpha) = -1$. Then $\theta(\tau) = \beta^\tau - \beta$, as desired.   ∎

Just as Hilbert's Theorem 90 (our Theorem 23.13) follows from Lemma 23.14, an additive analog of Theorem 90 can be deduced from Corollary 23.16 by the use of essentially the same proof. We leave this to the problems at the end of the chapter. Also in the problems we show how to compute norms in inseparable extensions.

23E

Suppose that $F \subseteq E$ is a finite degree extension of fields, and let $\mathcal{B} = \{\alpha_1, \alpha_2, \ldots, \alpha_n\}$ be an F-basis for E. Let $D \in M_n(F)$ be the matrix whose (i, j)-entry is equal to $T_{E/F}(\alpha_i\alpha_j)$.

If E is inseparable over F, then D is the zero matrix by Theorem 23.15(a). If E is separable over F, we shall see that D definitely is not the zero matrix and, even more surprising, its determinant is nonzero. We say that $\det(D)$ is the *discriminant* of the extension $F \subseteq E$ with respect to the basis $\mathcal{B}$, and we write $\det(D) = \Delta_{E/F}(\mathcal{B})$.

For example, suppose $E = F[\sqrt{d}]$, where $d \in F$ is not a square in F and $\text{char}(F) \neq 2$. Then $\mathcal{B} = \{1, \sqrt{d}\}$ is a basis for E over F, and in this case,

$$D = \begin{bmatrix} T(1) & T(\sqrt{d}) \\ T(\sqrt{d}) & T(d) \end{bmatrix} = \begin{bmatrix} 2 & 0 \\ 0 & 2d \end{bmatrix}$$

and $\Delta_{E/F}(\mathcal{B}) = 4d$. (We have written $T = T_{E/F}$, and we have used the fact that $T(\sqrt{d}) = \sqrt{d} + -\sqrt{d} = 0$ by Corollary 23.11.)

(23.17) THEOREM.   *Let $F \subseteq E$ be a finite degree separable extension, and let $\mathcal{B}$ be an F-basis for E. The following then hold.*

a. $\Delta_{E/F}(\mathcal{B}) \neq 0$.

b. *If $\mathcal{C}$ is another F-basis for E, then $\Delta_{E/F}(\mathcal{C}) = \delta^2 \Delta_{E/F}(\mathcal{B})$, where $\delta \in F$ is the determinant of the change-of-basis matrix from $\mathcal{C}$ to $\mathcal{B}$.*

Proof.   Define the map $\langle \cdot, \cdot \rangle : E \times E \to F$ by $\langle \alpha, \beta \rangle = T_{E/F}(\alpha\beta)$. We shall see that this defines a nondegenerate symmetric F-bilinear form on E. The symmetry is obvious, and since the trace map $T_{E/F}$ is F-linear, it is clear that the map $\langle \cdot, \cdot \rangle$ is F-bilinear. To see that it is nondegenerate, let $\gamma \in E$ with $T_{E/F}(\gamma) \neq 0$. (Such an element exists by Theorem 23.15(b).) Now if $\alpha \in E$ with $\alpha \neq 0$, then $\langle \alpha, \gamma/\alpha \rangle \neq 0$, and this establishes the nondegeneracy of the form.

The theorem now follows by standard linear algebra arguments. In particular, if P is the change-of-basis matrix from $\mathcal{C}$ to $\mathcal{B}$, then the ith row of P is the $\mathcal{B}$-coordinate row-vector of the ith element $\gamma_i \in \mathcal{C}$. Then PDP^t is the matrix with (i, j)-entry equal to $\langle \gamma_i, \gamma_j \rangle$. (We have written P^t to denote the transpose of the matrix P.) It follows that $\Delta_{E/F}(\mathcal{C}) = \det(PDP^t) = \det(P)^2 \det(D) = \det(P)^2 \Delta_{E/F}(\mathcal{B})$.   ∎

We should mention that in the case where E is a finite-degree extension of $\mathbb{Q}$, there always exists a special type of basis for E over $\mathbb{Q}$ called an "integral basis." (We define this in Chapter 28.) It is a fact that if $\mathcal{B}$ and $\mathcal{C}$ are any two integral bases for E over $\mathbb{Q}$, then the change-of-basis matrix P from $\mathcal{B}$ to $\mathcal{C}$ has entries in $\mathbb{Z}$. The same is true, of course, about P^{-1}, which is the change-of-basis matrix from $\mathcal{C}$ to $\mathcal{B}$. It follows that $\det(P)$ is in $\mathbb{Z}$, and $\det(P^{-1})$, its reciprocal, also lies in $\mathbb{Z}$. From this, we see that $\det(P) = \pm 1$ and thus $\det(P)^2 = 1$. It follows by Theorem 23.17 that $\Delta_{E/\mathbb{Q}}(\mathcal{B}) = \Delta_{E/\mathbb{Q}}(\mathcal{C})$, and this number, independent of the choice of integral basis, is often referred to simply as the "discriminant" of the field E. It turns out that this discriminant necessarily lies in $\mathbb{Z}$.

23F

The word "discriminant" is also used in a different, but not unrelated sense in field theory: the discriminant of a polynomial. Before we define this, we discuss some generalities.

Let $S = \{t_1, t_2, \ldots, t_n\}$ be a set of indeterminates, and let $R = F[S] = F[t_1, \ldots, t_n]$ be the corresponding polynomial ring over F. If $\sigma \in G = \mathrm{Sym}(S)$, then there is a unique F-linear automorphism (also called σ) of R that carries t_i to $(t_i)\sigma$ (viewing the t_i as elements of R). An element $f \in R$ that satisfies $f^\sigma = f$ for all $\sigma \in G$ is said to be a *symmetric polynomial*. Examples of these are the elementary symmetric functions s_r for $1 \leq r \leq n$ (as in Lemma 23.2) and the *power sum* polynomials $p_r = \sum (t_i)^r$ for integers $r \geq 0$.

If $f \in R$ and $\alpha_1, \alpha_2, \ldots, \alpha_n$ are not necessarily distinct elements of an extension field $E \supseteq F$, then we can substitute α_i for t_i in f to get an element of E. If f is symmetric, then this element, $f(\alpha_1, \ldots, \alpha_n)$, is independent of the numbering of the α_i.

(23.18) LEMMA.   *Let $S = \{t_1, \ldots, t_n\}$ as above, and define $v \in F[S]$ by*

$$v = \prod_{1 \le i < j \le n} (t_j - t_i).$$

Then $v^\sigma = \pm v$ for every $\sigma \in \mathrm{Sym}(S)$, and in particular, $d = v^2$ is symmetric. More precisely, we have $v^\sigma = -v$ if σ is an odd permutation and $v^\sigma = v$ if σ is even.

Proof.   The factors $(t_j - t_i)$ with $j > i$ are permuted by $\sigma \in S$, except that some of them may be mapped to the negatives of others. (This happens when $(t_j)\sigma = t_s$ and $(t_i)\sigma = t_r$ with $s < r$.)

   To prove the last statement, define the map $\theta : G \to \{1, -1\} \subseteq F$ such that $v^\sigma = \theta(\sigma)v$ for all $\sigma \in G$. If $\sigma, \tau \in G$, we have

$$\theta(\sigma\tau)v = v^{\sigma\tau} = (\theta(\sigma)v)^\tau = \theta(\sigma)v^\tau = \theta(\sigma)\theta(\tau)v,$$

and thus $\theta(\sigma\tau) = \theta(\sigma)\theta(\tau)$ and θ is a homomorphism of groups. Since $\{1, -1\}$ is abelian, θ is constant on conjugacy classes of G.

   Now let $\tau \in G$ be the transposition interchanging t_1 and t_2. Then τ permutes all factors of v other than $(t_2 - t_1)$, and it negates $(t_2 - t_1)$. It follows that $v^\tau = -v$, and so $\theta(\tau) = -1$. Since all transpositions of G are conjugate, θ takes the value -1 on each of them, and so if σ is a product of r transpositions, then $\theta(\sigma) = (-1)^r$, as required.   ∎

Note that since $v \neq -v$ if $\mathrm{char}(F) \neq 2$, we see that Lemma 23.18 provides an alternative proof of Theorem 6.8, which asserts that no permutation can be both even and odd.

   Now let $f \in F[X]$ have degree n. Choose some splitting field E for f over F, and factor $f(X) = a \prod(X - \alpha_i)$, where a is the leading coefficient. Then the *discriminant* of f in E is the quantity

$$\Delta(f) = \prod_{1 \le i < j \le n} (\alpha_j - \alpha_i)^2.$$

In other words, $\Delta(f) = d(\alpha_1, \ldots, \alpha_n)$, where d is as in Lemma 23.18. Since d is symmetric, $\Delta(f)$ is well defined; it is independent of the arbitrary numbering of the roots α_i of f in E.

   As an example, we compute $\Delta(f)$ when f is of degree 2. Writing $f(X) = aX^2 + bX + c$, we factor $f(X) = a(X - \alpha)(X - \beta)$ over some splitting field, and we observe that $-b/a = \alpha + \beta$ and $c/a = \alpha\beta$. Then

$$\Delta(f) = (\alpha - \beta)^2 = (\alpha + \beta)^2 - 4\alpha\beta = \frac{b^2 - 4ac}{a^2}.$$

(23.19) COROLLARY.   *Let $f \in F[X]$. Let E be a splitting field for f over F, and compute $\Delta(f)$ in E. Then*

a. $\Delta(f) \neq 0$ *iff f has distinct roots*,
b. $\Delta(f) \in F$, *and*
c. $\Delta(f)$ *is independent of the choice of E.*

Proof.   Part (a) is clear and (b) certainly holds if $\Delta(f) = 0$, so that to prove (b), we may assume that $\Delta(f) \neq 0$ and f has distinct roots. Then f is separable and E is Galois over F. Now $\mathrm{Gal}(E/F)$ permutes the roots of f in E, and since the polynomial d of Lemma 23.18 is symmetric, it follows that $\Delta(f) \in \mathrm{Fix}(\mathrm{Gal}(E/F)) = F$.

To prove (c) it suffices to note that if E_1 and E_2 are splitting fields for f over F, then they are F-isomorphic. Every F-isomorphism $E_1 \rightarrow E_2$ fixes $\Delta(f)$ by (b).   ∎

We can use the Vandermonde determinant to obtain a more convenient formula for the discriminant of a polynomial. If $t_1, t_2, \ldots, t_n$ are indeterminates, then the *Vandermonde matrix* $V = V(t_1, t_2, \ldots, t_n)$ is the $n \times n$ matrix with (i, j)-entry equal to $(t_j)^i$ for $0 \leq i \leq n - 1$ and $1 \leq j \leq n$. (Note that we have numbered the rows starting with zero.) Thus

$$V = \begin{bmatrix} 1 & 1 & \cdots & 1 \\ t_1 & t_2 & \cdots & t_n \\ t_1^2 & t_2^2 & \cdots & t_n^2 \\ \vdots & \vdots & \vdots & \vdots \\ t_1^{n-1} & t_2^{n-1} & \cdots & t_n^{n-1} \end{bmatrix}.$$

(23.20) LEMMA.   *Let* $R = F[t_1, \ldots, t_n]$, *where the* t_i *are indeterminates, and let* $V = V(t_1, \ldots, t_n) \in M_n(R)$ *be the Vandermonde matrix. Then*

$$\det(V) = \prod_{1 \leq i < j \leq n} (t_j - t_i).$$

Proof.   We compute $\det(V)$ using row and column operations, expansion by minors, and induction on n. (The result is trivial if $n = 1$ and is easily checked when $n = 2$.)

Perform the following row operations. Subtract t_1 times row $n - 2$ from row $n - 1$ (which is the last row). Then subtract t_1 times row $n - 3$ from row $n - 2$, and continue like this until row 1 has been modified. At this point, the (i, j)-entry (for $i \geq 1$) is $t_j^i - t_1 t_j^{i-1} = (t_j - t_1)t_j^{i-1}$. In particular, the first column consists of a 1 in position 0 followed by $n - 1$ zeros. It follows that $\det(V)$ is equal to the determinant of the $(n - 1) \times (n - 1)$ matrix W, which results by deleting the first row (numbered 0) and the first column from the matrix we have now.

The $(n - 1) \times (n - 1)$ matrix W has columns divisible by the quantities $(t_j - t_1)$ for $2 \leq j \leq n$. Factoring these out, we are left with the Vandermonde matrix $V(t_2, \ldots, t_n)$. The inductive hypothesis now yields

$$\det(V) = \prod_{j=2}^{n} (t_j - t_1) \cdot \prod_{2 \leq i < j \leq n} (t_j - t_i) = \prod_{1 \leq i < j \leq n} (t_j - t_i),$$

as required.

(23.21) LEMMA. *Let $f \in F[X]$ be monic, with $\deg(f) = n$, and suppose $f(X) = \prod(X - \alpha_i)$ in some splitting field $E \supseteq F$. Let $p_r = \sum(\alpha_i)^r$ for $r \geq 0$. (Thus p_r is the rth power-sum function of the roots of f.) Then $\Delta(f) = \det(P)$, where P is the $n \times n$ matrix with (i, j)-entry equal to p_{i+j} for $0 \leq i, j \leq n-1$.*

Proof. Let $V = V(\alpha_1, \alpha_2, \ldots, \alpha_n)$, the Vandermonde matrix, and note that $\Delta(f) = \det(V)^2 = \det(VV^T)$ by Lemma 23.20. Computing VV^T, we see that its (i, j)-entry is

$$\sum_{k=1}^{n}(\alpha_k)^i(\alpha_k)^j = p_{i+j},$$

and thus $VV^T = P$. ∎

The power-sums of the α_i can be computed (recursively) in terms of the elementary symmetric functions s_r of the α_i using the so-called Newton formulas:

$$p_r = \sum_{i=1}^{r-1}(-1)^{i+1}p_{r-i}s_i - (-1)^r r s_r.$$

(In these formulas, we set $s_k = 0$ for $k > n$.)

Although we do not prove the Newton formulas here, we show how to use them, together with Lemma 23.21, to express the discriminant of a polynomial in terms of its coefficients. We compute $\Delta(f)$ in the case where f is monic of degree 3 and the coefficient of X^2 vanishes. In general, the method is similar, although the computations can be quite tedious.

Suppose $f(X) = X^3 + bX + c$. By Lemma 23.2 we have $s_1 = 0$, $s_2 = b$, and $s_3 = -c$. The first Newton formula gives the obvious fact that $p_1 = s_1 = 0$. The second gives $p_2 = p_1 s_1 - 2s_2 = -2b$. (Note that the identity $p_2 = s_1^2 - 2s_2$ is obvious.) We have $p_3 = p_2 s_1 - p_1 s_2 + 3s_3 = -3c$ and $p_4 = p_3 s_1 - p_2 s_2 + p_1 s_3 - 4s_4 = 2b^2$. Since $p_0 = n = 3$, the matrix of Lemma 23.21 is

$$P = \begin{bmatrix} p_0 & p_1 & p_2 \\ p_1 & p_2 & p_3 \\ p_2 & p_3 & p_4 \end{bmatrix} = \begin{bmatrix} 3 & 0 & -2b \\ 0 & -2b & -3c \\ -2b & -3c & 2b^2 \end{bmatrix},$$

and $\Delta(f) = \det(P) = -4b^3 - 27c^2$.

We mentioned that the discriminant of a polynomial and the discriminant of a field extension (with respect to a basis) are related. The connection between $\Delta(f)$ and $\Delta_{E/F}(\mathcal{B})$ is given by the following theorem.

(23.22) THEOREM. *Let $E = F[\alpha]$, where α is separable over F, and let $f = \min_F(\alpha)$. Then*

$$\Delta(f) = \Delta_{E/F}(\mathcal{B}),$$

where $\mathcal{B} = \{1, \alpha, \alpha^2, \ldots, \alpha^{n-1}\}$ and $n = \deg(f)$.

Note that by the primitive element theorem (18.17), every finite-degree separable extension $F \subseteq E$ satisfies $E = F[\alpha]$ for some element α, necessarily separable over F, and thus Theorem 23.22 applies quite generally.

Proof of Theorem 23.22.   Let L be a splitting field for f over E, and write $f(X) = \prod(X - \alpha_i)$ with $\alpha_i \in L$ and $\alpha_1 = \alpha$. Note that L is a splitting field for f over F, and so L is Galois over F. Let $G = \mathrm{Gal}(L/F)$ and $H = \mathrm{Gal}(L/E) \subseteq G$, and let U be a set of representatives for the right cosets of H in G.

If $\sigma, \tau \in U$ and $\alpha^{\sigma} = \alpha^{\tau}$, then $\sigma\tau^{-1}$ fixes α. It follows that $\sigma\tau^{-1} \in \mathrm{Gal}(L/F[\alpha]) = H$, and this yields $\sigma \in H\tau$ and thus $\sigma = \tau$ (by the definition of U). In other words, distinct elements of U carry α to distinct images. Now α^{σ} is a root of f for each $\sigma \in U$, and we have

$$|U| = |G : H| = |E : F| = \deg(f) = n.$$

We deduce from this that as σ runs over U, we obtain all roots α_i of f in the form α^{σ}.

By Corollary 23.11, if $\beta \in E$ is arbitrary, then $T_{E/F}(\beta) = \sum \beta^{\sigma}$ for $\sigma \in U$. In particular,

$$T_{E/F}(\alpha^r) = \sum_{\sigma \in U}(\alpha^r)^{\sigma} = \sum_{\sigma}(\alpha^{\sigma})^r = \sum_{k=1}^{n} \alpha_k^r = p_r$$

for integers $r \geq 0$, where p_r is the rth power sum of the roots α_i.

By definition, $\Delta_{E/F}(\mathcal{B}) = \det(D)$, where (if we number the rows and columns of D from 0 to $n - 1$) the (i, j)-entry of D is

$$T_{E/F}(\alpha^i \alpha^j) = T_{E/F}(\alpha^{i+j}) = p_{i+j}.$$

Thus $D = P$, the matrix of Lemma 23.21, and $\Delta(f) = \det(P) = \det(D) = \Delta_{E/F}(\mathcal{B})$. ∎

The discriminant of a polynomial is occasionally useful for computing Galois groups.

(23.23) THEOREM.   *Assume that* $\mathrm{char}(F) \neq 2$ *and let* $f \in F[X]$ *be monic and have distinct roots in some splitting field E for f over F. Let Ω be the set of roots of f in E. Then the natural embedding of* $\mathrm{Gal}(E/F)$ *into* $\mathrm{Sym}(\Omega)$ *maps into* $\mathrm{Alt}(\Omega)$ *iff* $\Delta(f)$ *is a square in F.*

Proof.   Write $\Omega = \{\alpha_1, \alpha_2, \ldots, \alpha_n\}$ and let $\beta = v(\alpha_1, \alpha_2, \ldots, \alpha_n)$, where v is as in Lemma 23.18. If $\tau \in \mathrm{Gal}(E/F)$, then τ induces a permutation of Ω, and we let σ be the corresponding permutation of the indeterminates $t_1, t_2, \ldots, t_n$. Then

$$\beta^{\tau} = v(\alpha_1, \ldots, \alpha_n)^{\tau} = v^{\sigma}(\alpha_1, \ldots, \alpha_n) = \epsilon v(\alpha_1, \ldots, \alpha_n) = \epsilon\beta,$$

where $\epsilon = \pm 1$ by Lemma 23.18.

If $\text{Gal}(E/F)$ maps into $\text{Alt}(\Omega)$, then for each $\tau \in \text{Gal}(E/F)$, the corresponding permutation σ is even, and $\epsilon = 1$ by Lemma 23.18. Thus $\beta \in \text{Fix}(\text{Gal}(E/F))$ and $\Delta(f) = \beta^2$ is a square in F.

Otherwise, some $\tau \in \text{Gal}(E/F)$ corresponds to an odd permutation σ, and so $\tau(\beta) = -\beta$. Since $\text{char}(F) \neq 2$, we have $\tau(\beta) \neq \beta$ and so $\beta, -\beta \notin F$. These elements, however, are the only square roots of $\Delta(f)$ in E, and thus $\Delta(f)$ has no square root in F. ■

Consider, for example, the polynomial $X^3 - 7X + 7 \in \mathbb{Q}[X]$. It is irreducible by the Eisenstein criterion, and so its Galois group over $\mathbb{Q}$ is nontrivial. Also

$$\Delta(f) = -(4(-7)^3 + 27(7^2)) = 7^2(28 - 27) = 7^2,$$

a square in $\mathbb{Q}$. The Galois group is thus a subgroup (up to isomorphism) of the alternating group A_3 of order 3. It follows that the Galois group is cyclic of order 3.

Problems

23.1 Let V be an F-subspace of the space $M_n(F)$ of $n \times n$ matrices over F, and suppose that every nonzero matrix $A \in V$ is nonsingular.

a. Show that $\dim_F(V) \leq n$.

b. If $\dim_F(V) > 1$, show that there exists a field $E > F$ with $|E : F| \leq n$.

c. If there exists $E \supseteq F$ with $|E : F| = n$, show that it is possible to find V, as described, with $\dim_F(V) = n$.

HINT: For part (a), consider the map $\theta : V \to E^n$ mapping each matrix to its first row. For part (b), consider the characteristic polynomial of AB^{-1}, where $A, B \in V$ are distinct.

23.2 Let $F \subseteq K \subseteq E$, where E is of finite degree and separable over F. Prove the transitivity of the norm and trace maps. In other words, show the following for all elements $\alpha \in E$:

a. $N_{E/F}(\alpha) = N_{K/F}(N_{E/K}(\alpha))$

b. $T_{E/F}(\alpha) = T_{K/F}(T_{E/K}(\alpha))$

HINT: Use Galois groups.

23.3 Let $F \subseteq E$ be purely inseparable with $|E : F| = n < \infty$. Show that $N_{E/F}(\alpha) = \alpha^n$ for all $\alpha \in E$.

HINT: From Problem 19.13 we know that $\alpha^n \in F$.

23.4 Let $F \subseteq E$ with $|E : F| = n < \infty$. Assume that $N_{E/F}(\alpha) = \alpha^n$ for all $\alpha \in E$. Show that E is purely inseparable over F.

HINT: By Theorem 19.14 let $F \subseteq S \subseteq E$, where S is separable over F and E is purely inseparable over S. Let $S = F[\alpha]$, and note that $(\alpha^r)^s$ and $(\alpha^r + 1)^s$ both lie in F, where $r = |E : S|$ and $s = |S : F|$. Observe that $S = F[\alpha^r]$.

23.5 Generalize Problem 23.3 as follows. Let $F \subseteq S \subseteq E$, where S is separable over F and E is purely inseparable over S. If $\alpha \in E$, show that $N_{E/F}(\alpha) = N_{S/F}(\alpha^r)$, where $r = |E : S|$.

HINT: Use the fact that $\alpha^r \in S$ to show that $N_{E/F}(\alpha^r) = N_{S/F}(\alpha^r)^r$.

23.6 Show that Problem 23.2 would remain true if the separability hypothesis were dropped.

HINT: Note that we need to worry only about norms. Let $F \subseteq S \subseteq E$, with E purely inseparable over S and S separable over F. Let $U = \langle K, S \rangle$ and $V = K \cap S$. Compute $N_{E/K}$ and $N_{K/F}$ using Problem 23.5 with U and V, respectively, in the role of S. Use Lemma 19.20 to show that if $\beta \in S$, then $\min_V(\beta) = \min_K(\beta)$.

23.7 Let $F \subseteq E$ be finite fields with $|F| = q$ and $|E : F| = n$. Show that

$$N_{E/F}(\alpha) = \alpha^{(q^n-1)/(q-1)}$$

for all $\alpha \in E$.

23.8 In the situation of Problem 23.7, let $N : E^\times \to F^\times$ be the homomorphism induced by the norm map. By Problem 23.7 $\ker(N)$ is the subgroup of $E^\times$ of order $(q^n - 1)/(q - 1)$. Use Hilbert's Theorem 90 to give a different proof that $|\ker(N)| = (q^n - 1)/(q - 1)$.

HINT: Let $\mathrm{Gal}(E/F) = \langle \sigma \rangle$, and let $\theta : E^\times \to E^\times$ be defined by $\theta(\alpha) = \alpha^\sigma/\alpha$. What is $\ker(\theta)$? How large is the image $\mathrm{im}(\theta)$?

NOTE: Since $|\ker(N)| = (q^n - 1)/(q - 1)$, it follows that $|\mathrm{im}(N)| = (q^n - 1)/|\ker(N)| = q - 1$. Thus N is necessarily surjective. The norm map is usually not surjective for infinite fields.

23.9 Let $F \subseteq E$ be Galois and assume that $\mathrm{Gal}(E/F) = \langle \sigma \rangle$. Show that $\alpha \in E$ lies in $\ker(T_{E/F})$ iff $\alpha = \beta^\sigma - \beta$ for some element $\beta \in E$.

NOTE: This is the additive form of Hilbert's Theorem 90.

23.10 Let G be an arbitrary group, and suppose that $A \triangleleft G$, where A is abelian. Let $H \subseteq G$ be a complement for A in G. (This means that $HA = G$ and $H \cap A = 1$.) Note that conjugation defines an action of H on A. Now let K be any other complement for A in G. For each element $x \in H$, show that there is a unique element $a \in A$ such that $xa \in K$, and define the function $\theta : H \to A$ by setting $\theta(x) = a$. Show that θ is a crossed homomorphism. Conversely, show that every crossed homomorphism from H to A arises in this way from some choice of complement K.

23.11 In the situation of Problem 23.10, show that $K^g = H$ for some element $g \in G$ iff θ is a principal crossed homomorphism.

NOTE: It follows from Problem 23.11 that all complements K to A in G are conjugate to H iff the first cohomology group $H^1(H, A)$ is trivial.

23.12 Let Γ be a group and suppose $A \triangleleft \Gamma$ and A is abelian. Let $G = \Gamma/A$ and choose a set of representatives X for the cosets of A in Γ. For each element $g \in G$, let $x_g \in X$ be the element in g that lies in X. (Recall that g is a coset.) If $g, h \in G$, then $x_g x_h$ lies in the coset gh, and so $x_g x_h = x_{gh} a$ for some element $a \in A$. Define the function $\varphi : G \times G \to A$ by $\varphi(g, h) = a$.

a. Show that one can define an action of G on A by setting $a^g = a^{x_g}$.
b. Show that the map φ is a factor set; in other words, it is a 2-cocycle, an element of $Z^2(G, A)$.
c. Show that φ is a 2-coboundary iff there exists a subgroup $H \subseteq \Gamma$ with $HA = \Gamma$ and $H \cap A = 1$.

HINT: For part (c), note that if H exists, then $H = \{x_g \theta(g) \mid g \in G\}$ for some function $\theta : G \to A$.

NOTE: The subgroup H necessarily exists if $H^2(G, A) = 1$.

23.13 Let $f(X) = X^n - a \in \mathbb{Q}[X]$. Show that $\Delta(f) = \pm n^n a^{n-1}$, where the sign is positive if $n \equiv 1, 2 \bmod 4$ and is negative in the other two cases.

HINT: If $\epsilon = e^{2\pi i/n}$, observe that

$$\prod_{i \neq j}(\epsilon^i - \epsilon^j) = \prod_{i=0}^{n-1}\left(\epsilon^i \prod_{k=1}^{n-1}(1 - \epsilon^k)\right),$$

where on the left, i and j run over the set $\{0, 1, \ldots, n-1\}$. The product $\prod(1-\epsilon^k)$ can be evaluated by computing $h(1)$, where $h(X)$ is the polynomial $(X^n - 1)/(X - 1)$.

23.14 a. If $g(X) = (X - a)f(X)$, show that $\Delta(g) = f(a)^2 \Delta(f)$.
b. If $f(X) = X^{n-1} + X^{n-2} + \cdots + X + 1 \in \mathbb{Q}[X]$, compute $\Delta(f)$.

23.15 Let $E = \mathbb{Q}[\sqrt[p]{2}]$, where p is an odd prime. Let $\alpha \in E$ and let $d = \Delta(f)$, where $f = \min_{\mathbb{Q}}(\alpha)$. Show that if α is irrational, then $\sqrt{\epsilon p d}$ is rational, where $\epsilon = (-1)^{(p-1)/2}$.

23.16 Let $F \subseteq E$ be any finite-degree separable extension. Show that there exists a finite subset $S \subseteq F^\times$ such that for every element $\alpha \in E$, the quantity $s\Delta(f)$ is a square in F for some element $s \in S$, where $f = \min_F(\alpha)$.

23.17 Let $F \subseteq E$ be Galois and let $G = \mathrm{Gal}(E/F)$. Write $G = \{\sigma_1, \sigma_2, \ldots, \sigma_n\}$, where $n = |G|$. Let $\mathcal{B} = \{\alpha_1, \alpha_2, \ldots, \alpha_n\} \subseteq E$, and let B be the $n \times n$ matrix whose (i, j)-entry is $\sigma_i(\alpha_j)$.

a. If $\mathcal{B}$ is linearly dependent over F, show that $\det(B) = 0$.
b. If $\mathcal{B}$ is linearly independent over F (so that it is a basis for E), show that $\det(B)^2 = \Delta_{E/F}(\mathcal{B})$. (Thus $\det(B) \neq 0$ in this case.)

CHAPTER TWENTY-FOUR

Transcendental Extensions

24A

Suppose $F \subseteq E$ is an arbitrary field extension. Recall that an element $\alpha \in E$ is transcendental over F if $f(\alpha) \neq 0$ for all nonzero polynomials $f \in F[X]$. More generally, we say that a finite list $\alpha_1, \alpha_2, \ldots, \alpha_n$ of elements of E is *algebraically independent* over F if $f(\alpha_1, \ldots, \alpha_n) \neq 0$ for all nonzero polynomials $f \in F[X_1, \ldots, X_n]$. (Usually we speak loosely and say that "the elements $\alpha_1, \ldots, \alpha_n$ are algebraically independent," but we must remember that algebraic independence is not a property of the individual elements.)

An empty "list" is always algebraically independent, and a list consisting of a single element $\alpha \in E$ is algebraically independent over F precisely when α is transcendental over F. Also, the algebraic dependence or independence of a list is not affected by rearrangements. Finally, we observe that any list with repeats is automatically dependent. This is so because if $\alpha_j = \alpha_k$ with $j \neq k$, then the nonzero polynomial $f(X_1, \ldots, X_n) = X_j - X_k$ vanishes when the α_i are substituted for the X_i.

These remarks allow us to speak of algebraically independent *sets* instead of lists. To be precise, a (possibly infinite) subset $\mathcal{X} \subseteq E$ is *algebraically independent* over F if every finite list of distinct members of $\mathcal{X}$ is algebraically independent.

The reader should observe the close analogy between algebraic and linear independence. In both cases, a list is independent iff every nontrivial combination of the elements has nonzero value. In the linear case, of course, "combination" means "linear combination," and in the algebraic case, it refers to "polynomial combination." This analogy motivates much of the theory we are about to develop.

Examples of algebraically independent field elements are easy to find. Given F, let $E = F(T_1, T_2, \ldots, T_n)$, where the T_i are indeterminates. Thus E is the "rational function field" over F in n "variables"; it is the fraction field of the polynomial ring $F[T_1, \ldots, T_n]$. It is trivial to see that $T_1, T_2, \ldots, T_n$, viewed as elements of the field E, are algebraically independent. In some sense, sets of indeterminates are the only examples of algebraically independent elements. The next lemma makes this precise, but before stating it, we introduce a bit of notation.

If $F \subseteq E$ is a field extension and $\mathcal{X} \subseteq E$ is an arbitrary subset, we write $F(\mathcal{X})$ to denote the subfield of E generated over F by $\mathcal{X}$. This is the unique smallest subfield that contains both F and $\mathcal{X}$, and it can be constructed by taking the intersection of all such subfields. We write $F[\mathcal{X}]$ to denote the subring of E generated by F and $\mathcal{X}$. It consists of all polynomial expressions in finitely many of the elements of $\mathcal{X}$ and with coefficients in F. It should be clear that $F(\mathcal{X})$ is a fraction field for $F[\mathcal{X}]$.

(24.1) LEMMA. *Let $F \subseteq E$ and suppose the subset $\mathcal{X} \subseteq E$ is algebraically independent over F. Write $K = F(\mathcal{X})$. Then K is F-isomorphic to the rational function field in a set of indeterminates in bijective correspondence with $\mathcal{X}$.*

Proof. For each element $\alpha \in \mathcal{X}$, create a distinct indeterminate T_α, and let $R = F[T_\alpha \mid \alpha \in \mathcal{X}]$ be the polynomial ring in these indeterminates. (Although there may be infinitely many indeterminates, each polynomial $f \in R$ involves just finitely many of them, of course.) Evaluation of polynomials $f \in R$ by substituting α for T_α defines an F-homomorphism $\theta : R \to K$, and the image of θ is the subring $F[\mathcal{X}]$ of K. The algebraic independence of $\mathcal{X}$ tells us that $\ker(\theta)$ is trivial, and thus R is F-isomorphic to $F[\mathcal{X}]$. The result now follows because the rational function field in the T_α is the fraction field of R, and K is the fraction field of $F[\mathcal{X}]$. ∎

A field extension $F \subseteq E$ is called *purely transcendental* if E is generated by some subset $\mathcal{X}$, algebraically independent over F. Recall that a field extension $F \subseteq E$ is totally transcendental if the only elements of E that are algebraic over F are the elements of F. By Corollary 17.8, for example, if $E = F(\alpha)$, where α is transcendental over F, then E is totally transcendental over F. More generally, as we shall soon see, every purely transcendental extension is totally transcendental. The converse, however, is false.

The following lemma is useful for inductive arguments.

(24.2) LEMMA. *Let $F \subseteq E$ and suppose $\mathcal{X}$ is a subset of E that is algebraically independent over F. If $\alpha \in E - \mathcal{X}$, then the set $\mathcal{X} \cup \{\alpha\}$ is algebraically independent over F iff α is transcendental over the subfield $K = F(\mathcal{X})$.*

Proof. If α is algebraic over K, let $f(\alpha) = 0$ with $f \in K[X]$ and $f \neq 0$. Each coefficient of f has the form p/q, where $p, q \in F[\mathcal{X}]$. Clearing denominators, we can assume that all coefficients of f are polynomials in $\mathcal{X}$, and thus there exist elements $\alpha_1, \alpha_2, \ldots, \alpha_n \in \mathcal{X}$ and polynomials $p_i \in F[X_1, X_2, \ldots, X_n]$ such that

$$f(X) = p_r(\alpha_1, \ldots, \alpha_n)X^r + \cdots + p_0(\alpha_1, \ldots, \alpha_n),$$

where $p_r \neq 0$. Define $g \in F[X_1, \ldots, X_n, X]$ by

$$g(X_1, \ldots, X_n, X) = p_r(X_1, \ldots, X_n)X^r + \cdots + p_0(X_1, \ldots, X_n).$$

Thus $g \neq 0$ since $p_r \neq 0$, and yet

$$g(\alpha_1, \ldots, \alpha_n, \alpha) = f(\alpha) = 0 .$$

It follows that the list $\alpha_1, \ldots, \alpha_n, \alpha$ is not algebraically independent over F, and thus $\mathcal{X} \cup \{\alpha\}$ is dependent.

Conversely, suppose $\mathcal{X} \cup \{\alpha\}$ is dependent over F. Then there is some polynomial dependence among some finite subset of this set, and that subset must contain α since $\mathcal{X}$ is independent. We can therefore find $\alpha_1, \alpha_2, \ldots, \alpha_n \in \mathcal{X}$ and a nonzero polynomial $g \in F[X_1, \ldots, X_n, X]$ such that $g(\alpha_1, \ldots, \alpha_n, \alpha) = 0$. We can view $g \in F[X_1, \ldots, X_n][X]$, and since $g \neq 0$, it has a nonzero leading coefficient $p \in F[X_1, \ldots, X_n]$. Since the elements $\alpha_1, \ldots, \alpha_n$ are algebraically independent over F, we have $p(\alpha_1, \ldots, \alpha_n) \neq 0$, and the polynomial

$$f(X) = g(\alpha_1, \ldots, \alpha_n, X) \in K[X]$$

is therefore nonzero. Since $f(\alpha) = 0$, we see that α is algebraic over K.   ∎

(24.3) COROLLARY. *A purely transcendental extension is totally transcendental.*

Proof. Let $F \subseteq E$ and suppose that $E = F(\mathcal{X})$, where $\mathcal{X}$ is a subset algebraically independent over F. Suppose $\beta \in E$ is algebraic over F. Our goal is to show that $\beta \in F$. Observe that β is a quotient of two polynomial expressions in $\mathcal{X}$ and thus lies in $F[\mathcal{X}_0]$, where $\mathcal{X}_0$ is a finite subset of $\mathcal{X}$. It follows that we may assume that $\mathcal{X}$ is finite, and we show that $\beta \in F$ by induction on $n = |\mathcal{X}|$.

If $n = 0$, then $E = F$ and there is nothing to prove. Assuming $n > 0$, we let $\alpha \in \mathcal{X}$ and write $\mathcal{X}' = \mathcal{X} - \{\alpha\}$ and $K = F(\mathcal{X}')$. Then $E = K(\alpha)$ and α is transcendental over K by Lemma 24.2, since $\mathcal{X}' \cup \{\alpha\} = \mathcal{X}$ is independent over F.

Now $\beta \in E$ is algebraic over K, and so Corollary 17.8 tells us that $\beta \in K$. The inductive hypothesis then gives $\beta \in F$, as required.   ∎

We construct an example of a totally transcendental extension that is not purely transcendental. Let $K = \mathbb{Q}(T)$ for some indeterminate T, and let $f \in K[X]$ be the polynomial $f(X) = X^2 + (T^2 + 1)$. Note that f is irreducible over K by the Eisenstein criterion. (Observe that $T^2 + 1$ is a prime element in $\mathbb{Q}[T]$, which is a UFD.) Let $E = K[\alpha]$, where α is a root of f. We shall show that E is totally transcendental but not purely transcendental over $\mathbb{Q}$.

To show that E is totally transcendental, suppose $\beta \in E$ is algebraic over $\mathbb{Q}$. Since $|E : K| = 2$, we can write $\beta = a + b\alpha$ for some $a, b \in K$. Now f splits over K, and so E is Galois over K, and there exists $\sigma \in \mathrm{Gal}(E/K)$ with $\sigma(\alpha) = -\alpha$. Since β is algebraic, so too is $\sigma(\beta)$, and thus $(\beta - \sigma(\beta))/2 = b\alpha$ is also algebraic over $\mathbb{Q}$. It follows that $b^2\alpha^2 = -b^2(T^2 + 1)$ is algebraic over $\mathbb{Q}$, and hence it lies in $\mathbb{Q}$ since K is totally transcendental over $\mathbb{Q}$. We can write $b = g(T)/h(T)$ for polynomials $g, h \in \mathbb{Q}[T]$, and we get $g(T)^2(T^2 + 1) = rh(T)^2$ for some rational number r. If $g \neq 0$, the irreducible polynomial $T^2 + 1$ is a factor of $g(T)^2(T^2 + 1)$

with odd multiplicity, and yet its multiplicity in $rh(T)^2$ is even. This contradiction proves that $g = 0$, and so $b = 0$ and $\beta = a \in K$. It follows that $\beta \in \mathbb{Q}$ because K is totally transcendental.

To show that E is not purely transcendental over $\mathbb{Q}$, we appeal to Theorem 24.5, which is proved later. Because E is algebraic over $\mathbb{Q}(T)$, this result guarantees that no two elements of E can be algebraically independent over $\mathbb{Q}$. If E were purely transcendental, therefore, we would have $E = \mathbb{Q}(\gamma)$ for some element $\gamma \in E$, and we derive a contradiction from this. Because $\alpha, T \in E$, we can write $\alpha = p/q$ and $T = u/v$, where $p, q, u, v \in \mathbb{Q}[\gamma]$. Since $\alpha^2 + T^2 + 1 = 0$, this yields $p^2 v^2 + u^2 q^2 + v^2 q^2 = 0$.

Now γ must be transcendental over $\mathbb{Q}$ since $E \supseteq \mathbb{Q}$ is not an algebraic extension. We thus have $\mathbb{Q}[\gamma] \cong \mathbb{Q}[X]$, and we can view γ as an indeterminate and $p, q, u,$ and v as polynomials in this indeterminate. Since $q, v \neq 0$, each of these polynomials has only finitely many roots, and so we can choose $s \in \mathbb{Q}$ with $v(s)q(s) \neq 0$. If follows that $p(s)^2 v(s)^2 + u(s)^2 q(s)^2 + v(s)^2 q(s)^2$ is a positive rational number, and this contradicts $p^2 v^2 + u^2 q^2 + v^2 q^2 = 0$.

(24.4) LEMMA.   *Let $F \subseteq E$ be an arbitrary field extension. Then there exists an intermediate field K such that K is purely transcendental over F and E is algebraic over K. In particular, if $\mathcal{X} \subseteq E$ generates E over F, and $\mathcal{X}_0 \subseteq \mathcal{X}$ is a subset maximal with the property that it is algebraically independent over F, then the field $K = F(\mathcal{X}_0)$ has the desired properties.*

Proof.   Suppose $\mathcal{X}$ generates E over F. (If we are not actually given a generating set $\mathcal{X}$, we can certainly find one. We could take $\mathcal{X} = E$, for instance.)   Since to determine algebraic independence, we need only look at finitely many elements at a time, it is easy to see that Zorn's lemma applies to the poset of all algebraically independent subsets of $\mathcal{X}$. There definitely exists, therefore, a subset $\mathcal{X}_0 \subseteq \mathcal{X}$, as in the statement of the lemma, and we put $K = F(\mathcal{X}_0)$.

Since $\mathcal{X}_0$ is algebraically independent over F, we certainly know that K is purely transcendental over F, and it suffices to show that E is algebraic over K. The elements of $\mathcal{X} - \mathcal{X}_0$ generate E over K, and so it is enough to show that each of these elements is algebraic over K.

Let $\alpha \in \mathcal{X} - \mathcal{X}_0$, and note that by the maximality of $\mathcal{X}_0$, the set $\mathcal{X}_0 \cup \{\alpha\}$ is not algebraically independent. By Lemma 24.2 we see that α is algebraic over $F(\mathcal{X}_0) = K$, as required.   ■

Note that in the important case of Lemma 24.4, where E is finitely generated over F, we do not really need Zorn's lemma in the proof. If $\mathcal{X}$ is finite, then it is clear that the subset $\mathcal{X}_0$ exists and, of course, $\mathcal{X}_0$ is finite.

If $F \subseteq E$ is an arbitrary field extension, a subset $\mathcal{X} \subseteq E$ that is algebraically independent over F is said to be a *transcendence basis* for E over F if E is algebraic over $K = F(\mathcal{X})$. By Lemma 24.4, therefore, every field extension has a transcendence basis, and by the previous paragraph, every finitely generated field extension has a finite transcendence basis (of cardinality not exceeding that of the given generating set). Note that an extension is algebraic iff the empty set is a transcendence basis.

It is amusing to note that Lemmas 24.1 and 24.4 can be used to give a purely algebraic characterization of the complex numbers field $\mathbb{C}$. We can find K such that $\mathbb{Q} \subseteq K \subseteq \mathbb{C}$, where K is isomorphic to the field of rational functions in some cardinal number $\aleph$ of indeterminates, and $\mathbb{C}$ is algebraic over K. Since $\mathbb{C}$ is algebraically closed, it can be described (up to isomorphism) as the algebraic closure of the field of rational functions in $\aleph$ indeterminates. (Clearly, $\aleph \leq |\mathbb{C}|$, and it is not hard to show that $\aleph = |\mathbb{C}| = |\mathbb{R}|$.)

Recall that we explained that it was impossible to prove the fundamental theorem of algebra purely algebraically since we said that $\mathbb{C}$ is not defined algebraically. The observation of the previous paragraph does not contradict this; it uses the fact that $\mathbb{C}$ is algebraically closed to prove that the field described as "the algebraic closure of the field of rational functions over $\mathbb{Q}$ in $|\mathbb{C}|$ indeterminates" is actually isomorphic to $\mathbb{C}$.

Before returning to the general theory, we deduce one more interesting consequence about $\mathbb{C}$. It is clear that any permutation of the indeterminates determines an automorphism of a field of rational functions. It follows that $\mathrm{Aut}(K)$ is huge, and it contains elements of every finite order and of infinite order. Every automorphism of K extends to an automorphism of its algebraic closure $\mathbb{C}$. (This follows since isomorphic fields have isomorphic algebraic closures.) Therefore $\mathrm{Aut}(\mathbb{C})$ is also huge. It is a fact that we prove in Chapter 25, however, that no finite subgroup of $\mathrm{Aut}(\mathbb{C})$ can have order exceeding 2, and so no element can have finite order exceeding 2.

Ending our digression now, let us return to the analogy between vector spaces and field extensions, where linear independence of a set of vectors corresponds to algebraic independence of a set of field elements. The assertion for extension fields analogous to the statement that "vector w is a linear combination of vectors $v_1, v_2, \ldots, v_n$" is "$\alpha \in E$ is algebraic over $F(\alpha_1, \ldots, \alpha_n)$" and, similarly, the analog of "$v_1, v_2, \ldots, v_n$ span V" is "E is algebraic over $F(\alpha_1, \ldots, \alpha_n)$." (We stress that the correct analog is "algebraic over" and not "equal to.") The analog of "basis" for a vector space is "transcendence basis" for a field extension, and we would like to show that the cardinality of a transcendence basis is uniquely determined, so that we can get the field theory analog of "dimension." We need the following theorem for this.

(24.5) THEOREM. *Let $F \subseteq E$ and suppose E is algebraic over $F(\mathcal{X})$ for some finite subset $\mathcal{X} \subseteq E$. Then every subset $\mathcal{Y} \subseteq E$ with $|\mathcal{Y}| > |\mathcal{X}|$ is algebraically dependent over F.*

We need the "only if" part of the following lemma to prove Theorem 24.5, and the other half is used later.

(24.6) LEMMA. *Let $F \subseteq E$ and suppose $\mathcal{X}$ and $\mathcal{Y}$ are disjoint subsets of E, with $\mathcal{X}$ algebraically independent over F. Write $K = F(\mathcal{X})$. Then $\mathcal{X} \cup \mathcal{Y}$ is algebraically independent over F iff $\mathcal{Y}$ is algebraically independent over K.*

Note that Lemma 24.6 extends Lemma 24.2, which is the case where $|\mathcal{Y}| = 1$. It is the case of Lemma 24.6 where $|\mathcal{X}| = 1$ that we need for the proof of Theorem 24.5.

Proof of Lemma 24.6. Suppose first that $\mathcal{X} \cup \mathcal{Y}$ is independent. We may certainly assume that $\mathcal{Y}$ is finite, and we work by induction on $n = |\mathcal{Y}|$ to show that $\mathcal{Y}$ is independent over K. There is nothing to prove if $n = 0$, and so we assume that $n > 0$ and we choose $\alpha \in \mathcal{Y}$. Writing $\mathcal{Y}' = \mathcal{Y} - \{\alpha\}$, we know by the inductive hypothesis that $\mathcal{Y}'$ is algebraically independent over K, and thus by Lemma 24.2 it suffices to show that α is transcendental over $K(\mathcal{Y}')$.

Now $\mathcal{X} \cup \mathcal{Y}$ is independent over F, and thus by Lemma 24.2 we know that α is transcendental over $F(\mathcal{X} \cup \mathcal{Y}')$. Since this field is equal to $K(\mathcal{Y}')$, this completes the proof that $\mathcal{Y}$ is independent over K.

Conversely now, suppose that $\mathcal{Y}$ is algebraically independent over K. To show that $\mathcal{X} \cup \mathcal{Y}$ is independent over F, it is no loss to assume that $\mathcal{Y}$ is finite, and again we work by induction on $n = |\mathcal{Y}|$. If $n = 0$, there is nothing to prove, since by hypothesis, $\mathcal{X}$ is independent over F. We therefore assume that $n > 0$, and again choose $\alpha \in \mathcal{Y}$ and write $\mathcal{Y}' = \mathcal{Y} - \{\alpha\}$.

By the inductive hypothesis, $\mathcal{X} \cup \mathcal{Y}'$ is independent over F, and hence by Lemma 24.2 it suffices to show that α is transcendental over $F(\mathcal{X} \cup \mathcal{Y}') = K(\mathcal{Y}')$. This is true by Lemma 24.2, however, since $\mathcal{Y}$ is independent over K. ∎

Proof of Theorem 24.5. We have $|\mathcal{Y}| > 0$, and so we can choose $\beta \in \mathcal{Y}$. If β is algebraic over F, then certainly $\mathcal{Y}$ is dependent, and so we may assume that β is transcendental over F. Thus E is not algebraic over F, and $|\mathcal{X}| > 0$. We work by induction on $|\mathcal{X}|$.

If it happens that $\beta \in \mathcal{X}$, write $\mathcal{X}' = \mathcal{X} - \{\beta\}$ and $\mathcal{Y}' = \mathcal{Y} - \{\beta\}$, and let $K = F(\beta)$. Now E is algebraic over $F(\mathcal{X}) = K(\mathcal{X}')$, and since $|\mathcal{Y}'| > |\mathcal{X}'|$, the inductive hypothesis yields that $\mathcal{Y}'$ is algebraically dependent over K. Because $\mathcal{Y}$ is the disjoint union of $\{\beta\}$ and $\mathcal{Y}'$, and since $K = F(\beta)$, we conclude from Lemma 24.6 that $\mathcal{Y}$ cannot be independent over F, and we are done in this case.

If we are not so lucky as to have $\beta \in \mathcal{X}$, we show how to replace $\mathcal{X}$ with a new set $\mathcal{Z}$, where $|\mathcal{Z}| \le |\mathcal{X}|$ and E is algebraic over $F(\mathcal{Z})$, and such that $\beta \in \mathcal{Z}$. By the previous paragraph, this completes the proof. Since β is transcendental over F, the set $\{\beta\}$ is algebraically independent over F, and we let $\mathcal{Z} \supseteq \{\beta\}$ be a subset of $\mathcal{X} \cup \{\beta\}$, maximal with the property that it is independent over F.

Since β is algebraic over $F(\mathcal{X})$, Lemma 24.2 guarantees that $\mathcal{X} \cup \{\beta\}$ is not independent over F. It follows that $\mathcal{Z}$ is a proper subset of $\mathcal{X} \cup \{\beta\}$, and we have $|\mathcal{Z}| \le |\mathcal{X}|$, as required. To see that E is algebraic over $F(\mathcal{Z})$, we observe that Lemma 24.4 tells us that $F(\mathcal{X} \cup \{\beta\})$ is algebraic over $F(\mathcal{Z})$. Since E is algebraic over $F(\mathcal{X})$, it is certainly algebraic over $F(\mathcal{X} \cup \{\beta\})$, and it follows that E is algebraic over $F(\mathcal{Z})$, as claimed. Finally, $\beta \in \mathcal{Z}$ by construction, and the proof is complete. ∎

(24.7) COROLLARY. *Assume E has a finite transcendence basis over F. Then all transcendence bases for E over F have equal cardinality.*

Proof.   Let $\mathcal{X}$ be a transcendence basis with $|\mathcal{X}| < \infty$. Since E is algebraic over $F(\mathcal{X})$, Theorem 24.5 tells us that no algebraically independent set can have more than $|\mathcal{X}|$ elements (and, in particular, none is infinite). If $\mathcal{Y}$ is any other transcendence basis, therefore, we have $|\mathcal{Y}| \le |\mathcal{X}|$. Interchanging the two bases, we obtain the reverse inequality, and the result follows.   ∎

If $F \subseteq E$ and there is a transcendence basis for E over F consisting of $n < \infty$ elements, then by Corollary 24.7 every transcendence basis has cardinality n. In this case, we say that the *transcendence degree* of the extension is n, and we write $\mathrm{td}_F(E) = n$.

Recall that by Lemma 24.4 there always exists a transcendence basis. It is possible to extend Corollary 24.7 to show that even if they are infinite, two transcendence bases for the extension $F \subseteq E$ must have equal cardinality. One could then define $\mathrm{td}_F(E)$ to be this (possibly infinite) cardinal number, but we simply write $\mathrm{td}_F(E) = \infty$ if there is no finite transcendence basis.

Note that $\mathrm{td}_F(E) = 0$ iff E is algebraic over F. We claim that $\mathrm{td}_F(E)$ is finite iff E is algebraic over some intermediate field K, where K is finitely generated over F. If $\mathrm{td}_F(E) < \infty$, it is immediate from the definitions that K exists. Conversely, if we have K finitely generated over F, then K has a finite transcendence basis over F. If E is algebraic over K, this transcendence basis is also a transcendence basis for E over F.

(24.8) THEOREM. *Let $F \subseteq E \subseteq L$. Then $\mathrm{td}_F(L) = \mathrm{td}_F(E) + \mathrm{td}_E(L)$. In particular, the degree on the left is infinite iff one of the degrees on the right is infinite.*

Proof.   Suppose $\mathrm{td}_F(L) < \infty$. By Theorem 24.5 we get $\mathrm{td}_F(E) \le \mathrm{td}_F(L) < \infty$. Also, since any set of elements of L that is algebraically independent over E is automatically independent over F, we see that $\mathrm{td}_E(L) \le \mathrm{td}_F(L) < \infty$.

Conversely now, suppose $\mathrm{td}_F(E)$ and $\mathrm{td}_E(L)$ are finite, and let $\mathcal{X}$ and $\mathcal{Y}$ be transcendence bases for E over F and L over E, respectively. The elements of $\mathcal{Y}$ are transcendental over E, and thus certainly do not lie in E, and it follows that $\mathcal{X}$ and $\mathcal{Y}$ are disjoint. We will complete the proof by showing that $\mathcal{X} \cup \mathcal{Y}$ is a transcendence basis for L over F.

Since $\mathcal{Y}$ is algebraically independent over E, it is certainly independent over the smaller field $F(\mathcal{X})$. Also, $\mathcal{X}$ is independent over F, and so Lemma 24.6 guarantees that $\mathcal{X} \cup \mathcal{Y}$ is algebraically independent over F.

Finally, we need to show that L is algebraic over $K = F(\mathcal{X} \cup \mathcal{Y})$. Let $A = \{\gamma \in E \mid \gamma \text{ is algebraic over } K\}$ and note that $A \supseteq K$ is a field. Since $\mathcal{X}$ is a transcendence basis for E over F, every element of E is algebraic over $F(\mathcal{X}) \subseteq K$, and thus $E \subseteq A$. Therefore $E(\mathcal{Y}) \subseteq A$. However, L is algebraic over $E(\mathcal{Y})$, and so L is algebraic over A. Since A is algebraic over K, we are done.   ∎

24B

A striking consequence of some of the theory we have developed is the following theorem.

(24.9) THEOREM. *Let $F \subseteq L$ be a finitely generated field extension, and suppose E is any intermediate field. Then E is also finitely generated over F.*

Note that it is not generally true for algebraic structures that subobjects of finitely generated objects must be finitely generated. We need the following lemma .

(24.10) LEMMA. *Suppose that $F \subseteq E$ and that the subset $\mathcal{X} \subseteq E$ is algebraically independent over F. Let $F \subseteq A \subseteq E$, where A is a field that is algebraic over F. Then:*

   a. *$\mathcal{X}$ is algebraically independent over A.*
   b. *A is the full set of elements of $A(\mathcal{X})$ that are algebraic over F.*
   c. *If $|E : F(\mathcal{X})| < \infty$, then $|A : F| < \infty$.*

Proof.   To prove (a), we may assume that $|\mathcal{X}| = n < \infty$. Let m be the size of the largest subset of $\mathcal{X}$ that is algebraically independent over A. By Lemma 24.4 we have $m = \mathrm{td}_A(A(\mathcal{X}))$, and since $\mathrm{td}_F(A) = 0$, Theorem 24.8 gives $\mathrm{td}_F(A(\mathcal{X})) = m$. We can also use Theorem 24.8 to compute this transcendence degree in a different way. We have

$$m = \mathrm{td}_F(A(\mathcal{X})) = \mathrm{td}_F(F(\mathcal{X})) + \mathrm{td}_{F(\mathcal{X})}(A(\mathcal{X})).$$

However, $\mathrm{td}_F(F(\mathcal{X})) = |\mathcal{X}| = n$ since $\mathcal{X}$ is algebraically independent over F. We deduce that $m \geq n$, and thus $m = n$ and (a) follows.

    By (a), $A(\mathcal{X})$ is purely transcendental and thus is totally transcendental over A. If $\gamma \in A(\mathcal{X})$ is algebraic over F, it is certainly algebraic over A, and hence $\gamma \in A$. This proves (b).

    If $|A : F| = \infty$, then since A is algebraic over F, we have $A > F(\beta_1, \beta_2, \ldots, \beta_r)$ for any finite list $\beta_1, \beta_2, \ldots, \beta_r$ of elements of A. It follows that given any such list, we can append to it an element β_{r+1} such that

$$F(\beta_1, \ldots, \beta_r) < F(\beta_1, \ldots, \beta_r, \beta_{r+1}) < A.$$

We conclude that there exists an infinite ascending chain $F = A_0 < A_1 < \cdots$ of subfields of A, and writing $L_i = A_i(\mathcal{X})$, we have $F(\mathcal{X}) = L_0 \subseteq L_1 \subseteq \cdots$. All of these containments must be strict, however, since by part (b) applied to A_i, we have $L_i \cap A = A_i$, and hence if $L_i = L_j$, we deduce that $A_i = A_j$ and it follows that $i = j$.

    If $|E : F(\mathcal{X})| < \infty$, it is clearly impossible to have an infinite strictly ascending chain of intermediate fields, and this proves (c). ∎

Proof of Theorem 24.9.   Since L is finitely generated over F, we have $\mathrm{td}_F(L) < \infty$ by Lemma 24.4. Thus $\mathrm{td}_F(E) < \infty$ by Theorem 24.8, and we can choose a finite transcendence basis $\mathcal{X}$ for E over F, and we write $K = F(\mathcal{X})$. Now

$K \subseteq E \subseteq L$, and it suffices to show that E is finitely generated over K. We know that E is algebraic over K, and so our goal is to show that $|E : K| < \infty$.

Let $\mathcal{Y}$ be a transcendence basis for L over K. Since L is finitely generated over $K(\mathcal{Y})$ (by the same set that generates it over F) and L is algebraic over $K(\mathcal{Y})$, we see that $|L : K(\mathcal{Y})| < \infty$. We know that E is algebraic over K and $\mathcal{Y}$ is algebraically independent over K, and thus Lemma 24.10(c) applies. We deduce that $|E : K| < \infty$, as required.   ∎

24C

We can strengthen Lemma 24.10(c).

(24.11) LEMMA.   *In the situation of Lemma 24.10, we have $|A : F| \leq |E : F(\mathcal{X})|$.*

Proof.   Let $K = F(\mathcal{X})$. We may certainly assume that $|E : K| < \infty$, and so by Lemma 24.10(c) we have $|A : F| < \infty$, and we write $m = |A : F|$. If $m = 1$, there is nothing to prove, and so we assume $m > 1$ and work by induction on m.

Let $\beta \in A$ with $\beta \notin F$. By Lemma 24.10(a), $\mathcal{X}$ is algebraically independent over $F[\beta]$, and we see that $F[\beta](\mathcal{X}) = K[\beta]$. Since $|A : F[\beta]| < m$, we can apply the inductive hypothesis to the extension $F[\beta] \subseteq E$, and we deduce that $|A : F[\beta]| \leq |E : K[\beta]|$. To complete the proof, it suffices to show that $|F[\beta] : F| \leq |K[\beta] : K|$.

Let $f = \min_K(\beta)$ and $g = \min_F(\beta)$. Then $g \in K[X]$ and $g(\beta) = 0$, and we deduce that $f \mid g$. Every root of f in a splitting field, therefore, is a root of g and hence is algebraic over F. Since f is monic, it follows that the coefficients of f, which are elementary symmetric functions in the roots, are algebraic over F. But these coefficients lie in K, and K is purely (and hence totally) transcendental over F, and it follows that $f \in F[X]$. We conclude that $f = g$, and it follows that

$$|F[\beta] : F| = \deg(g) = \deg(f) = |K[\beta] : K|,$$

and the proof is complete.   ∎

Lemma 24.11 has a nice application, but to state it we need a definition. Suppose that $F \subseteq E$ and that K and L are intermediate fields. Then K and L are *linearly disjoint* over F provided that every F-linearly independent set of elements of either K or L remains linearly independent over L or K, respectively. Clearly, it is necessary to consider only finite F-linearly independent sets of elements.

It is true, but perhaps not so clear, that to verify that K and L are linearly disjoint over F, it suffices to check that F-linearly independent subsets of just one of the given fields are linearly independent over the other.

(24.12) LEMMA.   *Let $F \subseteq E$ with intermediate fields K and L. Suppose that every F-linearly independent subset of K is linearly independent over L. Then K and L are linearly disjoint over F.*

Proof.   Suppose $\beta_1, \beta_2, \ldots, \beta_r \in L$ are linearly independent over F. We need to show that these elements are linearly independent over K, and so we suppose that $\sum \alpha_i \beta_i = 0$ for some coefficients $\alpha_i \in K$.

Let $\gamma_1, \gamma_2, \ldots, \gamma_s$ be an F-basis for the F-span of $\alpha_1, \ldots, \alpha_r$ in K. We can thus write

$$\alpha_i = \sum_{j=1}^{s} a_{ij} \gamma_j \quad \text{for } 1 \le i \le r,$$

where $a_{ij} \in F$. Then

$$0 = \sum_i \alpha_i \beta_i = \sum_{i,j} a_{ij} \gamma_j \beta_i = \sum_{j=1}^{s} \gamma_j \left(\sum_{i=1}^{r} a_{ij} \beta_i \right).$$

The elements $\gamma_j \in K$ are linearly independent over F, and so by hypothesis, they are linearly independent over L. Since $\sum a_{ij} \beta_i \in L$, we conclude that

$$\sum_{i=1}^{r} a_{ij} \beta_i = 0 \quad \text{for } 1 \le j \le s,$$

and because the β_i are linearly independent over F, we deduce that all $a_{ij} = 0$. It follows that $\alpha_i = 0$ for all i, as required.   ∎

The promised application of Lemma 24.11 is the following theorem.

(24.13) THEOREM.   *Let $F \subseteq E$ and suppose that K and A are intermediate fields with K purely transcendental over F and A algebraic over F. Then K and A are linearly disjoint over F.*

We extract a part of the argument as a separate lemma that can be used in other situations where one wishes to prove linear disjointness.

(24.14) LEMMA.   *Let $F \subseteq E$ and suppose that K and A are intermediate fields with A algebraic over F. Assume that*

$$|B : F| \le |\langle K, B \rangle : K|$$

for every field B with $F \subseteq B \subseteq A$ and $|B : F| < \infty$. Then K and A are linearly disjoint over F.

Proof.   Let $\alpha_1, \alpha_2, \ldots, \alpha_n \in A$ be linearly independent over F. By Lemma 24.12 it suffices to show that these elements are linearly independent over K. Write $B = F[\alpha_1, \ldots, \alpha_n]$, so that $|B : F| < \infty$. Now $\alpha_1, \ldots, \alpha_n$ may not span B over F, but since they are linearly independent, we can append additional elements and assume that the α_i form a basis for B over F.

Let R denote the K-vector subspace of E spanned by $\alpha_1, \ldots, \alpha_n$. Of course, $R \subseteq \langle K, B \rangle$, and since the α_i span B over F, we have $B \subseteq R$. All of the products $\alpha_i \alpha_j$ lie in $B \subseteq R$, and hence R is a ring and we have $\langle K, B \rangle =$

$K[\alpha_1, \alpha_2, \ldots, \alpha_n] \subseteq R$. Thus $\langle K, B \rangle = R$, and the n elements α_i span $\langle K, B \rangle$ over K.

By hypothesis, however, $|\langle K, B \rangle : K| \geq |B : F| = n$, and so the spanning elements $\alpha_1, \alpha_2, \ldots, \alpha_n$ must be linearly independent over K. ∎

Proof of Theorem 24.13.   By Lemma 24.14 it is enough to show that $|B : F| \leq |\langle K, B \rangle : K|$ for all fields B such that $F \subseteq B \subseteq A$ and $|B : F| < \infty$. The desired inequality is immediate from Lemma 24.11. ∎

24D

We showed that a totally transcendental field extension $F \subseteq E$ need not be purely transcendental. It is harder to find examples if the condition that "E is totally transcendental over F" is strengthened to the assumption that "E is contained in some purely transcendental extension of F." In fact, if $\mathrm{td}_F(E) = 1$, no such example exists. (Recall that in our example where E was totally but not purely transcendental, the transcendence degree was 1.)

(24.15) THEOREM.   *Let $F \subseteq E \subseteq L$ and assume that L is purely transcendental over F. If $\mathrm{td}_F(E) = 1$, then $E = F(\beta)$ for some element $\beta \in E$.*

The case of Theorem 24.15 where $\mathrm{td}_F(L) = 1$ (and thus no hypothesis on $\mathrm{td}_F(E)$ is needed) is called Lüroth's theorem, and we state it separately.

(24.16) THEOREM (Lüroth).   *Let $L = F(\alpha)$, where α is transcendental over F, and suppose $F \subseteq E \subseteq L$. Then $E = F(\beta)$ for some element $\beta \in E$.*

Together with results from Chapter 17, Lüroth's theorem yields the following pretty result.

(24.17) COROLLARY.   *Let $F \subseteq E \subseteq L$ and assume that $L = F(\alpha)$ for some element $\alpha \in L$. Then $E = F(\beta)$ for some element $\beta \in E$.*

Proof.   If α is algebraic over F, then by Artin's theorem (17.11), there are just finitely many intermediate fields between F and L. We conclude that there are only finitely many between F and E, and the result follows by a second application of Theorem 17.11.

If α is transcendental over F, we are in the situation of Lüroth's theorem and we are done. ∎

Our strategy for proving Theorem 24.15 is to do the Lüroth case first. We need a technical lemma about polynomials.

(24.18) LEMMA.   *Let $p, q \in F[X]$ be relatively prime polynomials and, similarly, let $u, v \in F[Y]$ be relatively prime, and assume that u and v are not both constants. Write*

$$f(X, Y) = u(Y)p(X) - v(Y)q(X)$$

and assume that $f = gh$, where $g \in F[X, Y]$ and $h \in F[X]$. Then h is a constant.

Proof. We have

$$u(Y)p(X) - v(Y)q(X) = h(X)g(X, Y),$$

and we will show that h is a constant by proving that $h|p$ and $h|q$ in $F[X]$.

By hypothesis, at least one of u or v has positive degree. If $\deg(u) > 0$, choose an extension field $E \supseteq F$ in which u has a root α. Writing $\beta = v(\alpha) \in E$ and substituting α for Y, we deduce that $h|\beta q$ in $E[X]$. Now $\beta = v(\alpha) \neq 0$, since otherwise u and v would both be divisible by $\min_F(\alpha)$, and this is not the case because they are relatively prime. It follows that $h|q$ in $E[X]$ and thus also in $F[X]$. Therefore $h(X)$ divides $u(Y)p(X)$ in $F[X, Y]$, and so $h|p$ in $F(Y)[X]$ and thus also in $F[X]$. We are done in this case, and similar reasoning completes the proof if $\deg(v) > 0$. ∎

The following is an example of how Lemma 24.18 can be used.

(24.19) COROLLARY. Let $p, q \in F[X]$ be relative prime and write $d = \max(\deg(p), \deg(q))$. Let $E = F(\beta)$, where β is transcendental over F, and write $f = p - \beta q \in E[X]$. Then $\deg(f) = d$, and f is irreducible if $d > 0$.

Proof. First, note that since $\beta \notin F$, none of the nonzero coefficients of $p(X)$ can equal any coefficient of $\beta q(X)$. It is immediate from this that $\deg(f) = d$, and we assume that $d > 0$.

Since $E = F(\beta) \cong F(Y)$, where Y is an indeterminate, we may replace β by Y and show that the polynomial $f(X, Y) = p(X) - Yq(X)$ is irreducible in $F(Y)[X]$. If this polynomial reduces, we can apply Gauss's lemma (16.19) and write $f = gh$ for some polynomials $g, h \in F[X, Y]$, each of positive degree in X. Since f has degree 1 in Y, we see that one of g or h (say h) does not involve Y, and we have $h \in F[X]$. Now Lemma 24.18 applies with $u = 1$ and $v = Y$, and we conclude that h is a constant. This is a contradiction. ∎

Proof of Theorem 24.16. We may assume that $E > F$. We seek $\beta \in E$ such that $E = F(\beta)$, and so we start by choosing $\beta \in E - F$ arbitrarily. Note that β is transcendental over F, and we can write $\beta = p(\alpha)/q(\alpha)$ for polynomials $p, q \in F[X]$. Reducing the fraction if necessary, we may assume that p and q are relatively prime, and we write $d = \max(\deg(p), \deg(q))$. (We stress that this integer d depends on the choice of β.) Since $\beta \notin F$, we have $d > 0$, and so by Corollary 24.19 the polynomial $p(X) - \beta q(X)$ is irreducible of degree d in $F(\beta)[X]$.

Now α is a root of the polynomial $p - \beta q$, and therefore we conclude that $|L : F(\beta)| = |F(\beta)[\alpha] : F(\beta)| = d$. Since $F(\beta) \subseteq E$, we have $|L : E| \leq d$ and we write $|L : E| = n$. We thus have $|E : F(\beta)| = d/n$. Of course, n is independent of our choice of the element $\beta \in E - F$, and for each such choice we get an integer $d \geq n$. It suffices to find some element β such that $d = n$.

Since $|L : E| = n$, we can write $g = \min_E(\alpha)$, and we have $\deg(g) = n$. Because α is transcendental over F, we know that $g \notin F[X]$ and thus g has some coefficient $\beta \in E - F$. We will complete the proof by showing that for any such coefficient β, we get $d = n$.

Writing $\beta = p(\alpha)/q(\alpha)$ as above, we know that $p(X) - \beta q(X) \in E[X]$, and α is a root of this polynomial. Thus $g \mid (p - \beta q)$ in $E[X]$. Since $\beta = p(\alpha)/q(\alpha)$, we get

$$q(\alpha)p(X) - p(\alpha)q(X) = g(X)h(X)$$

for some polynomial $h \in L[X]$.

Replacing α by an indeterminate Y (as we may, because it is transcendental), we have $L = F(Y)$, and using Gauss's lemma (16.19), we get

$$(*) \qquad q(Y)p(X) - p(Y)q(X) = g_0(X, Y)h_0(X, Y),$$

where $g_0, h_0 \in F[X, Y]$ and where $g_0 = \gamma g$ for some element $\gamma \in L = F(Y)$. Since g is monic, we have $\gamma \in F[Y]$. Also, $\beta = p(Y)/q(Y)$ is a coefficient of g, and so $\beta\gamma$ is a coefficient of g_0, viewed in $F[Y][X]$. In particular, this tells us that $\beta\gamma \in F[Y]$.

Write $\gamma = r(Y)$ and $\beta\gamma = s(Y)$ for polynomials $r, s \in F[Y]$. Since $\beta = p(Y)/q(Y)$, we have $pr = qs$, and because p and q are coprime, we get $p \mid s$ and $q \mid r$. Thus

$$d = \max(\deg(p), \deg(q)) \leq \max(\deg(r), \deg(s)).$$

But $\gamma = r(Y)$ and $\beta\gamma = s(Y)$ both appear as coefficients of g_0, and thus the degree in Y of $g_0(X, Y)$ is at least $\max(\deg(r), \deg(s)) \geq d$.

Since the degree in Y on the left side of equation $(*)$ is clearly no more than d, it follows that h_0 does not involve Y. Lemma 24.18 thus applies, and we conclude that h_0 is constant. Also, the degree in Y of g_0 is d exactly.

Since $h_0(X, Y)$ is a constant, equation $(*)$ yields that $g_0(Y, X) = -g_0(X, Y)$, and thus the degrees in X and in Y of $g_0(X, Y)$ are equal. We know that the degree in Y of $g_0(X, Y)$ is equal to d, and on the other hand, the degree in X of g_0 is the same as that for g, namely n. Thus $d = n$, and the proof is complete.   ∎

24E

So far, we have proved Theorem 24.15 only in the case of Lüroth's theorem, where $\mathrm{td}_F(L) = 1$. We work now toward proving the next case, where $\mathrm{td}_F(L) = 2$. The general case then follows fairly easily. Our immediate goal, therefore, is to prove the following theorem.

(24.20) THEOREM. *Let $F \subseteq E \subseteq L = F(X, Y)$, where X and Y are indeterminates, and $\mathrm{td}_F(E) = 1$. Then $E = F(\beta)$ for some element $\beta \in E$.*

We need some preliminary results.

(24.21) LEMMA. *Let $f, g \in F[X, Y]$ be linearly independent over F. Then for all but finitely many positive integers n, the polynomials $f(X, X^n)$ and $g(X, X^n)$ in $F[X]$ are linearly independent over F.*

Proof. If two nonzero polynomials $u, v \in F[X]$ are linearly dependent over F, then they must have equal degrees and $\alpha v = \beta u$, where α and β are the leading coefficients of u and v, respectively. In view of this, it is reasonable to try to compute the degrees and leading coefficients of $f(X, X^n)$ and $g(X, X^n)$.
Write

$$f(X, Y) = p_r(X)Y^r + p_{r-1}(X)Y^{r-1} + \cdots + p_0(X)$$

and

$$g(X, Y) = q_s(X)Y^s + q_{s-1}(X)Y^{s-1} + \cdots + q_0(X),$$

where all $p_i, q_j \in F[X]$ and $p_r, q_s \neq 0$. Let α and β be the leading coefficients of p_r and q_s, and let u and v be the degrees of these polynomials. Write m to denote the maximum of the degrees of all the nonzero polynomials among the p_i and q_j.

Suppose $n > m$. The largest exponent on a power of X contributed to $f(X, X^n)$ by any of the terms $p_i(X)Y^i$ for $i < r$ is at most $m + in < rn$. It follows that the term αX^{u+rn} contributed by $p_r(X)Y^r$ is uncanceled and is the leading term of $f(X, X^n)$. Therefore, for $n > m$, the leading coefficient of $f(X, X^n)$ is α, and similarly, the leading coefficient of $g(X, X^n)$ is β. If $f(X, X^n)$ and $g(X, X^n)$ are linearly dependent with $n > m$, therefore, we have

$$\beta f(X, X^n) = \alpha g(X, X^n).$$

The polynomial $\beta f(X, Y) - \alpha g(X, Y)$ is nonzero, and so when it is viewed as lying in $F(X)[Y]$, it has only finitely many roots. Since we have seen that X^n is a root whenever $f(X, X^n)$ and $g(X, X^n)$ are dependent with $n > m$, there can be just finitely many such integers n. The result now follows. ∎

(24.22) LEMMA. *Let $\gamma \in F \subseteq F(Y)$, where Y is an indeterminate. Then there exists a subring $R \subseteq F(Y)$ and a homomorphism $\sigma : R \to F$ with the following properties*:

a. $F[Y] \subseteq R$
b. $\sigma(Y) = \gamma$
c. $\sigma(\alpha) = \alpha$ *for all* $\alpha \in F$
d. *If* $r \in R$ *and* $\sigma(r) \neq 0$, *then* $r^{-1} \in R$ *and* $\sigma(r^{-1}) = \sigma(r)^{-1}$.

Proof. Every element of $F(Y)$ has the form p/q for some $p, q \in F[Y]$. Let R be the set of elements that can be written in this way with $q(\gamma) \neq 0$. Because the sum, difference, and product of p_1/q_1 and p_2/q_2 can each be written with denominator $q_1 q_2$, it is clear that R is a ring.

Assertion (a) is obvious since the elements of $F[Y]$ can be written with denominator equal to 1.

We define σ by setting $\sigma(p/q) = p(\gamma)/q(\gamma)$ if $q(\gamma) \neq 0$. It is routine to check that σ is well defined and that it is a homomorphism, and assertions (b) and (c) are immediate.

To prove (d), let $r = p/q$ with $q(\gamma) \neq 0$ and $\sigma(r) \neq 0$. Then $p(\gamma) \neq 0$, and since $r^{-1} = q/p$ (in the field $F(Y)$), we have $r^{-1} \in R$. Also, $\sigma(r^{-1}) = q(\gamma)/p(\gamma) = \sigma(r)^{-1}$, as required. ∎

Proof of Theorem 24.20. By Theorem 24.9 we know that E is finitely generated over F. Now choose $\alpha \in E - F$. Then α is transcendental over F, and since $\mathrm{td}_F(E) = 1$, we see that E is algebraic over $F(\alpha)$. It is finitely generated over $F(\alpha)$, and so we can write $E = F(\alpha)[\beta_1, \beta_2, \ldots, \beta_r]$ for some $\beta_i \in E$.

Write $\alpha = f/g$ with $f, g \in F[X, Y]$, and note that f and g are linearly independent over F since $\alpha \notin F$. Thus $f(X, X^n)$ and $g(X, X^n)$ are linearly independent for infinitely many integers n. If we write $\beta_i = p_i/q_i$ with $p_i, q_i \in F[X, Y]$, we observe that each q_i, viewed as being a polynomial in Y with coefficients in $F(X)$, has just finitely many roots. We thus can choose a positive integer n so that if we write $\gamma = X^n$, we have $q_i(X, \gamma) \neq 0$ for $1 \leq i \leq r$. By Lemma 24.21 this can be done so that $f(X, \gamma)$ and $g(X, \gamma)$ are linearly independent over F.

We now apply Lemma 24.22 to the extension $F(X) \subseteq F(X, Y)$, and we obtain a certain subring R with $F(X)[Y] \subseteq R \subseteq F(X, Y)$ and a homomorphism $\sigma : R \to F(X)$ such that $\sigma(Y) = \gamma$ and σ is the identity on $F(X)$. Thus if $q \in F[X, Y]$, then $q \in R$ and $\sigma(q) = q(X, \gamma)$, and if this is nonzero, then $q^{-1} \in R$.

Since $\sigma(q_i) \neq 0$, we have $q_i^{-1} \in R$, and thus $\beta_i \in R$ for $1 \leq i \leq r$. Also, $\sigma(f)$ and $\sigma(g)$ are linearly independent over F, and in particular, $\sigma(g) \neq 0$, and so $\alpha \in R$ and $\sigma(\alpha) = \sigma(f)/\sigma(g) \notin F$ by linear independence. Thus $\sigma(\alpha) \in F(X) - F$, and so it is transcendental over F.

If h is any nonzero polynomial over F, then $h(\alpha) \in R$ and $\sigma(h(\alpha)) = h(\sigma(\alpha))$, and this is nonzero since $\sigma(\alpha)$ is transcendental. It follows that $h(\alpha)^{-1} \in R$, and so R contains the whole field $F(\alpha)$. Since we also know that $\beta_i \in R$ for each subscript i, we deduce that R contains $E = F(\alpha)[\beta_1, \beta_2, \ldots, \beta_r]$.

Now σ defines an F-homomorphism from E into $F(X)$, and because E is a field, this must be an isomorphism. We can apply Lüroth's theorem to $\sigma(E)$, and the result follows. ∎

Finally, we can prove our main result in this direction.

Proof of Theorem 24.15. We have $F \subseteq E \subseteq F(\mathcal{X})$, where $\mathrm{td}_F(E) = 1$ and $\mathcal{X}$ is algebraically independent over F. Suppose that $\mathcal{Y}$ is some finite subset of $\mathcal{X}$ such that $E \cap F(\mathcal{Y}) > F$. (Note that $\mathcal{Y}$ necessarily exists since $E > F$ and each element of E lies in $F(\mathcal{Y})$ for some finite subset $\mathcal{Y} \subseteq \mathcal{X}$.) Choose $\beta \in (E \cap F(\mathcal{Y})) - F$, and note that β is transcendental over F since L is purely and thus totally transcendental over F. Since $\mathrm{td}_F(E) = 1$, we conclude that E

is algebraic over $F(\beta)$, and it follows that the elements of E are algebraic over the field $F(\mathcal{Y})$. By Lemma 24.6 the set $\mathcal{X} - \mathcal{Y}$ is algebraically independent over $F(\mathcal{Y})$, and thus L is purely transcendental over $F(\mathcal{Y})$. We conclude that $E \subseteq F(\mathcal{Y})$.

By the previous paragraph, we may assume that $\mathcal{X}$ is finite, and we work by induction on $n = |\mathcal{X}|$. Since $E > F$, it is clear that $\mathcal{X}$ is nonempty, and we choose $\alpha \in \mathcal{X}$ and write $F^* = F(\alpha)$ and $E^* = E(\alpha)$. If $F^* \cap E > F$, then by the result of the first paragraph, we have $E \subseteq F^*$, and we are done by Lüroth's theorem. We may assume then that $F^* \cap E = F$ and hence $E^* > F^*$.

By Theorem 24.8 we have

$$\mathrm{td}_F(F^*) + \mathrm{td}_{F^*}(E^*) = \mathrm{td}_F(E^*) = \mathrm{td}_F(E) + \mathrm{td}_E(E^*).$$

Also $\mathrm{td}_{F^*}(E^*) \geq 1$ since L is purely transcendental over F^*. We have $\mathrm{td}_F(F^*) = 1$ and $\mathrm{td}_F(E) = 1$. Finally, $\mathrm{td}_E(E^*) \leq 1$ since E^* is generated by a single element over E. We conclude that $\mathrm{td}_{F^*}(E^*) = 1$.

Recall that L is purely transcendental over F^*, generated by the algebraically independent set $\mathcal{X} - \{\alpha\}$ of cardinality $n - 1$. By the inductive hypothesis, therefore, we have $E^* = F^*(\beta)$ for some element $\beta \in E^*$, and of course β is transcendental over $F^* = F(\alpha)$. It follows by Lemma 24.2 that α and β are algebraically independent over F. Since $F \subseteq E \subseteq E^* = F(\alpha, \beta)$, we are done by Theorem 24.20. (We can replace α and β by indeterminates X and Y by Lemma 24.1.) ∎

24F

Somewhat paradoxically perhaps, transcendental extensions are relevant to the question of which finite groups can be realized (up to isomorphism) as Galois groups of finite-degree extensions of the rationals. Our main result on this is the following theorem.

(24.23) THEOREM. *Suppose $\mathbb{Q} \subseteq F \subseteq E$, where E is Galois over F and F is purely transcendental and finitely generated over $\mathbb{Q}$. Then there exists a Galois extension $\mathbb{Q} \subseteq L$ such that $\mathrm{Gal}(L/\mathbb{Q}) \cong \mathrm{Gal}(E/F)$.*

This result is useful because it turns out that it is much easier to build extensions $E \supseteq F$ with prescribed Galois groups when F is a sufficiently large purely transcendental extension of $\mathbb{Q}$ than it is when $F = \mathbb{Q}$. Unfortunately, we are not able to give the full proof of Theorem 24.23 in this book since this result depends on the Hilbert irreducibility theorem, which we state without proof. We show how to obtain Theorem 24.23 from Hilbert's result, however, and we use Theorem 24.23 to prove that the symmetric group S_n occurs as a Galois group over $\mathbb{Q}$ for every positive integer n.

(24.24) LEMMA. *Let G be a finite group and let F be any field. Then there exist fields $E \supseteq L \supseteq F$, where L is finitely generated over F, and such that E is*

Galois over L with Galois group isomorphic to G. If $G = S_n$, this can be done in such a way that L is purely transcendental over F.

Proof.   By Cayley's theorem, it is no loss to assume that G is a subgroup of some symmetric group S_n, and we let $E = F(T_1, T_2, \ldots, T_n)$, where the T_i are indeterminates. Given any element $\sigma \in S_n$, we can uniquely extend σ to an F-automorphism of E, which we continue to call σ. We thus view $G \subseteq \text{Gal}(E/F)$, and we let $L = \text{Fix}(G)$. Then $F \subseteq L \subseteq E$, and Theorem 24.9 guarantees that L is finitely generated over F. By Theorem 18.20 we conclude that E is Galois over F, and $\text{Gal}(E/F) = G$.

To complete the proof, we must show that if G is the full symmetric group S_n, then L is purely transcendental over F. In fact, we will see that $L = F(s_1, s_2, \ldots, s_n)$, where the s_r are the elementary symmetric functions in the indeterminates $T_1, \ldots, T_n$. (Recall that s_r is the sum of all possible products of r distinct members of the set $\{T_1, \ldots, T_n\}$.) Once we show this, it follows by Lemma 24.4 that any maximally algebraically independent (over F) subset of $\{s_r \mid 1 \le r \le n\}$ is a transcendence basis for L over F, and thus also for E over F since E is algebraic over L. Since $\text{td}_F(E) = n$, we conclude that the whole set is algebraically independent, and thus L is purely transcendental over F.

To see that L is generated by the s_r over F, write

$$f(X) = \prod_{i=1}^{n}(X - T_i) \in E[X].$$

Note that f is invariant under G since each element $\sigma \in G$ merely permutes the factors, and thus $f \in L[X]$. We have

$$f(X) = X^n - s_1 X^{n-1} + s_2 X^{n-2} - \cdots + (-1)^n s_n ,$$

and so $s_r \in L$ for each r. Define the subfield $K = F(s_1, \ldots, s_n) \subseteq L$ and observe that $f \in K[X]$. Now f splits over E and E is generated over K by the roots $T_1, T_2, \ldots, T_n$ of f, and so E is a splitting field for f over K. By Problem 17.14 it follows that $|E : K| \le n!$, and since $|E : L| = |G| = |S_n| = n!$, we have $K = L$. (Alternatively, E is Galois over K and $\text{Gal}(E/K) \supseteq \text{Gal}(E/L) = G$. Since G already induces every possible permutation of $\{T_1, T_2, \ldots, T_n\}$ and since the action on this set is faithful, we have $\text{Gal}(E/K) = G$, and so $K = L$.)   ■

We extract a piece of the proof of Lemma 24.24 for future reference.

(24.25)   COROLLARY.   *Let $p \in F[T_1, \ldots, T_n]$ be a symmetric polynomial. (In other words, p is invariant under all permutations of the indeterminates T_i.) Then*

$$p = \frac{f(s_1, \ldots, s_n)}{g(s_1, \ldots, s_n)} ,$$

where the s_r are the elementary symmetric polynomials in the T_i, and f and g are suitably chosen polynomials over F in n indeterminates.

Proof.   In the notation of the last part of the proof of Lemma 24.24, we have $p \in L = K = F(s_1, \ldots, s_n)$.   ∎

As we shall see in Chapter 28, the polynomial g can always be taken to be the constant 1.

(24.26) COROLLARY.   *Given an integer $n \geq 1$, there exists a Galois extension $L \supseteq \mathbb{Q}$ with $\mathrm{Gal}(L/\mathbb{Q}) \cong S_n$.*   ∎

In order to state Hilbert's irreducibility theorem, which is the key to Theorem 24.23 (and thus to our proof of Corollary 24.26), we need to discuss the notion of "specialization." We met this idea in Lemma 24.22.

Let $E = F(T_1, T_2, \ldots, T_n)$ where the T_i are indeterminates, and choose an n-tuple $(\gamma_1, \gamma_2, \ldots, \gamma_n)$ of elements of F. If

$$R = \{p/q \in E \mid p, q \in F[T_1, \ldots, T_n] \text{ with } q(\gamma_1, \gamma_2, \ldots, \gamma_n) \neq 0\},$$

then $F[T_1, \ldots, T_n] \subseteq R \subseteq E$, and R is a subring. Furthermore, we have a homomorphism $R \to F$ defined by evaluation, setting T_i to γ_i. This map is called the *specialization* at the γ_i and the ring R is its *domain*. Note that the specialization is the identity map on F.

If F is infinite and $\alpha_1, \alpha_2, \ldots, \alpha_r$ is any finite list of elements of E, then it is possible to choose elements γ_i so that all of the elements α_j lie in the domain R of the corresponding specialization. To see this, write $\alpha_j = p_j/q_j$ with $p_j, q_j \in F[T_1, \ldots, T_n]$ and with $q_j \neq 0$, and write $q = \prod q_j$. Now q is a nonzero polynomial in n indeterminates, and since F is infinite, it is possible to choose an n-tuple of elements $\gamma_i \in F$ such that $q(\gamma_1, \ldots, \gamma_n) \neq 0$. (This can easily be proved by induction on n, but we omit the details.) It follows that all α_i lie in the domain of the corresponding specialization.

Given a specialization with domain R, we get a homomorphism $f \mapsto \overline{f}$ from $R[X]$ into $F[X]$. Furthermore, if $f \in E[X]$ is arbitrary and F is infinite, the above reasoning guarantees that we can find a specialization for which $f \in R[X]$ (and simultaneously such that any specified finite set of elements of E is contained in R).

Even if $f \in E[X]$ is irreducible, it would not be surprising to find that $\overline{f}$ reduces. (For example, if $n = 1$ and we take $f(X) = X^2 - T$, then f is irreducible, but $\overline{f}$ reduces if γ is a square in F.)

(24.27) THEOREM (Hilbert's irreducibility).   *Let f be an irreducible polynomial in $E[X]$, where $E = \mathbb{Q}(T_1, T_2, \ldots, T_n)$, and choose elements $\alpha_1, \alpha_2, \ldots, \alpha_r \in E$. Then there exists a specialization whose domain contains the coefficients of f and all α_i. This can be done so that the polynomial $\overline{f}$ is irreducible in $\mathbb{Q}[X]$.*

We omit the proof of this theorem.

Proof of Theorem 24.23.   We are given $\mathbb{Q} \subseteq F \subseteq E$ with E Galois over F and F purely transcendental and finitely generated over $\mathbb{Q}$. It is no loss to assume that $F = \mathbb{Q}(T_1, \ldots, T_n)$ for indeterminates T_i, and we write $G = \mathrm{Gal}(E/F)$.

Since E is Galois over F, we can write $E = F[\alpha]$ for some element $\alpha \in E$, and we put $f = \min_F(\alpha)$ with $d = \deg(f)$, so that $d = |E : F| = |G|$. Our strategy is to "capture" G in terms of polynomials, and then to specialize and recover G from the images of these polynomials.

Every element of E can be written in the form $u(\alpha)$ for some polynomial $u \in F[X]$ with $\deg(u) < d$ (or $u = 0$). For each element $\sigma \in G$, we choose $u_\sigma \in F[X]$ with $u_\sigma(\alpha) = \sigma(\alpha)$ and such that $\deg(u_\sigma) < d$.

We have, of course,

$$0 = \sigma(f(\alpha)) = f(\sigma(\alpha)) = f(u_\sigma(\alpha)),$$

and thus α is a root of the polynomial $h_\sigma(X) = f(u_\sigma(X))$. It follows that $f \mid h_\sigma$ for all elements $\sigma \in G$. Also, for $\sigma, \tau \in G$, we have

$$u_{\sigma\tau}(\alpha) = \tau(\sigma(\alpha)) = \tau(u_\sigma(\alpha)) = u_\sigma(\tau(\alpha)) = u_\sigma(u_\tau(\alpha)),$$

and hence α is a root of the polynomial $g_{\sigma,\tau}(X) = u_{\sigma\tau}(X) - u_\sigma(u_\tau(X))$. Thus $f \mid g_{\sigma,\tau}$ for all $\sigma, \tau \in G$. Finally, for $\sigma, \tau \in G$ with $\sigma \neq \tau$, we have $\sigma(\alpha) \neq \tau(\alpha)$ and so $u_\sigma \neq u_\tau$. Let $c_{\sigma,\tau} \in F$ be the reciprocal of the leading coefficient of $u_\sigma - u_\tau$. Thus $c_{\sigma,\tau}(u_\sigma - u_\tau)$ is monic.

Next we appeal to the Hilbert irreducibility theorem. We choose a specialization $R \to \mathbb{Q}$ such that $\overline{f} \in \mathbb{Q}[X]$ is irreducible. Of course, we are assuming that we have chosen the specialization so that the coefficients of f all lie in R. In addition, we may assume that all coefficients of the polynomials $u_\sigma, h_\sigma, g_{\sigma,\tau},$ h_σ/f and $g_{\sigma,\tau}/f$ lie in R, and that the elements $c_{\sigma,\tau}$ also are in R.

Since f is monic, we have $\deg(\overline{f}) = \deg(f) = d$. Also, for $\sigma, \tau \in G$, we have

$$\overline{h}_\sigma(X) = \overline{f}(\overline{u}_\sigma(X)),$$
$$\overline{g}_{\sigma,\tau}(X) = \overline{u}_{\sigma\tau}(X) - \overline{u}_\sigma(\overline{u}_\tau(X)),$$

and

$$\overline{f} \mid \overline{h}_\sigma \quad \text{and} \quad \overline{f} \mid \overline{g}_{\sigma,\tau}.$$

Furthermore, $\overline{c}_{\sigma,\tau}(\overline{u}_\sigma - \overline{u}_\tau)$ is monic when $\sigma \neq \tau$, and thus the polynomials $\overline{u}_\sigma$ are distinct as σ runs over G.

Now let $L = \mathbb{Q}[\beta]$, where β is a root of $\overline{f}$. Then $\overline{f} = \min_{\mathbb{Q}}(\beta)$ and $|L : \mathbb{Q}| = d = |G|$. For each element $\sigma \in G$, we define $\beta_\sigma = \overline{u}_\sigma(\beta) \in L$, and we note that since $\overline{f} \mid \overline{h}_\sigma$, we have

$$\overline{f}(\beta_\sigma) = \overline{f}(\overline{u}_\sigma(\beta)) = \overline{h}_\sigma(\beta) = 0$$

for all $\sigma \in G$, and thus β_σ is a root of $\overline{f}$. Also, the elements β_σ are distinct for distinct $\sigma \in G$ since if $\beta_\sigma = \beta_\tau$, then $\overline{u}_\sigma(\beta) - \overline{u}_\tau(\beta) = 0$, and so β is a root of the nonzero polynomial $\overline{u}_\sigma - \overline{u}_\tau$ of degree less than d. This is impossible since $\min_{\mathbb{Q}}(\beta) = \overline{f}$ has degree d.

We now have $|G| = d$ distinct roots β_σ of $\overline{f}$ in L, and we deduce that L is a splitting field for $\overline{f}$ over $\mathbb{Q}$ and so is Galois over $\mathbb{Q}$. For each element $\sigma \in G$, there exists $\sigma^* \in \mathrm{Gal}(L/\mathbb{Q})$ such that $\sigma^*(\beta) = \beta_\sigma$. (This is so since β and

β_σ are roots of the irreducible polynomial $\overline{f}$.) Furthermore, since $L = \mathbb{Q}[\beta]$, we see that σ^* is uniquely determined, and so $\sigma \mapsto \sigma^*$ is a well-defined map $G \to \text{Gal}(L/\mathbb{Q})$.

To see that this map is a homomorphism, we observe that $\overline{g}_{\sigma,\tau}(\beta) = 0$ since $\overline{f}|\overline{g}_{\sigma,\tau}$, and thus

$$(\sigma\tau)^*(\beta) = \beta_{\sigma\tau} = \overline{u}_{\sigma\tau}(\beta) = \overline{u}_\sigma(\overline{u}_\tau(\beta)) \,.$$

We also have

$$(\sigma^*\tau^*)(\beta) = \tau^*(\sigma^*(\beta)) = \tau^*(\overline{u}_\sigma(\beta)) = \overline{u}_\sigma(\tau^*(\beta)) = \overline{u}_\sigma(\overline{u}_\tau(\beta)) \,,$$

where the first equality is by our standing convention on function composition. Since β generates the field L over $\mathbb{Q}$, it follows that $(\sigma\tau)^* = \sigma^*\tau^*$, and our map is a group homomorphism. It is injective since the elements $\beta_\sigma = \sigma^*(\beta)$ are distinct as σ runs over G. We have now shown that G is isomorphically embedded in $\text{Gal}(L/\mathbb{Q})$, and the result follows since $|\text{Gal}(L/\mathbb{Q})| = |L : \mathbb{Q}| = d = |G|$. ∎

Problems

24.1 Suppose $E = F(\alpha, \beta, \gamma)$, with α, β, and γ algebraically independent over F.

a. Show that α/β, β/γ, and γ/α are algebraically dependent over F.

b. Show that any two of the elements in part (a) are algebraically independent over F.

24.2 In the situation of Problem 24.1, show that $\alpha\beta$, $\beta\gamma$, and $\gamma\alpha$ are algebraically independent over F.

HINT: Each of α, β, and γ is algebraic over the field generated by these elements.

24.3 In the situation of Problem 24.1, let $\sqrt{\alpha}$, $\sqrt{\beta}$, and $\sqrt{\gamma}$ be elements of some extension field $L \supseteq E$. Show that $\alpha\sqrt{\beta}$, $\beta\sqrt{\gamma}$, and $\gamma\sqrt{\alpha}$ are algebraically independent over F.

24.4 Let $F \subseteq K \subseteq E$ and suppose that $\alpha \in E$ is algebraic over F. Show that $N_{E/K}(\alpha)$ and $T_{E/K}(\alpha)$ are algebraic over F. (Assume that $|E : K|$ is finite.)

24.5 Let $K = F(T_1, T_2, \ldots, T_n)$, where the T_i are indeterminates. Suppose $f \in K[X]$ is monic, separable, and irreducible of prime degree, with discriminant $\Delta \in K$. Let $E = K[\alpha]$, where α is a root of f. Show that E is totally transcendental over F if Δ does not have the form ap^2/q^2, where $a \in F$ and $p, q \in F[T_1, \ldots, T_n]$.

HINT: If E is not totally transcendental over F, then $E = K[\beta]$, where β is algebraic. Use Theorems 23.17 and 23.22.

NOTE: If $F \subseteq K \subseteq E$, where E is algebraic over K and K is purely transcendental over F, then we say that E is a *regular* extension of K with respect to F if E is totally transcendental over F.

24.6 Let $F \subseteq E$ and suppose $\mathcal{X} \subseteq E$. Give the details of the Zorn's lemma argument (referred to in the proof of Lemma 24.4) and show that there exists a subset $\mathcal{X}_0 \subseteq \mathcal{X}$ maximal with the property that $\mathcal{X}_0$ is algebraically independent over F.

24.7 Let $\mathcal{X}$ be a transcendence basis for E over F and suppose that $\alpha \in E$. Show that for some finite subset $\mathcal{X}_0 \subseteq \mathcal{X}$, the element α is algebraic over $F(\mathcal{X}_0)$.

24.8 Let $F \subseteq E$ and suppose that $\mathcal{X} \subseteq E$ is a finite subset minimal with the property that E is algebraic over $F(\mathcal{X})$. Show that $\mathcal{X}$ is a transcendence basis for E over F.

24.9 Let $F \subseteq K \subseteq E$, where K is totally transcendental over F and E is Galois over K. Let A be the subfield consisting of those elements of E that are algebraic over F. Show that A is Galois over F.

24.10 Let $F \subseteq E$ and suppose that $\alpha, \beta \in E$ are algebraically independent over F. Show that $F(\alpha) \cap F(\beta) = F$.

HINT: Otherwise, $F(\alpha, \beta)$ is algebraic over the intersection.

24.11 Let $\mathbb{Q} \subseteq E$ and suppose that $\alpha, \beta \in E$ are transcendental but algebraically dependent over $\mathbb{Q}$. Prove that it can happen that $\mathbb{Q}(\alpha) \cap \mathbb{Q}(\beta) = \mathbb{Q}$.

HINT: Take $E = \mathbb{Q}(T)$ with $\alpha = T^2$ and $\beta = T - T^2$. Show that there exist $\sigma \in \mathrm{Gal}(E/\mathbb{Q}(\alpha))$ and $\tau \in \mathrm{Gal}(E/\mathbb{Q}(\beta))$ such that $\sigma \tau$ has infinite order.

24.12 Let $F \subseteq E$ with intermediate fields K and L. Assume that $K \cap L = F$ and that L is Galois over F. Show that K and L are linearly disjoint over F.

24.13 Let $F \subseteq E$ with intermediate fields K and L. Assume that K is totally transcendental over F and that L is algebraic and separable over F. Show that K and L are linearly disjoint over F.

24.14 Let $E = F(T)$, where T is an indeterminate and let $p, q \in F[T]$ be nonzero relatively prime polynomials. Write $d = \max(\deg(p), \deg(q))$ and suppose $d > 0$. Show that $|E : F(p/q)| = d$.

24.15 Let $E = F(T)$, where T is an indeterminate, and suppose $\sigma \in \mathrm{Gal}(E/F)$.
   a. Show that $\sigma(T) = (aT + b)/(cT + d)$ for some $a, b, c, d \in F$ with $ad - bc \neq 0$.
   b. If $a, b, c, d \in F$ with $ad - bc \neq 0$, show that there exists a unique element $\sigma \in \mathrm{Gal}(E/F)$ such that $\sigma(T) = (aT + b)/(cT + d)$.

HINT: For part (b), use the fact that if α and β are transcendental over F in fields $K, L \supseteq F$, then there exists a unique F-isomorphism $F(\alpha) \to F(\beta)$ that carries α to β.

24.16 Let $G = GL(2, F)$, the group of invertible 2×2 matrices over F. Let $E = F(T)$, where T is an indeterminate, and let $\theta : G \to \mathrm{Gal}(E/F)$ be the map that assigns to the matrix $\begin{bmatrix} a & c \\ b & d \end{bmatrix}$ the automorphism σ such that $\sigma(T) = (aT + b)/(cT + d)$. Show that θ is a homomorphism and that $\ker(\theta)$ is the set of nonzero scalar matrices $\begin{bmatrix} a & 0 \\ 0 & a \end{bmatrix}$.

NOTE: The group $GL(n, F)$ modulo the (central) subgroup of scalar matrices is denoted $PGL(n, F)$. Here, the "P" stands for "projective." By Problems 24.15 and 24.16 we conclude that $\text{Gal}(E/F) \cong PGL(2, F)$, where $E = F(T)$.

24.17 Let G be any finite subgroup of $PGL(2, \mathbb{Q})$. Show that G occurs as a Galois group over $\mathbb{Q}$.

CHAPTER TWENTY-FIVE

The Artin-Schreier Theorem

25A

Does it ever happen that the algebraic closure of a field F is a proper, but finite-degree extension of F? Of course this happens: $\mathbb{C}$ is the algebraic closure of $\mathbb{R}$ and $|\mathbb{C} : \mathbb{R}| = 2$. Remarkably, it is almost true that there are no other examples. In every case where this occurs, F looks a lot like $\mathbb{R}$, and its algebraic closure is an extension of degree 2 obtained by adjoining an element whose square is -1.

(25.1) THEOREM (Artin-Schreier). *Let $F \subseteq L$, where L is algebraically closed and $1 < |L : F| < \infty$. Then:*

a. *$|L : F| = 2$ and $L = F[i]$, where $i \in L$ and $i^2 = -1$.*
b. *If we write S to denote the set of nonzero squares in F, then S is closed under addition.*
c. *F is the disjoint union $S \mathbin{\dot\cup} \{0\} \mathbin{\dot\cup} -S$, where S is as in (b) and $-S$ is the set of negatives of elements of S.*

Note that in the situation of Theorem 25.1, conclusion (b) implies that 0 cannot be written as $1 + 1 + \cdots + 1$, and so $\operatorname{char}(F) = 0$. When $F = \mathbb{R}$, which is the case we know, S is exactly the set of positive real numbers, and conclusions (b) and (c) clearly hold.

We begin with a nearly trivial observation.

(25.2) LEMMA. *Suppose $F \subseteq L$, where L is algebraically closed and $|L : F| = n < \infty$. Then every irreducible polynomial in $F[X]$ has degree $\leq n$.*

Proof. Let $f \in F[X]$ be irreducible. Then f has a root $\alpha \in L$ and

$$\deg(f) = \deg(\min{}_F(\alpha)) = |F[\alpha] : F| \leq |L : F| = n,$$

as required. ∎

It follows that a field over which we can prove that there exist irreducible polynomials of arbitrarily large degree cannot have a finite-degree algebraically closed extension. We dispose first of the case where F has prime characteristic but is not perfect.

(25.3) LEMMA. *Suppose* char$(F) = p > 0$ *and* F *is not perfect. Then there exist irreducible polynomials in* $F[X]$ *that have degree* p^e *for each nonnegative integer* e.

Proof. Let $a \in F$ be an element that is not a pth power. We claim that $f(X) = X^{p^e} - a \in F[X]$ is irreducible whenever $e \geq 0$. To see this, let α be a root of f in some splitting field, and let $g = \min_F(\alpha)$. Then $0 = f(\alpha) = \alpha^{p^e} - a$ and so $a = \alpha^{p^e}$. Thus $f(X) = X^{p^e} - \alpha^{p^e} = (X - \alpha)^{p^e}$.

Now write $f = g_1 g_2 \cdots g_m$, where each factor $g_i \in F[X]$ is monic and irreducible. It follows that g_i is a power of $X - \alpha$, and so $g_i(\alpha) = 0$ and $g_i = g$. We conclude that $f = g^m$ and hence $p^e = \deg(f) = m \deg(g)$, and m is a power of p. Now $-a = f(0) = g(0)^m$ and hence $a = (-g(0))^m$. (Note that this holds even if m is even, since in that case, $p = 2$ and $-1 = 1$.)

We see that a is an mth power in F. Because a is not a pth power, p cannot divide m, and hence $m = 1$. This shows that f is irreducible, as required. ∎

(25.4) LEMMA. *Let* $F \subseteq L$, *where* L *is algebraically closed and* $1 < |L : F| < \infty$. *Then either* $|L : F| = 2$ *and* $L = F[i]$, *where* $i \in L$ *and* $i^2 = -1$, *or there exists a field* K *with* $F \subseteq K \subseteq L$ *such that the following hold:*

a. L *is Galois over* K,
b. $|L : K| = p$, *a prime, and*
c. -1 *is a square in* K.

Proof. By Lemma 25.2 the irreducible polynomials over F have bounded degree, and so by Lemma 25.3 either char$(F) = 0$ or F is perfect of prime characteristic. In either case, L is separable over F, and because L is algebraically closed, it is certainly normal over F. Thus L is Galois over F and over every intermediate field.

Now let $i \in L$ be a root of $X^2 + 1$. If $F[i] = L$, then $|L : F| = 2$ and there is nothing to prove. Assume, therefore, that $L > F[i]$ and let $G = \text{Gal}(L/F[i])$. Then $|G| > 1$ and we can choose a subgroup $P \subseteq G$ of prime order p. If we take $K = \text{Fix}(P)$, then (a) and (b) certainly hold, and statement (c) holds too since $i \in K$. ∎

By Lemma 25.4 it suffices to study Galois extensions of prime degree p. We consider first the case where the characteristic is different from p.

(25.5) LEMMA. *Let* $K \subseteq L$, *where* L *is algebraically closed, and suppose* $|L : K| = p$, *a prime. If* char$(L) \neq p$, *then* K *contains a primitive* pth *root of unity.*

Proof.   Because L is algebraically closed and of characteristic different from p, the polynomial $X^p - 1$ has p distinct roots in L. If $\epsilon \neq 1$ is one of these, then ϵ is a primitive pth root of unity, and so it is a root of $(X^p - 1)/(X - 1) \in K[X]$. Thus $|K[\epsilon] : K| = \deg(\min_K(\alpha)) \leq p - 1 < |L : K|$, and hence $K[\epsilon] < L$.

Since $K \subseteq L$ is an extension of prime degree, there are no proper intermediate fields. Thus $K[\epsilon] = K$ and $\epsilon \in K$. ∎

The following result is the key to the proof of Theorem 25.1(a) in the case where $\mathrm{char}(L) \neq |L : K|$.

(25.6) THEOREM.   *Let $K \subseteq E$ be any Galois extension of prime degree p, and assume that K contains a primitive pth root of unity. If $p = 2$, assume also that -1 is a square in K. Then there exists $\alpha \in E$ such that the polynomial $X^p - \alpha$ is irreducible in $E[X]$. In particular, E is not algebraically closed.*

Proof.   By Kummer's theorem (22.8), we have $E = K[\alpha]$ for some element $\alpha \in E$ with $\alpha^p \in L$. To prove that $X^p - \alpha$ is irreducible in $E[X]$, we let β be a root of this polynomial in some extension field of E. It suffices to show that $|E[\beta] : E| = p$. By the other direction of Kummer's theorem, we know that $|E[\beta] : E|$ divides p, and since p is prime, we may assume that $|E[\beta] : E| = 1$ (so that $\beta \in E$), and we work to derive a contradiction.

Write $\mathrm{Gal}(E/K) = \langle \sigma \rangle$ and note that $\beta^{p^2} = \alpha^p \in K$. Thus $\sigma(\beta)^{p^2} = \sigma(\beta^{p^2}) = \beta^{p^2}$, and so if we write $\delta = \sigma(\beta)/\beta \in E$, we have $\delta^{p^2} = 1$. Therefore δ^p is a pth root of unity, and so it lies in K. We have $\sigma(\delta)^p = \sigma(\delta^p) = \delta^p$, and thus setting $\epsilon = \sigma(\delta)/\delta$, we get $\epsilon^p = 1$. Since ϵ is a pth root of unity, we have $\epsilon \in K$.

Next we compute the effect of successive applications of σ to β. We have $\sigma(\beta) = \beta\delta$ and $\sigma^2(\beta) = \sigma(\beta\delta) = (\beta\delta)(\delta\epsilon) = \beta\delta^2\epsilon$. Another application of σ yields $\sigma^3(\beta) = \sigma(\beta\delta^2\epsilon) = (\beta\delta)(\delta\epsilon)^2\epsilon = \beta\delta^3\epsilon^3$. In general, we obtain

$$\sigma^i(\beta) = \beta\delta^i\epsilon^{i(i-1)/2},$$

as can easily be proved by induction on i. Since $\sigma^p = 1$, this gives

$$\beta = \sigma^p(\beta) = \beta\delta^p\epsilon^{p(p-1)/2}$$

and so

$$\delta^p\epsilon^{p(p-1)/2} = 1.$$

We claim that $\epsilon^{p(p-1)/2} = 1$ and thus $\delta^p = 1$. If p is odd, then $p(p-1)/2$ is a multiple of p, and since $\epsilon^p = 1$, the assertion follows. If $p = 2$, on the other hand, then $\delta^4 = 1$ and $\delta^2 = \pm 1$. By assumption in this case, -1 is a square in K, and we conclude that $\delta \in K$. It follows from this that $\epsilon = \sigma(\delta)/\delta = 1$, and here too the claim is established and $\delta^p = 1$.

Now $\sigma(\alpha) = \sigma(\beta^p) = \sigma(\beta)^p = \beta^p\delta^p = \beta^p = \alpha$. Since α generates E over K, this gives $\sigma = 1$, which contradicts our choice of σ as a generator of $\mathrm{Gal}(E/K)$. ∎

Next we do the characteristic p analog of Theorem 25.6.

(25.7) THEOREM. *Let $K \subseteq E$ be a Galois extension of prime degree p, and assume that $\text{char}(E) = p$. Then there exists $\alpha \in E$ such that the polynomial $X^p - X - \alpha$ is irreducible in $E[X]$. In particular, E is not algebraically closed.*

To prove this, we need an analog of Kummer's theorem (22.8).

(25.8) THEOREM. *Let $F < E$ and assume that $\text{char}(F) = p \neq 0$. Then the following are equivalent:*

i.   *E is Galois over F and $|E : F| = p$.*
ii. *$E = F[\alpha]$ for some element $\alpha \in E$ with $\alpha^p - \alpha \in F$.*

Proof.   Assume (i) and write $G = \text{Gal}(E/F)$. Since G has order p, there exists an isomorphism θ from G onto the additive subgroup of F generated by 1. Thus for $\sigma, \tau \in G$, we have

$$\theta(\sigma\tau) = \theta(\sigma) + \theta(\tau) = \theta(\sigma)^\tau + \theta(\tau)$$

since $\theta(\sigma) \in F$ is fixed by τ. (We have decided to use the exponential notation for field automorphisms.) In other words, θ is a crossed homomorphism from G into the additive group of E. By Corollary 23.16 the cohomology group $H^1(G, E^+)$ is trivial. Thus every crossed homomorphism from G into E^+ is principal, and there exists $\alpha \in E$ such that $\theta(\sigma) = \alpha^\sigma - \alpha$ for all elements $\sigma \in G$. In particular, since θ is injective, $\alpha^\sigma \neq \alpha$ if $\sigma \neq 1$, and so $\alpha \notin F$. Since $|E : F|$ is prime, we have $F[\alpha] = E$.

We need to show that $\alpha^p - \alpha \in F$, and so we compute $(\alpha^p - \alpha)^\sigma$ for a generator σ of G. We can choose σ so that $\theta(\sigma) = 1$, and this gives $\alpha^\sigma = \alpha + 1$. Thus

$$(\alpha^p - \alpha)^\sigma = (\alpha^\sigma)^p - \alpha^\sigma = (\alpha + 1)^p - (\alpha + 1) = \alpha^p - \alpha ,$$

where the last equality holds because we are in characteristic p. This proves that $\alpha^p - \alpha \in F$, as required.

Conversely now, assume (ii) and write $f(X) = X^p - X - a$, where $a = \alpha^p - \alpha \in F$. Thus $f \in F[X]$ and α is a root of f. Since $(\alpha + 1)^p - (\alpha + 1) = \alpha^p - \alpha = a$, we see that $\alpha + 1$ is another root for f, and continuing like this, we find p distinct roots $\alpha, \alpha + 1, \alpha + 2, \ldots, \alpha + p - 1$ for f in E. It follows that f is separable over F and that E is a splitting field for f over F. Therefore E is Galois over F.

Now E is generated over F by any one of the roots $\alpha + k$ of f, and it follows that $|E : F|$ is the degree of each irreducible factor of f. Since $\deg(f) = p$, we see that $|E : F|$ divides p, and because $|E : F| > 1$, we conclude that $|E : F| = p$.   ∎

Proof of Theorem 25.7.   By Theorem 25.8 we have $E = K[\beta]$ for some element β such that $\beta^p - \beta \in F$. Writing $\beta^p - \beta = b$, we see that since $|K[\beta] : K| = p$, we have $\min_K(\beta) = X^p - X - b$. We now set $\alpha = \beta^{p-1}b$, and we show that $f(X) = X^p - X - \alpha$ is irreducible in $E[X]$.

Let γ be a root of f in some extension field, and observe that by Theorem 25.8 we know that if $E[\gamma] > E$, then $|E[\gamma] : E| = p$ and so f is irreducible. We complete the proof by assuming that $\gamma \in E$ and deriving a contradiction.

Since $\gamma \in E$, we can write $\gamma = g(\beta)$ for some polynomial $g \in K[X]$ with $\deg(g) \leq p - 1$. Let $h \in K[X]$ be the polynomial whose coefficients are the pth powers of the corresponding coefficients of g. Then $g(X)^p = h(X^p)$ and hence

$$b\beta^{p-1} = \gamma^p - \gamma = g(\beta)^p - g(\beta) = h(\beta^p) - g(\beta).$$

However, $\beta^p = b + \beta$ and so

$$(*) \qquad\qquad b\beta^{p-1} = h(b + \beta) - g(\beta).$$

Since $1, \beta, \beta^2, \ldots, \beta^{p-1}$ are linearly independent over K, we can equate the coefficients of β^{p-1} on both sides of $(*)$.

Let a be the coefficient of X^{p-1} in $g(X)$ (allowing $a = 0$, of course). Then a^p is the coefficient of X^{p-1} in $h(X)$, and since $\deg(h) \leq p - 1$, we see that this is also the coefficient of β^{p-1} in the expansion of $h(b + \beta)$ in terms of powers of β. The coefficient of β^{p-1} in $h(b + \beta) - g(\beta)$ is therefore $a^p - a$. Since the coefficient of β^{p-1} on the left side of $(*)$ is b, this yields $b = a^p - a$. Thus a is a root of $X^p - X - b$, but this is a contradiction, since $a \in K$, and we saw in the first paragraph that this polynomial is irreducible over K.   ∎

We can assemble the pieces now and prove the first part of the Artin-Schreier theorem.

Proof of Theorem 25.1(a).   If the conclusion fails, Lemma 25.4 gives a subfield $K \subseteq L$ such that $|L : K| = p$, a prime; L is Galois over K; and -1 is a square in K. Also, since L is algebraically closed, Theorem 25.7 guarantees that char$(L) \neq p$, and Lemma 25.5 tells us that K contains a primitive pth root of unity. This contradicts Theorem 25.6.   ∎

(25.9) COROLLARY.   *Let L be any algebraically closed field. Then* Aut(L) *has no finite subgroup of order exceeding* 2.

Proof.   Let $G \subseteq$ Aut(L) be finite. If $F =$ Fix(G), then by Theorem 18.20(a) we know that $|L : F| = |G|$, and so by the first part of the Artin-Schreier theorem we have $|G| \leq 2$.   ∎

25B

We consider now the elements of a field F that can be written as sums of squares. In $\mathbb{R}$, of course, sums of squares are nonnegative, and so -1 is not a sum of squares. We abstract this property of the real numbers and say that a field F is *formally real* if -1 is not a sum of squares in F. A formally real field is *realclosed* if no proper algebraic extension is formally real.

Note that $\mathbb{R}$ and all of its subfields (such as $\mathbb{Q}$, for instance) are formally real. In fact, $\mathbb{R}$ is realclosed since (up to $\mathbb{R}$-isomorphism) the only proper algebraic extension of $\mathbb{R}$ is $\mathbb{C}$, and $\mathbb{C}$ certainly is not formally real.

If F is any field and $L \supseteq F$ is an algebraic closure for F, then every algebraic extension $E \supseteq F$ is F-isomorphic to a subfield of L. (This follows since, by Theorem 17.30, any algebraic closure of E is F-isomorphic to L.) We conclude that F is realclosed iff F is maximal among formally real subfields of L.

(25.10) COROLLARY. *If F is formally real, then there exists a realclosed field that contains F and is algebraic over F.*

Proof. Let L be an algebraic closure for F, and consider the poset of all formally real subfields of L containing F, ordered by containment. The hypotheses of Zorn's lemma are easily verified, and we obtain a maximal formally real subfield of L that contains F. This field is realclosed. ∎

Our principal result concerning realclosed fields is the following theorem.

(25.11) THEOREM. *A field F is realclosed iff it has a proper, finite degree extension that is algebraically closed.*

In other words, the realclosed fields are precisely the fields F that occur in the Artin-Schreier theorem (25.1). (We complete the proof of Theorem 25.1 along the way toward the proof of Theorem 25.11.)

(25.12) LEMMA *Let F be formally real. Then*

    a. *no sum of nonzero squares in F can be zero, and*
    b. *F has characteristic zero.*

Proof. If $\sum a_i^2 = 0$ with $a_1 \neq 0$, then summing over $i > 1$, we have $\sum(a_i/a_1)^2 = -1$. This is impossible because F is formally real, and this proves (a). Statement (b) follows since if F had nonzero characteristic p, then $1^2 + 1^2 + \cdots + 1^2 = 0$, where there are p terms in the sum. ∎

Our next result completes the proof of Theorem 25.1 and proves the "if" part of Theorem 25.11.

(25.13) THEOREM. *Let $F < L$, where L is algebraically closed and $|L : F| < \infty$, and let S be the set of squares in F. Then*

    a. *S is closed under addition,*
    b. *$F = S \,\dot{\cup}\, \{0\} \,\dot{\cup}\, -S$, and*
    c. *F is realclosed.*

Proof. By Theorem 25.1(a) we have $|L : F| = 2$ and $L = F[i]$, where $i^2 = -1$. The polynomial $X^2 + 1$ is thus irreducible in $F[X]$, and it follows that there exists $\tau \in \mathrm{Gal}(L/F)$ with $\tau(i) = -i$. Since $\{1, i\}$ is an F-basis for L, each element of L is uniquely of the form $a + bi$ with $a, b \in F$. For each such element, we have $\tau(a + bi) = a - bi$.

To prove (a), let $r, s \in F$ be nonzero. Note that $r^2 + s^2 \neq 0$, since otherwise $(r/s)^2 = -1$ and hence $i = \pm(r/s) \in F$, a contradiction. It suffices, therefore, to show that $r^2 + s^2$ is a square in F.

Since L is algebraically closed, we can find a square root $a + bi$ for $r + si$ in L (with $a, b \in F$). Then $(a + bi)^2 = r + si$, and application of τ yields $(a - bi)^2 = r - si$. Thus

$$r^2 + s^2 = (r + si)(r - si) = ((a + bi)(a - bi))^2 = (a^2 + b^2)^2 .$$

This is a square, as required.

Now $-1 \notin S$ since $i \notin F$, and thus -1 is not a sum of squares in F and F is formally real. Since L is not formally real and $|L : F| = 2$, we see that F is maximal among formally real subfields of its algebraic closure L, and it follows that F is realclosed and (c) is proved.

Certainly $0 \notin S$ and $0 \notin -S$. Also, if $S \cap -S$ were nonempty, we could choose nonzero $r, s \in F$ with $r^2 = -s^2$, and we have seen that this does not happen. The union in (b) is therefore disjoint, and what remains to be shown is that if $r \in F$ is nonzero, then either r or $-r$ is a square in F.

Since r is a square in L, we can choose $a, b \in F$ such that $(a + bi)^2 = r$. Thus $(a^2 - b^2) + 2abi = r$ and hence $2abi \in F$. Since $i \notin F$, we have $2ab = 0$. Because F is formally real, we have $\mathrm{char}(F) = 0$, and we deduce that either $a = 0$ or $b = 0$. If $b = 0$, then $r = a^2$ is a square in F; if $a = 0$, then $r = (bi)^2 = -b^2$ is the negative of a square. In either case, the proof is complete.   ∎

The following result completes the proof of Theorem 25.11.

(25.14) THEOREM. *Suppose F is realclosed, and let i be a root of the polynomial $X^2 + 1$ in some extension field of F. Then $L = F[i]$ is algebraically closed.*

The strategy for the proof of Theorem 25.14 is to use Theorem 22.17, which we used to prove that $\mathbb{C}$ is algebraically closed. We need to show (in the notation of Theorem 25.14) that L is closed under taking square roots, and that every polynomial of odd degree in $F[X]$ has a root in F. We work first to establish that L is square-root closed.

Since Theorem 25.14 asserts that conclusion (c) of Theorem 25.13 implies the hypotheses of that theorem, it should not be a surprise that (c) also implies conclusions (a) and (b) of Theorem 25.13. That fact is our next lemma. (Note that we could have proved this result earlier and then used it in the proof of Theorem 25.13. The direct proof of Theorem 25.13 seems preferable, however.)

(25.15) LEMMA. *Let S be the set of squares in the realclosed field F. Then S is closed under addition, and $F = S \; \dot\cup \; \{0\} \; \dot\cup \; -S$.*

Proof.   We will show that if $t \in F$ is any nonsquare, then it is possible to write $-1 = u + vt$, where $u, v \in F$ are sums of squares. We temporarily assume this and proceed with the rest of the proof.

To show that S is closed under addition, let $t = r^2 + s^2$, with $r, s \in F$. If t is not a square, then by the assumption of the first paragraph, we can write $-1 = u + vt$, where u and v are sums of squares in S. Since t is a sum of squares, it follows that $u + vt$ is also a sum of squares, and this is a contradiction since -1 is not a sum of squares in the formally real field F.

The squares of F are thus closed under addition, and so if t is any nonsquare in F, we can write $-1 = u + vt$, where u and v are actually squares. Now $v \neq 0$ since -1 is not a square in F, and this yields $-t = (1/v) + (u/v)$. Thus $-t$ is a sum of two squares and hence is a square. Since t is nonzero, we have $t \in -S$, and this shows that $F = S \cup \{0\} \cup -S$, as desired. This union is obviously disjoint since -1 is not a square.

We now prove the assertion of the first paragraph. If t is a nonsquare in F, let α be a root of the polynomial $X^2 - t$ in some extension field of F. Then $F[\alpha] > F$, and since F is realclosed, the field $F[\alpha]$ cannot be formally real, and -1 is a sum of squares in $F[\alpha]$. Since $|F[\alpha] : F| = 2$, every element of $F[\alpha]$ has the form $a + b\alpha$, where $a, b \in F$. Therefore, for some $a_i, b_i \in F$, we have

$$-1 = \sum (a_i + b_i\alpha)^2 = \sum a_i^2 + t \sum b_i^2 + 2\alpha \sum a_i b_i .$$

Now $\alpha \notin F$ and this forces $2 \sum a_i b_i = 0$. Thus $-1 = \sum a_i^2 + t \sum b_i^2 = u + vt$, where $u = \sum a_i^2$ and $v = \sum b_i^2$ are sums of squares, as desired. ∎

(25.16) LEMMA.   *Let F be realclosed and let $L = F[i]$, where $i^2 = -1$ as in Theorem 25.14. Then every element of L is a square in L.*

Proof.   First we observe that every element of F is a square in L. If $t \in F$ is not a square, then Lemma 25.15 guarantees that we can write $t = -s^2$ for some element $s \in F$. Thus $t = (si)^2$ is a square in L, as desired.

An arbitrary element of L has the form $a + bi$, with $a, b \in F$. Given $a, b \in F$, it suffices to produce elements $x, y \in L$ such that $(x + yi)^2 = a + bi$. (There is no need to require that $x, y \in F$, and it suffices to assume that $b \neq 0$.) Since

$$(x + yi)^2 = (x^2 - y^2) + 2xyi ,$$

it is enough to solve the simultaneous equations

$$x^2 - y^2 = a \qquad \text{and} \qquad 2xy = b$$

with $x, y \in L$.

By Lemma 25.15 we have $a^2 + b^2 = r^2$ for some element $r \in F$, and by the first paragraph we can find $x \in L$ with $x^2 = (a + r)/2$. If $x = 0$, then $r = -a$ and $a^2 = r^2 = a^2 + b^2$. In this case, $b = 0$, contrary to assumption.

Thus $x \neq 0$, and we set $y = b/2x$. Then $2xy = b$, and we need only show that $x^2 - y^2 = a$. We have

$$4x^2(x^2 - y^2) = (2x^2)^2 - b^2 = (a+r)^2 - b^2 .$$

Since $(a+r)^2 = a^2 + 2ar + (a^2 + b^2)$, we have

$$x^2 - y^2 = \frac{2a^2 + 2ar}{4x^2} = \frac{2a(a+r)}{2(a+r)} = a ,$$

as required.   ∎

To complete the proof of Theorem 25.14, we must show that every odd-degree polynomial over a realclosed field has a root. We need the following result.

(25.17) THEOREM.   *Let F be formally real and suppose that $F \subseteq E$, where $|E : F|$ is finite and odd. Then E is formally real.*

Proof.   Since there is nothing to prove in the case that $E = F$, we may assume that $n = |E : F| > 1$, and we work by induction on n. Let $\alpha \in E - F$. If $F[\alpha] < E$, then $|E : F[\alpha]|$ and $|F[\alpha] : F|$ are each smaller than n, and two applications of the inductive hypothesis complete the proof. We may therefore suppose that $F[\alpha] = E$.

Now suppose that E is not formally real, and write $-1 = \sum a_i^2$ with $a_i \in E$. For each subscript i, there exists a polynomial $g_i \in F[X]$ such that $a_i = g_i(\alpha)$ and $\deg(g_i) < n$. Write

$$h(X) = 1 + \sum g_i(X)^2 ,$$

and note that $h(\alpha) = 0$. If $f = \min_F(\alpha)$, we conclude that $f \,\big|\, h$.

Let $m = \max\{\deg(g_i)\}$ and consider the coefficient of X^{2m} in $h(X)$. This coefficient must be nonzero since a sum of nonzero squares in F cannot vanish because F is formally real. Thus $\deg(h) = 2m$ and $\deg(h/f) = 2m - n$ is odd. Choose an irreducible factor k of h/f of odd degree, and note that $\deg(k) \leq 2m - n < n$ since $m < n$.

Adjoin a root β of k to F to get a field $K = F[\beta] \supseteq F$. Then $|K : F| = \deg(k)$ is an odd number less than n, and hence by the inductive hypothesis, K is formally real. Since $k \,\big|\, h$, we have $h(\beta) = 0$, and thus $-1 = \sum g_i(\beta)^2$ is a sum of squares in K. This is a contradiction and completes the proof.   ∎

Proof of Theorem 25.14.   By Theorem 22.17 it suffices to verify that L is closed under taking square roots and that every odd degree polynomial $f \in F[X]$ has a root in F. The first requirement is satisfied because of Lemma 25.16.

To check the second condition, it is no loss to assume that f is irreducible. If we adjoin a root α of f to F to get $K = F[\alpha]$, then $|K : F|$ is odd, and so Theorem 25.17 implies that K is formally real. Since F is realclosed, however, we must have $K = F$, and so $\alpha \in F$.   ∎

Proof of Theorem 25.11. If F has a proper finite-degree extension L that is algebraically closed, then by Theorem 25.13(c) we know that F is realclosed. Conversely, if F is realclosed, then Theorem 25.14 guarantees that $L = F[i]$ is algebraically closed, where i is a root $X^2 + 1$. Since -1 is not a square in F, we have $1 < |L : F| < \infty$. ∎

<h1 style="text-align:center">25C</h1>

Which realclosed fields are algebraic over $\mathbb{Q}$? If $\mathbb{A}$ is the field of all algebraic numbers in $\mathbb{C}$, then $\mathbb{A}$ is algebraically closed and $\mathbb{A} \cap \mathbb{R}$ is the fixed field of complex conjugation acting on $\mathbb{A}$. Thus $|\mathbb{A} : \mathbb{A} \cap \mathbb{R}| = 2$, and so $\mathbb{A} \cap \mathbb{R}$ is realclosed. Our goal is to prove that up to isomorphism, $\mathbb{A} \cap \mathbb{R}$ is the unique algebraic (over $\mathbb{Q}$) realclosed field. This is in sharp contrast to the fact (which we do not prove here) that there are many isomorphism types of realclosed subfields $R \subseteq \mathbb{C}$ with $|\mathbb{C} : R| = 2$.

(25.18) THEOREM. *Let F be formally real and algebraic over $\mathbb{Q}$. Then there exists an isomorphism θ of F into $\mathbb{R} \cap \mathbb{A}$. If F is realclosed, then $\theta(F) = \mathbb{R} \cap \mathbb{A}$.*

To prove Theorem 25.18, we introduce the notion of an ordered field.

(25.19) DEFINITION. Let F be a field and assume that P is a subset of F closed under addition and multiplication, and such that $F = P \mathbin{\dot{\cup}} \{0\} \mathbin{\dot{\cup}} -P$, a disjoint union. Then F is *ordered*, and we write "$a < b$" if $b - a \in P$. The set P is the *positive set* of the order.

Sometimes we write "$b > a$" as a synonym for "$a < b$." Also, we write "$a \le b$" or "$b \ge a$" if either $a < b$ or $a = b$. Note that a field may have more than one subset P that defines an order and so, strictly speaking, we should write something like "$a <_P b$" to indicate the particular positive set P used to define the order. We omit this complication in our notation in the hope that no confusion will result.

We collect a number of elementary properties.

(25.20) LEMMA. *Let F be ordered.*

    a. *If $a < b$ and $b < c$, then $a < c$.*
    b. *If $a < b$, then $a + c < b + c$ for all $c \in F$.*
    c. *If $a < b$, then $-a > -b$.*
    d. *If $a < b$ and $c > 0$, then $ac > bc$.*
    e. *Given $a, b \in F$, exactly one of $a < b$, $a = b$, or $a > b$ is true.*
    f. *If $a > b > 0$, then $1/b > 1/a > 0$.*
    g. *$a^2 > 0$ for all $a \in F - \{0\}$.*
    h. *$\mathrm{char}(F) = 0$.*

Proof. Recalling that $x < y$ means that $y - x \in P$, we see that (b) and (c) are obvious, and (a) and (d) follow from the fact that P is closed under addition and multiplication, respectively. The "trichotomy law" (e) is just a restatement of the fact that $F = P \mathbin{\dot{\cup}} \{0\} \mathbin{\dot{\cup}} -P$.

If $a > 0$, then $a^2 > 0$ by (d), and if $a < 0$, then $-a > 0$ by (c), and so $a^2 = (-a)^2 > 0$, and this proves (g). In particular, $1 = 1^2 > 0$. If $a > 0$, we claim that $1/a > 0$. By (e), the only other possibility would be $1/a < 0$, and multiplication by a would yield $1 < 0$, a contradiction. Thus $1/a > 0$, proving part of (f). Now assuming $a > b > 0$, we have $ab > 0$ and hence $1/ab > 0$. Multiplication of the inequality $a > b$ by $1/ab$ yields $1/b > 1/a$, and (f) is proved.

By (g), we have $1 \in P$ and hence $1 + 1 + \cdots + 1 \in P$, and such a sum can thus never equal 0. This proves (h).   ∎

(25.21) LEMMA.   *Suppose F is ordered with positive set P. Then:*

a. *If $K \subseteq F$, then K is ordered with positive set $P \cap K$.*
b. *If $a, b \in K$, then $a < b$ in K iff $a < b$ in F.*
c. *If $K = \mathbb{Q}$, then the order inherited from F is the usual order on $\mathbb{Q}$.*

Proof.   It is clear that $K = (P \cap K) \,\dot{\cup}\, \{0\} \,\dot{\cup}\, -(P \cap K)$, and the first two statements follow.

To prove (c), we must show that $P \cap \mathbb{Q}$ is exactly the set of positive (in the usual sense) rational numbers. First observe that every positive integer lies in $P \cap \mathbb{Q}$ since if $n > 0$, we can write $n = 1^2 + \cdots + 1^2 \in P$ using Lemma 25.20(g). If r is a positive rational number, we can write $r = m/n$, where m and n are positive integers and hence lie in P. By Lemma 25.20(f) we have $1/n \in P$, and thus $r = m/n \in P$.

Conversely, if $r \in P \cap \mathbb{Q}$, then $r \neq 0$. Also, r cannot be negative, or else $-r$ is positive and thus lies in P, and hence $r \in -P$, which is not the case. It follows that r is positive.   ∎

The relevance of ordered fields is demonstrated in the following corollary.

(25.22) COROLLARY.   *A field is ordered iff it is formally real.*

Proof.   If F is ordered, then sums of squares in F are nonnegative by Lemma 25.20(g). In particular, $1 > 0$ and hence $-1 < 0$ by Lemma 25.20(c). This shows that -1 is not a sum of squares in F.

Conversely, assume F is formally real. By Corollary 25.10 we know that F is contained in some realclosed field, and by Lemma 25.21 it is enough to show that realclosed fields are ordered. If F is realclosed, its set S of nonzero squares serves as the positive set for an order. This set is clearly closed under multiplication, and the other conditions we need are given by Lemma 25.15.   ∎

Given an ordered field F, we know that F has characteristic zero and so we can assume that $\mathbb{Q} \subseteq F$. By Corollary 25.22 the order that $\mathbb{Q}$ inherits from F is the usual order on $\mathbb{Q}$, and we consider the question of where, in terms of the order on $\mathbb{Q}$, the irrational elements of F lie.

An ordered field F is said to be *archimedian* if for every element $a \in F$, there exists an integer $n \in \mathbb{Z}$ with $n \geq a$. (For example, $\mathbb{R}$ and all of its subfields are archimedian.) Thus a field is nonarchimedian precisely when it contains an element a larger than every integer, a so-called *infinite* element. If $a \in F$ is such an infinite element, then if we put $b = 1/a$, Lemma 25.20(f) tells us that $0 < b < 1/n$ for every positive integer n. Such an element b is said to be *infinitesimal*. Note that if $b \in F$ is infinitesimal, then $a = 1/b$ is infinite and F is nonarchimedian. In other words, we have the following lemma.

(25.23) LEMMA.   *An ordered field F is archimedian iff it contains no infinitesimal elements.*   ∎

(25.24) LEMMA.   *Let F be an ordered field and let I be its set of infinitesimal elements. Write $V = I \cup \{0\} \cup -I$. Then V is a $\mathbb{Q}$-vector subspace of F, and V is closed under multiplication.*

Proof.   We show first that V is closed under addition. Suppose $a, b \in V$. By negating a and b, if necessary, we can assume $a + b > 0$, and thus at least one of a or b (say a) is positive. Thus $a \in I$, and we want $a + b \in I$.

If $b < 0$, then $0 < a + b < a < 1/n$ for every positive integer n, and thus $a + b \in I$, as desired. If $b > 0$, then for every positive integer n, we have $a < 1/2n$ and $b < 1/2n$. Then

$$ 0 < a + b < a + \frac{1}{2n} < \frac{1}{2n} + \frac{1}{2n} = \frac{1}{n} $$

and $a + b \in I$.

To complete the proof, we must show that if either $r \in V$ or $r \in \mathbb{Q}$, then $rV \subseteq V$. Since $-V = V$, it is no loss to assume that $r > 0$, and it suffices to prove that $rI \subseteq I$. In fact, we will show that $rI \subseteq I$, when r is any positive noninfinite element of F.

Suppose $0 < r \leq m$, where $m \in \mathbb{Z}$, and let $a \in I$. Given a positive integer n, we have $a < 1/nm$ and so $ma < 1/n$. Also, $0 < ra \leq ma$, and it follows that $ra \in I$.   ∎

(25.25) THEOREM.   *Let F be an ordered field that is algebraic over $\mathbb{Q}$. Then F is archimedian.*

Proof.   If F is nonarchimedian, let $\beta \in F$ be infinitesimal and let $f = \min_{\mathbb{Q}}(\beta)$. Writing $f(X) = X^n + a_{n-1}X^{n-1} + \cdots + a_0$, we have $a_0 \neq 0$ since f is irreducible and $\beta \neq 0$. If V is as in Lemma 25.24, we have $\beta \in V$, and the lemma guarantees that

$$ \beta^n + a_{n-1}\beta^{n-1} + \cdots + a_1\beta \in V. $$

Since $f(\beta) = 0$, however, this quantity is equal to $-a_0$, and so $-a_0 \in V \cap \mathbb{Q} = \{0\}$. This is a contradiction.   ∎

(25.26) THEOREM.   *Let F be any archimedian ordered field. Then F is isomorphic to a subfield of $\mathbb{R}$.*

Once we prove Theorem 25.26, Theorem 25.18 will be immediate. We need the following lemma.

(25.27) LEMMA.   *Let F be an archimedian ordered field. If $a, b \in F$ with $a < b$, then there exists $r \in \mathbb{Q}$ such that $a < r < b$.*

Proof.   Since $b - a$ is positive but not infinitesimal, there exists a positive integer n such that $1/n < b - a$. Also, we can choose an integer m such that $a < m$ and $-a < m$. (Of course, $m > 0$.)

Now define $r_i = -m + i/n$ for integers $i \geq 0$. We have $r_0 = -m < a$ and $r_{2mn} = m > a$. We can thus find a subscript k such that $r_k \leq a$ and $r_{k+1} > a$. We have

$$a < r_{k+1} = r_k + \frac{1}{n} \leq a + \frac{1}{n} < a + (b - a) = b \,,$$

as required.   ∎

Proof of Theorem 25.26.   For each element $a \in F$, write $L(a) = \{r \in \mathbb{Q} \mid r < a\}$. Note that $L(a) \neq \varnothing$ since if $n \in \mathbb{Z}$ with $n > -a$ we have $-n < a$ and so $-n \in L(a)$. Also, $L(a)$ is bounded above in $\mathbb{Q}$ since if $a \leq m$ for $m \in \mathbb{Z}$, then every element $r \in L(a)$ satisfies $r < m$. There is therefore a uniquely defined real number $\sup(L(a))$, and we define our map $\theta : F \to \mathbb{R}$ by setting $\theta(a) = \sup(L(a))$.

Given $a, b \in F$, we will show that

$$L(a + b) = \{r + s \mid r \in L(a) \text{ and } s \in L(b)\} \,.$$

This clearly yields

$$\theta(a + b) = \sup(L(a + b)) = \sup(L(a)) + \sup(L(b)) = \theta(a) + \theta(b) \,,$$

and θ is an additive homomorphism.

If $r \in L(a)$ and $s \in L(b)$, then $r + s \leq a + s \leq a + b$, and $r + s \in L(a+b)$. Conversely, if $u \in L(a + b)$, we need to show that we can write $u = r + s$ with $r \in L(a)$ and $s \in L(b)$. Now $a + b > u$, and so we can choose $n \in \mathbb{Z}$ with $n > 0$ and $1/n < (a + b) - u$. By Lemma 25.27 choose a rational number r with $a - 1/n < r < a$. Then $r \in L(a)$, and we let $s = u - r$. It suffices to show that $s \in L(b)$, and since s is rational, we need only check that $s < b$. We have $r > a - 1/n$, and hence

$$s = u - r < u - a + \frac{1}{n} < u - a + (a + b - u) = b \,,$$

as desired.

The proof that θ is also a multiplicative homomorphism is similar but more complicated. First, we observe that it suffices to show that $\theta(ab) = \theta(a)\theta(b)$

for $a, b \in F$ with $a, b > 0$. This is so because we already know that θ is an additive homomorphism, and thus $\theta(0) = 0$ and $\theta(-x) = -\theta(x)$ for all $x \in F$.

Next we observe that for $0 < a \in F$, the set $L(a)$ contains some positive members. This is true because a is not infinitesimal, and so $1/n < a$ for some positive integer n. Writing $L^+(a) = \{r \in L(a) \mid r > 0\}$, we see that $\theta(a) = \sup(L(a)) = \sup(L^+(a))$ for $a > 0$. To prove that $\theta(ab) = \theta(a)\theta(b)$ for $a, b > 0$, therefore, it suffices to show that

$$L^+(ab) = \{rs \mid r \in L^+(a) \text{ and } s \in L^+(b)\}.$$

If $r \in L^+(a)$ and $s \in L^+(b)$, we have $rs > 0$ since $r > 0$ and $s > 0$. Also, since $r < a$ and $s > 0$, we get $rs < as$. Since $s < b$ and $a > 0$, we have $as < ab$ and hence $rs \in L^+(ab)$, as desired.

Conversely now, suppose $u \in L^+(ab)$. Since $ab - u$ is positive but not infinitesimal, we can find a positive integer n such that $ab - u > 1/n$. Now choose a positive integer m with $m > nab - 1$. Using n and m, we show how to choose $r \in L^+(a)$ and $s \in L^+(b)$ such that $rs = u$, and this will complete the proof.

We have $0 < m/(m+1) < 1$, and thus $0 < ma/(m+1) < a$ since $a > 0$. By Lemma 25.27 choose a rational number r with $ma/(m+1) < r < a$. Then $r \in L^+(a)$ and $s = u/r$ is positive and rational. It suffices to show that $s < b$.

Since $r > ma/(m+1)$, we have $1/r < (m+1)/ma = 1/a + 1/ma$. Recalling that $u < ab - 1/n$, we have

$$s = \frac{u}{r} < \left(ab - \frac{1}{n}\right)\left(\frac{1}{a} + \frac{1}{ma}\right),$$

and this gives

$$s < b + \frac{b}{m} - \frac{1}{nma} - \frac{1}{na} = b + \frac{nab - 1 - m}{nma} < b,$$

where the last inequality holds since $nab - 1 < m$ and $nma > 0$. ∎

Proof of Theorem 25.18.   We are given a formally real field F algebraic over $\mathbb{Q}$. By Corollary 25.22 we know that F is ordered, and by Theorem 25.25 it is archimedian. By Theorem 25.26, therefore, there exists an isomorphism θ of F into $\mathbb{R}$. Of course, $\theta(F)$ is algebraic over $\mathbb{Q}$, and so $\theta(F) \subseteq \mathbb{R} \cap \mathbb{A}$.

If F is realclosed, then so is $\theta(F)$, and thus $\theta(F)[i]$ is an algebraically closed subfield of $\mathbb{A}$ by Theorem 25.14. Since $\mathbb{A}$ is algebraic over $\mathbb{Q}$, it is also algebraic over $\theta(F)[i]$, and thus $\theta(F)[i] = \mathbb{A}$ and $|\mathbb{A} : \theta(F)| = 2$. We have $\theta(F) \subseteq \mathbb{R} \cap \mathbb{A} < \mathbb{A}$, and it follows that $\theta(F) = \mathbb{R} \cap \mathbb{A}$. ∎

We close this chapter with a somewhat surprising corollary.

(25.28) COROLLARY.   *Let $\mathbb{Q} \subseteq E$ be a Galois extension, and let $\mathcal{R}$ be the collection of intermediate fields that happen to be formally real.   Then $\mathrm{Gal}(E/\mathbb{Q})$ transitively permutes the maximal elements of $\mathcal{R}$.*

Proof.   We may assume that $E \subseteq \mathbb{C}$. If $F \in \mathcal{R}$, then by Theorem 25.18 there exists an isomorphism θ from F into $\mathbb{R}$. If $\alpha \in F$, then $\theta(\alpha)$ is a root of $\min_{\mathbb{Q}}(\alpha)$, and this polynomial splits over E since E is normal over $\mathbb{Q}$. It follows that $\theta(F) \subseteq E \cap \mathbb{R}$.

By the normality of E over $\mathbb{Q}$, the isomorphism θ between the intermediate fields F and $\theta(F)$ extends to an element of $\mathrm{Gal}(E/\mathbb{Q})$. It follows that each orbit of the action of the Galois group on $\mathcal{R}$ contains some subfield of $E \cap \mathbb{R}$. Since $E \cap \mathbb{R}$ is itself an element of $\mathcal{R}$, however, it follows that the orbit of $E \cap \mathbb{R}$ under the action of $\mathrm{Gal}(E/\mathbb{Q})$ is the unique orbit of maximal elements of $\mathcal{R}$.   ∎

Problems

25.1   Suppose that F is a field with the property that the irreducible polynomials in $F[X]$ have bounded degree. Show that an algebraic closure of F has finite degree over F.

   HINT: The finite-degree separable extensions of F have bounded degree over F.

25.2   Suppose F contains a primitive pth root of unity, where p is prime, and if $p = 2$, assume that F contains a primitive 4th root of unity. Suppose that some element of F is not a pth power in F. Show that for every positive integer e, there exists an extension $E \supseteq F$ with $|E : F| = p^e$.

25.3   In a field F, let us write $S_k(F)$ to denote the set of all elements of F that can be written as a sum of k squares in F. The identity

$$(a^2 + b^2)(c^2 + d^2) = (ac - bd)^2 + (ad + bc)^2$$

shows that $S_2(F)$ is closed under multiplication.
   a. If $u \in S_2(F)$ and $u \neq 0$, show that $1/u \in S_2(F)$.
   b. Show that if n is as small as possible such that $-1 \in S_n(F)$, then n is even.

25.4   Show that the hypotheses of Zorn's lemma are satisfied in the proof of Corollary 25.10.

25.5   Let F be formally real and suppose that $F \subseteq E$ is a Galois extension. If K is an intermediate field maximal with the property of being formally real, show that $|E : K| \leq 2$.

25.6   Let E be the splitting field for $X^4 - 2$ over $\mathbb{Q}$ in $\mathbb{C}$, and let $F = \mathbb{Q}[\sqrt{2}] \subseteq E$. Show that $\mathrm{Gal}(E/F)$ does not transitively permute the maximally formally real intermediate fields.

25.7   Let F be formally real and let S be the set of nonzero squares in F. We know that if F is realclosed, then S is closed under addition. Show that the converse of this statement is false.

HINT: Construct fields $F_i \subseteq \mathbb{R}$ as follows. Let $F_0 = \mathbb{Q}$, and for $i > 0$, put $F_i = F_{i-1}[U_i]$, where $U_i = \{\alpha \in \mathbb{R} \mid \alpha^2 \in F_{i-1}\}$. Let $F = \bigcup F_i$. To prove that F is not realclosed, show that $\sqrt[3]{2} \notin F$.

25.8 If F is formally real, show that $F(X)$ is formally real, where X is an indeterminate.

25.9 Let $F = \mathbb{Q}(X)$, where X is an indeterminate, and let $L \supseteq F$ be an algebraic closure. For $i = 1, 2, 3, \ldots$, let $\alpha_i \in L$ be a square root of $X - i$. Write $F_n = F[\alpha_1, \alpha_2, \ldots, \alpha_n]$ and note that F_n is a uniquely determined subfield of L even though the α_i are only determined up to sign.

  a. Show that for each $n > 0$, the field F_n is isomorphic to a subfield of $\mathbb{R}$ and hence is formally real.
  b. Let $K = \bigcup F_n$ and show that K is formally real.
  c. Show that K has no archimedian order.

HINT: To prove part (a) choose a transcendental real number $r > n$ and let $\beta_i = \sqrt{r - i} \in \mathbb{R}$ for $1 \le i \le n$. Show that $F_n \cong \mathbb{Q}(r, \beta_1, \beta_2, \ldots, \beta_n)$.

NOTE: We know that an ordered field algebraic over $\mathbb{Q}$ must be archimedian. Here we have constructed a nonarchimedian ordered field K that just fails to be algebraic since its transcendence degree over $\mathbb{Q}$ is 1.

25.10 Let K and L be ordered fields, and suppose θ is an isomorphism of K into L. Let "$<$" and "$\prec$" be orders on K and L, respectively, and suppose that $\alpha, \beta \in K$ with $\alpha < \beta$.

  a. If K and L are realclosed, show that $\theta(\alpha) \prec \theta(\beta)$.
  b. Show that the conclusion of part (a) does not hold in general.

NOTE: By part (a) a realclosed field has a unique order.

25.11 Let $E \supseteq \mathbb{R}$ and suppose that K is a realclosed subfield of E.

  a. If θ is an isomorphism from K into $\mathbb{R}$, show that $\theta(\alpha) = \alpha$ if $\alpha \in \mathbb{R} \cap K$.
  b. If $K > \mathbb{R}$, show that K has no archimedian order.
  c. Let E be an algebraic closure of $\mathbb{R}(X)$. Show that E has an automorphism of order 2 whose fixed field K is nonarchimedian.

HINT: If $\theta(\alpha) = \beta \ne \alpha$ in part (a), choose a rational number between α and β.

NOTE: The field E in part (c) is actually isomorphic to $\mathbb{C}$. It follows that $\mathbb{C}$ has an automorphism of order 2 with a nonarchimedian fixed field.

25.12 Let F be an ordered field and let V be as in Lemma 25.24. Let

$$R = \{a \in F \mid -n < a < n \text{ for some } n \in \mathbb{Z}\}.$$

Show that R is a ring and that V is an ideal of R.

25.13 Let $\mathbb{Q} \subseteq E$ with E Galois over $\mathbb{Q}$, and let $\mathcal{R}$ be the set of formally real intermediate fields. Assume that $E \notin \mathcal{R}$. Show that $\mathrm{Gal}(E/\mathbb{Q})$ has a conjugacy class Σ of involutions such that for $F \subseteq E$, we have $F \in \mathcal{R}$ iff $F \subseteq \mathrm{Fix}(\sigma)$ for some $\sigma \in \Sigma$.

CHAPTER TWENTY-SIX

Ideal Theory

26A

The goal of this chapter is to study the properties of ideals in general commutative rings. The results we obtain are useful in the study of the deeper properties of noetherian rings, which we undertake in the next chapter. As always in this book, the word "ring" means "ring with 1," and we continue to assume without further mention that our rings are commutative.

Recall from Chapter 16 that a proper ideal P of a ring R is a *prime* ideal of R if, whenever $ab \in P$ for elements $a, b \in R$, we have that either $a \in P$ or $b \in P$. Useful alternative characterizations of primality are given by the following lemma.

(26.1) LEMMA. *Let P be a proper ideal of R. Then the following are equivalent:*

i.   *P is prime.*
ii. *If $AB \subseteq P$ for ideals A and B, then either $A \subseteq P$ or $B \subseteq P$.*
iii. *There do not exist ideals $A > P$ and $B > P$ such that $AB \subseteq P$.*

Recall that given two additive subgroups $A, B \subseteq R$, their product AB is defined to be the additive subgroup generated by all elements of the form ab with $a \in A$ and $b \in B$. It follows that AB is the set of all finite sums of such products ab. If either A or B is an ideal, then AB is automatically an ideal. (Note that this latter observation depends on the commutativity of R.)

Proof of Lemma 26.1. Suppose that P is prime and that $AB \subseteq P$ for ideals A and B of R. If $B \nsubseteq P$, choose $b \in B - P$, and note that $ab \in AB \subseteq P$ for all $a \in A$. Since P is prime and $b \notin P$, this forces $a \in P$ and hence $A \subseteq P$, and this shows that (i) implies (ii).

Since (ii) trivially implies (iii), we complete the proof by assuming (iii) and showing that P is prime. Let $ab \in P$ with $a, b \in R$, and suppose that neither

418

a nor *b* lies in *P*. We derive a contradiction by constructing ideals $A > P$ and $B > P$ and showing that $AB \subseteq P$. Let

$$A = (a) + P \quad \text{and} \quad B = (b) + P,$$

where, as usual, (a) and (b) are the principal ideals generated by *a* and *b*. Then $A, B > P$ since $a, b \notin P$.

Typical elements of *A* and *B* have the forms $ax + u$ and $by + v$, respectively, where $x, y \in R$ and $u, v \in P$. We have

$$(ax + u)(by + v) = (ab)(xy) + u(by) + (ax)v + uv,$$

and this lies in *P* since ab, u, and v all are in *P*. It follows that *P* contains all sums of products of this type, and hence $AB \subseteq P$. ∎

It should be noted that the proof of Lemma 26.1 depends on the commutativity of *R*. In fact, in the general case, condition (ii) can hold without *P* being prime (according to our definition). For this reason, condition (ii) of Lemma 26.1 is taken as the definition of primeness in noncommutative ring theory. By Lemma 26.1 we could, of course, have used this as the definition in the commutative case, too.

We saw in Chapter 16 that maximal ideals are always prime. More generally, we have the following lemma.

(26.2) LEMMA. *Let M be a multiplicative system in the ring R, and let $P \subseteq R$ be an ideal maximal with the property that $P \cap M = \varnothing$. Then P is prime.*

Recall that a *multiplicative system* in *R* is a subset that contains 1, does not contain 0, and is closed under multiplication. If *M* is any multiplicative system, then $M \cap (0) = \varnothing$ and so there certainly exists an ideal disjoint from *M*. One can apply Zorn's lemma (11.17) to conclude that there exist ideals that are maximal with this property. We say that such ideals are *maximally disjoint* from *M*. In fact, Zorn's lemma implies that if $M \cap A = \varnothing$ for some ideal *A*, then there exists an ideal $P \supseteq A$ such that *P* is maximally disjoint from *M* (and hence *P* is prime by Lemma 26.2).

Proof of Lemma 26.2. Since $1 \in M$, we have $1 \notin P$ and $P < R$. By Lemma 26.1, therefore, it suffices to show that it is impossible to have ideals $A > P$ and $B > P$ with $AB \subseteq P$.

If $A > P$, then $A \cap M \neq \varnothing$ by the maximality of *P*, and we can choose $a \in A \cap M$. Similarly, if $B > P$, we can find $b \in B \cap M$, and thus $ab \in AB \cap M$ because *M* is closed under multiplication. If $AB \subseteq P$, this gives a contradiction since $P \cap M = \varnothing$. ∎

Note that a special case of Lemma 26.2 is when $M = \{1\}$. In this case, the ideals maximally disjoint from *M* are simply the maximal proper ideals; they are, in other words, the maximal ideals of *R*. We thus recover the fact we already knew: maximal ideals are prime.

By the Zorn's lemma argument to which we referred previously, every nonzero ring has maximal ideals and thus has prime ideals. (Note that if R is a field, then (0) is its only maximal ideal.) We write $\mathrm{Spec}(R)$ to denote the collection of all prime ideals of R. This (nonempty) set is called the prime *spectrum* of R.

(26.3) THEOREM. *The set* $\mathrm{nil}(R)$ *of all nilpotent elements of R is an ideal. It is the intersection of all prime ideals of R.*

Proof. Let $a \in R$ be nilpotent and suppose that $P \in \mathrm{Spec}(R)$. Since $a^n = 0$ for some integer $n > 0$, we have $a^n \in P$, and because P is prime, it follows that $a \in P$. This shows that $\mathrm{nil}(R) \subseteq \bigcap \mathrm{Spec}(R)$.

Conversely, suppose $a \in \bigcap \mathrm{Spec}(R)$. If a is not nilpotent, then zero does not lie in the set

$$M = \{a^i \mid 0 \le i \in \mathbb{Z}\},$$

and hence M is a multiplicative system in R. By Zorn's lemma we can choose an ideal P of R maximal such that $P \cap M = \varnothing$. By Lemma 26.2 we know that P is prime and hence $a \in P$. Since also $a \in M$, this is a contradiction. ∎

For noncommutative rings, the set of nilpotent elements is not in general an ideal (and is not even closed under addition). For example, in the 2×2 matrix ring over an arbitrary ring, the matrices

$$\begin{bmatrix} 0 & 1 \\ 0 & 0 \end{bmatrix} \quad \text{and} \quad \begin{bmatrix} 0 & 0 \\ 1 & 0 \end{bmatrix}$$

are nilpotent, but their sum is not.

We wish to generalize $\mathrm{nil}(R)$. If $A \subseteq R$ is any ideal, we define the *radical* of A to be the set

$$\sqrt{A} = \{r \in R \mid r^n \in A \text{ for some positive integer } n\}.$$

Note that $\mathrm{nil}(R) = \sqrt{(0)}$, and in general, $\sqrt{A}$ is the inverse image in R of $\mathrm{nil}(R/A)$. In particular, $\sqrt{A}$ is an ideal of R containing A.

If $A \subseteq P \in \mathrm{Spec}(R)$, then the ideal P/A of R/A is a prime ideal of R/A. Conversely, every ideal of R/A has the form I/A for some unique ideal I of R containing A, and it is routine to check that if I/A is prime, then I is prime. This yields the following lemma.

(26.4) LEMMA. *We have*

$$\mathrm{Spec}(R/A) = \{P/A \mid P \in \mathrm{Spec}(R) \text{ and } P \supseteq A\}$$

for all ideals A of R. ∎

(26.5) COROLLARY. *Let $A \subseteq R$ be an ideal. Then $\sqrt{A}$ is the intersection of the prime ideals of R that contain A.*

Proof.   Let $\mathcal{P} = \{P \in \mathrm{Spec}(R) \mid P \supseteq A\}$, the set of prime ideals of R that contain A. Then $\mathrm{Spec}(R/A) = \{P/A \mid P \in \mathcal{P}\}$ by Lemma 26.4. By Theorem 26.3 we therefore have

$$\sqrt{A}/A = \mathrm{nil}(R/A) = \bigcap \mathrm{Spec}(R/A) = \left(\bigcap \mathcal{P} \right)/A \, ,$$

and it follows that $\sqrt{A} = \bigcap \mathcal{P}$, as required.   ∎

Note that if we follow the usual convention that the intersection of the empty collection of subsets of a set is the whole set, then Corollary 26.5 remains true even if $A = R$. We can use Corollary 26.5 to describe those ideals of R that occur as radicals. These are called *radical ideals*.

(26.6) COROLLARY.   *Let A be an ideal of R. Then the following are equivalent*:

i. $A = \sqrt{B}$ *for some ideal B of R.*
ii.   *A is an intersection of some prime ideals of R.*
iii. $A = \sqrt{A}$.

Proof.   Corollary 26.5 shows that (i) implies (ii). Assuming (ii) now, suppose that $A = \bigcap \mathcal{X}$, where $\mathcal{X}$ is some subset of $\mathrm{Spec}(R)$. If $r \in \sqrt{A}$, then $r^n \in A$ for some integer $n > 0$, and hence $r^n \in P$ for all $P \in \mathcal{X}$. Since P is prime, it follows that $r \in P$ for all $P \in \mathcal{X}$, and thus $r \in A$. This shows that $\sqrt{A} \subseteq A$, and (iii) follows. That (iii) implies (i) is trivial.   ∎

Often, we do not need to intersect *all* prime ideals that contain an ideal A of R in order to get $\sqrt{A}$. For instance, if $A \subseteq P < Q$, where $P, Q \in \mathrm{Spec}(R)$, then Q can safely be deleted from any set $\mathcal{P}$ of primes such that $\bigcap \mathcal{P} = \sqrt{A}$. (This is clear if $P \in \mathcal{P}$. We leave the general argument to the problems at the end of the chapter.) It is tempting to conjecture at this point that $\sqrt{A}$ is the intersection of all primes minimal with respect to the property of containing A. These are said to be the primes *minimal over* A, and they are also called the *isolated* primes of A. (We use both names, as convenient.) Note that if A is prime, it is minimal over itself.

It should be obvious that for any ideal A, the isolated primes of A are exactly the isolated primes of $\sqrt{A}$. What is not obvious, however, is that A has any isolated primes at all. If A is a proper ideal, then it is surely contained in a prime ideal, but it is conceivable that there is no minimal prime over it. If that happened, it certainly would not be true that $\sqrt{A}$ was the intersection of the isolated primes of A.

We could avoid this difficulty if we were willing to assume the minimal condition (or, equivalently, the descending chain condition) on the partially ordered set of all prime ideals of R. (As we shall see in Chapter 27, if we assume that R is noetherian, so that its ideals satisfy the *ascending* chain condition, then the descending chain condition on prime ideals is a surprising consequence.)

Actually, we do not need to make any assumption at all. The proof of this fact involves an unusual application of Zorn's lemma.

(26.7) THEOREM.   *Let $A \subseteq R$ be an ideal. Then every prime ideal of R containing A contains some prime minimal over A.*

Proof.   Fix some prime ideal Q containing A and let $\mathcal{P} = \{P \in \mathrm{Spec}(R) \mid A \subseteq P \subseteq Q\}$. Clearly $\mathcal{P}$ is nonempty, and we use Zorn's lemma to show that $\mathcal{P}$ contains a minimal element. We do this by viewing $\mathcal{P}$ as a poset with respect to "reverse inclusion." In other words, if $U, V \in \mathcal{P}$, we write (for this proof only) $U \leq V$ when $U \supseteq V$. If we can find a maximal element of $\mathcal{P}$ with respect to "$\leq$," that will be our desired minimal element.

Since $\mathcal{P}$ is nonempty, it suffices to show that every linearly ordered subset $\mathcal{L} \subseteq \mathcal{P}$ has an upper bound in $\mathcal{P}$. In other words, we are assuming that $\mathcal{L}$ is a collection of prime ideals containing A and contained in Q and such that if $U, V \in \mathcal{L}$, then either $U \subseteq V$ or $V \subseteq U$. We seek some prime ideal D with $A \subseteq D \subseteq Q$ such that $D \geq P$ (which means that $D \subseteq P$) for all $P \in \mathcal{L}$. (Recall that we do not require $D \in \mathcal{L}$.)

Simply take $D = \bigcap \mathcal{L}$. It is trivial to check that D is an ideal, and of course $A \subseteq D \subseteq Q$. To complete the proof, we must show that D is prime. Suppose, then, that $xy \in D$ with $x, y \in R$. If $x \notin D$, then $x \notin U$ for some ideal $U \in \mathcal{L}$, and similarly, if $y \notin D$, then $y \notin V$ for some $V \in \mathcal{L}$. Thus neither x nor y lies in $U \cap V$, although $xy \in D \subseteq U \cap V$. This is a contradiction since $U \cap V$ is either U or V and so is prime. Thus at least one of x or y must lie in D, which is therefore prime.   ∎

(26.8) COROLLARY.   *If A is an ideal of R, then $\sqrt{A}$ is the intersection of the isolated primes of A.*   ∎

When can we compute $\sqrt{A}$ by intersecting just finitely many prime ideals?

(26.9) COROLLARY.   *If A is an ideal of R, then the following are equivalent*:

i.   *A has just finitely many isolated primes.*
ii.   *$\sqrt{A}$ is an intersection of some finite collection of prime ideals.*

Furthermore, every isolated prime of A occurs in every finite collection of primes with intersection equal to $\sqrt{A}$.

Proof.   That (i) implies (ii) is immediate from Corollary 26.8, and it suffices to prove the final sentence of the corollary in order to show that (ii) implies (i). Suppose, therefore, that

$$\sqrt{A} = P_1 \cap P_2 \cap \cdots \cap P_n \,,$$

where the P_i are prime, and let P be any isolated prime of A. Then

$$P \supseteq \sqrt{A} = P_1 \cap \cdots \cap P_n \supseteq P_1 P_2 \cdots P_n \,.$$

By repeated use of Lemma 26.1(ii), it follows that P contains at least one of the ideals P_i. Since $P_i \supseteq A$ and P is minimal over A, we deduce that $P = P_i$, as required.   ∎

As we shall see in the problems at the end of this chapter, it is not always true that there are just finitely many primes minimal over an ideal. This does hold in noetherian rings, however. In fact, it suffices to assume merely that the set of radical ideals of R satisfies the maximal condition.

(26.10) THEOREM. *Suppose that the set of radical ideals of R satisfies the maximal condition. Then every ideal of R has just finitely many isolated primes.*

Proof. Let $\mathcal{X}$ be the set of radicals of all ideals of R that have infinitely many isolated primes. If we assume that the conclusion is false, then the set $\mathcal{X}$ is nonempty, and by the maximal condition, we can choose a maximal element $U \in \mathcal{X}$. Since $U = \sqrt{I}$ for some ideal I that has infinitely many isolated primes, we see that U itself has infinitely many isolated primes. In particular, U is not prime.

By Lemma 26.1 it follows that there exist ideals $A > U$ and $B > U$ with $AB \subseteq U$. Since $\sqrt{A} \supseteq A > U$, the maximality of U guarantees that $\sqrt{A}$ does not lie in $\mathcal{X}$, and hence A has just finitely many isolated primes, and similarly for B. Since U has infinitely many isolated primes, we can choose a prime ideal P isolated for U but not isolated for either A or B. We have $AB \subseteq U \subseteq P$, and thus one of A or B (say A) is contained in P.

Now $U < A \subseteq P$ and P is a minimal prime over U. It follows that P is a minimal prime over A, and this is a contradiction. ∎

It occurs surprisingly often that an ideal maximal with respect to some property can be proved to be prime. (Observe that Lemma 26.2 provides an example of this phenomenon.) If we examine the proof of Theorem 26.10, we see that our contradiction arose from the existence of an ideal U that was maximal with respect to a certain property and yet was not prime. Another example of the heuristic principle "maximal implies prime" is given by the following lemma.

(26.11) LEMMA. *Suppose U is maximal among ideals of R that are not principal. Then U is prime.*

Proof. If not, then by Lemma 26.1 there exist ideals $A, B > U$ with $AB \subseteq U$. Since $A > U$, the maximality of U guarantees that A is principal, and we write $A = (a)$. Now let $I = \{r \in R \mid ar \in U\}$ and observe that I is an ideal. Since $aB \subseteq AB \subseteq U$, we see that $B \subseteq I$, and so $I > U$ and I is principal, say $I = (x)$.

Since $U \subseteq A = (a)$, an arbitrary element $u \in U$ has the form $u = ar$ for some element $r \in R$. Thus $r \in I$ by the definition of I, and it follows that $r = sx$ for some element $s \in R$. We now have $u = ar = axs \in (ax)$ and $U \subseteq (ax)$.

The reverse containment follows since $x \in I$, and so $ax \in U$ by the definition of I. We have now shown that $U = (ax)$, and this contradicts the assumption that U is not principal. ∎

(26.12) THEOREM. *If every prime ideal of R is principal, then every ideal of R is principal.*

Proof. Suppose that the set $\mathcal{X}$ of nonprincipal ideals of R is nonempty. If we can show that $\mathcal{X}$ satisfies the hypotheses of Zorn's lemma (when partially ordered by inclusion), then $\mathcal{X}$ contains a maximal element. That ideal will be prime and hence principal by Lemma 26.11, and that contradiction will prove the result.

To verify the hypothesis of Zorn's lemma, suppose $\mathcal{L} \subseteq \mathcal{X}$ is linearly ordered. We need to produce a member $U \in \mathcal{X}$ such that $U \supseteq I$ for all $I \in \mathcal{L}$. In fact, we can take $U = \bigcup \mathcal{L}$. That U is an ideal is an immediate consequence of the fact that $\mathcal{L}$ is linearly ordered, and so to show that $U \in \mathcal{X}$, it suffices to prove that U is nonprincipal.

Suppose, therefore, that $U = (u)$. Since $u \in U = \bigcup \mathcal{L}$, it follows that $u \in I$ for some ideal $I \in \mathcal{L}$. But then

$$U \supseteq I \supseteq (u) = U ,$$

and it follows that $I = (u)$ is principal. This is a contradiction since $I \in \mathcal{L} \subseteq \mathcal{X}$. ∎

One way to interpret Theorem 26.12 is that $\mathrm{Spec}(R)$ is rich enough so that information about the structure of prime ideals is sufficient to imply results about all ideals. A similar result, which we defer to the problems at the end of Chapter 27, is that if every prime ideal of R is finitely generated (as an R-module), then every ideal is finitely generated.

We mention one more result about prime ideals in general.

(26.13) THEOREM. *Let $P_1, P_2, \ldots, P_n$ be a finite collection of prime ideals of R, and suppose that A is an ideal of R such that $A \subseteq \bigcup P_i$. Then A is contained in one of the ideals P_i.*

Proof. We may assume that A is not contained in the union of any proper subset of $\{P_1, P_2, \ldots, P_n\}$, and our task is to show that $n = 1$. Since each ideal P_i is assumed to be essential, we can choose $a_i \in A$ such that $a_i \in P_i$ and $a_i \notin P_j$ for $j \neq i$.

Now assume $n > 1$ and put

$$b_i = \prod_{j \neq i} a_j$$

for $1 \leq i \leq n$. Certainly, $b_i \in A$ and $b_i \in P_j$ for $i \neq j$. However, $b_i \notin P_i$ or else, since P_i is prime, some factor a_j of b_i (with $j \neq i$) would have to belong to P_i, and this would contradict the choice of a_j.

Let $b = \sum b_i$. Then $b \in A$ and so b lies in some P_i by hypothesis. Since $b_j \in P_i$ for $j \neq i$, we have

$$b_i = b - \sum_{j \neq i} b_j \in P_i ,$$

and this is a contradiction. ∎

26B

An ideal $Q < R$ is a *primary* ideal if, whenever $ab \in Q$, we have either $a \in Q$ or $b \in \sqrt{Q}$. Clearly, prime ideals are primary, and it is easy to find examples of primary ideals that are not prime. If p is a prime number in $\mathbb{Z}$, for example, then of course (p) is a prime ideal in $\mathbb{Z}$, and we claim that (p^n) is primary for integers $n \geq 1$. To see this, note that if $ab \in (p^n)$ but $a \notin (p^n)$, then $p^n | (ab)$ but $p^n \nmid a$, and this forces $p | b$ (by unique factorization in $\mathbb{Z}$). Thus $p^n | b^n$ and so $b^n \in (p^n)$ and $b \in \sqrt{(p^n)}$.

Notice that if Q is primary but not prime, then there exist $a, b \in R$ with $ab \in Q$ but $a \notin Q$ and $b \notin Q$. It follows that both $a \in \sqrt{Q}$ and $b \in \sqrt{Q}$.

(26.14) LEMMA.   *Let Q be an ideal of R.*

    a. *If Q is primary, then $\sqrt{Q}$ is prime.*
    b. *If $\sqrt{Q}$ is maximal, then Q is primary.*

Proof.   For (a) we suppose that Q is primary. Then $1 \notin Q$ and so $1 \notin \sqrt{Q}$ and $\sqrt{Q}$ is proper. We assume that $ab \in \sqrt{Q}$ and $a \notin \sqrt{Q}$, and we prove that $b \in \sqrt{Q}$. Since $ab \in \sqrt{Q}$, we have $(ab)^n \in Q$ for some positive integer n. Then $a^n b^n \in Q$, but $a^n \notin Q$ since $a \notin \sqrt{Q}$. Since Q is primary, this forces $b^n \in \sqrt{Q}$, and hence $b^{nm} \in Q$ for some positive integer m. Therefore $b \in \sqrt{Q}$, and $\sqrt{Q}$ is prime, as required.

Now suppose that $\sqrt{Q}$ is maximal. Since $\sqrt{Q} < R$, certainly $Q < R$, and so to show that Q is primary, it suffices to assume that $ab \in Q$ and $b \notin \sqrt{Q}$ and to prove that $a \in Q$.

Since $\sqrt{Q} + (b) > \sqrt{Q}$, the maximality of $\sqrt{Q}$ forces $\sqrt{Q} + (b) = R$, and we can write $1 = u + br$, where $u \in \sqrt{Q}$ and $r \in R$. Now $u^n \in Q$ for some integer n, and we have

$$1 = 1^n = (u + br)^n = u^n + bx$$

for some element $x \in R$. Thus $a = a1 = au^n + (ab)x \in Q$, and this completes the proof of (b).   ∎

Since maximal ideals are prime, we see that statements (a) and (b) of Lemma 26.14 are almost converses of each other. This suggests the question of whether or not either of (a) or (b) can be strengthened to be the full converse of the other. We ask, in other words, if either of the following is true.

a′. If Q is primary, then $\sqrt{Q}$ is maximal.
b′. If $\sqrt{Q}$ is prime, then Q is primary.

Neither of these statements is correct. For a counterexample to (a′), let Q be any nonmaximal prime. Then Q is certainly primary, and $\sqrt{Q} = Q$ is not maximal. It is a little harder to see that (b′) is false; we leave this to the problems at the end of the chapter.

(26.15) COROLLARY. *Let M be a maximal ideal of R. Then M^n is primary for integers $n \geq 1$.*

Proof. Observe that $R > \sqrt{M^n} \supseteq M$, and so $\sqrt{M^n} = M$ and Lemma 26.14(b) applies. ∎

In the ring $\mathbb{Z}$, we saw that (p^n) is primary for prime numbers p. This is a special case of Corollary 26.15 since (p) is maximal in this case. It is not true in general that powers of prime ideals are always primary.

By Lemma 26.14(a) the radical of a primary ideal Q is some prime ideal P, and we often say that Q *belongs* to P in this case. The following easy result is needed in Chapter 27.

(26.16) LEMMA. *The intersection of any finite collection of primary ideals all belonging to the same prime P is also a primary ideal belonging to P.*

Proof. It suffices to prove this for intersections of two primary ideals. Suppose, then, that Q_1 and Q_2 are primary and $\sqrt{Q_1} = P = \sqrt{Q_2}$. We show first that $\sqrt{Q_1 \cap Q_2} = P$. If $x \in P$, then $x^n \in Q_1$ and $x^m \in Q_2$ for positive integers n and m. If $k = \max(n, m)$, then $x^k \in Q_1 \cap Q_2$, and this shows that $P \subseteq \sqrt{Q_1 \cap Q_2}$. The reverse inclusion is immediate.

Now suppose that $ab \in Q_1 \cap Q_2$ but that $a \notin Q_1 \cap Q_2$. Then $a \notin Q_1$ (say) and hence

$$b \in \sqrt{Q_1} = P = \sqrt{Q_1 \cap Q_2}.$$

This shows that $Q_1 \cap Q_2$ is primary. ∎

26C

We limit our attention in this section to the case where R is a domain. We recall from Chapter 16 (specifically, Theorems 16.23 and 16.24) that if M is a multiplicative system in a domain R, then R can be embedded in a unique (up to isomorphism) unitary overring S (sometimes called the *localization* of R at M) in such a way that the elements of M are invertible in S and that every element $s \in S$ has the form $s = rm^{-1}$ for some elements $r \in R$ and $m \in M$.

An important case of this situation is where M has the form $R - P$, the complement of the prime ideal P in R. (Note that the complement of the prime ideal is always a multiplicative system.) In this case, we write $S = R_P$, and we refer to R_P (in conflict with the usage in the previous paragraph) as the *localization* of R at P. (We shall henceforth use the word "localization" in the latter sense only; in the general case, we refer to S as the result of "adjoining inverses" for M.)

We use the process of localization to study the ideal theory of noetherian rings in Chapter 27, and so we need some general (and nearly trivial) results relating the ideals of a domain R to the the ideals of an overring S obtained by adjoining inverses for a multiplicative system $M \subseteq R$.

In this situation, if J is an ideal of S, then clearly $J \cap R$ is an ideal of R. We refer to this ideal as the *contraction* of J, and we write $J \cap R = J^C$. If I is an ideal of R, then IS is the ideal of S generated by I. It is called the *expansion* of I and is denoted I^E. A slightly more convenient description of I^E is given by the following lemma.

(26.17) LEMMA. *In the above situation,*

$$I^E = \{um^{-1} \mid u \in I \text{ and } m \in M\}.$$

Proof.   Every element of the form um^{-1} with $u \in I$ and $m \in M$ certainly lies in $IS = I^E$. A typical element of I^E, on the other hand, is a finite sum of elements of the form us, with $u \in I$ and $s \in S$. We can write $s = rm^{-1}$ with $r \in R$ and $m \in M$, and thus $us = (ur)m^{-1}$. Since $ur \in I$, we see that it suffices to prove that the set $\{um^{-1} \mid u \in I, m \in M\}$ is closed under addition.

If $u_1, u_2 \in I$ and $m_1, m_2 \in M$, we have

$$u_1 m_1^{-1} + u_2 m_2^{-1} = (u_1 m_2 + u_2 m_1)(m_1 m_2)^{-1},$$

and since $u_1 m_2 + u_2 m_1 \in I$ and $m_1 m_2 \in M$, the result follows.   ∎

The following collection of simple results allows us to relate the ideal theory of a domain to the ideal theory of its localizations. As we shall see in Chapter 27, one can sometimes prove theorems about the ideals of a ring by working in its localizations.

(26.18) LEMMA.   *Let R, M, and S be as above.*

   a. *The maps C and E preserve containments.*
   b. *$I \subseteq I^{EC}$ and $J = J^{CE}$ for ideals I of R and J of S.*
   c. *If $J_1 < J_2$ are ideals of S, then $J_1^C < J_2^C$.*
   d. *If I is primary in R and $I \cap M = \varnothing$, then $I = I^{EC}$.*
   e. *If I is prime or primary in R and $I \cap M = \varnothing$, then I^E is, respectively, prime or primary in S.*
   f. *If J is prime or primary in S, then J^C is, respectively, prime or primary in R.*
   g. *If $M = R - I$, where I is prime in R, then I^E is the unique maximal ideal of the localization $S = R_I$.*

We remark that it is property (g) that (at least partially) explains the use of the word "localization." A commutative ring is said to be a *local* ring if it has a unique maximal ideal.

Proof of Lemma 26.18.   Statement (a) and the first part of (b) are immediate from the definitions, as is the containment $J^{CE} \subseteq J$ for ideals J of S. To complete the proof of (b), let $x \in J$ and write $x = rm^{-1}$ for $r \in R$ and $m \in M$. Then $r = xm \in J \cap R = J^C$, and so $x \in rS \subseteq J^C S = J^{CE}$, as required.

If $J_1 < J_2$ are ideals of S, then $J_1^C \subseteq J_2^C$ by (a). If equality were to occur here, we would have the contradiction $J_1 = J_1^{CE} = J_2^{CE} = J_2$ by (b). This proves (c).

In the situation of (d), we certainly have $I \subseteq I^{EC}$ by (b). If $x \in I^{EC}$, then $x \in R$, and by Lemma 26.17 we can write $x = um^{-1}$ with $u \in I$ and $m \in M$. Then $xm = u \in I$ but $m \notin \sqrt{I}$ since the powers of m lie in M, which is disjoint from I. Since I is primary, this forces $x \in I$, as required.

To prove (e), we observe first that $I^E < S$, since by (d) we have $I^{EC} = I < R = S^C$. Now let $xy \in I^E$ with $x, y \in S$, and write $x = am^{-1}$ and $y = bn^{-1}$ with $a, b \in R$ and $m, n \in M$. Then $ab = xymn \in I^E \cap R = I^{EC} = I$, from (d). If $x \notin I^E$, then certainly $a \notin I$, and thus since I is primary, we have $b^k \in I$ for some integer k, and $k = 1$ if I is prime. Then $y^k = b^k(n^{-1})^k \in I^E$, and so I^E is primary and is prime if I is prime.

For (f), we note that $J^C < R$ since $1 \notin J$. Suppose that $ab \in J^C$ with $a, b \in R$ and that $a \notin J^C$. Then $a \notin J$, but $ab \in J^C \subseteq J$, and thus $b^k \in J$ and $k = 1$ if J is prime. Since $b^k \in R$, we have $b^k \in J^C$, as required.

Finally, in the situation of (g), we have $I^E < S$ (by (e), for instance). We need to show that $J \subseteq I^E$ for every ideal $J < S$. Now J contains no invertible elements of S and so $J \cap M = \varnothing$. It follows that $J^C \subseteq R - M = I$, and thus $J = J^{CE} \subseteq I^E$ by (a) and (b). ∎

As examples of how Lemma 26.18 can be used, we give two consequences.

(26.19) COROLLARY. *Let $R, M,$ and S be as above. If R is noetherian or artinian, then so is S.*

Proof.   We consider the noetherian case; the proof when R is artinian is very similar. Suppose $J_1 \subseteq J_2 \subseteq J_3 \subseteq \cdots$ is an ascending chain. If there are infinitely many strict containments, then by Lemma 26.18(c) the chain

$$J_1^C \subseteq J_2^C \subseteq \cdots$$

would have infinitely many strict containments. This cannot happen since R is noetherian. ∎

Our second corollary remedies the difficulty that powers of nonmaximal prime ideals are not necessarily primary. Suppose P is a prime in a domain R. Working in the localization $S = R_P$, we define the *symbolic powers* $P^{(n)}$ of P for integers $n \geq 1$ by

$$P^{(n)} = ((P^E)^n)^C .$$

Note that these are ideals of R.

(26.20) COROLLARY. *Let R be a domain and P a prime ideal of R. Then the symbolic powers $P^{(n)}$ are primary ideals belonging to P, and we have $P^{(n)} \subseteq P^{(m)}$ if $n \geq m$.*

Proof.   By Lemma 26.18(g) we see that P^E is a maximal ideal in $S = R_P$, and thus $(P^E)^n$ is primary in S by Corollary 26.15. It follows by Lemma 26.18(f) that $P^{(n)} = ((P^E)^n)^C$ is primary in R. Also, if $n \geq m$, then clearly $(P^E)^n \subseteq (P^E)^m$ and so $P^{(n)} \subseteq P^{(m)}$ by Lemma 26.18(a).

We need to show that $\sqrt{P^{(n)}} = P$. Since $(P^E)^n \subseteq P^E$, we have

$$P^{(n)} = ((P^E)^n)^C \subseteq P^{EC} = P$$

by Corollary 26.19(a) and (d), and it follows since P is prime that $\sqrt{P^{(n)}} \subseteq P$. To obtain the reverse containment, let $x \in P$. Then $x \in P^E$ and so

$$x^n \in (P^E)^n \cap R = ((P^E)^n)^C = P^{(n)}.$$

Thus $x \in \sqrt{P^{(n)}}$, as required. ∎

Problems

26.1 Recall that an *idempotent* is a nonzero element of a ring equal to its own square. Show that if R contains an idempotent different from 1, then every prime ideal of R contains an idempotent.

26.2 Given a positive integer n, find the unique positive integer m such that $\sqrt{(n)} = (m)$.

26.3 Show that

$$\sqrt{AB} = \sqrt{A \cap B} = \sqrt{A} \cap \sqrt{B}$$

for ideals A and B of R.

26.4 Say that a subset $X \subseteq \mathrm{Spec}(R)$ is *closed* if there exists an ideal $A \subseteq R$ such that

$$X = \{P \in \mathrm{Spec}(R) \mid P \supseteq A\}.$$

a. Show that if $X, Y \supseteq \mathrm{Spec}(R)$ are closed, then so is $X \cup Y$.
b. Show that the intersection of an arbitrary collection of closed subsets of $\mathrm{Spec}(R)$ is closed.
c. If $\mathcal{X}$ is a collection of closed subsets of $\mathrm{Spec}(R)$ with the property that every intersection of finitely many members of $\mathcal{X}$ is nonempty, show that $\bigcap \mathcal{X}$ is nonempty.

NOTE: Problem 26.4 shows how to view $\mathrm{Spec}(R)$ as a compact topological space. Observe that the only points that are closed (as singleton sets) are the maximal ideals.

26.5 Let R be the ring of all continuous real-valued functions on the closed interval $[0, 1]$. (Multiplication and addition are pointwise.)

a. If $a \in [0, 1]$, let $M_a = \{f \in R \mid f(a) = 0\}$. Show that M_a is a maximal ideal of R.
b. Show that R contains some nonmaximal prime ideals.
c. Show that the M_a are all of the maximal ideals of R.

HINT: For part (b), consider functions that vanish in a neighborhood of a point. For part (c), if M is a maximal ideal of R unequal to any M_a, consider the collection of sets $X_f = \{x \in [0, 1] \mid f(x) \neq 0\}$ for $f \in M$. Use the Heine-Borel theorem.

26.6 Let R be the ring of all real valued functions on some arbitrary set X. (Again, addition and multiplication are pointwise.) For $x \in X$, let $e_x \in R$ be the function with value 1 at x and 0 elsewhere.

    a. Show that each prime of R fails to contain at most one of the elements e_x.

    b. Show for $f \in R$ that $\sqrt{(f)}$ is an intersection of finitely many primes iff f vanishes at just finitely many points.

26.7 In R, suppose that there are just finitely many minimal primes. (Minimal primes, of course, are just primes minimal over (0).) Show that the union of the set of minimal primes is

$$\{r \in R \mid rs \text{ is nilpotent for some nonnilpotent element } s \in R\}.$$

26.8 Show that R has just finitely many maximal ideals iff there is some finite collection of ideals whose union is the set of all nonunits of R.

26.9 Show that every nonzero primary ideal in $\mathbb{Z}$ has the form (p^n), where p is prime and $n \geq 1$ is an integer.

26.10 Let $R = F[X, Y]$, where F is a field, and write $\overline{R} = R/(XY)$.

    a. Show that (X) is prime in R and that its image $I = (X)/(XY)$ is prime in $\overline{R}$.

    b. Show that in any ring, if P is prime, then $\sqrt{P^n} = P$ for positive integers n. In $\overline{R}$, in particular, $\sqrt{I^2} = I$.

    c. Show that I^2 is not primary.

    HINT: For part (c), note that $\overline{X}\,\overline{Y} \in I^2$, where $\overline{X} = X + (XY)$ and $\overline{Y} = Y + (XY)$. Show that $\overline{X} \notin I^2$ and that $\overline{Y} \notin I$.

26.11 Let P be a prime ideal of the domain R and let $S = R_P$, the localization at P. Fix an integer $n \geq 1$.

    a. Show that $P^n \subseteq P^{(n)}$, the symbolic power.

    b. Show that $(P^n)^E = (P^E)^n$, where E is the expansion map from ideals of R to ideals of S.

    c. If P is maximal in R, show that $P^{(n)} = P^n$.

26.12 Let U and V be radical ideals of a domain R. Assume that for every choice of prime $P \in \text{Spec}(R)$, we have $U^E = V^E$ in R_P. Show that $U = V$.

26.13 Let M be a multiplicative system in domain R. If $I \subseteq R$ is an ideal with just finitely many isolated primes, show that I^E has just finitely many isolated primes in the ring S obtained from R by adjoining inverses for M.

26.14 Let R be a subring of $\mathbb{Q}$. Show that R is the result of adjoining inverses to $\mathbb{Z}$ for the multiplicative system $M = \{b \in \mathbb{Z} \mid b^{-1} \in R\}$.

26.15 A ring R is said to be *von Neumann regular* if for every element $a \in R$, there exists $b \in R$ with $a = a^2 b$.

a. Show that domain R is a field iff it is von Neumann regular.

b. In general, show that R is von Neumann regular iff every ideal is a radical ideal.

NOTE: In the noncommutative case, the defining equation for von Neumann regularity is $a = aba$.

26.16 Ideals A and B of R are *comaximal* if $A + B = R$. (Note that distinct maximal ideals are necessarily comaximal.) Suppose that ideals $A_1, A_2, \ldots, A_n$ of R are pairwise comaximal. Show that

$$\prod_{i=1}^{n} A_i = \bigcap_{i=1}^{n} A_i.$$

HINT: First do the case where $n = 2$. For the general case, show that $\prod_{i<n} A_i$ is comaximal with A_n.

26.17 Let $A_1, A_2, \ldots, A_n$ be pairwise comaximal ideals of R. Show that

$$\frac{R}{A_1 A_2 \cdots A_n} \cong \frac{R}{A_1} \oplus \frac{R}{A_2} \oplus \cdots \oplus \frac{R}{A_n}.$$

HINT: To prove that the natural map from R to the right side of the above expression is surjective, find an element of R that maps to 1 in R/A_i and to 0 in R/A_j for $j \neq i$. Use the hint to Problem 26.16.

NOTE: This problem is a very general form of the Chinese remainder theorem.

26.18 If A and B are ideals of R, we write $(A : B)$ to denote the set $\{r \in R \mid rB \subseteq A\}$. Show that $(A : B)$ is an ideal. If A is primary and $B \not\subseteq A$, show that $(A : B)$ is primary and that

$$\sqrt{(A : B)} = \sqrt{A}.$$

26.19 Let M be a right R-module and let $\mathcal{X}$ be the collection of ideals I of R such that $mI = 0$ for some nonzero element $m \in M$. Show that every maximal element of $\mathcal{X}$ is a prime ideal.

26.20 Suppose $I \subseteq R$ is maximal among nonprime proper ideals. Let $S = R/I$.

a. Show that every nonzero proper ideal of S is maximal.

b. Show that S has at most two nonzero proper ideals.

HINT: For part (a), show that S/T is a field for nonzero proper ideals T of S. Use Problem 26.15.

26.21 Suppose $\sqrt{A}$ and $\sqrt{B}$ are comaximal for ideals A and B of R. Show that A and B are comaximal.

26.22 Let A be an ideal of R and suppose that Q is a prime ideal containing A that is not minimal over A. If $\mathcal{P}$ is a collection of prime ideals such that $\bigcap \mathcal{P} = \sqrt{A}$, let $\mathcal{P}' = \mathcal{P} - \{Q\}$. Show that $\bigcap \mathcal{P}' = \sqrt{A}$.

CHAPTER TWENTY-SEVEN

Noetherian Rings

27A

Recall from Chapter 11 that an abelian group with operators is *noetherian* if its poset of "operator subgroups" satisfies the ascending chain condition (ACC) or, equivalently, the maximal condition. By Theorem 11.4 an abelian group with operators is noetherian iff every operator subgroup is finitely generated.

If R is a commutative ring, viewed as a module over itself, then the submodules (operator subgroups) are precisely the ideals. To say that an ideal I of R is finitely generated means that there is some finite set $\{x_1, x_2, \ldots, x_n\}$ of elements of I such that I is the smallest ideal that contains these elements. Since the set $Rx_1 + Rx_2 + \cdots + Rx_n$ of all R-linear combinations of the x_i is certainly an ideal contained in I, we must have

$$I = Rx_1 + Rx_2 + \cdots + Rx_n .$$

The following is therefore a consequence of the results of Chapter 11.

(27.1) COROLLARY. *Let R be a commutative ring. Then the following are equivalent*:

   i.   *The ideals of R satisfy the ACC.*
   ii. *The ideals of R satisfy the maximal condition.*
   iii. *The ideals of R all have the form $Rx_1 + Rx_2 + \cdots + Rx_n$ for finite sets of elements $x_1, x_2, \ldots, x_n \in R$.* ∎

In this chapter we study commutative noetherian rings, which are the rings that satisfy the conditions of Corollary 27.1. (We will not henceforth explicitly mention the assumption that our rings are all commutative.) Note that the ideals for which $n = 1$ in Corollary 27.1(iii) are exactly the principal ideals, and so the noetherian condition on a ring can be viewed as a generalization of the condition that the ring is a PIR.

Our first significant result is the Hilbert basis theorem, which provides a convenient source of examples of noetherian rings.

(27.2) THEOREM (Hilbert basis).   *Let R be noetherian. Then the polynomial ring $R[X]$ is also noetherian.*

The word "basis" in the name of this theorem refers to the elements x_i of condition (iii) in Corollary 27.1. Of course, these elements need not actually be a basis in the modern sense of that word. It may happen that coefficients $r_i \in R$ exist such that $r_1 x_1 + r_2 x_2 + \cdots + r_n x_n = 0$ without having all $r_i = 0$. In fact, most ideals of most noetherian rings do not have a true basis; they are not, in other words, *free* R-modules.

Before proving Theorem 27.2, we mention a few consequences.

(27.3) COROLLARY.   *Let R be noetherian and let $X_1, X_2, \ldots, X_n$ be distinct indeterminates. Then the polynomial ring $R[X_1, X_2, \ldots, X_n]$ is noetherian. In particular, if F is any field, then $F[X_1, X_2, \ldots, X_n]$ is a noetherian ring.*

Proof.   Work by induction on n, using the obvious isomorphism

$$R[X_1, X_2, \ldots, X_i] \cong R[X_1, X_2, \ldots, X_{i-1}][X_i].$$

For the last statement, simply note that F is noetherian since its ideal set is finite.   ∎

(27.4) COROLLARY.   *Let $R \subseteq S$, where R is noetherian and S is a unitary overring. Suppose S is finitely generated as a ring over R. Then S is noetherian.*

Proof.   Suppose R and $s_1, s_2, \ldots, s_n \in S$ generate S as a ring. Since $R[s_1, s_2, \ldots, s_n]$ is a subring of S, we have $S = R[s_1, s_2, \ldots, s_n]$, and this is the image of the polynomial ring $R[X_1, X_2, \ldots, X_n]$ under the evaluation homomorphism $X_i \mapsto s_i$. Since homomorphic images of noetherian rings are noetherian, the result follows.   ∎

As an example of how Corollary 27.4 might be used, we observe, for example, that the subring $\mathbb{Z}[\sqrt{3}, \sqrt{5}, i] \subseteq \mathbb{C}$ is noetherian. We also mention that although $F[X]$ is a principal ideal ring if F is a field, neither $F[X, Y]$ nor $\mathbb{Z}[X]$ enjoys this property. By Corollary 27.3 each of these rings is noetherian, however.

Proof of Theorem 27.2.   Let $I \subseteq R[X]$ be an ideal. Our goal is to find a finite "basis" (generating set) for I over $R[X]$.

For integers $n \geq 0$, we let I_n be the subset of I consisting of all polynomials $f \in I$ such that for all integers $k > n$, the coefficient of X^k in f is 0. (In other words, I_n consists of 0 together with all polynomials $f \in I$ such that $\deg(f) \leq n$.) Since I is an additive group that is closed under multiplication by elements of R, it is easy to see that I_n is an R-submodule of I for each integer $n \geq 0$.

~~We write $A_n \subseteq R$ to denote the set of all elements~~ that occur as the coefficient of X^n in some polynomial $f \in I_n$. Since I_n is an R-module, it follows that A_n is an ideal of R.

Since $IX \subseteq I$, we see that $I_n X \subseteq I_{n+1}$ for $n \geq 0$, and this yields

$$A_0 \subseteq A_1 \subseteq A_2 \subseteq \cdots .$$

Because R is noetherian, there is some subscript m such that

$$A_m = A_{m+1} = A_{m+2} = \cdots ,$$

and we can choose finite generating sets S_i for the ideals A_i for $0 \leq i \leq m$. For each element $s \in S_i$, choose a polynomial $f \in I_i$ such that s is the coefficient of X^i in f, and let $\mathcal{X}$ be the (finite) set of all the polynomials chosen this way.

Of course, $\mathcal{X} \subseteq I$, and so if J is the ideal of $R[X]$ generated by $\mathcal{X}$, we have $J \subseteq I$. We complete the proof by showing that $J = I$. If this is false, choose $g \in I - J$ with the smallest possible degree, and let $t = \deg(g)$ so that $g \in I_t$.

Let a be the leading coefficient of g and note that $a \in A_t$. Writing $u = \min(m, t)$, we have $A_u = A_t$, and so we can express a in terms of the generating set S_u for A_u. We have

$$a = r_1 a_1 + r_2 a_2 + \cdots + r_k a_k ,$$

where $r_i \in R$ and $a_i \in S_u$ for $1 \leq i \leq k$. Now let $f_i \in \mathcal{X}$ be the polynomial we selected in I_u such that the coefficient of X^u in f_i is a_i. Let

$$f = r_1 f_1 + r_2 f_2 + \cdots + r_k f_k ,$$

and note that $f \in J$ has degree u and that the leading coefficient of f is a. Now $X^{t-u} f \in J$, and this polynomial has the same degree t and the same leading coefficient a as does g. It follows that either $g - X^{t-u} f = 0$ or $\deg(g - X^{t-u} f) < t$. In either case, we deduce that $g - X^{t-u} f \in J$ and thus $g \in J$. This is a contradiction and completes the proof. $\blacksquare$

27B

Now that we know how to construct some noetherian rings, we will start to investigate their properties and especially their ideal theory. We record first a result that was essentially proved in Chapter 26.

(27.5) COROLLARY. *Assume that R is noetherian. Then each ideal I of R has only finitely many isolated primes.*

Proof. By Theorem 26.10 we need only verify the maximal condition on radical ideals. In fact, we have the maximal condition on all ideals. $\blacksquare$

It follows that the radical ideals of the noetherian ring R are exactly the ideals that can be constructed as finite intersections of prime ideals. This suggests the question of which ideals of R can be obtained as intersections of finitely many primary ideals. The answer is all.

(27.6) THEOREM (Lasker-Noether). *Let I be an ideal of the noetherian ring R. Then*

$$I = \bigcap_{i=1}^{t} Q_i$$

for some collection of primary ideals Q_i of R with distinct radicals.

It is instructive to examine the case where $R = \mathbb{Z}$. If $I \subseteq \mathbb{Z}$ is a nonzero proper ideal, we can write $I = (n)$ for some integer $n > 1$. We factor $n = p_1^{e_1} p_2^{e_2} \cdots p_t^{e_t}$ with distinct primes p_i and positive exponents e_i. Writing $Q_i = (p_i^{e_i})$, we see that Q_i is primary and $I = \bigcap Q_i$. (Note that the Lasker-Noether theorem also holds in the two excluded cases: $I = (0)$ and $I = R$. For the latter, simply take the empty collection of primary ideals.)

The reader may feel that it would be more natural and more interesting to write $(n) = \prod(p_i^{e_i})$ since this ideal factorization would then reflect the factorization of individual elements of $\mathbb{Z}$. The Lasker-Noether theorem may seem somewhat deficient because it gives a decomposition of an arbitrary ideal as an intersection rather than as a product of primary ideals.

Actually, there is a close connection between these two types of decomposition. If we have $I = \bigcap Q_i$ and the radicals $P_i = \sqrt{Q_i}$ are distinct and they happen to be maximal ideals, then by Problem 26.21 the Q_i are pairwise comaximal. By Problem 26.16 we have $\bigcap Q_i = \prod Q_i$ in this case. In other words, in a ring (such as $\mathbb{Z}$) where all nonzero primes are maximal, the Lasker-Noether theorem does indeed give a factorization of an arbitrary nonzero ideal as a product of primary ideals.

As we shall see in Chapter 29, if R is a noetherian domain in which every nonzero prime ideal is maximal and in which a certain additional condition holds, then every nonzero ideal factors uniquely as a product of powers of prime ideals. For these rings (called "Dedekind domains"), the ideals behave almost exactly as they do in $\mathbb{Z}$.

Observe that the Lasker-Noether theorem implies that if a proper ideal I (of a noetherian ring) cannot be written as an intersection of two strictly larger ideals, then I must itself be primary. Our first (and the major) step in the proof of Theorem 27.6 is to give an independent proof of this "corollary."

(27.7) LEMMA. *Let R be noetherian and suppose that $I < R$ is an ideal. If I is not of the form $I = A \cap B$ for ideals $A, B > I$, then I is primary.*

Proof. Let $xy \in I$ with $x \notin I$. Our task is to show that $y \in \sqrt{I}$, and so we assume that $y^n \notin I$ for all $n \geq 1$. Let

$$A_n = \{ r \in R \mid ry^n \in I \}$$

and note that A_n is an ideal of R. We have

$$A_1 \subseteq A_2 \subseteq A_3 \subseteq \cdots ,$$

and obviously $I \subseteq A_1$ and $x \in A_1$. By the ACC, we can choose n so that $A_n = A_{2n}$, and we write $A = A_n$ and $B = I + (y^n)$. Now $x \in A_1 \subseteq A$, and so $A > I$ since $x \notin I$. Also, $y^n \notin I$, and so $B > I$, as required.

We obtain our contradiction by showing that $A \cap B = I$. Since $A \cap B \supseteq I$, it is sufficient to take $z \in A \cap B$ and show that $z \in I$.

Since $z \in B$, we can write $z = u + ry^n$ for some elements $u \in I$ and $r \in R$. Also, because $z \in A = A_n$, we have $zy^n \in I$. But

$$zy^n = uy^n + ry^{2n} ,$$

and since $uy^n \in I$, this gives $ry^{2n} \in I$ and hence $r \in A_{2n} = A_n$. Thus $ry^n \in I$, and it follows that $z = u + ry^n \in I$, as required.   ∎

Proof of Theorem 27.6.   We prove that every ideal I of R can be written as a finite (possibly empty) intersection of primary ideals. If this is false, then the collection of counterexample ideals is nonempty, and hence by the maximal condition, there exists an ideal I, maximal with the property of not being an intersection of primaries. Then $I < R$ (since R is the empty intersection) and I is surely not primary. By Lemma 27.7 we can write $I = A \cap B$ with $A > I$ and $B > I$.

Neither A nor B is a counterexample since each is larger than I, and thus each of A and B is a finite intersection of primary ideals. It follows that $I = A \cap B$ is a finite intersection of primaries, and this is a contradiction.

We must now show that primary ideals with intersection equal to some given ideal I can always be chosen to have distinct radicals. This is easy: write I as a finite intersection of arbitrary primary ideals and group together those primaries with equal radicals. By Lemma 26.16 the intersections of the ideals in each batch are primary, and they have distinct radicals.   ∎

Before we address the uniqueness questions associated with the Lasker-Noether theorem, we present an application.

(27.8) THEOREM (Krull intersection).   *Let A be an ideal of the noetherian ring R and write*

$$D = \bigcap_{n=1}^{\infty} A^n .$$

Then:

a. $DA = D$.
b. $D(1 - a) = (0)$ *for some element* $a \in A$.
c. $D = (0)$ *if R is a domain and A is proper.*

We need a preliminary result.

(27.9) LEMMA.   *Let R be noetherian and suppose that $I \subseteq R$ is an ideal. Then for some positive integer n, we have $(\sqrt{I})^n \subseteq I$.*

Proof.   Write $\sqrt{I} = Rx_1 + Rx_2 + \cdots + Rx_t$ and note that since $x_i \in \sqrt{I}$, we can write $x_i^{e_i} \in I$ for suitable integers $e_i \geq 1$. Now write $n = \sum e_i$ and note that any product of n elements of $\sqrt{I}$ can be written as an R-linear combination of elements of the form

$$x = x_1^{u_1} x_2^{u_2} \cdots x_t^{u_t},$$

where $\sum u_i = n$. It follows that $u_i \geq e_i$ for some subscript i, and thus x is a multiple of $x_i^{e_i}$ and hence lies in I. Thus $(\sqrt{I})^n \subseteq I$.   ∎

We cannot resist digressing to give an alternative, less computational (and less direct) proof of Lemma 27.9. It is based on the following lemma.

(27.10) LEMMA.   *Let $I \subseteq R$ be an ideal, where R is noetherian. Then I contains some finite product of prime ideals, each containing I.*

Proof.   If the result is false, let I be maximal among counterexample ideals. Note that $I \neq R$ since (by convention) R is the empty product of primes. Also I is not prime, and thus there exist ideals $A > I$ and $B > I$ with $AB \subseteq I$.

Since neither A nor B can be a counterexample, we have

$$A \supseteq \prod P_i \quad \text{and} \quad B \supseteq \prod Q_j$$

for primes $P_i \supseteq A$ and $Q_j \supseteq B$. Then $I \supseteq AB \supseteq \prod P_i \prod Q_j$, and $P_i \supseteq I$ and $Q_j \supseteq I$ for all subscripts i, j. Thus I is not a counterexample ideal after all.   ∎

Alternative Proof of Lemma 27.9.   Choose primes $P_i \supseteq I$ with

$$I \supseteq \prod_{i=1}^{n} P_i.$$

Since $P_i \supseteq \sqrt{I}$, it follows that

$$I \supseteq (\sqrt{I})^n,$$

as required.   ∎

Proof of Theorem 27.8(a).   Since D is an ideal, we have $DA \subseteq D$, and so we must show that $D \subseteq DA$. By the Lasker-Noether theorem, the ideal DA is an intersection of primary ideals, and thus it suffices to show that $D \subseteq Q$ whenever Q is a primary ideal containing DA.

If $D \nsubseteq Q$, then there exists $d \in D$ with $d \notin Q$, and since $da \in Q$ for all $a \in A$, it follows (since Q is primary) that $A \subseteq \sqrt{Q}$. By Lemma 27.9, therefore, $A^n \subseteq Q$ for some integer n. This is a contradiction since $D \subseteq A^n$ and yet $D \nsubseteq Q$.   ∎

In order to complete the proof of the Krull intersection theorem (27.8), we need a very general lemma that can be stated conveniently in the language of modules.

(27.11) LEMMA. *Suppose that M is a finitely generated R-module and that $MA = M$ for some ideal $A \subseteq R$. Then $M(1-a) = 0$ for some element $a \in A$.*

Proof. Let $m_1, m_2, \ldots, m_n$ generate M. Thus each element $m \in M$ has the form

$$m = m_1 r_1 + m_2 r_2 + \cdots + m_n r_n$$

with coefficients $r_i \in R$. If $a \in A$, we have

$$ma = m_1(r_1 a) + \cdots + m_n(r_n a),$$

and because $r_i a \in A$, we see that ma is an A-linear combination of the generators m_i. It follows that every element of $M = MA$ is such a linear combination. We can thus write

$$m_j = \sum_{i=1}^{n} m_i a_{ij} \quad \text{for } 1 \leq j \leq n,$$

where the coefficients $a_{ij} \in A$. We can rewrite these equations in matrix language as $\mathbf{m} = \mathbf{m}X$, where $\mathbf{m}$ is the "row vector" $(m_1, m_2, \ldots, m_n)$ and $X = [a_{ij}]$ is an $n \times n$ matrix over R with entries in the ideal A.

Write $Y = I - X$, where I is the $n \times n$ identity matrix. We thus have $\mathbf{m}Y = 0$ and hence $\mathbf{m}YZ = 0$, where Z is the "classical adjoint matrix" of Y. (The (i, j)-entry of Z, therefore, is the cofactor of Y corresponding to position (j, i); it is $(-1)^{i+j}$ times the determinant that results from the deletion of the jth row and ith column of Y.)

By the standard expansion-by-minors computation of determinants (which works as well over commutative rings as over fields), we have $YZ = \det(Y)I$. The equation $\mathbf{m}YZ = 0$ thus yields $\mathbf{m}\det(Y) = 0$, and hence writing $b = \det(Y)$, we see that b annihilates each entry m_j in the row vector $\mathbf{m}$. Since b annihilates the generators of M, we conclude that $Mb = 0$.

Finally, since $b = \det(Y) = \det(I - X)$ and all entries of X lie in A, we see that $b = 1 - a$ for some element $a \in A$. ∎

The completion of the proof of Theorem 27.8 is now a triviality.

Proof of Theorem 27.8(b, c). Since D is an ideal of the noetherian ring R, it is a finitely generated R-module. We have $DA = D$ by (a), and so by Lemma 27.11 we have $D(1 - a) = 0$ for some element $a \in A$, and this proves (b). If A is proper, then $1 \notin A$ and hence $1 - a \neq 0$. If R is a domain, therefore, it follows that $D = 0$. ∎

27C

In order to address the question of uniqueness in the Lasker-Noether theorem, we say that the equation

$$I = \bigcap_{i=1}^{t} Q_i$$

represents a *standard* Lasker-Noether decomposition of I if the ideals Q_i are primary with distinct radicals, and no proper subset of $\{Q_1, Q_2, \ldots, Q_t\}$ has intersection equal to I. (In other words, the intersection is *irredundant*.)   We know that a standard Lasker-Noether decomposition exists for every ideal of a noetherian ring. Simply apply Theorem 27.6 and choose a primary decomposition of I where the number t of primary ideals is as small as possible. (Note, however, that the definition of a "standard" decomposition does not require the minimality of t.)

(27.12) THEOREM.   *Let*

$$\bigcap_{i=1}^{s} U_i = I = \bigcap_{i=1}^{t} V_i$$

be two standard Lasker-Noether decompositions for an ideal I of a noetherian ring. Then $s = t$ and (after renumbering the V_i if necessary) we have $\sqrt{U_i} = \sqrt{V_i}$ for all i.

Note that we did not say (and it is not, in general, true) that $U_i = V_i$; only the radicals of the primary ideals are uniquely determined. These radicals are called the *associated* primes of I and (as we shall see) they include the isolated primes of I.

In order to prove Theorem 27.12, we need to discuss the "prime annihilators" of an R-module. If M is any R-module and $x \in M$, we write

$$\text{ann}(x) = \{r \in R \mid xr = 0\},$$

the *annihilator* of x. It is easy to see that $\text{ann}(x)$ is always an ideal of R. If it happens that $\text{ann}(x)$ is a prime ideal for some element $x \in M$, we say that it is a *prime annihilator* of M. These prime annihilators are also called the *associated primes* of M, and the set of them is denoted $\text{ass}(M)$. There is an inconsistent (though standard) use of language here since, as we prove in Theorem 27.13, the associated primes of an ideal I (in the Lasker-Noether sense) turn out to be exactly the associated primes (prime annihilators) of the module R/I (and not the prime annihilators of the module I).

The following result includes Theorem 27.12.

(27.13) THEOREM.   *Suppose that*

$$I = \bigcap_{i=1}^{t} Q_i$$

is a standard Lasker-Noether decomposition for the ideal I in the noetherian ring R. Then

$$\{\sqrt{Q_i} \mid 1 \le i \le t\} = \text{ass}(R/I).$$

Proof.   We begin with a bit of notation. Write $\overline{R} = R/I$, and for $r \in R$, let $\overline{r}$ denote the image of r under the canonical homomorphism $R \to \overline{R}$. Thus $\overline{r} = I + r$ and $\text{ann}(\overline{r}) = \{x \in R \mid rx \in I\}$. Also, write $P_i = \sqrt{Q_i}$.

We claim that if $r \in R$ but $r \notin Q_i$, then $\text{ann}(\bar{r}) \subseteq P_i$. To see this, write $A = \text{ann}(\bar{r})$ and note that $rA \subseteq I \subseteq Q_i$. Since $r \notin Q_i$ and Q_i is primary, we deduce that $A \subseteq \sqrt{Q_i} = P_i$, as desired.

Suppose $P \in \text{ass}(\bar{R})$. Thus P is prime and $P = \text{ann}(\bar{r})$ for some element $r \in R$. Since $P < R$, we know that $\bar{r} \neq 0$ and so $r \notin I$. It follows that there exist subscripts j with $r \notin Q_j$, and we write

$$B = \prod_{r \notin Q_j} Q_j .$$

Thus $rB \subseteq Q_i$ for all subscripts i and hence $rB \subseteq \bigcap Q_i = I$, and this tells us that $B \subseteq \text{ann}(\bar{r}) = P$.

Since P is prime and B was defined as a product, it follows that P contains one of the factors of B. Thus $Q_j \subseteq P$ for some subscript j such that $r \notin Q_j$. Since P is a prime ideal containing Q_j, we have

$$P_j = \sqrt{Q_j} \subseteq P .$$

We saw earlier, however, that $\text{ann}(\bar{r}) \subseteq P_j$ whenever $r \notin Q_j$. This gives the reverse containment and proves that every associated prime of $\bar{R}$ is one of the ideals P_j.

Conversely now, we show that each of the ideals P_j has the form $\text{ann}(\bar{r})$ for some $r \in R$. Since P_j is certainly prime, this proves that $P_j \in \text{ass}(\bar{R})$, as required. Fix j and let

$$D = \bigcap_{i \neq j} Q_i .$$

Since the Q_i constitute a standard (irredundant) Lasker-Noether decomposition of I, it follows that $D > I$. By Lemma 27.9 we know that Q_j contains some power of P_j, and thus

$$D P_j^n \subseteq D \cap Q_j = I$$

for some integer $n \geq 0$. Choose n as small as possible and note that $n > 0$ since $D \not\subseteq I$. Because $D P_j^{n-1} \not\subseteq I$, we can choose $d \in D P_j^{n-1}$ with $d \notin I$. Since $d \in D$, we have $d \notin Q_j$, and as we observed earlier, this implies that $\text{ann}(\bar{d}) \subseteq P_j$. Also, $dP_j \subseteq D P_j^n \subseteq I$ and so $P_j \subseteq \text{ann}(\bar{d})$. Thus $P_j = \text{ann}(\bar{d})$, and the proof is complete. ∎

(27.14) LEMMA. *Let I be an ideal of a noetherian ring R. Then each isolated prime of I is one of the associated primes of I.*

Proof. Let P be an isolated prime of I and let $I = \bigcap Q_i$ be a standard Lasker-Noether decomposition. Then

$$\prod Q_i \subseteq \bigcap Q_i = I \subseteq P$$

and hence $Q_i \subseteq P$ for some subscript i. This yields

$$I \subseteq Q_i \subseteq \sqrt{Q_i} \subseteq P ,$$

and since P is a minimal prime over I, we have $P = \sqrt{Q_i}$. Thus P is one of the associated primes of I, as desired.   ∎

Associated primes of I that are not isolated are said to be *embedded*. Frequently, an ideal has no embedded primes, but one can see that embedded primes can exist by considering a prime ideal P of R for which some power P^n is not primary. (See Problem 26.10, for example.) In this case, P is the unique isolated prime of P^n, but since P^n is not primary, there must be more than one primary ideal occurring in a standard Lasker-Noether decomposition for this ideal, and thus there is more than one associated prime. It follows that an embedded prime must exist in this case.

27D

We examine the set $\text{ass}(M)$ of prime annihilators of an R-module M a bit further.

(27.15) LEMMA. *Let R be noetherian and suppose that M is an R-module. If $0 \neq m \in M$, then $\text{ann}(m)$ is contained in some prime annihilator of M.*

Proof. Let $A = \text{ann}(m)$ and write

$$\mathcal{A} = \{\text{ann}(x) \mid 0 \neq x \in M \text{ and } \text{ann}(x) \supseteq A\}.$$

Since $A \in \mathcal{A}$, we know that $\mathcal{A}$ is nonempty, and hence the maximal condition enables us to choose a maximal element $P \in \mathcal{A}$. It is not surprising that P turns out to be prime, and this proves the lemma.

Write $P = \text{ann}(x)$ with $0 \neq x \in M$. Since $x \cdot 1 = x \neq 0$, we see that $1 \notin P$ and P is proper. It suffices to show, therefore, that there do not exist ideals $U, V > P$ with $UV \subseteq P$.

Since $U > P = \text{ann}(x)$, we have $xU \neq 0$, and we can choose $y \in xU$ with $y \neq 0$. Then $yV \subseteq xUV \subseteq xP = 0$ and so $\text{ann}(y) \supseteq V > P \supseteq A$. Since $y \neq 0$, we have $\text{ann}(y) \in \mathcal{A}$, and this contradicts the maximality of P.   ∎

Lemma 27.15 suggests that the set $\text{ass}(M)$ is "large" since it contains supersets for the annihilators $\text{ann}(m)$ for every nonzero element $m \in M$. Nevertheless, for some modules M, we know that $\text{ass}(M)$ is finite. By Theorem 27.13, for instance, this is true when $M = R/I$ for some ideal $I \subseteq R$ (for noetherian rings R). Note that $\overline{R} = R/I$ is a finitely generated module; it is generated by the single element $\overline{1}$. In fact, every finitely generated module over a noetherian ring has a finite set of prime annihilators.

(27.16) THEOREM. *Let R be noetherian and let M be a finitely generated R-module. Then $\text{ass}(M)$ is a finite set.*

(27.17) LEMMA. *Let R be noetherian and let M be an R-module. Then the following are equivalent:*

i.   *M is finitely generated.*
ii. *Every submodule of M is finitely generated.*
iii. *M is noetherian. (The set of submodules of M satisfies the maximal condition.)*

Proof.   The equivalence of (ii) and (iii) is given by Theorem 11.4 (and does not depend on the assumption that R is noetherian). That (i) implies (iii) is a consequence of Theorem 12.19, and it is obvious that (ii) implies (i).   ∎

Proof of Theorem 27.16.   Suppose ass(M) is infinite, where M is a finitely generated R-module, and let

$$\mathcal{M} = \{N \subseteq M \mid \text{ass}(M/N) \text{ is infinite}\}\,.$$

Since $0 \in \mathcal{M}$, we know that $\mathcal{M}$ is nonempty, and because M is noetherian by Lemma 27.17, we can choose a maximal element $N \in \mathcal{M}$. Writing $X = M/N$, we have ass(X) is infinite, but ass(X/Y) is finite for all nonzero submodules $Y \subseteq X$.

Now ass(X) is an infinite set of prime ideals of the noetherian ring R, and so we can choose a maximal element P of this set. Then $P = \text{ann}(y)$ for some element $y \in X$, and $y \neq 0$ since $P < R$. Let

$$Y = \{x \in X \mid xP = 0\}\,.$$

Then Y is a submodule of X, and $Y \neq 0$ since $y \in Y$.

Now let $Q \in \text{ass}(X)$ with $Q \neq P$. We will show that $Q \in \text{ass}(X/Y)$ and thus

$$|\text{ass}(X)| \leq 1 + |\text{ass}(X/Y)| < \infty\,.$$

This is the contradiction we seek.

Let $z \in X$ be such that $Q = \text{ann}(z)$; write $\bar{z} = z + Y \in X/Y$ and let $A = \text{ann}(\bar{z})$. Since $zQ = 0$, we certainly have $\bar{z}Q = 0$ and so $Q \subseteq A$. We will show that $Q = A$, and since Q is prime, this will give $Q \in \text{ass}(X/Y)$, as required. To complete the proof, therefore, we need to show that $A \subseteq Q$.

Since $\bar{z}A = 0$, we have $zA \subseteq Y$ and thus $zAP \subseteq YP = 0$. It follows that $AP \subseteq \text{ann}(z) = Q$. Since $Q \neq P$, we know (by the maximality of P) that $P \not\subseteq Q$, and hence since Q is prime, we have $A \subseteq Q$, as required.   ∎

27E

It is a remarkable fact about noetherian rings that the set Spec(R) of prime ideals satisfies a strong form of the *descending* chain condition.

(27.18) THEOREM.   *Let P be a prime ideal of the noetherian ring R. Then there exists a nonnegative integer n (depending on P) such that if*

$$P = P_0 > P_1 > P_2 > \cdots > P_m$$

is any strictly descending chain of primes of R, then $m \leq n$.

If P is a prime ideal of a ring R, we define the *height* of P, denoted ht(P), to be the largest integer m such that there exists a chain $P = P_0 > \cdots > P_m$ of prime ideals of R. If there is no bound to the lengths of such chains, we write ht(P) $= \infty$. (Thus the isolated primes of the ideal 0, for instance, have height 0, and in $\mathbb{Z}$, the prime ideal (p) has height 1, where p is a prime number.)

Theorem 27.18 asserts that the heights of primes in noetherian rings are always finite. We actually prove more, since the following refinement of Theorem 27.18 gives an explicit bound.

(27.19) THEOREM (Krull).   *Let P be a prime ideal of the noetherian ring R. If P is a minimal prime over an n-generator ideal, then ht(P) $\leq n$. In particular, if P is generated by n elements, then ht(P) $\leq n$, and in all cases, ht(P) is finite.*

The first (and major) step in proving Theorem 27.19 is the following "principal ideal" theorem. This is the case of Theorem 27.19 where $n = 1$.

(27.20) THEOREM (Principal ideal).   *Let R be a noetherian ring and suppose P is a prime ideal of R minimal over some principal ideal. Then ht(P) ≤ 1.*

We need some preliminary results.

(27.21) LEMMA.   *Let P be a prime ideal of the noetherian domain R. Then the symbolic powers $P^{(n)}$ for $n \geq 1$ satisfy*

$$\bigcap_{n=1}^{\infty} P^{(n)} = 0.$$

Proof.   Write $S = R_P$, the localization, and recall that

$$P^{(n)} = ((P^E)^n)^C = (P^E)^n \cap R.$$

Then $\bigcap P^{(n)} = \left(\bigcap (P^E)^n\right) \cap R$. Because S is a noetherian domain by Corollary 26.19, the Krull intersection theorem (27.8) applies and yields $\bigcap (P^E)^n = 0$. The result follows.   ∎

(27.22) LEMMA.   *Let M be a maximal ideal of R, where R is noetherian. Then R/M^n is an artinian R-module for integers $n \geq 0$.*

Proof.   We work by induction on n. The result is trivial when $n = 0$, and so we suppose that $n > 0$ and we write $X = R/M^n$ and $Y = M^{n-1}/M^n$. Then $Y \subseteq X$ and $X/Y \cong R/M^{n-1}$ is artinian by the inductive hypothesis. By Theorem 11.6, in order to show that X is artinian, it suffices to show that Y is artinian.

Since $YM = 0$, we can view Y as an (R/M)-module. (Define $y(r + M) = yr$, and observe that this does not depend on the choice of the coset representative r.) Note that the R-submodules of Y are (R/M)-submodules, and thus to show that Y is artinian as an R-module, it suffices to check that it is artinian as an

(R/M)-module. But R/M is a field, and so we must show that Y is finite dimensional as a vector space over this field.

Because R is noetherian, the ideal M^{n-1} is finitely generated, and thus $Y = M^{n-1}/M^n$ is finitely generated as an R-module. It is therefore finitely generated as an (R/M)-space and hence is finite dimensional. ∎

Proof of Theorem 27.20. Let P be a minimal prime over $I = (a)$. We want to show that if Q and U are prime ideals of R such that $P > Q \supseteq U$, then $Q = U$. In this situation, let $\overline{R} = R/U$ and write $\overline{P}, \overline{I}$, and $\overline{Q}$ to denote the images of P, I, and Q under the canonical homomorphism $R \to \overline{R}$. Now $\overline{I}$ is a principal ideal of $\overline{R}$, and it is trivial to check that $\overline{P}$ is a prime ideal minimal over $\overline{I}$. We also know that $\overline{Q} < \overline{P}$ and that $\overline{Q}$ is prime, and our goal is to show that $\overline{Q} = 0$. We can now replace R by $\overline{R}$ and assume that R is a domain. We have a prime ideal $Q < P$, and we work to show that $Q = 0$.

Now that we have reduced the problem to the case where R is a domain, we show that we can make another reduction and assume that P is the unique maximal ideal of R. For this purpose, we use the technique of localization discussed in Chapter 26.

In the localization $S = R_P$, we have prime ideals $Q^E < P^E$. (These ideals are prime by Lemma 26.18(e) and the containment is strict by Lemma 26.18(d).) Also, S is a domain since 0^E is prime by Lemma 26.18(e). We have that $I^E = IS = aS$ is principal and $I^E \subseteq P^E$. In fact, P^E is a minimal prime over I^E since if $I^E \subseteq V < P^E$ for some prime ideal V of S, we would have

$$I \subseteq I^{EC} \subseteq V^C < P^{EC} = P$$

by Lemma 26.18(a, c, d). Since V^C is prime in R by Lemma 26.18(f), this contradicts the choice of P. Note also that S is noetherian by Corollary 26.19.

At this point, we can replace I, P, Q, and R by I^E, P^E, Q^E, and $S = R_P$. The hypotheses are still satisfied, and since $Q \subseteq Q^E$, it suffices to show that $Q^E = 0$. Since P^E is the unique maximal ideal of S, we have shown that it is no loss to assume that P is the unique maximal ideal of R.

Now P contains every proper ideal of R, and hence P is the unique prime minimal over I. It follows that $P = \sqrt{I}$, and hence by Lemma 27.9 we have $P^m \subseteq I$ for some integer m. The R-module R/P^m is artinian by Lemma 27.22, and thus its homomorphic image R/I is artinian. Writing $I \cap Q = J$, we see that

$$\frac{Q}{J} \cong \frac{I+Q}{I} \subseteq \frac{R}{I},$$

and thus Q/J is an artinian R-module.

Consider now the symbolic powers $Q^{(n)}$ of Q. These form a descending chain, and thus

$$\frac{Q^{(1)}+J}{J} \supseteq \frac{Q^{(2)}+J}{J} \supseteq \frac{Q^{(3)}+J}{J} \supseteq \cdots$$

is a descending chain of R-submodules of Q/J. This chain must eventually stabilize, and so there exists an integer n such that

$$Q^{(n)} + J = Q^{(k)} + J$$

for all integers $k \geq n$.

For $k \geq n$, therefore, we have

$$Q^{(n)} \subseteq Q^{(k)} + J \subseteq Q^{(k)} + I,$$

and so if $x \in Q^{(n)}$, we can write $x = y + ra$ for some elements $y \in Q^{(k)}$ and $r \in R$. Since $Q^{(k)} \subseteq Q^{(n)}$, this gives $ra \in Q^{(n)}$.

Now Corollary 26.20 tells us that $Q^{(n)}$ is a primary ideal belonging to Q. However, $I \not\subseteq Q$ since P is minimal over I and $P > Q$, and thus $a \notin Q = \sqrt{Q^{(n)}}$. It follows that $r \in Q^{(n)}$. Thus $x \in Q^{(k)} + aQ^{(n)}$, and we have

$$Q^{(n)} = Q^{(k)} + aQ^{(n)}.$$

We deduce that the R-module $M = Q^{(n)}/Q^{(k)}$ satisfies $Ma = M$, and so $MI = M$. By Lemma 27.11 we conclude that $M(1 - z) = 0$ for some element $z \in I$. (This requires that M be finitely generated. That hypothesis is satisfied, however, since $Q^{(n)}$ is finitely generated as an ideal of R.)

Because $z \in I \subseteq P$, we see that $1 - z \notin P$, and thus the principal ideal $(1 - z)$ cannot be proper. (Recall that P is the unique maximal ideal of R.) Thus

$$0 = M(1 - z) = MR = M,$$

and so $Q^{(n)} = Q^{(k)}$ for all $k \geq n$. By Lemma 27.21, however,

$$0 = \bigcap_{i \geq 1} Q^{(i)} = Q^{(n)}.$$

But 0 is a prime ideal since R is a domain, and this yields

$$0 = \sqrt{0} = \sqrt{Q^{(n)}} = Q,$$

as required.   ∎

To prove the general case of Theorem 27.19, we need the following lemma.

(27.23) LEMMA.   *Let P be a prime ideal of the noetherian ring R, and assume that $\mathrm{ht}(P) \geq k$ for some integer $k \geq 1$. Let $\mathcal{P}$ be a finite collection of prime ideals of R, none of which contains P. Then there exist primes P_i with*

$$P = P_0 > P_1 > \cdots > P_k$$

and such that P_{k-1} is contained in no member of $\mathcal{P}$.

Proof.   By definition of the height of P, we can certainly find prime ideals P_i such that

$$P = P_0 > P_1 > \cdots > P_k,$$

and we show by induction on n that for $n < k$, the primes P_i can be chosen so that P_n is not contained in any member of $\mathcal{P}$. For $n = 0$, this is just our hypothesis, and so we assume that $n > 0$ and that the ideals P_i have been chosen so that P_{n-1} is not contained in a member of $\mathcal{P}$.

By Theorem 26.13, we know that $P_{n-1} \not\subseteq \bigcup \mathcal{P}$, and so we can choose $a \in P_{n-1}$ such that a is contained in no member of $\mathcal{P}$. (Note that we may assume that $a \notin P_n$, or else we have nothing more to prove.)

The ideal $(P_{n+1} + (a))/P_{n+1}$ is principal in the ring R/P_{n+1}, and $\mathrm{ht}(P_{n-1}/P_{n+1}) \geq 2$ in this ring. It follows by the principal ideal theorem that P_{n-1}/P_{n+1} is not a minimal prime over any principal ideal, and thus in R there exists a prime Q such that

$$P_{n-1} > Q \supseteq P_{n+1} + (a) > P_{n+1},$$

where the last containment is strict since $a \notin P_{n+1}$. We can thus replace P_n by Q. ∎

Proof of Theorem 27.19.   Since the prime ideal P is minimal over itself and it is finitely generated by some number n of generators, we see that it suffices to prove the first statement. We thus assume that P is a minimal prime over an n-generator ideal I, and we work by induction on n to show that $\mathrm{ht}(P) \leq n$. We may certainly assume that $n > 0$, and so we can choose an ideal $J \subseteq I$ with fewer than n generators and such that I/J is a principal ideal of R/J.

Let $\mathcal{P}$ be the (finite) set of isolated primes for J. By the inductive hypothesis, the primes in $\mathcal{P}$ all have heights at most $n-1$, and so we may assume that $P \notin \mathcal{P}$. Since $P \supseteq J$, it follows that P cannot be contained in any member of $\mathcal{P}$.

Now suppose we have prime ideals P_i such that

$$P = P_0 > P_1 > \cdots > P_m.$$

Our goal is to show that $m \leq n$, and so we can suppose that $m > 0$. By Lemma 27.23 we may assume that P_{m-1} is contained in no member of $\mathcal{P}$.

We will show that P is a minimal prime over $J + P_{m-1}$. Assuming this for the moment, we deduce that P/P_{m-1} is a minimal prime over the $(n-1)$-generator ideal $(J + P_{m-1})/P_{m-1}$ of R/P_{m-1}. By the inductive hypothesis, therefore, $\mathrm{ht}(P/P_{m-1}) \leq n-1$, and this implies that $m-1 \leq n-1$ and thus $m \leq n$, as desired.

We may assume now that there exists a prime ideal Q_1 of R with

$$P > Q_1 \supseteq J + P_{m-1},$$

and we work to obtain a contradiction.

Since $P_{m-1} \subseteq Q_1$, we cannot have $Q_1 \in \mathcal{P}$. Since $Q_1 \supseteq J$, however, there must exist a prime ideal Q_2 of R with

$$P > Q_1 > Q_2 \supseteq J$$

and hence $\mathrm{ht}(P/J) \geq 2$ in R/J. We know, however, that P/J is a minimal prime over the principal ideal I/J of R/J, and by Theorem 27.20 this implies that $\mathrm{ht}(P/J) \leq 1$. This is the desired contradiction. ∎

27F

We know that for every prime ideal P of a noetherian ring R, the height $\text{ht}(P)$ is a well-defined nonnegative integer. We say that the *Krull dimension* of R, denoted simply $\dim(R)$, is the maximum of $\text{ht}(P)$ as P runs over $\text{Spec}(R)$. We write $\dim(R) = \infty$ if there is no bound to the heights of the primes.

If R is a field, then $\dim(R) = 0$. If R is a PID but is not a field, then $\dim(R) = 1$ since by Lemma 16.9 every nonzero prime ideal of R is maximal. In particular, if F is a field, then $\dim(F[X]) = 1$. We close this chapter with a generalization of this fact.

(27.24) THEOREM. *Let R be a noetherian domain with $\dim(R) = k$. Then $\dim(R[X]) = k + 1$.*

We need a preliminary result.

(27.25) LEMMA. *Let $Q < P$ be prime ideals in $R[X]$, where R is a domain. If $P \cap R = 0$, then $Q = 0$.*

Proof. We introduce denominators for the multiplicative system $M = R - \{0\}$ in $R[X]$. This yields the ring $S = F[X]$, where F is the field of fractions of R. Since $Q < P$ are primes in $R[X]$ and $P \cap M = varnothing = Q \cap M$, we deduce from Lemma 26.18(e) that Q^E and P^E are prime ideals of $F[X]$.

If $Q \neq 0$, then $Q^E \neq 0$ and hence Q^E is a maximal ideal of $F[X]$. (This is because $F[X]$ is a PID.) It follows that $Q^E = P^E$ and

$$P = P^{EC} = Q^{EC} = Q$$

by Lemma 26.18(d). This is a contradiction, and so $Q = 0$.   ∎

Proof of Theorem 27.24. We begin with a general remark. If S is a noetherian domain with a nonzero prime ideal I, then

$$\dim(S) \geq \dim(S/I) + 1.$$

To see this, note that if

$$Q_0 > Q_1 > \cdots > Q_n = 0$$

is a chain of prime ideals of S/I, and P_i is the full inverse image of Q_i in S, then

$$P_0 > P_1 > \cdots > P_n = I > 0$$

is a longer chain of prime ideals of S.

In our situation, this gives $\dim(R[X]) \geq k+1$ since we can take $S = R[X]$ and we let I be the kernel of the "evaluate at 0" homomorphism from $R[X]$ onto R. (Note that $S/I \cong R$, and so I is prime since R is a domain.)

Conversely, suppose that P is a prime of (the necessarily noetherian ring) $R[X]$. If $\text{ht}(P) = m$, we must show that $m \leq k + 1$. We can assume that $k < \infty$, and we work by induction on k. If $k = 0$, then R is a field, and we know that $\dim(R[X]) = 1$ in this case. We may therefore suppose that $k \geq 1$.

Let

$$P = P_0 > P_1 > \cdots > P_m$$

be a chain of prime ideals of $R[X]$. Write $Q = P_{m-1} \cap R$ and observe that Q is prime in R. We let $\theta : R[X] \to (R/Q)[X]$ be the surjection obtained by reading the coefficients of polynomials modulo Q, and we write $I = \ker(\theta)$. Thus $I = Q[X] \subseteq P_{m-1}$.

Suppose $Q \neq 0$. By the first paragraph of the proof, therefore, we have $\dim(R/Q) \leq \dim(R) - 1 = k - 1$, and so by the inductive hypothesis

$$\dim\left(\frac{R[X]}{I}\right) = \dim\left(\frac{R}{Q}[X]\right) = \dim\left(\frac{R}{Q}\right) + 1 \leq k\,.$$

Since

$$\frac{P_0}{I} > \frac{P_1}{I} > \cdots > \frac{P_{m-1}}{I}$$

is a chain of prime ideals in $R[X]/I$, this gives $m - 1 \leq k$ and $m \leq k + 1$, as required.

We may assume, therefore, that no matter how we construct the sequence

$$P = P_0 > P_1 > \cdots > P_m\,,$$

we always have $P_{m-1} \cap R = 0$.

Because $k \geq 1$, we may assume that $m \geq 2$, and we derive a contradiction. Since $P_{m-1} > 0$, Lemma 27.25 guarantees that $P_{m-2} \cap R \neq 0$, and we choose $a \in P_{m-2} \cap R$ with $a \neq 0$. Since $\text{ht}(P_{m-2}) \geq 2$, the principal ideal theorem tells us that P_{m-2} cannot be a minimal prime over the principal ideal (a) in $R[X]$. It follows that there exists a prime P^* of $R[X]$ with

$$P_{m-2} > P^* \supseteq (a) > 0\,,$$

and so we can replace P_{m-1} by P^* and P_m by 0. Since $0 \neq a \in P^* \cap R$, this is our desired contradiction. ∎

(27.26) COROLLARY. *Let F be a field. Then*

$$\dim(F[X_1, X_2, \ldots, X_n]) = n\,,$$

where the X_i are indeterminates. ∎

Problems

27.1 Let $P_1, P_2, \ldots, P_t$ be the isolated primes of some ideal I in the noetherian ring R.

a. Prove that there exist nonnegative integers e_i such that

$$I \supseteq P_1^{e_1} P_2^{e_2} \cdots P_t^{e_t}.$$

b. Prove that the exponents e_i in part (a) are all positive.

27.2 Let I be an ideal with no embedded primes in a noetherian ring R.

a. Suppose that $r \notin I$ but that $rP_1 \subseteq I$ and $rP_2 \subseteq I$ for prime ideals $P_1, P_2 \supseteq I$. Show that $P_1 = P_2$.

b. Now assume that $P_1, P_2, \ldots, P_t$ are the isolated primes of I. Show that there are unique integers e_i such that

$$I \supseteq P_1^{e_1} P_2^{e_2} \cdots P_t^{e_t}$$

and

$$I \not\supseteq P_1^{f_1} P_2^{f_2} \cdots P_t^{f_t}$$

if any $f_i < e_i$.

27.3 Let I be a principal ideal of a domain R. Show that 0 is the only finitely generated prime ideal properly contained in I.

NOTE: If R is noetherian, this is an immediate consequence of the principal ideal theorem.

27.4 Let M be a finitely generated R-module, where R is noetherian. Suppose I is an ideal of R such that for each element $a \in I$, there exists a nonzero element $x \in M$ such that $xa = 0$. Show that $xI = 0$ for some nonzero element $x \in M$.

27.5 Let $I \subseteq R$ be an ideal, where R is noetherian.

a. Show that I is primary iff $|ass(R/I)| = 1$.

b. If $ass(R/I) = \{P\}$, show that $(R/I)P^n = 0$ for some positive integer n.

27.6 Let R be noetherian and suppose that M is an R-module such that $ass(M) = \{P\}$.

a. If $x \in M$, show that $xP^n = 0$ for some integer n.

b. If M is finitely generated, show that $MP^n = 0$ for some n.

HINT: Note that $xR \cong R/I$ as R-modules, where $I = ann(x)$.

27.7 Let $N < M$ be R-modules. We say that N is a *primary* submodule of M if whenever $xr \in N$ for $x \in M$ and $r \in R$, we have that either $x \in N$ or $Mr^n \subseteq N$ for some integer $n \geq 0$.

a. If $M = R$, show that the primary submodules are exactly the primary ideals.

b. If R is noetherian and M is finitely generated, show that $N \subseteq M$ is primary iff $|ass(M/N)| = 1$.

HINT: In the situation of part (b), suppose that N is primary in M and let $P, Q \in \text{ass}(M/N)$. Show that $P^n \subseteq Q$ for some integer n.

27.8 Let A and B be ideals of the noetherian ring R. Show that

$$B \cap A^n \subseteq BA$$

for some positive integer n.

HINT: Consider a Lasker-Noether decomposition for BA.

27.9 Let $N \subseteq M$ be R-modules and suppose that A is an ideal of R.
  a. For $x \in M$, let $B = \{r \in R \mid xr \in N\}$ and note that B is an ideal. Show that $xA \cap N = x(A \cap B)$.
  b. If R is noetherian and $x \in M$, show that $xA^n \cap N \subseteq NA$ for some integer $n \geq 0$.
  c. If R is noetherian and M is finitely generated, show that $MA^n \cap N \subseteq NA$ for some integer $n \geq 0$.

HINT: If part (c) is false, choose $N \subseteq M$ maximal such that there exists an ideal $U \subseteq R$ for which

$$MU^n \cap N \not\subseteq NU$$

for all $n \geq 0$. Now let $x \in M - N$, and show that for every ideal $V \subseteq R$, there exists $n \geq 0$ such that

$$MV^n \cap N \subseteq NV + (xV \cap N).$$

Apply this with V taken to be some suitable power of U.

NOTE: The result of part (c) is called the Artin-Rees theorem.

27.10 Let A and B be ideals of R, where A is finitely generated and R/A and R/B are both noetherian or both artinian. Show that R/AB is noetherian or artinian, respectively.

HINT: View A/AB as a finitely generated R/B module and deduce that it is a noetherian or artinian R-module.

27.11 Let R be noetherian but not artinian. Show that R has a homomorphic image S such that S is a nonartinian domain with the property that S/I is artinian whenever I is a nonzero ideal of R.

27.12 Suppose that every prime ideal of R is finitely generated. Show that R is noetherian.

HINT: Choose $I \subseteq R$ maximal such that I is not finitely generated. Let $A, B > I$ with $AB \subseteq I$, and use Problem 27.10 to deduce that I/AB is finitely generated. Note also that AB is finitely generated.

27.13 Let R be noetherian. Show that a radical ideal can have no embedded primes.

27.14 Let Z be the set of zero divisors of the noetherian ring R. (An element $z \in R$ is a *zero divisor* if there exists $x \in R$ with $zx = 0$ but $x \neq 0$.) Show that Z is the union of a finite collection of prime ideals.

CHAPTER TWENTY-EIGHT

Integrality

28A

If $F \subseteq E$ is an extension of fields, then the elements of E most closely tied to F are those that are algebraic over F, the roots of nonzero polynomials $f \in F[X]$. If we generalize to unitary extensions of commutative rings, the appropriate analog of "algebraic" is "integral." (Recall that a ring extension is *unitary* if the rings have the same unity element.)

(28.1) DEFINITION. Let $R \subseteq S$ be a unitary extension. An element $s \in S$ is *integral* over R if $f(s) = 0$ for some monic polynomial $f \in R[X]$.

Each element $r \in R$ is a root of the monic polynomial $X - r \in R[X]$, and so it is integral over R. If R is a field, then an element $s \in S$ is integral over R iff it is algebraic. To see this, note that if $0 \neq f \in R[X]$ and $s \in S$ is a root of f, then s is also a root of the monic polynomial $r^{-1}f$, where r is the leading coefficient of f.

In the general situation, not every element $s \in S$ that is algebraic over R is integral. We shall see, for example, that the only elements of $\mathbb{Q}$ that are integral over $\mathbb{Z}$ are the elements of $\mathbb{Z}$. (In other words, a rational number is integral over $\mathbb{Z}$ iff it is integral in the usual sense of being a member of the ring $\mathbb{Z}$, of integers.) If $\alpha \in \mathbb{Q}$ is not integral, we can clear denominators; we can choose a nonzero element $m \in \mathbb{Z}$ such that $m\alpha$ is integral. The following lemma can be viewed as a generalization of this observation.

(28.2) LEMMA. *Let $R \subseteq S$ be a unitary ring extension, and suppose that $s \in S$ is algebraic over R. Then rs is integral over R, where r is the leading coefficient of any nonzero polynomial $f \in R[X]$ such that $f(s) = 0$.*

Proof. Suppose that $f(s) = 0$, where
$$f(X) = a_n X^n + a_{n-1} X^{n-1} + \cdots + a_0 .$$

Substitution of s and multiplication by $(a_n)^{n-1}$ yield

$$(a_n s)^n + a_{n-1}(a_n s)^{n-1} + a_n a_{n-2}(a_n s)^{n-2} + \cdots + (a_n)^{n-1} a_0 = 0,$$

and hence $g(a_n s) = 0$, where $g \in R[X]$ is the monic polynomial given by

$$g(X) = X^n + a_{n-1}X^{n-1} + a_n a_{n-2}X^{n-2} + \cdots + a_0(a_n)^{n-1}.$$

It follows that $a_n s$ is integral over R.   ∎

We consider now the extension $\mathbb{Z} \subseteq \mathbb{C}$. The complex numbers that are integral over $\mathbb{Z}$ are called *algebraic integers* or sometimes just *integers*. To avoid confusion with the ordinary integers, the elements of $\mathbb{Z}$ are often referred to as "rational integers" since (as we have not yet proved) they are precisely the rational algebraic integers. Of course, algebraic integers are algebraic, but not all complex numbers algebraic over $\mathbb{Q}$ are integers. Some examples of nonrational algebraic integers are $\sqrt[n]{k}$ and $e^{2\pi i k/n}$, where n and k are appropriate rational integers. (These are roots of $X^n - k$ and $X^n - 1$, respectively.)

(28.3) LEMMA. *Let R be a UFD with field of fractions K. Then the elements of K integral over R are precisely the elements of R.*

Proof.   Since the elements of R are surely integral over R, it suffices to prove that if $\alpha \in K$ is integral over R, then $\alpha \in R$.

Write $\alpha = a/b$, where $a, b \in R$. Since R is a UFD, we may assume that a and b have no common prime divisor, and we show that b has no prime divisors at all; it is a unit. This proves that $\alpha = ab^{-1} \in R$, as required.

Since $\alpha = a/b$ is integral over R, we can write

$$\left(\frac{a}{b}\right)^n + r_{n-1}\left(\frac{a}{b}\right)^{n-1} + \cdots + r_0 = 0,$$

with $r_i \in R$. This yields

$$a^n = -b(r_{n-1}a^{n-1} + r_{n-2}ba^{n-2} + \cdots + r_0 b^{n-1}),$$

and hence every prime divisor of b divides a^n. Since no prime divisor of b divides a, unique factorization shows that b has no prime divisors.   ∎

As promised, we have the following corollary .

(28.4) COROLLARY. *The algebraic integers in $\mathbb{Q}$ are exactly the ordinary integers.*   ∎

If $R \subseteq S$ is a unitary extension, we say that R is *integrally closed in S* if the only elements of S integral over R are the elements of R. If R is a domain that is integrally closed in its field of fractions, we simply say that R is *integrally closed*. The assertion of Lemma 28.3, therefore, is that UFDs are always integrally closed.

The converse of this is not true. Consider $R = \{a + b\sqrt{-5} \mid a, b \in \mathbb{Z}\} \subseteq \mathbb{C}$, for example. By Problem 16.2 the domain R is not a UFD, but as we shall see, it is integrally closed.

We can combine Lemmas 28.2 and 28.3 to obtain a result that for many readers should be familiar from high school algebra. When $R = \mathbb{Z}$ and $K = \mathbb{Q}$, this is called the "rational root theorem."

(28.5) COROLLARY.   *Let K be the field of fractions of R, where R is a UFD. Suppose that $f \in R[X]$ is nonzero and that $f(\alpha) = 0$, with $\alpha \in K$. If we write $\alpha = a/b$, where $a, b \in R$ have no common prime divisor, then a divides the constant term of f and b divides the leading coefficient.*

Proof.   Let $u, v \in R$ be, respectively, the leading coefficient and constant term of f. Now $u\alpha$ is integral over R by Lemma 28.2, and so by Lemma 28.3 we have $u\alpha \in R$. Thus b divides ua in R, and hence b divides u since a and b are coprime.

If $v = 0$, then a divides v and there is nothing more to prove. Assuming that $v \neq 0$ and writing $n = \deg(f)$, we see that the polynomial $X^n f(1/X)$ has leading coefficient v and root $1/\alpha = b/a$. The result now follows from the previous paragraph. ∎

28B

In a field extension, the algebraic elements form an intermediate field. The corresponding result here is the following theorem.

(28.6) THEOREM.   *Let $R \subseteq S$ be a unitary extension of rings, and let*

$$T = \{s \in S \mid s \text{ is integral over } R\}.$$

Then T is a ring, and T is integrally closed in S.

The keys to proving Theorem 28.6 are the following two results, which concern the R-module structure of a unitary overring of R. These lemmas are the integrality-theory analogs of the fact that in a field extension, the algebraic elements are exactly those that lie in intermediate fields that are finite dimensional over the ground field.

(28.7) LEMMA.   *Let $R \subseteq S$ be a unitary ring extension, and suppose that S is finitely generated as an R-module. Then every element of S is integral over R.*

Proof.   Let $x_1, x_2, \ldots, x_n$ be generators for S as an R-module, and let $s \in S$. For each subscript j, we have $x_j s \in S$, and so we can write

$$x_j s = \sum_{i=1}^{n} x_i a_{ij},$$

where $a_{ij} \in R$. Writing A for the matrix $[a_{ij}]$ and $\mathbf{x}$ for the row vector $(x_1, \ldots, x_n)$, we have $\mathbf{x}s = \mathbf{x}A$, and so $\mathbf{x}(sI - A) = 0$, where I is the $n \times n$

identity matrix. Let $d = \det(sI - A)$. Then $d = f(s)$, where f is the characteristic polynomial of the matrix A. If we can show that $d = 0$, then since $f \in R[X]$ is monic, the proof will be complete.

We will show that $\mathbf{x}d = 0$. This implies that $x_i d = 0$ for all subscripts i and hence $Sd = 0$. Because $1 \in S$, this yields $d = 0$, as desired. To prove that d annihilates the row vector $\mathbf{x}$, we write B to denote the "classical adjoint" of the matrix $sI - A$. (This is the same trick that was used in the proof of Lemma 27.11, where a bit more detail was given.) Then $(sI - A)B = dI$ and so $0 = \mathbf{x}(sI - A)B = \mathbf{x}d$, as required.   ∎

Conversely, every element of $S \supseteq R$ that is integral over R lies in some module-finite intermediate ring. In fact, given any finite set of integral elements, there exists a module-finite ring that contains all of them. This is the content of the following result.

(28.8) LEMMA. *Let $R \subseteq S$ be a unitary ring extension. If $S = R[s_1, s_2, \ldots, s_n]$, where each s_i is integral over R, then S is finitely generated as an R-module.*

Proof. Since s_i is integral over R, there is a monic polynomial over R that has s_i as a root. It follows that for suitable integers $e_i > 0$, we can write $s_i^{e_i}$ as an R-linear combination of lower powers of s_i. Now let

$$\mathcal{X} = \{s_1^{m_1} s_2^{m_2} \cdots s_n^{m_n} \mid 0 \le m_i < e_i \text{ for } 1 \le i \le n\}.$$

The set $\mathcal{X}$ is clearly finite, and for most elements $x \in \mathcal{X}$, we have $xs_i \in \mathcal{X}$. The exception is when the exponent on s_i in x is exactly equal to $e_i - 1$. In that case, although xs_i need not lie in $\mathcal{X}$, it is an R-linear combination of $\mathcal{X}$. It follows that the finitely generated R-module $R\mathcal{X}$ is closed under multiplication by s_i for $1 \le i \le n$. Thus $R\mathcal{X}$ is an S-submodule of S; in other words, it is an ideal. Since $1 \in \mathcal{X}$, we deduce that $S = R\mathcal{X}$, and S is a finitely generated R-module.   ∎

Proof of Theorem 28.6. To see that T is a ring, we must show that $u \pm v$ and uv lie in T for $u, v \in T$. These elements lie in $R[u, v]$, which by Lemma 28.8 is finitely generated as an R-module. By Lemma 28.7, however, all of the elements of $R[u, v]$ are integral over R and hence lie in T. We conclude that T is a ring, as desired.

Suppose now that $s \in S$ is integral over T. Then there exists a monic polynomial $f \in T[X]$ with $f(s) = 0$. By Lemma 28.8 there is a subring $T_0 \subseteq T$ that contains the (finitely many) coefficients of f and that is finitely generated as an R-module. Also, s is integral over T_0, and so by Lemma 28.8 again, $T_0[s]$ is finitely generated as a T_0-module. If $\mathcal{X}$ is a finite generating set for T_0 as an R-module and $\mathcal{Y}$ is a finite generating set for $T_0[s]$ as a T_0-module, then the finite set $\mathcal{X}\mathcal{Y}$ is a generating set for $T_0[s]$ as an R-module. By Lemma 28.7, we conclude that s is integral over R and hence $s \in T$. Thus T is integrally closed in S, as required.   ∎

An important situation in the theory of integrality is where R is an integrally closed domain, and we consider the set S (which we now know to be a ring) of elements integral over R in some extension field E of the field of fractions of R. (For example, if $R = \mathbb{Z}$ and $E = \mathbb{C}$, then S is the ring of all algebraic integers.)

(28.9) COROLLARY.   *Let R be a domain with fraction field F, and suppose $F \subseteq E$ is an algebraic field extension. Let S be the ring of elements of E integral over R. Then S is an integrally closed domain with fraction field E.*

Proof.   By Theorem 28.6 we know that S is a ring integrally closed in E, and as a subring of a field, it is certainly a domain. What remains is to show that E is the fraction field of S. By Lemma 28.2, however, if $\alpha \in E$ is arbitrary, then since α is algebraic over F, we can clear denominators in the minimal polynomial $\min_F(\alpha)$ to produce a polynomial $g \in R[X]$ with $g(\alpha) = 0$. By Lemma 28.2 we know that if r is the leading coefficient of g, then $r\alpha \in S$, and this gives $\alpha = s/r$ for some element $s \in S$. Since $r \in R \subseteq S$, we see that α is a quotient of elements of S, as required.   ∎

In the situation of Corollary 28.9, each element $s \in S$, being algebraic over F, is a root of a monic irreducible polynomial $\min_F(s) \in F[X]$. Also, because s is integral, it is a root of some monic polynomial with coefficients in R. This suggests the question of whether or not $\min_F(s)$ might have coefficients in R. It does if R is integrally closed.

(28.10) LEMMA.   *In the situation of Corollary 28.9, if $s \in S$, then the images of s under the Galois group $\mathrm{Gal}(E/F)$ are integral over R, and all of the roots of the minimal polynomial $\min_F(s)$ in any extension field of E are also integral. If R is integrally closed, then the coefficients of $\min_F(s)$ all lie in R.*

Proof.   By the definition of integrality, we know that there exists a monic polynomial $g \in R[X]$ with $g(s) = 0$. Since the Galois group permutes the roots of g in E, the first assertion is clear. Let K be a splitting field for $f = \min_F(s)$ over E, and let α be a root of f in K. Then $f \mid g$ in $F[X]$, and it follows that $g(\alpha) = 0$ and hence α is integral over R.

We can factor $f(X) = \prod(X - \alpha_i)$, and we now know that all α_i are integral over R. Since the coefficients of f are sums of products of the α_i or the negatives of such sums, it follows from the fact that the integral elements form a ring that all of the coefficients of f are integral over R. These coefficients lie in F, and so if R is integrally closed in its quotient field F, then the coefficients are in R, as required.   ∎

We investigate the case where $R = \mathbb{Z}$ and $F = \mathbb{Q}$. In general, it is quite difficult to compute explicitly the ring S of algebraic integers in a finite degree extension $E \supseteq \mathbb{Q}$. In the case where the degree is 2, however, the solution to this problem is within reach, and we present it now. If $|E : \mathbb{Q}| = 2$, then by the quadratic formula, it is easy to see that $E = \mathbb{Q}[\sqrt{d}]$ for some rational integer d. We may clearly assume that d is square free; it is a product of distinct primes or the negative of such a product. A general element $\alpha \in E$ has the form $\alpha = u + v\sqrt{d}$, where $u, v \in \mathbb{Q}$, and our task is to find necessary and sufficient conditions on u and v so that α is integral.

Suppose first that α is integral and write $\beta = u - v\sqrt{d}$. Note that the automorphism of E that carries $\sqrt{d}$ to $-\sqrt{d}$ takes α to β, and so β is integral by Lemma 28.10. By Theorem 28.6, $\alpha + \beta = 2u$ is integral, and since $2u \in \mathbb{Q}$, we see that $2u \in \mathbb{Z}$ since $\mathbb{Z}$ is integrally closed. We can thus write $u = y/t$, where $y \in \mathbb{Z}$ and $t \in \{1, 2\}$, and where $t = 1$ if y is even. We also know that $\alpha\beta = u^2 - dv^2$ is integral and hence it lies in $\mathbb{Z}$. It follows that $(tv)^2 d \in \mathbb{Z}$. Since d is square free, we see from this that $tv \in \mathbb{Z}$. (If tv had a nontrivial denominator when written as a reduced fraction, then the square of that denominator would have to divide d.) We can thus write $v = z/t$ with $z \in \mathbb{Z}$. Now $u^2 - dv^2 = (y^2 - dz^2)/t^2$, and so if $t = 2$ (so that y is odd), then $y^2 - dz^2$ must be divisible by 4. Thus d and z must both be odd. In this case, $y^2 \equiv 1 \equiv z^2 \bmod 4$ and hence $d \equiv 1 \bmod 4$.

If d is even, therefore, or $d \equiv 3 \bmod 4$, then $t = 1$ and the integers in E all have the form $y + z\sqrt{d}$, with $y, z \in \mathbb{Z}$. If $d \equiv 1 \bmod 4$, on the other hand, then the integers in E necessarily have the form $(y + z\sqrt{d})/2$, where $y, z \in \mathbb{Z}$ are either both even or both odd. (Note that we have considered all possible cases for d.) To see that all of the numbers we have just described actually are integers, we observe that since $\sqrt{d}$ is integral, the numbers $y + z\sqrt{d}$ with $y, z \in \mathbb{Z}$ are integral by Theorem 28.6. What remains is to prove that $\alpha = (y + z\sqrt{d})/2$ is integral when $d \equiv 1 \bmod 4$ and $y, z \in \mathbb{Z}$ are both odd. In this case, we let $\delta = (1 + \sqrt{d})/2$. Then $\alpha - \delta$ has the form $u + v\sqrt{d}$, with $u, v \in \mathbb{Z}$, and so is integral. It thus suffices to show that δ is integral when $d \equiv 1 \bmod 4$, and to see this we observe that δ is a root of $X^2 - X + (1-d)/4$, which is a monic polynomial in $\mathbb{Z}[X]$.

In particular, the ring $R = \{a + b\sqrt{-5} \mid a, b \in \mathbb{Z}\}$ is the full ring of algebraic integers in $\mathbb{Q}[\sqrt{-5}]$, and so it is integrally closed. As we claimed earlier, this ring is an example of an integrally closed domain that is not a UFD.

28C

Let $R \subseteq S$ be a unitary extension of rings. We say that S is *integral over* R or is an *integral extension* of R if every element of S is integral over R. In this situation, it turns out that the ideal theory of S is closely related to that of R, especially when we consider prime ideals. Given a prime Q of S, it is a triviality to see that $Q \cap R$ is a prime ideal of R, and we say that Q *lies over* $Q \cap R$. Conversely, given a prime P of R, we ask whether there necessarily exists a prime Q of S that lies over it. The answer is yes. In fact, something a bit stronger is true.

(28.11) THEOREM. *Let S be integral over R and suppose P is a prime ideal of R. If A is an ideal of S such that $A \cap R \subseteq P$, then there exists a prime ideal Q of S such that $Q \supseteq A$ and $Q \cap R = P$*

Proof. Define the subset $M = R - P$ of R. Since P is prime, we see that M contains 1, does not contain zero, and is closed under multiplication. In other words, M is a multiplicative system in R and hence also in S. Also, $A \cap M = \varnothing$. By Lemma 26.2 and the discussion following its statement, we can use Zorn's lemma to choose an ideal Q of S that contains A and is maximal with the property that $Q \cap M = \varnothing$, and this ideal is necessarily prime in S. Certainly, $R \cap Q \subseteq P$, and our task is to show equality here.

Let $\pi \in P$. We derive a contradiction from the assumption that $\pi \notin Q$, and this will complete the proof. Since $Q + \pi S > Q$, the maximality of Q guarantees that $(Q + \pi S) \cap M$ is nonempty, and thus we can write $m = q + \pi s$ for some elements $m \in M, q \in Q$, and $s \in S$. In particular, $m \equiv \pi s \bmod Q$. (In other words, m and πs have equal images under the canonical homomorphism $S \to S/Q$.) Since S is integral over R, we can find some monic polynomial over R with s as a root, and we can thus write

$$s^n + r_{n-1}s^{n-1} + r_{n-2}s^{n-2} + \cdots + r_0 = 0,$$

with $r_i \in R$.

We multiply through by π^n and read the result mod Q. Since $\pi s \equiv m$, this gives

$$m^n + \pi(r_{n-1}m^{n-1} + \pi r_{n-2}m^{n-2} + \cdots + \pi^{n-1}r_0) \equiv 0 \bmod Q,$$

and so $m^n + \pi r \in Q$ for some element $r \in R$. Thus $m^n + \pi r \in R \cap Q \subseteq P$. Since $\pi r \in P$, this gives $m^n \in P$, and this is a contradiction because $m^n \in M$ and $M \cap P = \varnothing$. ∎

(28.12) COROLLARY (Lying over).   *Let S be integral over R and suppose P is a prime ideal of R. Then there exists a prime ideal Q of S such that $Q \cap R = P$.*

Proof.   Apply Theorem 28.11 with $A = 0$. ∎

By LO (the common abbreviation for the "lying over" theorem), we know that the intersection map $\mathrm{Spec}(S) \to \mathrm{Spec}(R)$ is surjective. Several different primes of S often lie over the same prime of R, and hence this map is usually not injective. Nevertheless, there is information available about the set of primes lying over a given prime, and so the failure of injectivity is not totally out of control.

(28.13) THEOREM.   *Let $R \subseteq S$ be an integral extension of rings, and suppose that $Q \subseteq A$ are ideals of S, where Q is prime and $Q \cap R = A \cap R$. Then $Q = A$.*

Proof.   We obtain a contradiction by assuming that $A > Q$. Choose $s \in A$ with $s \notin Q$, and note that since s is integral over R, we can write $f(s) = 0 \in Q$ for some monic polynomial $f \in R[X]$. Choose a monic polynomial $g \in R[X]$ of minimal possible degree m such that $g(s) \in Q$. Writing $g(s) = q$, we have

$$s^m + r_{m-1}s^{m-1} + \cdots + r_0 = q \in Q$$

for some choice of coefficients $r_i \in R$. Since $s \in A$ and $Q \subseteq A$, this gives

$r_0 \in R \cap A = R \cap Q$, and hence

$$s(s^{m-1} + r_{m-1}s^{m-2} + \cdots + r_1) = q - r_0 \in Q .$$

But Q is prime and $s \notin Q$, and we deduce that

$$s^{m-1} + r_{m-1}s^{m-2} + \cdots + r_1 \in Q ,$$

and this contradicts the minimality of m.   ∎

(28.14) COROLLARY (Incomparability). *Suppose that $R \subseteq S$ is an integral extension and that Q_1 and Q_2 are prime ideals of S that lie over the same prime P of R. If $Q_1 \neq Q_2$, then $Q_1 \nsubseteq Q_2$ and $Q_2 \nsubseteq Q_1$.*

Proof. If $Q_1 \subseteq Q_2$, then since Q_1 is prime, we get $Q_1 = Q_2$ by Theorem 28.13. This is a contradiction, and we get a similar contradiction if $Q_2 \subseteq Q_1$.   ∎

From INC (the incomparability theorem), we can get information relating chains of prime ideals in R and S.

(28.15) COROLLARY. *Let $R \subseteq S$ be an integral extension of rings. If $Q_0 < Q_1 < \cdots < Q_n$ is a properly increasing chain of prime ideals of S, then $P_0 < P_1 < \cdots < P_n$ is a properly increasing chain of prime ideals of R, where $P_i = Q_i \cap R$.*

Proof. We have already observed that $P_i = Q_i \cap R$ is prime in R, and it is clear that $P_i \subseteq P_j$ when $i < j$. We now apply INC to see that we cannot have $P_i = P_j$ if $i < j$.   ∎

Corollary 28.15 suggests the question of whether or not R could possibly have a longer proper chain of prime ideals than the longest such chain in S. In fact, it cannot. To prove this, we need the following result.

(28.16) COROLLARY (Going up). *Let $R \subseteq S$ be an integral extension, and suppose $P_1 \subseteq P_2$ are prime ideals of R. If Q_1 is a prime of S lying over P_1, then there exists a prime ideal Q_2 of S such that Q_2 lies over P_2 and $Q_2 \supseteq Q_1$.*

Proof. Apply Theorem 28.11 to the prime P_2, taking $A = Q_1$.   ∎

(28.17) COROLLARY. *Suppose $R \subseteq S$ is integral and let $P_1 < P_2 < \cdots < P_n$ be a properly increasing chain of prime ideals of R. Then there exists a (necessarily properly) increasing chain of primes $Q_1 < Q_2 < \cdots < Q_n$ of S such that Q_i lies over P_i.*

Proof. Apply GU (Corollary 28.16) repeatedly.   ∎

28D

We wish now to consider integral extensions of rings in the context of finite-degree field extensions. Our basic situation is where R is an integrally closed domain with fraction field F, and where S is the full ring of elements integral over R in the field $E \supseteq F$, where $|E : F|$ is finite. (Note that by Corollary 28.9, E is the field of fractions of S.) Of course, S is an integral extension of R, and so LO, INC, and GU hold, but more is true in this context. To avoid restating all of these hypotheses in the results that follow, we simply say that (F, E, R, S) satisfy the *standard hypotheses* for integers in field extensions.

We consider first the case where E is Galois over F. In this case, the Galois group $G = \mathrm{Gal}(E/F)$ induces a group of automorphisms of S by Lemma 28.10. It follows that G permutes the prime ideals of S, and in fact if Q is a prime of S lying over $P \subseteq R$, then since $\sigma \in G$ fixes P, we see that $\sigma(Q)$ also lies over P.

(28.18) THEOREM. *Let (F, E, R, S) satisfy the standard hypotheses, and assume in addition that E is Galois over F with Galois group G. If P is a prime ideal of R, then there are just finitely many primes of S lying over P, and these are permuted transitively by G.*

Proof. Let Q lie over P and let $Q = Q_1, Q_2, \ldots, Q_n$ be the distinct images of Q under the elements of G. (Note that since $|G| = |E : F|$ is finite, we do have a finite list of ideals here.) Now let Q^* be any prime of S lying over P. We must show that $Q^* = Q_i$ for some subscript i. If $s \in Q^*$, consider the product $a = \prod \sigma(s)$ as σ runs over the elements of G. Since each element $\sigma(s)$ is integral over R, so is their product a. Observe, however, that a is G-invariant and hence lies in F. Since R is integrally closed in F, this gives $a \in R$.

The factor s of a lies in the ideal Q^* and hence $a \in Q^* \cap R = P \subseteq Q$. Because Q is a prime ideal of S, some factor $\sigma(s)$ of a must lie in Q and hence $s \in \sigma^{-1}(Q) = Q_i$ for some subscript i. This shows that Q^* is contained in the union $\bigcup Q_i$, and by Theorem 26.13 we deduce that Q^* must be contained in one of the ideals Q_i. Finally, since Q^* and Q_i both lie over P, we get $Q^* = Q_i$ by INC (28.14). ∎

In view of GU (28.16), it is natural to ask whether the containments can be reversed so that a "going down" theorem can be proved. Examples show that GD does not hold in general, although it can be proved under our standard hypotheses for integers in field extensions. We prove only the case where $F \subseteq E$ is a separable extension since that is an easy consequence of Theorem 28.18.

(28.19) COROLLARY (Going down). *Let (F, E, R, S) satisfy the standard hypotheses and assume that E is separable over F. Let $P_1 \supseteq P_2$ be prime ideals of R and suppose that Q_1 is a prime of S lying over P_1. Then there exists a prime Q_2 of S such that Q_2 lies over P_2 and $Q_2 \subseteq Q_1$.*

Proof.   Let $L \supseteq E$ be Galois over F. (For example, we could take L to be a splitting field for a polynomial over F whose roots contain a generating set for E over F.) Let T be the ring of integers over R in L, and choose a prime U_1 of T lying over Q_1.

Now let V_2 be any prime of T lying over P_2, and apply GU to find a prime V_1 of T containing V_2 and lying over P_1. By Theorem 28.18 we have $\sigma(V_1) = U_1$ for some element $\sigma \in \mathrm{Gal}(L/F)$. Now define $U_2 = \sigma(V_2)$ and note that U_2 lies over P_2 and $U_2 \subseteq U_1$. The ideal $Q_2 = S \cap U_2$ is now easily seen to have the desired properties.   ∎

We have seen that if S is an integral extension of R, then we get some control over chains of primes in S from the corresponding information in R. This suggests that we might be able to control arbitrary chains of ideals. If R is noetherian, for instance, we would like to prove that S is noetherian. As we did with GD, we prove this in the separable case of our standard hypotheses. In fact, even more is true in this situation. Not only does S satisfy the ACC for ideals (that is, S-submodules), but it is even noetherian as an R-module, which is a much stronger condition. In particular, it is a finitely generated R-module, thus providing a sort of weak converse for Lemma 28.7.

(28.20) THEOREM.   *Let (F, E, R, S) satisfy the standard hypotheses for integers in field extensions and assume that E is separable over F. Then S is contained in some finitely generated R-submodule of E. If R is noetherian, then S is finitely generated and is a noetherian R-module.*

Proof.   Let $n = |E : F|$ and note that since E is a finite degree separable extension of F, the primitive element theorem (18.17) guarantees that $E = F[\alpha]$ for some element $\alpha \in E$. By Lemma 28.2 there exists $r \in R$ such that $r\alpha$ is integral over R, and so $r\alpha \in S$. Since $r\alpha$ also generates E over F, we may replace α by $r\alpha$ and assume that $\alpha \in S$.

Let M be the R-submodule of E generated by $1, \alpha, \alpha^2, \ldots, \alpha^{n-1}$. Can we prove that $S \subseteq M$? Each element $s \in S$ can be written in the form $s = g(\alpha)$ for some unique polynomial $g \in F[X]$ with $\deg(g) < n$. If, for some choice of $s \in S$, it happens that $g \in R[X]$, this would imply that $s \in M$, as desired. It is too much to expect, however, that this holds for every element $s \in S$, and we prove instead that there exists a fixed nonzero element $d \in F$ such that $dg \in R[X]$ for every choice of $s \in S$. This implies that $dS \subseteq M$ and hence $S \subseteq (1/d)M$. Since $(1/d)M$ is the R-module generated by the elements α^i/d for $0 \le i < n$, this proves that S is contained in a finitely generated R-module.

Let $L \supseteq E$ be a splitting field for $f = \min_F(\alpha)$ over F, and let $\alpha = \alpha_1, \alpha_2, \ldots, \alpha_n$ be the (necessarily distinct) roots of f. Fix $s \in S$ and let $g \in F[X]$ be as above, so that $g(\alpha) = s$ and $\deg(g) < n$. Set $s_i = g(\alpha_i)$ and note that the Galois group of L over F permutes the α_i transitively, and hence it also permutes the s_i transitively. Since $s_1 = s$, it follows that the elements $s_i \in L$ are all integral over R.

Write
$$g(X) = a_0 + a_1 X + a_2 X^2 + \cdots + a_{n-1} X^{n-1}.$$

Then
$$\sum_{i=0}^{n-1} a_i \alpha_j^i = s_j \text{ for } 1 \le j \le n,$$

and we can write these equations in matrix form as $\mathbf{a}A = \mathbf{s}$ for row vectors $\mathbf{s} = (s_1, \ldots, s_n)$ and $\mathbf{a} = (a_0, \ldots a_{n-1})$. The coefficient matrix A has (i, j)-entry equal to α_j^i, and so it is the Vandermonde matrix $V(\alpha_1, \ldots, \alpha_n)$, as in Lemma 23.20.

Our goal is to find some nonzero element $d \in F$, independent of the choice of s, such that all entries of $\mathbf{a}d$ lie in R. We claim that we can take $d = \Delta(f)$, the discriminant. By Corollary 23.19 we know that $0 \neq d \in F$, and since the entries of $\mathbf{a}d$ lie in F, it suffices to show that they are integral over R. To see this, multiply the equation $\mathbf{a}A = \mathbf{s}$ by the classical adjoint matrix B of A. This gives $\mathbf{a}v = \mathbf{s}B$, where $v = \det(A)$. By Lemma 23.20 we have $d = v^2$ and thus $\mathbf{a}d = \mathbf{s}Bv$. Since all α_i are integral over R by Lemma 28.10, it follows that v and all entries in B are integral over R, and thus all entries in the row vector $\mathbf{s}Bv$ are integral, as required.

We have now shown that S is contained in a finitely generated R-submodule of E. By Theorem 12.19, if R is noetherian, then this is a noetherian module and S is finitely generated. ■

We mention two applications of Theorem 28.20 to the case where $F = \mathbb{Q}$.

(28.21) COROLLARY. *Let E be a finite-degree field extension of $\mathbb{Q}$. Then there exists a basis $\mathcal{B}$ for E over $\mathbb{Q}$ consisting of algebraic integers and having the property that every algebraic integer in E is a $\mathbb{Z}$-linear combination of $\mathcal{B}$.*

A basis $\mathcal{B}$ as in Corollary 28.21 is said to be an *integral basis* for E. If $\mathcal{B}$ and $\mathcal{C}$ are two integral bases for the same field E, we see that the change-of-basis matrices from $\mathcal{B}$ to $\mathcal{C}$ and $\mathcal{C}$ to $\mathcal{B}$ have entries in $\mathbb{Z}$. Since these matrices are inverses of each other, each must have determinant ± 1, and it follows by Theorem 23.17 that the discriminants $\Delta_{E/\mathbb{Q}}(\mathcal{B})$ and $\Delta_{E/\mathbb{Q}}(\mathcal{C})$ must be equal. This quantity depends only on E, therefore, and it is called the *discriminant* of E. This discriminant lies in $\mathbb{Q}$, of course, and it is easy to see that it must be an algebraic integer. It follows that the discriminant of the field E is a rational integer.

Proof of Corollary 28.21. Let S be the ring of algebraic integers in E. Thus $(\mathbb{Q}, E, \mathbb{Z}, S)$ satisfies our standard hypotheses, and since E is certainly separable over $\mathbb{Q}$, Theorem 28.20 applies. Because $\mathbb{Z}$ is noetherian, we deduce that S is finitely generated as a $\mathbb{Z}$-module, and because S is a domain, it is torsion free as a $\mathbb{Z}$-module. By Theorem 16.28, however, finitely generated torsion-free modules over PIDs are free, and thus there is a $\mathbb{Z}$-basis $\mathcal{B}$ for S. To complete the proof, it suffices to show that $\mathcal{B}$ is also a $\mathbb{Q}$-basis for E.

Let $s_1, s_2, \ldots, s_n$ be distinct members of $\mathcal{B}$ and suppose that $\sum a_i s_i = 0$, where the coefficients a_i lie in $\mathbb{Q}$. We can find a "common denominator" $d \in \mathbb{Z}$ such that d is nonzero and each product da_i lies in $\mathbb{Z}$. Then $\sum (da_i)s_i = 0$, and since the $\mathcal{B}$ is a $\mathbb{Z}$-basis, we conclude that $da_i = 0$ for each subscript i. It follows that $a_i = 0$ for each i, and $\mathcal{B}$ is linearly independent over $\mathbb{Q}$.

If $\alpha \in E$, then by Lemma 28.2 there exists a nonzero rational integer n such that $n\alpha \in S$. We can thus write $n\alpha = \sum z_i s_i$ for some choice of basis elements $s_i \in \mathcal{B}$ and coefficients $z_i \in \mathbb{Z}$. Thus $\alpha = \sum (z_i/n)s_i$ is a $\mathbb{Q}$-linear combination of $\mathcal{B}$, as required.   ∎

In Chapter 29 we study a class of rings called *Dedekind domains*. These are noetherian integrally closed domains in which every nonzero prime ideal is maximal. We have now accumulated enough theory to see that Dedekind domains occur naturally.

(28.22) COROLLARY.   *Let E be a finite-degree extension of $\mathbb{Q}$. Then the ring S of algebraic integers in E is a Dedekind domain.*

Proof.   Corollary 28.9 tells us that S is integrally closed. Since every nonzero prime ideal of $\mathbb{Z}$ is maximal, it follows by Corollary 28.15 that S cannot have prime ideals Q_1 and Q_2 such that $0 < Q_1 < Q_2$, and this implies that every nonzero prime of S is maximal. Finally, since $\mathbb{Z}$ is noetherian, Theorem 28.20 guarantees that S is a finitely generated $\mathbb{Z}$-module, and so of course it is a noetherian ring.   ∎

28E

We devote the rest of this chapter to two applications of integrality theory. The first of these is to the determination of Galois groups and the second is to "pure" finite group theory.

Suppose $f \in \mathbb{Z}[X]$ is monic of degree n and has distinct roots, and let E be a splitting field for f over $\mathbb{Q}$. We know that the Galois group $G = \mathrm{Gal}(E/Q)$ can be embedded into the symmetric group on n symbols (the roots of f in E), and if f is irreducible in $\mathbb{Q}[X]$, then G is a transitive subgroup. In general, however, the task of determining G up to group isomorphism can be quite difficult.

Consider the polynomial

$$f(X) = X^7 + 2X^5 + X^4 + 4X^3 + X^2 + 3X - 2.$$

If we read the coefficients of f modulo various primes p, we get polynomials $\overline{f} \in (\mathbb{Z}/p\mathbb{Z})[X]$, and the trick is to use some of these infinitely many polynomials to obtain information about the unknown Galois group G. If we take $p = 11$, for instance, we find that $\overline{f}$ is irreducible. (One can use the Berlekamp algorithm discussed in Chapter 21 to establish this. Similar algorithms are also included as parts of a number of different commercially available symbolic algebra software packages.) It is easy to see that the irreducibility of f mod p guarantees that f is irreducible in $\mathbb{Q}[X]$. (If f had a nontrivial factorization over $\mathbb{Q}$, it would have a corresponding factorization over $\mathbb{Z}$ by Gauss's lemma (16.19), and this would yield a proper factorization over $\mathbb{Z}/p\mathbb{Z}$ for every prime p.) We now know that the group G of the above polynomial is transitive of degree 7.

Since the fact that $\overline{f}$ is irreducible when $p = 11$ gives useful information about G, it should not be too surprising that we get something interesting even when $\overline{f}$ reduces. What is striking is just how much we can deduce in the (common) case where $\overline{f}$ factors into distinct irreducibles. If we take $p = 5$, for instance, the computer tells us (or the skeptical reader can check) that

$$\overline{f}(X) = (X^3 + X + 1)(X^2 + 2)(X + 1)(X + 4)$$

is the factorization of $\overline{f}$ into irreducible polynomials. Since the factors are distinct and have degrees 3, 2, 1, and 1, our theorem, stated below, asserts that G contains an element that is the disjoint product of a 3-cycle, a 2-cycle, and two 1-cycles. It is a fact (from the theory of permutation groups) that the only transitive group of degree 7 that has an element with this cycle structure is the whole symmetric group. The Galois group of our polynomial is therefore determined. In particular, we deduce that f is not solvable by radicals.

(28.23) THEOREM. *Let $f \in \mathbb{Z}[X]$ be monic and irreducible over $\mathbb{Q}$ and let p be a prime number. Write $F = \mathbb{Z}/p\mathbb{Z}$ and let $\overline{f}$ denote the natural image of f in $F[X]$. Factor $\overline{f} = f_1 f_2 \ldots f_r$, where the polynomials f_i are monic and irreducible over F, and assume they are distinct. Let $G = \mathrm{Gal}(E/\mathbb{Q})$, where E is a splitting field for f over $\mathbb{Q}$, and view G as a permutation group on the set of roots of f in E. Then G contains some element that can be written as a product of disjoint cycles of lengths $d_1, d_2, \ldots, d_r$, where d_i is the degree of f_i.*

As we shall see, there are only finitely many "bad" primes p for which the irreducible factors of $\overline{f}$ fail to be distinct. For each of the infinitely many remaining primes, Theorem 28.23 gives a little information about the group G. Each time we choose a prime, we get to see the cycle structure of one element of G. In some sense, this element is randomly chosen. It is a deep theorem, well beyond what we can prove here, that in the "long run" we get to see each element of G equally often. In other words, as we apply Theorem 28.23 for a large number of primes, the different cycle structures appear with frequencies proportional to the numbers of elements of G that have each structure.

Suppose, for example, that f has degree 7 and that we have applied Theorem 28.23 for 100 different primes. If it turned out every time that the cycle structure was either seven 1-cycles or one 7-cycle, it would be reasonable to suspect that these are the only cycle structures in G and thus G is cyclic of order 7. Although in this case we may be confident that we know G, we have not, of course, *proved* anything. In fact, the only thing that G can be proved to be, using only Theorem 28.23, is the full symmetric group.

Proof of Theorem 28.23. Write $|E : \mathbb{Q}| = m$ and let R be the ring of elements of E integral over $\mathbb{Z}$. By Corollary 28.21 there exists an integral basis $\mathcal{B}$ for E, and we have $|\mathcal{B}| = m$. The elements of R are exactly the $\mathbb{Z}$-linear combinations of $\mathcal{B}$, and the additive group of R is thus the direct sum of the m cyclic subgroups generated by the elements of $\mathcal{B}$. It follows that the additive group of the ring R/pR is isomorphic to the direct sum of m cyclic groups of order p, and we conclude that $|R/pR| = p^m$.

Next we consider the prime ideals $P_1, P_2, \ldots, P_t$ of R that lie over the principal maximal ideal (p) of $\mathbb{Z}$. (Note that by Theorem 28.18 there are just finitely many of these, and they are transitively permuted by the Galois group G.) Also, by Corollary 28.15 it follows that each P_i is maximal in R. Let $D = \bigcap P_i$ and consider R/D. First note that $p \in D$, and so R/D is a homomorphic image of R/pR, and by the previous paragraph, this ring has finite order p^m. It follows that $|R/D| \leq p^m$.

We claim that R/D is isomorphic to the (external) direct sum S of the t rings R/P_i. (This is a version of the Chinese remainder theorem.) We certainly have a map $\theta : R \to S$ that maps $r \in R$ to the t-tuple $(x_1, x_2, \ldots, x_t)$, where x_i is the natural image of r in R/P_i. Clearly, $\ker(\theta) = D$, and to see that θ is surjective, it suffices to find subsets $U_i \subseteq R$ such that U_i maps onto R/P_i and U_i maps to 0 in R/P_j for $j \neq i$.

Take $U_i = \prod P_j$, where the product is taken over all subscripts $j \neq i$. Since $U_i \subseteq P_j$ for $j \neq i$, we certainly have that U_i maps to 0 in R/P_j. To compute the image of U_i in R/P_i, we observe that $U_i \not\subseteq P_i$. This is because P_i is prime, and none of the factors P_j of U_i can be contained in P_i since these ideals are distinct and the P_j are maximal. Since P_i is maximal, it follows that $P_i + U_i = R$, and thus U_i maps onto R/P_i as desired. It follows that θ is surjective and $R/D \cong S$, as claimed.

Now write $P = P_1$ and let K be the quotient field R/P. Note that since $pR \subseteq P$ and $|R/pR| = p^m$, we certainly know that K is a finite field of characteristic p, and we write $|K| = p^k$. Also, since the elements of G define automorphisms of the ring R and permute the P_i transitively, we see that all of the fields R/P_i are isomorphic and thus $|R/D| = |S| = |K|^t$. This gives $p^m \geq |R/D| = |K|^t = p^{kt}$ and hence $m \geq kt$.

Now let $\alpha_1, \alpha_2, \ldots, \alpha_n$ be the (necessarily distinct) roots of f in E, and note that since f is monic, these all lie in R. We use overbars to denote the canonical homomorphism of R onto $R/P = K$. (Note that this does not conflict with our earlier use of overbars to represent the homomorphism $\mathbb{Z} \to \mathbb{Z}/p\mathbb{Z} = F$, since if we view F as being contained in K, then the map on $\mathbb{Z}$ is just the restriction of the map on R.) We have

$$f(X) = (X - \alpha_1)(X - \alpha_2) \cdots (X - \alpha_n),$$

and so

$$f_1(X) f_2(X) \cdots f_r(X) = \overline{f}(X) = (X - \overline{\alpha_1})(X - \overline{\alpha_2}) \cdots (X - \overline{\alpha_n}).$$

Since each polynomial f_i is irreducible over the finite field F, it has no repeated roots. Also, these monic irreducible polynomials are distinct by assumption, and it follows that no two of them can have a common root. We conclude that the n elements $\overline{\alpha_i}$ of K are distinct.

Now let $H \subseteq G$ be the stabilizer of P in the action of G on the set $\{P_1, P_2, \ldots, P_t\}$. Then $|G : H| = t$, and since $|G| = |E : \mathbb{Q}| = m$, we have $|H| = m/t \geq k$. Also, since H acts on R and stabilizes P, we see that each element $\sigma \in H$ induces an automorphism $\hat{\sigma}$ of the field $K = R/P$.

We claim that this map $\sigma \mapsto \hat{\sigma}$ is injective. If $\hat{\sigma}$ is trivial, then for $1 \leq i \leq n$, we have

$$\overline{\alpha_i} = \hat{\sigma}(\overline{\alpha_i}) = \overline{\sigma(\alpha_i)} = \overline{\alpha_j},$$

where we have written $\alpha_j = \sigma(\alpha_i)$. (This is valid since the elements of G permute the roots of f.) Since we know that the elements $\overline{\alpha_i}$ are distinct for distinct subscripts i, it follows that $\alpha_i = \alpha_j$. We have now shown that σ fixes all α_i, and thus $\sigma = 1$.

Since the map $H \to \text{Aut}(K)$ is injective, we have $k \leq |H| \leq |\text{Aut}(K)| = k$. This forces equality, and thus H induces the full automorphism group of K, which is a cyclic group of order k.

Let $\sigma \in H$ induce a generator τ of $\text{Aut}(K)$. Then σ and τ induce corresponding permutations of the α_i and the $\overline{\alpha_i}$, respectively. Necessarily, however, $\langle \tau \rangle = \text{Aut}(K)$ acts transitively on the set of roots of each of the irreducible polynomials f_i, and so its cycle decomposition corresponds exactly to the various degrees of these polynomials. This is exactly what we wanted to prove. ∎

We mentioned that in the situation of Theorem 28.23, there can be at most finitely many primes p for which the polynomial $\overline{f}$ has repeated irreducible factors. To see why this is so, recall that we saw in the proof that the elements $\overline{\alpha_i}$ are exactly the roots of $\overline{f}$ in R/P. If $\overline{f}$ has repeated factors, then the elements $\overline{\alpha_i}$ are not distinct and $\prod(\overline{\alpha_i} - \overline{\alpha_j})^2 = 0$, where the product is taken over all pairs i, j with $i < j$. In other words, the discriminant $\Delta(f) = \prod(\alpha_i - \alpha_j)^2 \in P$. Since the α_i are integral and $\Delta(f) \in \mathbb{Q}$, we see that $\Delta(f) \in \mathbb{Z}$ and hence $\Delta(f) \in \mathbb{Z} \cap P = (p)$. This shows that the only "bad" primes, where there are repeated factors, are prime divisors of the rational integer $\Delta(f)$. It follows that there are just finitely many of these.

28F

We close this chapter with an application of the theory of integrality to finite group theory. In Chapter 15 we developed some of the basic ingredients of the character theory of finite groups, and we used that theory to prove the purely group theoretic theorem of Frobenius. When the results of Chapter 15 are augmented by a little of the theory of algebraic integers, we get a machine powerful enough to prove the following celebrated theorem.

(28.24) THEOREM (Burnside). *Let G be a finite group whose order is divisible by no more than two different prime numbers. Then G is solvable.*

The proof of Theorem 28.24 that we present is essentially the original one of Burnside, but we mention that there now exist purely group theoretic proofs of this result, independent of character theory. These character-free arguments, however, are quite complex and subtle and are nowhere nearly so accessible as the proof we present. (To date, the theorem of Frobenius proved in Chapter 15 has resisted all efforts to find a character-free proof.)

It is a triviality to reduce the proof of Burnside's theorem to an assertion about simple groups.

(28.25) THEOREM. *Let G be a finite simple group with order divisible by no more than two primes. Then G has prime order.*

Proof of Theorem 28.24 (assuming Theorem 28.25).   Each composition factor of G is a simple group whose order divides that of G and hence has no more than two prime divisors. By Theorem 28.25, therefore, these composition factors all have prime order, and hence G is solvable.   ∎

Now that the classification of finite simple groups is complete (or at least is alleged to be so by the experts), it has become common to prove theorems about finite groups by reducing the question to some assertion about simple groups, and then verifying this assertion case by case for the groups on the list of all finite simple groups. Although Burnside's proof of Theorem 28.24 also proceeds by reduction to simple groups, the similarity ends there. The proof of Theorem 28.25 that we are about to present does not rely on the classification or on a case-by-case analysis. The key step is the following result.

(28.26) THEOREM. *Let G be a finite simple group that has a nonidentity conjugacy class of prime-power size. Then G has prime order.*

Proof of Theorem 28.25 (assuming Theorem 28.26).   Let P be a nontrivial Sylow p-subgroup of G, and let $1 \neq x \in \mathbf{Z}(P)$. Since $P \subseteq \mathbf{C}_G(x)$, it follows that $|G : \mathbf{C}_G(x)|$ is a power of the one prime q different from p that can divide $|G|$. The conjugacy class of x in G, therefore, has q-power size, and the result follows via Theorem 28.26.   ∎

Although, as we mentioned earlier, Burnside's theorem can now be proved without characters, no one has yet found a character-free proof of Theorem 28.26.

At last, characters come into the picture, and we assume from now on that the reader is familiar with Chapter 15. The following result is the key step in Burnside's argument. Both its proof and its application to the proof of Theorem 28.26 depend on integrality. Recall that for characters χ of G, we have defined $\mathbf{Z}(\chi) = \{g \in G \mid |\chi(g)| = \chi(1)\}$. By Lemma 15.17 this is a normal subgroup of G.

(28.27) THEOREM. *Let $x \in G$ and $\chi \in \mathrm{Irr}(G)$, where G is a finite group, and assume that $\chi(1)$ is relatively prime to the size of the conjugacy class of x. If $x \notin \mathbf{Z}(\chi)$, then $\chi(x) = 0$.*

Theorem 28.27 is especially useful when G is nonabelian simple since in that case, $\mathbf{Z}(\chi)$ is almost always trivial.

(28.28) LEMMA. *If G is a nonabelian simple group, then $\mathbf{Z}(\chi) = 1$ for all nonprincipal irreducible characters χ of G.*

Proof. Let $\chi \in \mathrm{Irr}(G)$ and write $Z = \mathbf{Z}(\chi) \triangleleft G$. Since G is simple, we have $Z = 1$ or $Z = G$. We assume that $Z = G$ and we show that χ is the principal character. (In other words, it is the constant function with value 1.)

By the definition of $\mathbf{Z}(\chi)$, we have $|\chi(g)| = \chi(1)$ for all $g \in G$, and hence $[\chi, \chi] = \chi(1)^2$, where $[\cdot, \cdot]$ denotes the character inner product, as in Chapter 15. Since χ is irreducible, however, we have $[\chi, \chi] = 1$ and hence $\chi(1) = 1$, and it follows that χ is a homomorphism from G into the multiplicative group of the complex numbers. Since G is nonabelian simple and $\mathbb{C}^\times$ is abelian, it follows that the only possible homomorphism is $\chi(g) = 1$ for all $g \in G$. Thus χ is the principal character, as required. ∎

There are two essentially different ways in which integrality comes into character theory. The first of these, which enables us to deduce Theorem 28.26 from Theorem 28.27, is very simple.

(28.29) LEMMA. *The values of characters of finite groups are algebraic integers.*

Proof. By Lemma 15.13 and the remarks following its proof, character values are sums of roots of unity. Since roots of unity are clearly integral, the result follows by Theorem 28.6. ∎

Proof of Theorem 28.26 (assuming Theorem 28.27). Let $x \in G$ lie in a conjugacy class whose size is a power of the prime p, and suppose $x \neq 1$. Then $\rho(x) = 0$, where ρ is the regular character of G. Now by the discussion following Corollary 15.6, we know that $\rho = \sum \chi(1)\chi$, where the sum is taken over all $\chi \in \mathrm{Irr}(G)$. It follows that

$$0 = \rho(x) = 1 + \sum_\psi \psi(1)\psi(x) + \sum_\eta \eta(1)\eta(x),$$

where ψ runs over all irreducible characters of degree divisible by p and η runs over all nonprincipal irreducible characters with degrees not divisible by p. (The first term in the above sum, the quantity 1, is the contribution to $\rho(x)$ that comes from the principal character.)

Now suppose that $\eta \in \mathrm{Irr}(G)$ is nonprincipal and that p does not divide $\eta(1)$. Then $\eta(1)$ is relatively prime to the size of the class of x, and Theorem 28.27 applies. Assuming that G is nonabelian, we conclude from Lemma 28.28 that $\mathbf{Z}(\eta) = 1$ and hence $x \notin \mathbf{Z}(\eta)$. By Theorem 28.27, therefore, $\eta(x) = 0$.

If $\psi \in \mathrm{Irr}(G)$ with $\psi(1)$ divisible by p, then since $\psi(x)$ is an algebraic integer by Lemma 28.29, we see that the quantity $\psi(1)\psi(x)$ has the form $p\beta$, where β is an algebraic integer. It follows that we can write

$$\sum_{\psi} \psi(1)\psi(x) = p\alpha \,,$$

where ψ runs over irreducible characters with degree divisible by p, and α is some algebraic integer. This yields $0 = 1 + p\alpha$, and so $\alpha = -1/p$, a rational number. This is a contradiction, however, since we know that all rational algebraic integers lie in $\mathbb{Z}$ and yet $-1/p \notin \mathbb{Z}$.

This contradiction arose from the assumption that the group G is nonabelian. Thus G is abelian and hence, being simple, has prime order.   ∎

What remains now is to prove Theorem 28.27. For this purpose, we define the functions $\omega_\chi : G \to \mathbb{C}$ for characters $\chi \in \mathrm{Irr}(G)$. We set

$$\omega_\chi(g) = \frac{\chi(g)|K|}{\chi(1)} \,,$$

where K is the conjugacy class of g in G.

The explanation for this somewhat peculiar definition is the following lemma.

(28.30) LEMMA.   *Let $\chi \in \mathrm{Irr}(G)$ and suppose that $\mathcal{X}$ is a representation affording χ. Let $g \in G$ and write $\hat{K}$ to denote the element in the group algebra $\mathbb{C}G$ that is the sum of the elements in the conjugacy class K of g in G. Then*

$$\mathcal{X}(\hat{K}) = \omega_\chi(g) \cdot I \,,$$

where I is the $\chi(1) \times \chi(1)$ identity matrix.

Proof.   Since $\hat{K}$ is central in $\mathbb{C}G$, we see that $\mathcal{X}(\hat{K})$ is central in the $\mathcal{X}(\mathbb{C}G)$, which is the full ring of $\chi(1) \times \chi(1)$ matrices over $\mathbb{C}$. (See Lemma 15.10.) It follows that $\mathcal{X}(\hat{K}) = \omega I$ for some scalar $\omega \in \mathbb{C}$. Computation of traces gives

$$\chi(1)\omega = \mathrm{tr}(\omega I) = \mathrm{tr}(\mathcal{X}(\hat{K})) = |K|\chi(g)$$

since $\mathrm{tr}(\mathcal{X}(x)) = \chi(g)$ for each of the $|K|$ conjugates x of g in G. Solving for ω, we get $\omega = \omega_\chi(g)$, as required.   ∎

Now we come to the second way that integrality comes into character theory.

(28.31) THEOREM.   *Let $\chi \in \mathrm{Irr}(G)$. Then $\omega_\chi(g)$ is an algebraic integer for all elements $g \in G$.*

Proof.   Let M be the $\mathbb{Z}$-submodule of $\mathbb{C}$ generated by the values of the function $\omega_\chi(g)$ as g runs over G. We will show that M is closed under multiplication, and the result follows by Theorem 28.7. It suffices, therefore, to show for $x, y \in G$ that $\omega_\chi(x)\omega_\chi(y)$ can be written as a $\mathbb{Z}$-linear combination of $\omega_\chi(g)$ for $g \in G$.

Write X and Y to denote the conjugacy classes of x and y in G, and observe that working in $\mathbb{C}G$ and using notation as in Lemma 28.30, we can write $\hat{X}\hat{Y}$ as a $\mathbb{Z}$-linear combination of conjugacy class sums. (The coefficient of the class sum $\hat{K}$ in this product is the number of ways that an element $g \in K$ can be written as a product of an element of X with an element of Y.)

Let $\mathcal{X}$ be a representation affording χ as in Lemma 28.30. Then $\mathcal{X}(\hat{X})\mathcal{X}(\hat{Y}) = \mathcal{X}(\hat{X}\hat{Y})$ is a $\mathbb{Z}$-linear combination of the matrices $\mathcal{X}(\hat{K})$ as K runs over the various conjugacy classes of G. The result now follows by Lemma 28.30. ∎

Proof of Theorem 28.27.   Write $f = \chi(1)$ and $n = |K|$, where K is the conjugacy class of X in G. Then $\omega_\chi(x) = n\chi(x)/f$ is an algebraic integer by Theorem 28.31. Since n and f are coprime, we can choose rational integers k and l such that $kf + ln = 1$, and this gives

$$\frac{\chi(x)}{f} = (kf + ln)\frac{\chi(x)}{f} = k\chi(x) + l\omega_\chi(x),$$

and this is an algebraic integer. We write $\alpha = \chi(x)/f$.

By Lemma 15.13 we know that $\chi(x)$ is a sum of f roots of unity, and these are all equal since we are assuming that $x \notin \mathbf{Z}(\chi)$. Thus $|\chi(x)| < f$, and it follows that $|\alpha| < 1$. Also $|\alpha^\sigma| \le 1$ for all $\sigma \in \mathrm{Gal}(E/\mathbb{Q})$, where E is any extension of the field $\mathbb{Q}[\alpha]$ in $\mathbb{C}$. (This is so because $\chi(x)^\sigma$ is also a sum of f roots of unity.)

Now take $E \supseteq \mathbb{Q}[\alpha]$ to be Galois over $\mathbb{Q}$. (Actually, $\mathbb{Q}[\alpha]$ is already Galois, but we do not need this fact.) The product of the α^σ as σ runs over $\mathrm{Gal}(E/\mathbb{Q})$ is rational and is an algebraic integer (since α is integral), and so it lies in $\mathbb{Z}$. The absolute value of this product is less than 1, however, and it follows that the product equals 0. We deduce that $\alpha = 0$, and thus $\chi(x) = 0$, as required. ∎

Problems

28.1 Let $R \subseteq S \subseteq T$ be unitary extensions of rings. Suppose that S is integral over R and that $t \in T$ is integral over S. Show that t is integral over R.

28.2 Let $\omega = e^{2\pi i/3}$ and show that the algebraic integers in $\mathbb{Q}[\omega]$ are exactly the $\mathbb{Z}$-linear combinations of 1 and ω.

NOTE: In fact, if ϵ is any complex root of unity, then $\mathbb{Z}[\epsilon]$ is exactly the ring of integers in $\mathbb{Q}[\epsilon]$.

28.3 Let α be an algebraic integer and let $f \in \mathbb{Q}[X]$ be the minimal polynomial of α over $\mathbb{Q}$. Show that $1/\alpha$ is integral iff $f(0) = \pm 1$.

28.4 Let α be a nonzero algebraic integer. Show that there exists a unique largest rational integer n such that α/n is integral. If also α/m is integral, show that m divides n.

28.5 Let $R < S$ be a proper unitary extension of rings, and assume that R is noetherian and is integrally closed in S. If $I \subseteq R$ is an ideal of S, show that there exists $d \in S$ such that $d \neq 0$ and $dI = 0$.

HINT: Use the fact that I is finitely generated as an R-module to write a matrix equation expressing the fact that $Is \subseteq I$ for $s \in S$. If we take s non-integral over R, this guarantees the nonvanishing of a certain determinant.

28.6 Suppose $\mathbb{Z} \subseteq T \subseteq \mathbb{C}$, where T is a ring. Assume that there exists $n \in \mathbb{Z}$ with $n \neq 0$ and such that nT consists entirely of algebraic integers. Show that T consists entirely of algebraic integers.

HINT: If $s \in T$ is nonintegral, let R be the ring of integers in $\mathbb{Q}[s]$ and let $S = R[s]$. Apply Problem 28.5 with $I = \{r \in R \mid rS \subseteq R\}$.

28.7 Let $F \subseteq E$ be fields and suppose that $\alpha_1, \ldots, \alpha_n \in E$. Define $f \in E[X]$ by $f(X) = \prod(X - \alpha_i)$, and write

$$f(X) = X^n + c_{n-1}X^{n-1} + c_{n-2}X^{n-2} + \cdots + c_1 X + c_0.$$

a. Show that $F[c_0, \ldots, c_{n-1}] \subseteq F[\alpha_1, \ldots, \alpha_n]$ is an integral extension of rings.
b. If the elements $\alpha_1, \ldots, \alpha_n$ are algebraically independent over F, show that
$$F(c_0, \ldots, c_{n-1}) \cap F[\alpha_1, \ldots, \alpha_n] = F[c_0, \ldots, c_{n-1}].$$

HINT: For part (b), show that the c_i are algebraically independent.

NOTE: When this problem is combined with Problem 18.25 or with Corollary 24.25, it shows that every symmetric polynomial in n indeterminates is a polynomial in the elementary symmetric polynomials. The earlier results gave only that it was a rational function in the elementary symmetric polynomials.

28.8 Let $E \supseteq \mathbb{Q}$ be a Galois extension with $E \subseteq \mathbb{C}$ and suppose that $a > 0$ is a real number. Consider the set S of all algebraic integers $\alpha \in E$ such that $|\alpha^\sigma| \leq a$ for all $\sigma \in \mathrm{Gal}(E/\mathbb{Q})$. Show that S is a finite set.

28.9 Let $E \supseteq \mathbb{Q}$ be a Galois extension with $E \subseteq \mathbb{C}$. Suppose that $\mathrm{Gal}(E/\mathbb{Q})$ is abelian and let $\alpha \in E$ be an algebraic integer with $|\alpha| = 1$. Show that α is a root of unity.

HINT: Show that all the powers of α lie in the set S of Problem 28.8 (taking $a = 1$).

NOTE: The assumption that the Galois group is abelian is necessary. Consider, for instance, the field $E = \mathbb{Q}[i, \sqrt[4]{2}]$, the splitting field of $X^4 - 2$ over $\mathbb{Q}$. Take $\alpha = (\sqrt{2} - 1)((\sqrt{2} - 1) + (2\sqrt[4]{2})i)$ and note that $|\alpha| = 1$. To see that α is not a root of unity, observe that there exists $\sigma \in \mathrm{Gal}(E/\mathbb{Q})$ such that $(\sqrt[4]{2})^\sigma = i\sqrt[4]{2}$ and $i^\sigma = i$. Compute that α^σ does not have absolute value equal to 1.

28.10 Let R be an integrally closed domain and suppose that $M \subseteq R$ is a multiplicative system in R. Let $S = R_M$, the localization of R at M. Show that S is an integrally closed domain.

HINT: View S as a subring of the fraction field of R.

28.11 Let d be a square-free rational integer. Compute the discriminant of the field $\mathbb{Q}[\sqrt{d}]$.

28.12 Let $\chi \in \mathrm{Irr}(G)$, where G is a finite group. Show that

$$\sum_x \chi(x)\overline{\omega_\chi(x)} = \frac{|G|}{\chi(1)},$$

where x runs over a set of representatives for the conjugacy classes of G. Deduce that the degree $\chi(1)$ divides $|G|$ for all irreducible characters χ of G.

28.13 Let G be a finite group and write S to denote the set of prime numbers p such that there exists an irreducible character of G with degree divisible by p. If $|S| \le 2$ and G is simple, show that G has prime order.

HINT: If $p \in S$, let x be a nontrivial element in the center of a Sylow p-subgroup of G, and mimic the proof of Theorem 28.26.

NOTE: Since S is a subset of the set of prime divisors of $|G|$ by Problem 28.12, we see that the assertion of this problem generalizes Theorem 28.26.

28.14 Let G be nonabelian of order p^3, where p is a prime. Compute the degrees of all the irreducible characters of G, and deduce that G has exactly $p^2 + p - 1$ conjugacy classes.

28.15 Let χ be a character of a finite group G and suppose that $|\chi(g)| = 1$ for some element $g \in G$. Show that $\chi(g)$ is a root of unity.

CHAPTER TWENTY-NINE

Dedekind Domains

29A

Let E be a finite-degree field extension of $\mathbb{Q}$, and let R be the ring of algebraic integers in E. Although R may be viewed as a kind of generalization of the ring $\mathbb{Z}$ of rational integers, and indeed R enjoys many of the familiar properties of $\mathbb{Z}$, there is one important respect in which R can differ from the ordinary integers: unique factorization can (and usually does) fail. We saw in Chapter 28, for instance, that the ring of integers in $\mathbb{Q}[\sqrt{-5}]$ is not a UFD.

In the latter part of the nineteenth century, attempts were made to prove the so-called "last theorem" of Fermat using algebraic integers. (Recall that Fermat's claim was that no positive integers x, y, and z can satisfy the equation $x^n + y^n = z^n$ for integers $n > 2$. As of 1992, it remained unknown whether or not this is true.) These attempts failed precisely because unique factorization does not hold in rings R of algebraic integers in finite degree extensions of $\mathbb{Q}$. Dedekind was able to show, however, that although unique factorization of numbers (ring elements) can fail, there is another sort of "ideal number" for which unique factorization always holds. Some of these ideal numbers corresponded to genuine numbers and the rest were "ideal" (in the sense of being not "actual"). Dedekind was unable, however, to use his modified version of unique factorization to prove Fermat's famous assertion.

Dedekind's ideal numbers are what are called simply "ideals" today, and those that corresponded to numbers were the principal ideals. The rings for which Dedekind's unique factorization of ideals is valid are now called "Dedekind domains," and as we saw in Corollary 28.22, this class of rings includes the rings of integers in finite-degree extensions of $\mathbb{Q}$. We repeat the definition here.

(29.1) DEFINITION. A domain is a *Dedekind domain* if it is noetherian and integrally closed and has the property that every nonzero prime ideal is maximal.

The most familiar examples of Dedekind domains are principal ideal domains (PIDs).

474

(29.2) LEMMA. *Let R be a PID. Then R is a Dedekind domain.*

Proof. Since every ideal of R is generated by a single element, the ideals are certainly finitely generated, and so R is noetherian. We know by Theorem 16.10 that R is a UFD, and so it is integrally closed by Lemma 28.3. Finally, the nonzero prime ideals of R are maximal by Lemma 16.9. ∎

We may view Dedekind domains as being only very slight generalizations of PIDs. For example, it turns out (as we shall see) that ideals of Dedekind domains never require more than two generators.

Recall that if A and B are any two additive subgroups of a ring, then AB is defined to be the additive subgroup generated by the set of all products ab, where $a \in A$ and $b \in B$. In other words, the elements of AB are all finite sums of products of elements of A with elements of B. Of course, A^n denotes the product of n copies of A for $n > 0$. With this notation, we can state the unique factorization theorem.

(29.3) THEOREM. *Let R be a Dedekind domain and suppose that $A \subseteq R$ is a nonzero ideal. Then there exist a unique finite set $\mathcal{X}$ of nonzero prime ideals of R and a unique positive-integer valued function e defined on $\mathcal{X}$ such that*

$$A = \prod_{P \in \mathcal{X}} P^{e(P)}.$$

Given a domain R, which we do not yet assume to be Dedekind, it turns out to be convenient to consider a kind of generalized ideal.

(29.4) DEFINITION. Let R be a domain with field F of fractions. An R-submodule $M \subseteq F$ is said to be a *fractional ideal* of R if there exists some nonzero element $\alpha \in F$ such that $M\alpha \subseteq R$.

Note that an ideal of R (in the usual sense) is automatically a fractional ideal since we can take $\alpha = 1$ in Definition 29.4. Also, any finitely generated R-submodule of the field F of fractions of R is a fractional ideal. To see this, let $x_1, \ldots, x_n$ be generators and write $x_i = a_i/b_i$, with $a_i, b_i \in R$ and $b_i \neq 0$. Set $\alpha = \prod b_i$ and observe that $\alpha \neq 0$. If x is any R-linear combination of the x_i, then multiplication by α clears denominators and we have $x\alpha \in R$, as claimed. In particular, the one-generator submodules of F are fractional ideals; these are the *principal* fractional ideals. (The reader should check that the ideals of R that are principal as fractional ideals are exactly the principal ideals.)

If M is any R-submodule of F and $0 \neq \alpha \in R$, then $M\alpha$ is also a submodule and $M\alpha \cong M$ as R-modules. If M is a fractional ideal, therefore, then M is module-isomorphic to an ordinary ideal of R. For noetherian rings, it follows that fractional ideals are finitely generated, and we have proved the following lemma .

(29.5) LEMMA. *Let R be a noetherian domain. Then the fractional ideals of R are exactly the finitely generated R-submodules of the fraction field of R.* ∎

Products of fractional ideals are again fractional ideals, since if $A\alpha \subseteq R$ and $B\beta \subseteq R$, then $(AB)(\alpha\beta) \subseteq R$. Another useful way to construct new fractional ideals from old is as follows. If A is a nonzero fractional ideal of R, we write A^{-1} to denote $\{\alpha \in F \mid A\alpha \subseteq R\}$, where as usual, F is the field of fractions of R. (For example, $R^{-1} = R$ and $A^{-1} \supseteq R$ for ideals $A \subseteq R$.) It is a triviality that A^{-1} (pronounced "A-inverse") is an R-submodule of F. Furthermore, A^{-1} is also a fractional ideal since $\beta A^{-1} \subseteq R$ for any nonzero element $\beta \in A$. Also, if $A \neq 0$ is a fractional ideal, then Definition 29.4 guarantees that $A^{-1} \neq 0$ and so $(A^{-1})^{-1}$ is defined. (Caution: It is not always true that $(A^{-1})^{-1} = A$.)

Consider the set $\mathcal{F} = \mathcal{F}(R)$ of nonzero fractional ideals for a domain R. We have an associative commutative multiplication on $\mathcal{F}$ and also a unary operation called "inverse." Note that $R \in \mathcal{F}$, and R acts as an identity element in $\mathcal{F}$. A fractional ideal $A \in \mathcal{F}$ is said to be *invertible* if there exists $B \in \mathcal{F}$ such that $AB = R$. We stress that although every element of $\mathcal{F}$ has an inverse, it is not obvious (and it is not in general true) that every nonzero fractional ideal is invertible. What is true is that if $A, B \in \mathcal{F}$ satisfy $AB = R$, then necessarily $B = A^{-1}$. To see this, note that $B \subseteq A^{-1}$ by the definition of A^{-1}, and also $A^{-1} = A^{-1}R = A^{-1}AB \subseteq RB = B$. Note that in this case where A is invertible, we have $AA^{-1} = R$ and $(A^{-1})^{-1} = A$.

The reader has probably guessed that in some important cases (if not always) the set $\mathcal{F}(R)$ forms a (necessarily abelian) group. It should be clear that $\mathcal{F}$ is a group precisely when every nonzero fractional ideal is invertible. It is one of the key results in this subject that this happens iff R is a Dedekind domain.

(29.6) THEOREM.   *Let R be a domain. Then the following are equivalent:*

   i.   *R is a Dedekind domain.*
  ii.   *Every nonzero ideal of R is invertible.*
 iii.   *$\mathcal{F}(R)$ is an abelian group.*

For any domain R, it is easy to show that the nonzero principal fractional ideals are invertible, and in fact, they always form a group. The fact that (i) implies (ii) in Theorem 29.6 thus shows another way in which Dedekind domains behave as generalized PIDs. This part of Theorem 29.6 is also crucial for our proof of the unique factorization theorem (29.3).

(29.7) LEMMA.   *Let R be a noetherian domain in which all nonzero prime ideals are maximal. (For example, R may be a Dedekind domain.) Let $0 < A \subseteq P$ be ideals of R, where P is prime, and write $U = \{x \in R \mid xP \subseteq A\}$. Then U is an ideal of R and $U > A$.*

Proof.   It is clear that U is an ideal and obviously $U \supseteq A$. We work to find elements of U that are not in A.

     Since R is noetherian, Lemma 27.10 applies, and we can find prime ideals $P_1, P_2, \ldots, P_n$, each containing A and such that $\prod P_i \subseteq A$. Choose these ideals so that n is as small as possible. (Note that the P_i need not be distinct.) Since $A \subseteq P$ and P is prime, we deduce that $P_i \subseteq P$ for some subscript i, and we may assume that $P_n \subseteq P$. But P_n is nonzero since it contains A, and thus P_n is a maximal ideal by hypothesis. It follows that $P_n = P$ and thus $P_1 P_2 \cdots P_{n-1} \subseteq U$. By the minimality of n, this product is not contained in A.   ∎

(29.8) LEMMA.   *Let R be a Dedekind domain. If $P \subseteq R$ is a nonzero prime ideal, then P is invertible.*

Proof.   Clearly, $P^{-1} \supseteq R$, and we begin by showing that this containment is proper. (Note that this must be true if we expect to prove that $P P^{-1} = R$.) Let $a \in P$ be nonzero and apply Lemma 29.7 to the principal ideal generated by a. We get an ideal $U > (a)$ such that $UP \subseteq (a)$.

Since $UP \subseteq aR$, we have $(1/a)UP \subseteq R$ (working in the fraction field of R) and hence $(1/a)U \subseteq P^{-1}$. But $U \not\subseteq (a)$ and thus $(1/a)U \not\subseteq R$, and it follows that $P^{-1} \not\subseteq R$, as claimed.

Now $P = PR \subseteq P P^{-1}$. Because $P P^{-1} \subseteq R$ is an ideal and P is maximal, there are just two possibilities: either $P P^{-1} = R$, as desired, or $P P^{-1} = P$. We assume the latter and derive a contradiction.

If $\alpha, \beta \in P^{-1}$, then $P\alpha\beta \subseteq P P^{-1}\beta = P\beta \subseteq R$. Thus $\alpha\beta \in P^{-1}$ and P^{-1} is closed under multiplication. In other words, P^{-1} is a subring of F. Since the ring P^{-1} is a fractional ideal of the noetherian ring R, it is finitely generated as an R-module, and Lemma 28.7 tells us that its elements are all integral over R. The ring R is Dedekind, however, and so it is integrally closed in F. This is a contradiction since $P^{-1} > R$.   ∎

We need one more lemma to assist with the proof that (iii) implies (i) in Theorem 29.6.

(29.9) LEMMA.   *Let A be an invertible fractional ideal of a domain R. Then A is finitely generated as an R-module.*

Proof.   Since $AA^{-1} = R$, we have $1 \in AA^{-1}$, and we can write $1 = a_1\alpha_1 + a_2\alpha_2 + \cdots + a_n\alpha_n$, where $a_i \in A$ and $\alpha_i \in A^{-1}$ for all subscripts i. We claim that the elements a_i form a generating set for A. To see this, we show how to write an arbitrary element of A as an R-linear combination of the a_i.

Let $b \in A$ and observe that $b = 1b = a_1(\alpha_1 b) + \cdots + a_n(\alpha_n b)$. The coefficients $\alpha_i b$ lie in $A^{-1}A = R$, as required.   ∎

Proof of Theorem 29.6.   To prove that (i) implies (ii), we assume that R is Dedekind. If (ii) fails, choose a nonzero ideal $I \subseteq R$, maximal with the property that it is not invertible. By Lemma 29.8 we know that I is not prime, and thus there exist ideals $A > I$ and $B > I$ with $AB \subseteq I$. We have $IB^{-1} \subseteq BB^{-1} \subseteq R$, and thus IB^{-1} is an ideal. By the maximality of I, we see that B is invertible, and thus $I < A = ABB^{-1} \subseteq IB^{-1}$. It follows that IB^{-1} is invertible and we have $IB^{-1}C = R$, where $C = (IB^{-1})^{-1}$. Since $I(B^{-1}C) = R$, we see that I is invertible, and this is a contradiction.

We now assume (ii). Thus nonzero ideals are invertible, and to prove that $\mathcal{F}(R)$ is a group, we must show that an arbitrary nonzero fractional ideal is invertible. If $A \in \mathcal{F}$, choose a nonzero element $\alpha \in A^{-1}$. Then $A\alpha \subseteq R$ and $A\alpha$ is an ideal. By (ii), $A\alpha$ is invertible and thus $A\alpha B = R$, where $B = (A\alpha)^{-1}$. Since $A(\alpha B) = R$, we see that A is invertible.

Assuming (iii) now, we have $AA^{-1} = R$ for all nonzero ideals A of R. It follows by Lemma 29.9 that these ideals are all finitely generated and hence R is noetherian. To see that R is integrally closed in its fraction field F, suppose $\alpha \in F$ is integral over R. By Lemma 28.8 the ring $R[\alpha]$ is finitely generated as an R-module, and thus it is a fractional ideal. Since $R[\alpha]$ is closed under multiplication, however, we have $R[\alpha]R[\alpha] = R[\alpha]$. Multiplication by $R[\alpha]^{-1}$ gives $R[\alpha] = R$, and thus $\alpha \in R$, as required.

Finally, we must show that the nonzero prime ideals of R are maximal. Suppose, then, that P is a prime ideal and $0 < P < A$, where A is an ideal. We observe that $PA^{-1} \subseteq R$. Thus PA^{-1} is an ideal and we have $(PA^{-1})A \subseteq P$. Since P is prime and $A \not\subseteq P$, we deduce that $PA^{-1} \subseteq P$. Multiplication by AP^{-1} gives $R \subseteq A$, and hence $A = R$, as required. ∎

At last, we are ready to prove the unique factorization theorem.

Proof of Theorem 29.3. We work first to show existence. We must prove that every nonzero ideal A of R can be written as a product of prime ideals, and so if this is false, choose A maximal among ideals that cannot be so written. Since by convention the empty product is R itself, we have $A < R$, and we can choose a maximal ideal P that contains A.

We have $A = AR \subseteq AP^{-1} \subseteq PP^{-1} \subseteq R$, and thus AP^{-1} is an ideal of R that contains A. Because the fractional ideals form a group and $P \neq R$, we know that $A \neq AP^{-1}$ and thus $A < AP^{-1}$. It follows by the maximality of A that we can write $AP^{-1} = P_1 P_2 \cdots P_n$ for some choice of (not necessarily distinct) prime ideals P_i. This yields $A = P P_1 P_2 \cdots P_n$, as required.

We now know that every ideal A of R can be written in the form

$$A = \prod_{P \in \mathcal{X}} P^{e(P)}$$

for some finite set $\mathcal{X}$ of nonzero prime ideals and some choice of positive exponents. For convenience, we extend the exponent function $e(P)$ by defining $e(P) = 0$ for primes $P \notin \mathcal{X}$, and we view the product as being taken over the possibly infinite set of all nonzero primes of R. We show by induction on $m = \sum e(P)$ that if also

$$A = \prod P^{f(P)}$$

for some nonnegative valued function f with nonzero values on just finitely many primes, then $e = f$. This will complete the proof of the theorem.

If $m = 0$, then e is identically zero and $A = R$. If $f(P) > 0$, however, then $A \subseteq P$, and this contradiction implies that f is identically zero, as required. Suppose now that $m > 0$ and choose Q so that $e(Q) > 0$. Then $A \subseteq Q$, and since Q is prime, this implies that $f(Q) > 0$. Since $A \subseteq Q$, we see that AQ^{-1} is an ideal and we have

$$\prod P^{e'(P)} = AQ^{-1} = \prod P^{f'(P)} \,,$$

where we have obtained the functions e' and f' by decrementing the values of e and f by one at Q and leaving all of the other values unchanged. Now $m' = \sum e'(P) = m - 1$, and so the inductive hypothesis gives $e' = f'$, and hence $e = f$.   ■

Let $\mathcal{P}$ denote the set of all nonzero prime ideals of the Dedekind domain R, and write $\mathcal{G}(\mathcal{P})$ to denote the group (with respect to pointwise addition) of all rational-integer-valued functions with finite support on $\mathcal{P}$. (This means that we are considering only those functions whose value is nonzero on at most finitely many elements of the set $\mathcal{P}$.) There is a natural map θ from $\mathcal{G}(\mathcal{P})$ into the group $\mathcal{F}(R)$ of nonzero fractional ideals of R. We simply set

$$\theta(f) = \prod_{P \in \mathcal{P}} P^{f(P)} \,,$$

and note that this makes sense because this is essentially a finite product. The content of Theorem 29.3 is precisely that θ defines a bijection from the set of nonnegative-valued elements of $\mathcal{G}(\mathcal{P})$ onto the set of nonzero ideals of R. The following result extends this.

(29.10) COROLLARY.   *The map θ defined above is an isomorphism from the group $\mathcal{G}(\mathcal{P})$ onto the group $\mathcal{F}(R)$ of all nonzero fractional ideals of R.*

It should be clear that up to isomorphism, the group $\mathcal{G}(\mathcal{P})$ depends on only the cardinality $|\mathcal{P}|$, and not at all on any of the deeper structure of the ring R. In fact, this group is exactly the restricted direct sum of $|\mathcal{P}|$ copies of $\mathbb{Z}$. Because of this, Corollary 29.10 is perhaps a bit disappointing. It tells us that the isomorphism type of the group of nonzero fractional ideals of R is not a very strong invariant; it gives no more information than the cardinality of $\mathcal{P}$.

Proof of Corollary 29.10.   Since θ is easily seen to be a group homomorphism, it suffices to check that it is both injective and surjective. We have observed that by Theorem 29.3 the image of θ contains all nonzero ordinary ideals and that θ is injective when restricted to the set of nonnegative functions.

If A is a nonzero fractional ideal, then $A\alpha \subseteq R$ for some element α in the fraction field. Writing $\alpha = r/s$, with $r, s \in R$, we see that $Ar \subseteq R$, and hence Ar is an ideal and lies in the image of θ. The principal ideal (r) of R also lies in the image of θ, and since $A(r)$ is in the image, it follows that A is in the image, too. The homomorphism θ is thus surjective.

Now suppose $\theta(f) = 0$. We can write $f = g - h$, where $g, h \in \mathcal{G}(\mathcal{P})$ and g and h are nonnegative valued. Since $\theta(g) = \theta(h)$, we have $g = h$ and hence $f = 0$.   ■

29B

We wish to argue now that Dedekind domains are "almost" principal ideal domains. One possible interpretation of this assertion is that "many" of the ideals are principal. One can look, for instance, at the subgroup of $\mathcal{F}(R)$ consisting of principal fractional ideals, and try to show that its index is small. It is a theorem of number theory (which we do not prove here) that if R is the ring of integers in a finite-degree extension of $\mathbb{Q}$, then this index (called the *class number* of the field) is finite. This does not hold in general, however. What is true is that every ideal of a Dedekind domain can be "closely approximated" by a principal ideal. To make this precise, we introduce some notation.

If R is a Dedekind domain and $A \subseteq R$ is an ideal, we write $f_A(P)$ to denote the exponent on the nonzero prime P in the unique prime factorization of A. Thus f_A is a nonnegative-integer-valued function on the set $\mathcal{P}$ of nonzero primes of R. Note that by the unique factorization theorem, the function f_A completely determines A. (In fact, in the notation of Corollary 29.10, we have $A = \theta(f_A)$.)

(29.11) THEOREM (Finite approximation). *Let R be a Dedekind domain and suppose $A \subseteq R$ is an ideal. Given any finite set $\mathcal{X}$ of nonzero primes of R, we can find a principal ideal B of R such that $f_A(P) = f_B(P)$ for all $P \in \mathcal{X}$.*

(29.12) COROLLARY. *A Dedekind domain with just finitely many prime ideals is a PID.* ∎

The key ingredient in the proof of the finite approximation theorem is the following version of the "Chinese remainder" theorem. (This is essentially Problem 26.17.)

(29.13) LEMMA. *Let R be an arbitrary commutative ring, and suppose $U_1, U_2, \ldots, U_n$ are ideals of R such that $U_i + U_j = R$ whenever $i \neq j$. Define the map $\varphi : R \to (R/U_1) \times \cdots \times (R/U_n)$ by setting the ith component of $\varphi(x)$ to be the natural image of x in R/U_i. Then φ is a surjective homomorphism.*

Proof. The map φ is clearly a homomorphism, and so it suffices to find subsets $V_i \subseteq R$ such that V_i maps to zero in R/U_j when $i \neq j$, and V_i maps onto R/U_i. In fact, we can take $V_i = \prod U_j$, where j runs over all values different from i.

Clearly, $V_i \subseteq U_j$ when $j \neq i$, and so V_i maps to zero in R/U_j, as required. To see that V_i maps onto R/U_i, it suffices to check that $R = U_i + V_i$. If this is false, however, then $U_i + V_i$ is contained in some maximal ideal P of R, and we have

$$\prod_{j \neq i} U_j = V_i \subseteq P .$$

Since P is prime, this yields $U_j \subseteq P$ for some subscript $j \neq i$. Since also $U_i \subseteq P$ and $U_i + U_j = R$, this is a contradiction. ∎

We need one more preliminary result.

(29.14) LEMMA.   *Let A be an ideal of the Dedekind domain R. If P is a nonzero prime of R, then $f_A(P)$ is the largest integer e such that $A \subseteq P^e$.*

Proof.   Write $f = f_A(P)$. Since P^f is a factor of A, we certainly have $A \subseteq P^f$. We need to show that if $A \subseteq P^e$, then $e \le f$ or, equivalently, P^e is a factor of A. Write $B = A(P^e)^{-1}$ and note that $B \subseteq R$, so that B is an ideal. Since $A = BP^e$, the result follows.   ∎

Proof of Theorem 29.11   For each prime $P \in \mathcal{X}$, write $T_P = P^{f_A(P)}$ and $U_P = PT_P$. By Lemma 29.14 we have $A \not\subseteq U_P$, and so we can choose elements a_P in A with $a_P \notin U_P$. We wish now to apply Lemma 29.13 (the Chinese remainder theorem) to the finite collection of ideals U_P for $P \in \mathcal{X}$. We need to verify that $U_P + U_Q = R$ if P and Q are distinct nonzero primes. To see this, observe that by primeness, P is the only maximal ideal that can contain a power of P, and similarly, Q is the only maximal ideal that contains a power of Q. Since $P \neq Q$, this forces $U_P + U_Q = R$, and Lemma 29.13 does apply.

Since φ maps R onto the direct product of the rings R/U_P, we can find $b \in R$ such that for all $P \in \mathcal{X}$, the image of b in R/U_P is the same as the image of a_P in R/U_P.

We claim that the principal ideal $B = (b)$ has the desired properties. By Lemma 29.14, therefore, we must show for each $P \in \mathcal{X}$ that $B \subseteq T_P$, but that B is not contained in any higher power of P. Now $a_P \in A \subseteq T_P$ and $b \equiv a_P \bmod U_P$. Since $U_P \subseteq T_P$, it follows that $B = (b) \subseteq T_P$, as required. Finally, if B were contained in a higher power of P, we would have $b \in B \subseteq U_P$, and thus $a_P \in U_P$ since $a_P \equiv b \bmod U_P$. This is a contradiction.   ∎

Another interesting application of the finite approximation theorem is the following corollary.

(29.15) COROLLARY.   *Let I be a nonzero proper ideal of a Dedekind domain R. Then the ring R/I is a principal ideal ring.*

Proof.   Every ideal of R/I has the form A/I for some ideal A of R that contains I. Let $\mathcal{X}$ be the set of primes P of R for which $f_I(P) > 0$. Since this set is finite, we can choose a principal ideal B of R such that $f_B(P) = f_A(P)$ for all $P \in \mathcal{X}$. We claim that $B + I = A$. It follows that A/I is the image of the principal ideal B under the natural homomorphism $R \to R/I$, and thus A/I is principal.

We will show that $f_{B+I}(Q) = f_A(Q)$ for all nonzero primes Q of R, and this proves that $B + I = A$, as claimed. First suppose $Q \notin \mathcal{X}$. Then $f_I(Q) = 0$ and so $I \not\subseteq Q$. It follows that neither A nor $B + I$ is contained in Q, and hence $f_{B+I}(Q) = 0 = f_A(Q)$. Finally, suppose that $Q \in \mathcal{X}$ and let $e = f_A(Q) = f_B(Q)$. Then B and A are both contained in P^e, and so $B + I \subseteq B + A \subseteq P^e$. But $B + I$ cannot be contained in any higher power of Q since not even B is contained in such a power. It follows that $f_{B+I}(Q) = e$, and the proof is complete.   ∎

It is perhaps surprising that the converse of Corollary 29.15 is also true. If R is a domain with the property that every proper homomorphic image is a principal ideal ring, then R must be Dedekind. We leave the proof of this to the problems at the end of the chapter.

We mentioned earlier that every ideal of a Dedekind domain can be generated by two elements. We can now prove this, and in fact we get more: one of the two elements is arbitrary, provided it is nonzero.

(29.16) COROLLARY. *Let A be an ideal of a Dedekind domain R. Given any nonzero element $a \in A$, it is possible to find a second element $b \in A$ such that A is generated by a and b.*

Proof. Let $I = (a)$. By Corollary 29.15, R/I is a PIR and so A/I is a principal ideal of R/I. If the coset $I + b$ is a generating element for A/I, then a and b generate A. ∎

29C

We saw in Corollary 28.22 that the ring of integers in any finite degree field extension of $\mathbb{Q}$ is a Dedekind domain. More generally, we have the following theorem.

(29.17) THEOREM. *Suppose $F \subseteq E$ is a finite-degree field extension, where F is the fraction field of a Dedekind domain R. Then the ring S of elements of E integral over R is a Dedekind domain with fraction field E.*

Of the three defining properties of a Dedekind domain: noetherian, integrally closed, and nonzero primes maximal, the only one that is hard to establish for the ring S of Theorem 29.17 is that it is noetherian. If E happens to be separable over F, then Theorem 28.20 guarantees that S is noetherian as an R-module, and thus it is certainly noetherian as a ring. In the inseparable case, S may not be a noetherian R-module, but it is always a noetherian ring. This is what we intend to establish here.

First, however, we dispose of the easy part of Theorem 29.17.

(29.18) LEMMA. *In the situation of Theorem 29.17, the domain S is integrally closed and E is its fraction field. Also, all nonzero primes of S are maximal.*

Proof. That S is integrally closed with fraction field E is given by Corollary 28.9. To see that the nonzero prime ideals of S are maximal, suppose to the contrary that P is a prime that is neither zero nor maximal. If Q is a maximal ideal of S that contains P, then $0 < P < Q$ is a strictly increasing chain of prime ideals. By INC (see Corollary 28.14), we get a strictly increasing chain of three prime ideals of R, and this contradicts the fact that all nonzero primes of R are maximal. ∎

We need two preliminary results to prove that S is noetherian. The first of these is quite general.

(29.19) LEMMA. *Let M be a module for a commutative ring R. Suppose that $A \subseteq R$ is maximal among ideals of R such that there exists an R-submodule $N \subseteq M$ for which N/NA is not noetherian as an R-module. Then A is prime.*

Proof. Certainly $A \neq R$, and so if A is not prime, there exist ideals $U, V > A$ such that $UV \subseteq A$. Since $U > A$, the maximality of A yields that N/NU is noetherian. Similarly, since $V > A$, we see that $NU/(NU)V$ is noetherian, and it follows by Theorem 11.6 that N/NUV is noetherian. Since $A \supseteq UV$, we have $NA \supseteq NUV$ and so N/NA is a homomorphic image of N/NUV. It follows that N/NA is noetherian, and this is a contradiction. $\blacksquare$

(29.20) LEMMA. *Let A be a proper ideal of a Dedekind domain R with fraction field F. If $\alpha_1, \alpha_2, \ldots, \alpha_n \in F$ are not all zero, then there exists an element $\beta \in F$ such that $\beta\alpha_i \in R$ for all i, but not all $\beta\alpha_i$ lie in A.*

Proof. Let U be the R-submodule of F generated by the α_i, and note that since U is finitely generated, it is a nonzero fractional ideal. By Theorem 29.6 we have $UU^{-1} = R \not\subseteq A$, and so we can choose $\beta \in U^{-1}$ with $\beta U \not\subseteq A$. Since $\beta U \subseteq R$, the result follows. $\blacksquare$

Proof of Theorem 29.17. By Lemma 29.18 it suffices to show that S is a noetherian ring. There is nothing to prove if S is a noetherian R-module, and so we suppose that it is not. The set $\mathcal{Y}$ of ideals $A \subseteq R$ with the property that there exists an R-submodule $N \subseteq S$ such that N/NA is not noetherian as an R-module is therefore nonempty. (We can take $N = S$ to see that $0 \in \mathcal{Y}$.) Since R is noetherian, we choose (and fix) a maximal element $A \in \mathcal{Y}$. By Lemma 29.19, A is prime and thus either A is a maximal ideal of R or $A = 0$.

Suppose first that A is maximal and write $K = R/A$, a field. Fix $N \subseteq S$ such that N/NA is not noetherian as an R-module. We can view N/NA also as an R/A-module, and we observe that the R-submodules of N/NA are exactly the same subsets that are the R/A-submodules, and these fail to satisfy the ACC. It follows that N/NA has infinite dimension as a vector space over $R/A = K$.

Let $n = |E : F|$ and choose $n + 1$ elements of N/NA that are linearly independent over K. Let $v_0, v_1, \ldots, v_n \in N$ be representatives for these cosets of NA, and note that these elements of E must be linearly dependent over F. We can thus find elements $\alpha_i \in F$ that are not all zero and such that

$$(*) \qquad \sum_{i=0}^{n} \alpha_i v_i = 0.$$

By Lemma 29.20 there exists $\beta \in F$ such that $\beta\alpha_i \in R$ for every subscript i, but not all of these elements lie in A. Multiplying equation $(*)$ by β, we can assume that $\alpha_i \in R$ for all i but that not all α_i lie in A. Using overbars to denote the image of an element α of R in R/A and also to denote the image of an element v of N in N/NA, we have $\overline{\alpha v} = \overline{\alpha}\,\overline{v}$ for all $\alpha \in R$ and $v \in N$. If we now read equation $(*)$ modulo NA, we get

$$\sum_{i=0}^{n} \overline{\alpha_i}\, \overline{v_i} = 0.$$

Since not all α_i lie in A, we see that not all of the coefficients $\overline{\alpha_i}$ are zero, and this contradicts the linear independence of the $\overline{v_i}$ over K.

We are left with the case where $A = 0$. Suppose that I is any nonzero ideal of S. By Theorem 28.13 we conclude that $I \cap R > 0 = A$, and thus $S/S(I \cap R)$ is a noetherian R-module. Since $S(I \cap R) \subseteq I$, we conclude that S/I is a noetherian R-module, and thus it is a noetherian ring. This implies that a strictly ascending chain of ideals of S must terminate after at most finitely many steps beyond any nonzero ideal I. In other words, every ascending chain of ideals of S terminates after finitely many steps, and S is noetherian. ∎

29D

We now consider modules over a Dedekind domain R. We are primarily interested in *torsion-free* modules, that is, modules M with the property that if $m \in M$ and $r \in R$ with $mr = 0$, then either $m = 0$ or $r = 0$. We already have a large supply of such modules: the fractional ideals. From the point of view of module theory, we do not wish to distinguish between isomorphic modules, and so we should first decide when two fractional ideals are isomorphic.

(29.21) LEMMA. *Let $\theta : A \to B$ be an R-module homomorphism, where A and B are nonzero fractional ideals of a domain R with fraction field F. Then there exists an element $\alpha \in F$ such that $(x)\theta = \alpha x$ for all $x \in A$. In particular, $A \cong B$ iff $B = \alpha A$ for some nonzero element $\alpha \in F$.*

Proof. Fix a nonzero element $a \in A$ and let $\alpha = (a)\theta/a \in F$. We have $(a)\theta = \alpha a$, and we wish to compute $(x)\theta$ for an arbitrary element x of A. Given $x \in A$, choose $r, s \in R$ such that $x/a = r/s$. This gives

$$(x)\theta s = (xs)\theta = (ar)\theta = (a)\theta r = \alpha a r = \alpha x s ,$$

and we conclude that $(x)\theta = \alpha x$, as required.

If θ is surjective, then $B = A\theta = \alpha A$, and conversely, multiplication by any nonzero element $\alpha \in F$ certainly defines an isomorphism of A onto αA, as required. ∎

If A is an ideal of a domain R, it is easy to see that A is invertible iff αA is invertible, where α is any nonzero element of the fraction field F of R. It follows that whether or not A is invertible depends only on its isomorphism class as an R-module. In fact, it turns out that there is a very simple module theoretic characterization of invertibility.

(29.22) THEOREM. *Let A be a nonzero ideal of a domain R. Then A is invertible iff it is a projective R-module.*

(29.23) COROLLARY.   *A domain is Dedekind iff every ideal is projective.*

Proof.   By Theorem 29.6 we know that a domain is Dedekind iff every nonzero ideal is invertible. The result then follows by Theorem 29.22.   ∎

Recall from Chapter 14 that projective modules are a generalization of free modules. Since a principal ideal of a domain R is module-isomorphic to the regular module, it is free. In fact, it is easy to see that R is a PID iff every ideal is free. Corollary 29.23, therefore, shows another way in which Dedekind domains can be viewed as generalizations of PIDs.

Proof of Theorem 29.22.   By Corollary 14.22 there exists a homomorphism π from some free module M onto A. If we assume that A is projective, then Lemma 14.28 gives an injective homomorphism $\varphi : A \to M$ such that $\varphi\pi$ is the identity map on A. (Remember that the composition $\varphi\pi$ means φ first, then π.)

Let B be an R-basis for M. Each element $m \in M$ can be expressed uniquely as an R-linear combination of B with at most finitely many nonzero coefficients. In particular, if $a \in A$, then $a\varphi$ can be so expressed. For $x \in B$, we define a map $\varphi_x : A \to R$ by setting $a\varphi_x$ to be the coefficient of x in the expansion of $a\varphi$ in terms of B. It is easy to see that φ_x is a module homomorphism, and for $a \in A$, we have $\sum x(a\varphi_x) = a\varphi$, where the sum runs over $x \in B$. (We observe that this sum makes sense since at most finitely many of the elements $a\varphi_x$ are nonzero.)

Pick $a \in A$ and define for each basis element x the quotient $\alpha_x = (a\varphi_x)/a \in F$, where F is the fraction field of R. We claim that $\alpha_x \in A^{-1}$. To see this, let $b \in A$ and note that $(a\varphi_x)b = (ab)\varphi_x = (b\varphi_x)a$. Division by a then yields $\alpha_x b = b\varphi_x \in R$, as required. Because just finitely many α_x are nonzero, we can compute

$$\sum_{x\in B}(x\pi)\alpha_x = \frac{1}{a}\sum_{x\in B}(x\pi)(a\varphi_x) = \frac{1}{a}\Big(\sum_{x\in B}x(a\varphi_x)\Big)\pi = \frac{1}{a}(a\varphi\pi) = 1.$$

Thus $1 \in AA^{-1}$ and it follows that $AA^{-1} = R$, as required.

Conversely, assume that A is invertible. Then $1 \in AA^{-1}$ and we can write

$$1 = a_1\alpha_1 + a_2\alpha_2 + \cdots + a_n\alpha_n$$

for some choice of elements $a_i \in A$ and $\alpha_i \in A^{-1}$. Let $M = R \oplus \cdots \oplus R$ be the direct sum of n copies of the regular module. (In earlier chapters, we wrote $R^\bullet$ to denote a ring R viewed as the regular R-module, but we have decided that in the present context, it is clearer not to use the dots.) Then M is a free module, and we let $\pi : M \to A$ be the module homomorphism defined by

$$(x_1, \ldots, x_n)\pi = \sum_i x_i a_i .$$

We also construct a map $\varphi : A \to M$ by

$$a\varphi = (a\alpha_1, \ldots, a\alpha_n).$$

Note that each component $a\alpha_i$ lies in R since $\alpha_i \in A^{-1}$, and so $a\varphi$ lies in M, as required.

Now

$$a\varphi\pi = \sum_i a\alpha_i a_i = a \sum_i \alpha_i a_i = a1 = a\,,$$

and so $\varphi\pi$ is the identity map on A. It follows from this that φ is injective, and a routine argument yields that M is the direct sum of $\ker(\pi)$ and $A\varphi$. Thus A is isomorphic to the direct summand $A\varphi$ of the free module M, and this implies that A is projective by Theorem 14.26.   ∎

29E

We close this chapter with a classification of finitely generated torsion-free modules over Dedekind domains. The first step is to construct some modules. If A is a nonzero ideal of R and n is a positive integer, we write $S_n(A)$ to denote the R-module defined as the external direct sum $R \oplus R \oplus \cdots \oplus R \oplus A$, where there is a total of n summands. It should be clear that $S_n(A)$ is a finitely generated torsion-free R-module.

(29.24) THEOREM.   *Suppose R is a Dedekind domain. Then:*

    a. *Every finitely generated torsion-free R-module is isomorphic to $S_n(A)$ for some integer n and nonzero ideal A.*

    b. *$S_n(A) \cong S_m(B)$ iff $n = m$ and $A \cong B$.*

The reader may have noticed that we could have built other finitely generated torsion-free R-modules by taking finite direct sums of arbitrary ideals of the Dedekind domain R. (In our construction of $S_n(A)$, we exercised arbitrariness only once, since $n - 1$ of the direct summands were taken to be the regular module.) By the theorem, then, if $A_1, \ldots, A_m$ are nonzero ideals, the module $A_1 \oplus \cdots \oplus A_m$ must be isomorphic to $S_n(A)$ for some integer n and nonzero ideal A. It is not hard to guess that $n = m$, but what is A? There is a very satisfying answer.

(29.25) THEOREM.   *Let $A_1, \ldots, A_n$ be nonzero ideals of a Dedekind domain R. Then we have*

$$A_1 \oplus A_2 \oplus \cdots \oplus A_n \cong S_n(A_1 A_2 \cdots A_n)\,.$$

We need to develop a little general theory. If M is a torsion-free module for a domain R, we say that a subset $\mathcal{X} \subseteq M$ is a *weak generating set* for M if for every element $m \in M$, there exists a nonzero element $r \in R$ such that $mr \in \mathcal{X}R$, the set of R-linear combinations of $\mathcal{X}$. (Note that in the case where R is a field, this yields $m \in \mathcal{X}R$, and so a weak generating set is the same as a generating set or spanning set over a field.) Of course, a finitely generated module necessarily has a finite weak generating set. In general, if M has a finite weak generating set, then the cardinality of the smallest such set is called the *rank* of M. Since we are assuming that M is torsion free, we see that the rank cannot be zero if M is nonzero. As an example, we mention that any nonzero element of an ideal of R forms a weak generating set. Nonzero ideals, therefore, always have rank 1.

Let $\mathcal{X} = \{m_1, \dots, m_n\}$ be a weak generating set for M. If $\sum m_i r_i = 0$, with $r_i \in R$ and (say) $r_n \neq 0$, then $\mathcal{Y} = \mathcal{X} - \{m_n\}$ is also a weak generating set. To see this, note that $m_n r_n \in \mathcal{Y}R$, and so if $mr \in \mathcal{X}R$, then $mr r_n \in \mathcal{Y}R$. It follows that if $\mathcal{X}$ is a minimal weak generating set for M, then $\mathcal{X}$ is R-linearly independent, and so every member of $\mathcal{X}R$ has uniquely determined coefficients.

We now begin work toward a proof of our classification theorem (29.24).

(29.26) LEMMA. *Let M be a nonzero finitely generated torsion-free R-module, where R is a domain. Then there exists an R-module homomorphism of M onto a nonzero ideal of R such that the kernel of this homomorphism has smaller rank than does M.*

Proof.   Since M is finitely generated, we can choose a finite weak generating set $\mathcal{X}$ of M with minimum possible size. Since M is nonzero and torsion free, $\mathcal{X}$ is nonempty, and we fix $x \in \mathcal{X}$. We see that for each element $u \in \mathcal{X}R$, there is a uniquely determined coefficient of x that occurs in the expression of u as an R-linear combination of $\mathcal{X}$. If we write $(u)\theta$ to denote this coefficient, we observe that $\theta : \mathcal{X}R \to R$ is an R-module homomorphism.

We now define a map $\lambda : M \to F$, where F is the fraction field of R. If $m \in M$, choose nonzero $r \in R$ with $mr \in \mathcal{X}R$, and set $(m)\lambda = (mr)\theta/r$. To see that this is well defined, suppose that also $ms \in \mathcal{X}R$ with $0 \neq s \in R$. Then $(mr)\theta s = (mrs)\theta = (ms)\theta r$ and hence $(mr)\theta/r = (ms)\theta/s$, as required. This map is additive since if $n \in M$ and $nt \in \mathcal{X}R$ with $0 \neq t \in R$, then $(m+n)rt \in \mathcal{X}R$ and

$$(m+n)\lambda = \frac{((m+n)rt)\theta}{rt} = \frac{(mrt)\theta}{rt} + \frac{(nrt)\theta}{rt} = (m)\lambda + (n)\lambda \,.$$

Also, if $a \in R$, then $(ma)r \in \mathcal{X}R$ and

$$(ma)\lambda = \frac{(mar)\theta}{r} = \frac{(mr)\theta a}{r} = (m)\lambda a \,.$$

It follows that λ is an R-module homomorphism, and hence $M\lambda$ is an R-submodule of F. Since M is finitely generated, $M\lambda$ is finitely generated also, and hence $M\lambda$ is a fractional ideal. It is nonzero because $(x)\lambda = 1$.

The fractional ideal $M\lambda$ is isomorphic to some nonzero ordinary ideal A, and this gives a homomorphism of M onto A. The kernel of this map is exactly $\ker(\lambda)$, which is equal to the set of elements of M that have a multiple in $\mathcal{Y}R$, where $\mathcal{Y} = \mathcal{X} - \{x\}$. This submodule thus has rank smaller than $|\mathcal{X}|$, which is the rank of M. ∎

(29.27) LEMMA. *Every finitely generated torsion-free module for a Dedekind domain is isomorphic to a finite direct sum of ideals.*

Proof. Work by induction on the rank of the module M. By Lemma 29.26 we have a surjection $\pi : M \to A$, where A is a nonzero ideal and the rank of $\ker(\pi)$ is smaller than that of M. Since A is a projective module, it follows from Lemma 14.28 that M is the direct sum of $\ker(\pi)$ and an isomorphic copy of A. The inductive hypothesis applies to $\ker(\pi)$, and the result follows. ∎

By Theorem 16.28 we know that finitely generated torsion-free modules over a PID are free. The following corollary gives the corresponding result for Dedekind domains, where we replace "free" by "projective." The argument here also provides another proof for Theorem 16.28.

(29.28) COROLLARY. *If R is a Dedekind domain, then every finitely generated torsion-free R-module is projective. If R is a PID, then these modules are free.*

Proof. By Corollary 29.23 the ideals of R are projective modules, and if R is a PID, they are module isomorphic to the regular module. Since direct sums of projectives are projective (by Corollary 14.29) and direct sums of copies of R are free, the result follows by Lemma 29.27. (Note that Lemma 29.27 applies in the case where R is a PID, since by Lemma 29.2, PIDs are Dedekind domains.) ∎

We need one more ingredient to get Theorem 29.24(a). This lemma is also the key step in the proof of Theorem 29.25.

(29.29) LEMMA. *Let A and B be nonzero ideals of a Dedekind domain R. Then $A \oplus B \cong R \oplus AB$.*

Proof. We consider first the special case where $A + B = R$. Define the map $\pi : A \oplus B \to R$ by $(a, b) \mapsto a + b$ for $(a, b) \in A \oplus B$. This is clearly an R-module homomorphism of $A \oplus B$ onto R, and since R is a projective R-module, Lemma 14.28 guarantees that $A \oplus B \cong R \oplus N$, where $N = \ker(\pi)$. Clearly, $N = \{(x, -x) \mid x \in A \cap B\}$ and so $N \cong A \cap B$. We claim that $A \cap B = AB$ when $A + B = R$, and this completes the proof in this case.

Certainly, $AB \subseteq A \cap B$, and so we need to prove the reverse inclusion. Since $1 \in R = A + B$, we can write $1 = a + b$ for some elements $a \in A$ and $b \in B$. If $d \in A \cap B$, we have $d = da + db$, and we note that each term da and db lies in AB. It follows that $d \in AB$, as required.

To complete the proof in the general case, it suffices to find ideals isomorphic to A and B whose sum is R. We seek, therefore, nonzero elements $\alpha \in A^{-1}$ and $\beta \in B^{-1}$ such that $\alpha A + \beta B = R$. Choose a nonzero element $\alpha \in A^{-1}$ arbitrarily. Now B^{-1} is an R-module isomorphic to an ideal of R, and since R is Dedekind, it follows by Corollary 29.16 that B^{-1} can be generated by two elements, one of which is arbitrary. Since $0 < \alpha A B^{-1} \subseteq B^{-1}$, we can choose an element $\beta \in B^{-1}$ such that $B^{-1} = \alpha A B^{-1} + \beta R$. It follows that $R = B^{-1}B = (\alpha A B^{-1} + \beta R)B = \alpha A + \beta B$, as required. (Note that it is always true that $(X + Y)Z = XZ + YZ$ for additive subgroups X, Y and Z of a ring.) ∎

We can now prove Theorem 29.25 and the existence part of Theorem 29.24.

Proof of Theorem 29.25   By Lemma 29.29 we have $A_1 \oplus A_2 \cong R \oplus A_1 A_2$. This, together with a second application of Lemma 29.29, yields

$$A_1 \oplus A_2 \oplus A_3 \cong R \oplus A_1 A_2 \oplus A_3 \cong R \oplus R \oplus A_1 A_2 A_3.$$

The result follows by continuing this process, using Lemma 29.29 repeatedly. ∎

Proof of Theorem 29.24(a).   By Lemma 29.27 a finitely generated torsion-free R-module is isomorphic to a direct sum of ideals. The result follows by Theorem 29.25. ∎

What remains now is the proof of Theorem 29.24(b). For this, we need to analyze R-module homomorphisms between direct sums of ideals. We obtain a result that is entirely analogous to the familiar result that a linear transformation between two row vector spaces over a field is given by matrix multiplication. The following result can also be viewed as a generalization of Lemma 29.21.

(29.30) LEMMA.   *Let $A_1, \ldots, A_m$ and $B_1, \ldots, B_n$ be nonzero ideals of a domain R with fraction field F. Write*

$$M = A_1 \oplus \cdots \oplus A_m \quad and \quad N = B_1 \oplus \cdots \oplus B_n,$$

and let $\theta : M \to N$ be an R-module homomorphism. Then there exists a unique $m \times n$ matrix X over F such that $\mathbf{a}X = \mathbf{b}$, whenever $\mathbf{a}\theta = \mathbf{b}$, where $\mathbf{a} = (a_1, \ldots, a_m) \in M$ and $\mathbf{b} = (b_1, \ldots, b_n) \in N$.

We say that the unique matrix X in Lemma 29.30 *represents* the module homomorphism θ. Note that if φ is a homomorphism from N to $P = C_1 \oplus \cdots \oplus C_k$, another direct sum of ideals, and if Y represents φ, then XY is the unique matrix representing the map $\theta\varphi : M \to P$.

Proof of Lemma 29.30.   Let $\mathbf{e}_1, \ldots, \mathbf{e}_m$ be the "standard basis" in the row space F^m. The row vector $\mathbf{e}_i$, therefore, has the entry 1 in position i and all other coordinates are 0. Fix nonzero elements $a_i \in A_i$ and define the row vector $\mathbf{x}_i \in F^n$ by $\mathbf{x}_i = ((\mathbf{e}_i a_i)\theta)/a_i$. Let X be the matrix whose ith row is $\mathbf{x}_i$, for $1 \le i \le m$.

If $u_i \in A_i$ is arbitrary, we compute that

$$(\mathbf{e}_i u_i)\theta a_i = (\mathbf{e}_i u_i a_i)\theta = (\mathbf{e}_i a_i)\theta u_i = \mathbf{x}_i a_i u_i = \mathbf{e}_i X a_i u_i = (\mathbf{e}_i u_i)X a_i,$$

and it follows that the equation $\mathbf{a}\theta = \mathbf{a}X$ holds whenever $\mathbf{a}$ has the form $\mathbf{e}_i u_i$. Since every member of M is a sum of such vectors, we have $\mathbf{a}\theta = \mathbf{a}X$ for all $\mathbf{a} \in M$.

To prove uniqueness, it suffices to show that if Y is an $m \times n$ matrix such that $\mathbf{a}Y = 0$ for all $\mathbf{a} \in M$, then $Y = 0$. The row vectors $\mathbf{a} \in M$, however, span the whole row space F^m, and so we conclude that $\mathbf{e}_i Y = 0$ for all i. Since $\mathbf{e}_i Y$ is exactly the ith row of Y, we have $Y = 0$.   ∎

We can now prove a result that includes Theorem 29.24(b) but is more general since it does not require R to be Dedekind.

(29.31) THEOREM (Kaplansky).   *Let R be any domain and assume that A_i and B_j are nonzero ideals for $1 \le i \le m$ and $1 \le j \le n$. Let M and N be the direct sums of these ideals as in Lemma 29.30, and define the ideals $P = \prod A_i$ and $Q = \prod B_j$. If $M \cong N$, then $m = n$ and $P \cong Q$. In fact, if θ is any isomorphism from M onto N, then $P\delta = Q$, where δ is the determinant of the matrix representing θ.*

Proof.   Let θ be an isomorphism from M onto N, and let X be the $m \times n$ matrix that represents θ as in Lemma 29.30. Similarly, let Y be the $n \times m$ matrix that represents θ^{-1}. By the uniqueness in Lemma 29.30, the identity homomorphisms on M and N are represented only by appropriately sized identity matrices, and it follows that XY and YX are both identity matrices. By elementary linear algebra, this implies that $m = n$.

Now write x_{ij} to denote the (i, j)-entry of the matrix X. The fact that

$$(0, \ldots, 0, a_i, 0, \ldots, 0)X \in B_1 \oplus \ldots \oplus B_n$$

for elements $a_i \in A_i$ tells us that the product $a_i x_{ij}$ lies in B_j. It follows that if x is any product of n entries of X, chosen one from each row and column, then $Px \subseteq Q$.

Since $\delta = \det X$ is a sum and difference of products x, as above, it follows that $P\delta \subseteq Q$ and, similarly, $Q\gamma \subseteq P$, where $\gamma = \det Y$. Now $XY = I$ and this implies that $\gamma\delta = 1$. Thus $Q = Q\gamma\delta \subseteq P\delta$. We conclude that $P\delta = Q$ and hence $P \cong Q$.   ∎

Problems

29.1 Let $R = F[X, Y]$, where F is a field, and define the ideal $A = XR + YR$. Compute A^{-1} and show that $(A^{-1})^{-1} \ne A$.

29.2 Show that a Dedekind domain is a PID iff it is a UFD.

HINT: Show that the maximal ideals of a Dedekind UFD are principal.

29.3 Let R be a domain and suppose that $A \subseteq R$ is an ideal maximal with the property that $A^{-1} \not\subseteq R$. Show that A is prime.

29.4 Let R be a domain with the property that every nonzero ideal is a product of maximal ideals. Show that R is Dedekind.

HINT: Show that every nonzero ideal is invertible. Use the fact that nonzero principal ideals are invertible.

29.5 Let M be a multiplicative system in a Dedekind domain R. Show that the ring $S = R_M$, obtained by adjoining inverses for M to R, is a Dedekind domain. If M is the complement of a nonzero prime, show that S is a PID.

HINT: See Problem 28.10.

29.6 Let R be the ring of all algebraic integers in $\mathbb{C}$, and suppose that A is a proper ideal of R. Suppose $\alpha_1, \ldots, \alpha_n$ are algebraic numbers that are not all zero. Show that there exists some algebraic number β such that all $\beta\alpha_i$ are integral but not all lie in A.

29.7 Suppose that at most finitely many of the prime ideals of a Dedekind domain R are nonprincipal. Show that R is a PID.

29.8 Let R be a Dedekind domain and write $\mathcal{P}$ to denote the set of nonzero primes of R. As in Section 29B, let f_A denote the nonnegative-integer-valued function on $\mathcal{P}$ associated with the ideal A of R.

   a. If A and B are ideals, show that $f_{A+B}(P) = \min(f_A(P) + f_B(P))$.
   b. Find and prove a corresponding formula for $f_{A \cap B}$.
   c. Prove that $AB = (A + B)(A \cap B)$ for ideals A and B of R.

29.9 Show that the primary ideals of a Dedekind domain are exactly the powers of primes.

29.10 Let R be a domain, and let M be a torsion-free R-module of finite rank n. Show that every R-linearly independent subset of M contains no more than n members.

29.11 Let R be a domain with fraction field F. Show that if M is a torsion-free R-module of finite rank n, then M is isomorphic to a submodule of the row space F^n, viewed as an R-module.

29.12 Let M be a module for a domain R. The *torsion* submodule of M is the set $T = \{m \in M \mid mr = 0 \text{ for some nonzero element } r \in R\}$.

   a. Show that T is a submodule and that M/T is torsion free.
   b. If R is Dedekind and M is finitely generated, show that T is a direct summand of M.
   c. In the situation of part (b), show that $Ta = 0$ for some nonzero element $a \in R$.

29.13 Suppose that $Q < P$ are prime ideals of a ring R. If P/PQ is a principal ideal of R/PQ, show that $Q = PQ$. If R is a noetherian domain, deduce that $Q = 0$.

HINT: Write $P = PQ + aR$ for some element $a \in R$, and consider the ideal $X = \{x \in R \mid ax \in Q\}$. Note that $PX \subseteq Q$.

29.14 Let R be a domain for which every proper homomorphic image is a principal ideal ring. Show that R is Dedekind.

HINT: By Problem 29.13 all nonzero primes of R are maximal. Let P be a nonzero prime and work to show that P is invertible. Once this is done, show that all nonzero ideals of R are invertible using the relevant argument in the proof of Theorem 29.6. To show that P is invertible, write $P = P^2 + aR$. If P is not invertible, show that $P + aP^{-1} < R$ and deduce that $aP^{-1} \subseteq P$. By Lemma 29.7 let U be an ideal such that $UP \subseteq aP^{-1}$ with $U \nsubseteq aP^{-1}$, and obtain a contradiction by computing UP^2.

29.15 Let R be the ring of integers in a Galois extension $E \supseteq \mathbb{Q}$. If p is a prime number, show that there exist distinct prime ideals of $P_1, P_2, \ldots, P_t$ of R such that $pR = (P_1 P_2 \cdots P_t)^e$ for some positive integer e.

CHAPTER THIRTY

Algebraic Sets and the Nullstellensatz

30A

The two main areas of "application" of commutative ring theory are algebraic number theory and algebraic geometry, each of which is a discipline to which numerous books and many lifetimes of research have been devoted. Because of the limitations of space and, more significant, the limitations in the expertise of the author, all we can hope to do here is to give a small sample of each of these subjects. In the previous two chapters, we attempted to do this for number theory, and here in the final chapter of this book, we present a tiny taste of algebraic geometry.

Suppose we are given a polynomial in several variables over a field. For example, we might consider the polynomial $f(X, Y) = Y - X^2$ over $\mathbb{Q}$. The graph of this polynomial is its zero-set: the set of points (x, y) (on an appropriate piece of graph paper) such that $f(x, y) = 0$. Observe that the coordinates of points on the graph paper are not restricted to lie in the coefficient field of the polynomial; they come from the extension field $\mathbb{R} \supseteq \mathbb{Q}$. In the case of our example, of course, the graph is a parabola, and we can ask such "geometric" questions as whether or not the graph crosses itself, or whether or not there is a line it crosses in as many as three points. In essence, algebraic geometry is the study of such geometric questions about the graphs of polynomials.

We do not wish to be limited to two dimensions. If we take a polynomial in three indeterminates and graph it on three-dimensional "graph paper," in general we get a surface. If we wish to study curves in 3-space, we could take two polynomials and consider their common zero-set. (For instance, the line through the points $(1, 1, 1)$ and $(0, 1, 2)$ is the common zero-set of the polynomials $X + Y + Z - 3$ and $X + Z - 2$.) We consider, therefore, common zero-sets of sets of polynomials.

In general, let $F \subseteq E$ be fields and fix an integer $n > 0$. Write $R = F[X_1, \ldots, X_n]$, the polynomial ring in n indeterminates. For our "graph paper," we take the space E^n of all n-tuples over E. If $\mathbf{x} = (x_1, \ldots, x_n) \in E^n$ and $f \in R$, we write $f(\mathbf{x})$ to denote $f(x_1, \ldots, x_n)$. For arbitrary subsets $Q \subseteq R$, we define

$$\text{Zer}(Q) = \{\mathbf{x} \in E^n \mid f(\mathbf{x}) = 0 \text{ for all } f \in Q\},$$

the common zero-set of Q. A subset of E^n of the form $\text{Zer}(Q)$ for some subset $Q \subseteq R$ is called an *algebraic set*. (More precisely, it is an algebraic set over F in E^n.) It is these algebraic sets that are the primary objects of study in algebraic geometry. Note that the whole set E^n and the empty set are algebraic sets since they are, respectively, $\text{Zer}(\{0\})$ and $\text{Zer}(R)$. In the case where $E = F$, every singleton set $\{\mathbf{x}\}$ for $\mathbf{x} \in E^n$ is also algebraic; it is the zero-set of the collection $Q = \{X_i - x_i \mid 1 \le i \le n\}$, where $\mathbf{x} = (x_1, \dots, x_n)$.

(30.1) LEMMA. *The collection of algebraic subsets of E^n is closed under finite unions and arbitrary intersections.*

Proof.   Suppose $\mathcal{X}$ is a collection of subsets of R and write $Q = \bigcup \mathcal{X}$. Then $\mathbf{x} \in E^n$ lies in $\text{Zer}(Q)$ iff $\mathbf{x}$ is a zero for all polynomials in all the sets that constitute $\mathcal{X}$. In other words, $\text{Zer}(Q) = \bigcap\{\text{Zer}(S) \mid S \in \mathcal{X}\}$. It follows that arbitrary intersections of algebraic sets are algebraic. To complete the proof, it suffices to show that a union of two algebraic sets is again algebraic. To see this, note that $\text{Zer}(Q_1) \cup \text{Zer}(Q_2) = \text{Zer}(Q)$, where Q is the set of all products of a polynomial from Q_1 and one from Q_2. This is so because the polynomial fg vanishes at $\mathbf{x}$ iff either f or g vanishes at $\mathbf{x}$.   ∎

It follows that we can make E^n into a topological space by taking the algebraic sets to be the closed sets. (This is called the *Zariski* topology for E^n.) Although topological considerations form an important part of algebraic geometry, we do not pursue that aspect of the subject any further.

There is something unsatisfactory about what we have done. If we take $F = \mathbb{Q}$ and $E = \mathbb{R}$ as before, we see that the polynomial $X^2 + Y^2 + 1$ has an empty graph, whereas the graph of the very similar polynomial $X^2 + Y^2 - 1$ is a circle. This difference is because our graph paper is inadequate; the field E is too small. Had we taken $E = \mathbb{C}$, for instance, there would not have been such a dramatic difference between the two graphs. In order to avoid this difficulty, we usually assume that the field E is algebraically closed.

As we saw in the previous paragraph, if $F \subseteq L \subseteq E$, then we cannot necessarily recover an algebraic set in E^n by looking only at points whose coordinates lie in L. (The zero-set of $X^2 + Y^2 + 1$ in $\mathbb{C}^2$ has no real points at all, for example.) One of our major results is that if L contains all elements of E algebraic over F, where E is algebraically closed, then the intersection of any algebraic set in E^n with L^n contains enough information to recover the full algebraic set in E^n. In short, an algebraic set is determined by its algebraic points.

(30.2) THEOREM.   *Let $F \subseteq L \subseteq E$, where E is algebraically closed and L is the algebraic closure of F in E. If A is an algebraic set over F in L^n, then there exists a unique algebraic set B over F in E^n such that $B \cap L^n = A$.*

(30.3) COROLLARY.   *Let $S \subseteq E^n$ be a nonempty algebraic set over F, where E is algebraically closed. Then S contains a point with all coordinates algebraic over F.*

Proof.   Let L be the algebraic closure of F in E, as in Theorem 30.2. We must show that $S \cap L^n$ is nonempty. If this is false, then S and $\varnothing$ are two different algebraic sets in E^n that have the same intersection with L^n. This contradicts Theorem 30.2. ∎

How might we try to prove Theorem 30.2? One strategy might be to try to discover the subset of R for which A is the zero-set, and then to show that B must be the zero-set in E^n for that same collection of polynomials. If $A \subseteq E^n$ is an arbitrary subset (not necessarily algebraic), we write

$$\text{Pol}(A) = \{f \in R \mid f(\mathbf{x}) = 0 \text{ for all } \mathbf{x} \in A\},$$

where, as usual, $R = F[X_1, \ldots, X_n]$.

(30.4) LEMMA.   *If A is an algebraic set, then $A = \text{Zer}(\text{Pol}(A))$.*

We should mention that the functions Zer and Pol between subsets of R and subsets of E^n form a Galois connection. Lemma 30.4 says that $\text{Zer}(Q) = \text{Zer}(\text{Pol}(\text{Zer}(Q)))$ for arbitrary subsets $Q \subseteq R$. Although this is one of the general trivialities about Galois connections, we spare the reader the necessity of reviewing this material by giving the simple direct proof.

Proof of Lemma 30.4.   The containment $A \subseteq \text{Zer}(\text{Pol}(A))$ is obvious for all subsets $A \subseteq E^n$. We are assuming that $A = \text{Zer}(Q)$ for some subset $Q \subseteq R$, and so it follows that $\text{Pol}(A) = \text{Pol}(\text{Zer}(Q)) \supseteq Q$. Since $\text{Pol}(A) \supseteq Q$, we have $\text{Zer}(\text{Pol}(A)) \subseteq \text{Zer}(Q) = A$, as required. ∎

Note that $\text{Pol}(A)$ is an ideal of the polynomial ring R. It follows by Lemma 30.4 that every algebraic set has the form $\text{Zer}(I)$ for some ideal I of R.

(30.5) COROLLARY.   *Every algebraic set has the form $\text{Zer}(Q)$ for some finite set Q of polynomials of R.*

Proof.   If A is an algebraic set, then $A = \text{Zer}(I)$ for some ideal I. Since R is a noetherian ring by the Hilbert basis theorem, the ideal I has a finite generating set Q, and it is clear that $\text{Zer}(Q) = \text{Zer}(I) = A$. ∎

Not every ideal has the form $\text{Pol}(A)$ for a subset $A \subseteq E^n$. To see this, note that if $I = \text{Pol}(A)$ and $f \in R$ has the property that $f^k \in I$ for some integer k, then for $\mathbf{x} \in A$, we have $f(\mathbf{x})^k = 0$ and hence $f(\mathbf{x}) = 0$. It follows that $f \in I$ and hence $I = \sqrt{I}$. In other words, only radical ideals can occur as $\text{Pol}(A)$. This argument also shows that if $I \subseteq R$ is any ideal, then $\text{Pol}(\text{Zer}(I)) \supseteq \sqrt{I}$.

(30.6) THEOREM (Hilbert Nullstellensatz).   *Let $E \supseteq F$, where E is algebraically closed. Computing zero-sets in E^n, we have $\text{Pol}(\text{Zer}(I)) = \sqrt{I}$ for all ideals $I \subseteq R = F[X_1, \ldots, X_n]$.*

We mention that the German word "Nullstellensatz" means "zero location theorem." This is the essential ingredient in the proof of Theorem 30.2.

Proof of Theorem 30.2.   For the purposes of this proof, we write Zer_L and Zer_E to denote the zero-set functions in L^n and E^n, respectively. Let $I = \text{Pol}(A)$ and define $B = \text{Zer}_E(I)$. Then $B \cap L^n = \text{Zer}_L(I) = A$ by Lemma 30.4.

   Now let C be any algebraic set in E^n with $C \cap L^n = A$, and write $J = \text{Pol}(C)$. Since $C = \text{Zer}_E(J)$ by Lemma 30.4, we have $A = C \cap L^n = \text{Zer}_L(J)$. Thus

$$I = \text{Pol}(A) = \text{Pol}(\text{Zer}_L(J)) = \sqrt{J}$$

by the Nullstellensatz (30.6) applied in L. (This is valid since L is algebraically closed because it is the algebraic closure of F in the algebraically closed field E.) Now we apply the Nullstellensatz in E. This gives

$$J = \text{Pol}(C) = \text{Pol}(\text{Zer}_E)(J) = \sqrt{J},$$

and so $I = J$. It follows that $B = \text{Zer}_E(I) = \text{Zer}_E(J) = C$, as required. ∎

We did not use the full strength of the Nullstellensatz in the proof of Theorem 30.2; we needed only the fact that $\text{Pol}(\text{Zer}(I))$ is determined by I, and not that it is the radical. The following Nullstellensatz corollary uses the fact that we actually get the radical of I. An even stronger form of this result (which requires only that the field E be infinite) can be proved by induction on n, without the Nullstellensatz.

(30.7) COROLLARY.   *If E is algebraically closed, then no nonzero polynomial in $R = E[X_1, \ldots, X_n]$ can vanish on all of E^n.*

Proof.   If f vanishes on E^n, then since $E^n = \text{Zer}(0)$, we have $f \in \text{Pol}(\text{Zer}(0)) = \sqrt{0} = 0$. ∎

The Nullstellensatz is easily seen to imply that if I is a proper ideal of R, then $\text{Zer}(I)$ is nonempty. (This follows because $\text{Pol}(\text{Zer}(I)) = \sqrt{I} < R$.) We reverse the logic, however, and prove this "corollary" first, and then derive the Nullstellensatz from it. We begin by proving a theorem of Zariski that, although it seems purely algebraic and does not mention algebraic sets, is essentially equivalent to this assertion that zero-sets of proper ideals are nonempty.

(30.8) THEOREM (Zariski).   *Let $F \subseteq E$ be an extension of fields, and assume that there exist $\alpha_1, \ldots, \alpha_n \in E$ such that $E = F[\alpha_1, \ldots, \alpha_n]$. Then E is algebraic over F.*

In other words, a field extension that is finitely generated as a ring extension is algebraic. Note that this extension, being algebraic and finitely generated, necessarily has finite degree. Conversely, every finite-degree field extension is obviously finitely generated as a ring.

Observe that we already know Zariski's theorem in the case where $n = 1$. If $F[\alpha]$ is a field, then certainly α must be algebraic because otherwise $F[\alpha]$ would be isomorphic to the polynomial ring $F[X]$, which is not a field.

Proof of Theorem 30.8.   As we have observed, we know the result to be true when $n = 1$, and so we assume that $n > 1$ and proceed by induction on n. Let R be the subring $F[\alpha_n] \subseteq E$, and let K be the field of fractions of R in E. Then $E = K[\alpha_1, \ldots, \alpha_{n-1}]$ and so E is algebraic over K by the inductive hypothesis. If α_n is algebraic over F, then $K = R$ is algebraic over F, and hence E is algebraic over F, as desired. We may assume, therefore, that α_n is transcendental over F, and we work to get a contradiction.

Since α_i is algebraic over the fraction field K of R for $1 \le i \le n$, it is automatically algebraic over R. (Just clear denominators from the minimal polynomial of α_i over K.) It follows by Lemma 28.2 that $a_i \alpha_i$ is integral over R for some nonzero element $a_i \in R$. Let $a = \prod a_i$, and note that $0 \ne a \in R$ and $a\alpha_i$ is integral over R for all subscripts i.

We claim that for every element $\beta \in E$, there exists some nonnegative integer exponent m such that $a^m \beta$ is integral over R. To see this, observe that the set of elements $\beta \in E$ for which this assertion is true is closed under addition and multiplication. This set includes all α_i and it obviously includes F since $F \subseteq R$. It follows that it includes the whole ring $F[\alpha_1, \ldots, \alpha_n] = E$, as claimed.

Since $R = F[\alpha_n]$ and α_n was assumed to be transcendental over F, we know that R is isomorphic to $F[X]$, and we identify R with this polynomial ring. Since the polynomial $Xa + 1$ is nonzero (where a is as in the previous paragraphs and so is a polynomial over F), we have $1/(Xa + 1) \in K \subseteq E$ and so $a^m/(Xa + 1)$ is integral over R for some exponent m. The polynomial ring R is a UFD, however, and so by Lemma 28.3 it is integrally closed in K. Thus $a^m/(Xa + 1)$ lies in R and a^m is a multiple of $Xa + 1$. Since $Xa + 1$ has positive degree, it has a root δ in some extension field of F, and it follows that $a^m(\delta) = 0$. Thus $a(\delta) = 0$ and evaluation of $Xa + 1$ at δ gives 1. This is a contradiction since δ is a root of $Xa + 1$.   ∎

We can now prove the assertion we mentioned earlier as a consequence of the Nullstellensatz.

(30.9) COROLLARY.   *Assume the usual notation, with the field E algebraically closed. If I is a proper ideal of R, then $\mathrm{Zer}(I)$ is nonempty.*

Proof.   Since I is proper, we can choose a maximal ideal M of R with $M \supseteq I$. Then $\overline{R} = R/M$ is a field. Since the nonzero elements of F are invertible in R, they cannot lie in a proper ideal. It follows that $F \cap M = 0$, and the image $\overline{F}$ of F in $\overline{R}$ is isomorphic to F. We identify $\overline{F}$ with F via the natural map.

Since $R = F[X_1, \ldots, X_n]$, we have $\overline{R} = F[\alpha_1 \ldots, \alpha_n]$, where $\alpha_i = \overline{X}_i$, the image of X_i under our homomorphism. By Zariski's theorem (30.8), $\overline{R}$ is algebraic over F, and so we can identify $\overline{R}$ with a subfield of the algebraically closed field E.

We claim that $(\alpha_1, \ldots, \alpha_n) \in E^n$ lies in $\mathrm{Zer}(I)$. To see this, let $f \in I$. Then

$$f(\alpha_1, \ldots, \alpha_n) = f(\overline{X_1}, \ldots, \overline{X_n}) = \overline{f(X_1, \ldots, X_n)} = \overline{f} = 0\,,$$

where the last equality holds since $f \in I \subseteq M$. ∎

We are now ready to prove Hilbert's Nullstellensatz.

Proof of Theorem 30.6.   Let $f \in R$ and assume that $f \in \mathrm{Pol}(\mathrm{Zer}(I))$. Our task is to show that some power of f lies in I. To do this, we introduce a new indeterminate T and we write $S = R[T]$. Let $I[T]$ denote the ideal of S that consists of those polynomials in T for which all coefficients lie in I, and define the ideal J of S by setting $J = I[T] + (1 - Tf)S$. Suppose J is a proper ideal of $S = F[X_1, X_2, \ldots, X_n, T]$. By Corollary 30.9 (applied in the case of $n + 1$ indeterminates) there exist elements $\alpha_1, \ldots, \alpha_n, \beta \in L$ such that every member of J vanishes when evaluated at this $(n + 1)$-tuple.

Since $I \subseteq J$, it follows that the point $\mathbf{x} = (\alpha_1, \ldots, \alpha_n) \in E^n$ lies in $\mathrm{Zer}(I)$. But $f \in \mathrm{Pol}(\mathrm{Zer}(I))$, and so we deduce that f vanishes at $\mathbf{x}$. This is impossible, however, since $1 - Tf \in J$, and thus $1 - \beta f(\mathbf{x}) = 0$. This contradiction shows that $J = S$.

Since $1 \in J$, we can write $1 = u(T) + (1 - Tf)v(T)$, where $u \in I[T]$ and $v \in R[T]$. Viewing this as a polynomial identity in $R[T]$, we can deduce valid equations in any unitary overring R^* of R by substituting any element of R^* for T. We apply this in the case where R^* is the field of fractions of R, and we substitute $1/f$ for T. (We may certainly assume that $f \neq 0$, since otherwise $f \in I$ and there is nothing to prove.) Carrying out this substitution, we obtain $u(1/f) = 1$.

If we write $u(T) = a_n T^n + a_{n-1}T^{n-1} + \cdots + a_0$, with all $a_i \in I$, we have

$$\frac{a_n}{f^n} + \frac{a_{n-1}}{f^{n-1}} + \cdots + a_0 = 1\,,$$

and so

$$a_n + a_{n-1}f + \cdots + a_0 f^n = f^n\,.$$

Since all $a_i \in I$, we conclude that $f^n \in I$, as required. ∎

30B

We continue to consider fields $F \subseteq E$ and we write $R = F[X_1, \ldots, X_n]$, as usual. If E is algebraically closed, the Nullstellensatz gives us a correspondence between the radical ideals of R (including R itself) and the algebraic subsets of E^n.

(30.10) COROLLARY. *The maps* Zer *and* Pol *define inverse bijections between the set of radical ideals of R and the set of algebraic subsets of E^n, where E is algebraically closed.*

Proof.   Certainly Zer maps radical ideals to algebraic sets, and we have seen that
Pol maps algebraic sets to radical ideals. It suffices, therefore, to show that
$\text{Zer}(\text{Pol}(A)) = A$ and $\text{Pol}(\text{Zer}(I)) = I$ for algebraic sets A and radical ideals
I. The first of these equations is exactly Lemma 30.4, and if I is a radical ideal,
then the Nullstellensatz gives $\text{Pol}(\text{Zer}(I)) = \sqrt{I} = I$, as required.   ∎

Since the radical ideals of R are exactly the intersections of the prime ideals,
we see that prime ideals play a special role. This suggests the question of which
algebraic sets correspond to prime ideals.

A nonempty algebraic set is *irreducible* if it is not the union of two proper
algebraic subsets. For example, the zero-set of the polynomial $X^2 - Y^2$ is the union
of two lines, the zero sets of $X - Y$ and $X + Y$, and so it is not irreducible. In the
literature, irreducible algebraic sets are also called *varieties*, although sometimes the
word "variety" is used to mean an algebraic set in general, one that is not necessarily
irreducible. It is exactly the irreducible algebraic sets that correspond to prime
ideals under the bijections of Corollary 30.10. (Note that although we used the
Nullstellensatz to establish Corollary 30.10, it is not needed for the following two
results.)

(30.11) THEOREM.   *Let $A \subseteq E^n$ be an algebraic set. Then A is irreducible iff*
$\text{Pol}(A)$ *is prime.*

Proof.   Let $I = \text{Pol}(A)$ and suppose that I is prime. In particular, $I < R$ and so A
is nonempty. If $A = B \cup C$, where B and C are algebraic sets, we must show
that either $B = A$ or $C = A$. Write $U = \text{Pol}(B)$ and $V = \text{Pol}(C)$, and note that
every polynomial in UV vanishes on $B \cup C = A$. Thus $UV \subseteq \text{Pol}(A) = I$, and
since I is prime, it must contain one of U or V. Suppose $U \subseteq I$. By Lemma
30.4 we have

$$B = \text{Zer}(\text{Pol}(B)) = \text{Zer}(U) \supseteq \text{Zer}(I) = \text{Zer}(\text{Pol}(A)) = A .$$

It follows that $A = B$, as required.

Now suppose that A is irreducible and again write $I = \text{Pol}(A)$. In particular,
A is nonempty and so I is proper. To show that I is prime, we suppose that
$fg \in I$ for polynomials f and g. We have

$$\text{Zer}(f) \cup \text{Zer}(g) = \text{Zer}(fg) \supseteq \text{Zer}(I) = \text{Zer}(\text{Pol}(A)) = A$$

by Lemma 30.4. Thus $A = B \cup C$, where $B = A \cap \text{Zer}(f)$ and $C = A \cap \text{Zer}(g)$.
By Lemma 30.1 both B and C are algebraic sets, and hence by the irreducibility
of A, we may suppose that $B = A$. Since $B \subseteq \text{Zer}(f)$, we have $f \in \text{Pol}(B) =
\text{Pol}(A) = I$, as required.   ∎

Just as the zero-set of $X^2 - Y^2$ is the union of two lines, each of which is an
irreducible algebraic set, we have the following theorem.

(30.12) THEOREM.   *Every algebraic set is a finite union of irreducible algebraic
sets.*

Proof.   If $A_1 > A_2 > \cdots$ is a properly decreasing chain of algebraic sets, write $I_i = \mathrm{Pol}(A_i)$. Then clearly $I_1 \subseteq I_2 \subseteq \cdots$, and these containments must be proper since $A_n = \mathrm{Zer}(I_n)$ by Lemma 30.4. Since R is noetherian by the Hilbert basis theorem, this chain of ideals must terminate, and we deduce that the set of algebraic subsets of E^n satisfies the DCC and hence also the minimal condition.

   If the theorem is false, choose an algebraic set A minimal such that it is not a finite union of irreducible algebraic sets. In particular, A is nonempty since the empty set is the union of the empty collection, and A is not irreducible. We can thus write $A = B \cup C$, where $B < A$ and $C < A$ are algebraic sets. By the minimality of A, each of B and C is a finite union of irreducible algebraic sets, and thus A is, too. This is a contradiction.   ∎

Unlike the previous two elementary results, which were about algebraic sets but did not depend on the Nullstellensatz, our next result is nongeometric but uses the Nullstellensatz heavily.

(30.13) THEOREM.   *Let R be a polynomial ring in several variables over an arbitrary field. Then every radical ideal (and, in particular, every prime ideal) of R is an intersection of maximal ideals.*

(30.14) LEMMA.   *Let $F \subseteq E$ be fields, where E is algebraic over F. If $\mathbf{x} \in E^n$, then $\mathrm{Pol}(\mathbf{x})$ is a maximal ideal of $R = F[X_1, \ldots, X_n]$.*

Proof.   Write $\mathbf{x} = (\alpha_1, \ldots, \alpha_n)$ and consider the evaluation map $\psi : R \to E$ defined by setting $\psi(f) = f(\mathbf{x})$ for $f \in R$. Then ψ is a homomorphism and $\ker(\psi) = \mathrm{Pol}(\mathbf{x})$. The image of ψ is $F[\alpha_1, \ldots, \alpha_n]$, which is a field since the elements α_i of E are all algebraic over F. Since $R/\mathrm{Pol}(\mathbf{x})$ is a field, the ideal $\mathrm{Pol}(\mathbf{x})$ must be maximal.   ∎

Proof of Theorem 30.13.   We have $R = F[X_1, \ldots, X_n]$ and we let E be the algebraic closure of F. Suppose that I is a radical ideal of R and let $A = \mathrm{Zer}(I)$. We have

$$I = \sqrt{I} = \mathrm{Pol}(A) = \bigcap \{\mathrm{Pol}(\mathbf{x}) \mid \mathbf{x} \in A\}$$

by the Nullstellensatz. The result follows since all of the ideals $\mathrm{Pol}(\mathbf{x})$ are maximal by Lemma 30.14.   ∎

The correspondence given in Corollary 30.10 (when E is algebraically closed) tells us that the maximal ideals of R correspond to the minimal (nonempty) algebraic sets. Of course, these are irreducible (this corresponds to the fact that maximal ideals are prime). What else can we say about minimal nonempty algebraic sets? Note that we are not assuming that E is algebraically closed in the following theorem.

(30.15) THEOREM.   *Suppose E is algebraic over F. Then every point of E^n lies in a unique minimal algebraic set, and these minimal algebraic subsets of*

E^n are all finite. If $E = F$, then each minimal algebraic set consists of just a single point.

Proof. Let $\mathbf{x} \in E^n$ and write $M = \text{Pol}(\mathbf{x})$. Then M is a maximal ideal of $R = F[X_1, \ldots, X_n]$ by Lemma 30.14. Writing $A = \text{Zer}(M)$, we have $\mathbf{x} \in A$ and we claim that A is a minimal algebraic set. To see this, suppose B is a nonempty algebraic set and $B \subseteq A$. Then

$$R > \text{Pol}(B) \supseteq \text{Pol}(A) = \text{Pol}(\text{Zer}(M)) \supseteq M,$$

and it follows that $\text{Pol}(B) = M$ since M is maximal. By Lemma 30.4 we have $B = \text{Zer}(\text{Pol}(B)) = \text{Zer}(M) = A$, and so A is minimal, as claimed. It is the unique minimal algebraic set that contains $\mathbf{x}$ because if $\mathbf{x} \in C$, where C is any algebraic set, then $A \cap C$ is a nonempty algebraic set contained in A and hence $A = A \cap C \subseteq C$.

To see that A is finite, write $\mathbf{x} = (\alpha_1, \ldots, \alpha_n)$ and note that each coordinate α_i is algebraic over F. Let f_i be the minimal polynomial of α_i over F, and view $f_i(X_i)$ as an element of the ring R. Then $f_i(\mathbf{x}) = 0$ and so $f_i \in M$. Since $A = \text{Zer}(M)$, it follows that $f_i(\mathbf{y}) = 0$ for all $\mathbf{y} \in A$. If $\mathbf{y} = (\beta_1, \ldots, \beta_n)$, we see that $f_i(\beta_i) = 0$, and thus there are just finitely many possibilities for β_i; it must be one of the roots of f_i. It follows that there are just finitely many possibilities for points $\mathbf{y} \in A$.

If $F = E$, then the polynomials f_i of the previous paragraph all have degree 1, and so $\beta_i = \alpha_i$ and $\mathbf{x}$ is the unique point in A. ∎

(30.16) COROLLARY. *If F is algebraically closed, then the map*

$$\mathbf{x} \mapsto \text{Pol}(\mathbf{x})$$

defines a bijection from F^n onto the set of all maximal ideals of $R = F[X_1, \ldots, X_n]$.

Proof. Since $\text{Pol}(\mathbf{x})$ is a maximal ideal of R by Lemma 30.14, we do have a map. By Theorem 30.15 we know that $\{\mathbf{x}\}$ is an algebraic set for $\mathbf{x} \in F^n$, and Lemma 30.4 thus gives $\{\mathbf{x}\} = \text{Zer}(\text{Pol}(\mathbf{x}))$. It follows that our map is injective.

To prove surjectivity, we appeal to Corollary 30.9 of Zariski's theorem. If M is a maximal ideal of R, then $\text{Zer}(M)$ is nonempty, and we can choose $\mathbf{x} \in \text{Zer}(M)$. Then $M \subseteq \text{Pol}(\mathbf{x})$, and since M is maximal, we have $M = \text{Pol}(\mathbf{x})$, as required. ∎

The situation when $E = F$ is algebraically closed can be clarified further by describing the map Zer within the ring R.

(30.17) COROLLARY. *Let $R = F[X_1, \ldots, X_n]$, where F is algebraically closed. If $Q \subseteq R$ is an arbitrary subset, then $\text{Zer}(Q)$ is exactly the set of those points in F^n that correspond under the bijection of Corollary 30.16 to maximal ideals of R containing Q.*

Proof. It suffices to observe that a point **x** lies in Zer(Q) iff $Q \subseteq$ Pol(**x**).   ■

30C

We close with an application of our results.

(30.18) THEOREM. *Let $R = F[X_1, \ldots, X_n]$, where F is algebraically closed. Suppose that $Q \subseteq R$ is a subset that contains fewer than n polynomials, and suppose that* Zer(Q) *is nonempty. Then* Zer(Q) *is infinite.*

To demonstrate the significance of Theorem 30.18, we use it to prove the following result in linear algebra.

(30.19) THEOREM. *Let F be an algebraically closed field and suppose that V is an F-subspace of the vector space of all $m \times m$ matrices over F. Let $n = \dim_F(V)$ and assume either that $n > m$ or that $n = m$ and V consists entirely of singular matrices. Then V contains a nonzero nilpotent matrix.*

Proof. Let $M_1, M_2, \ldots, M_n$ be a basis for V and write $R = F[X_1, \ldots, X_n]$. We can view the matrix

$$M = X_1 M_1 + X_2 M_2 + \cdots + X_n M_n$$

as being an $m \times m$ matrix over R. Its characteristic polynomial $p(T) = \det(TI - M)$ lies in $R[T]$, and we write

$$p(T) = T^m + f_{m-1}T^{m-1} + f_{m-2}T^{m-2} + \cdots + f_1 T + f_0,$$

where $f_j \in R$ for $0 \le j \le m-1$. In particular, $\det(M) = \pm f_0$. For each point $\mathbf{x} = (\alpha_1, \ldots, \alpha_n) \in F^n$, we define the matrix $M(\mathbf{x}) \in V$ by setting $X_i = \alpha_i$ in the expression for M above. Observe that the characteristic polynomial of $M(\mathbf{x})$ can be obtained from $p(T)$ by substituting α_i for X_i in the polynomials $f_j(X_1, \ldots, X_n)$, for $0 \le j \le m-1$. In particular, writing $\mathbf{0} = (0, \ldots, 0)$, we see that $M(\mathbf{0})$ is the zero matrix with characteristic polynomial T^m. It follows that $f_j(\mathbf{0}) = 0$ for all subscripts j.

Suppose we could find some point $\mathbf{x} \in F^n$ with $\mathbf{x} \ne \mathbf{0}$ and such that $f_j(\mathbf{x}) = 0$ for all j. Then the characteristic polynomial of $M(\mathbf{x})$ would be T^m, and so $M(\mathbf{x})$ would be nilpotent. Since $\mathbf{x} \ne \mathbf{0}$, we would have $M(\mathbf{x}) \ne 0$ because of the linear independence of the matrices M_i, and we would be done.

If $n > m$, let $Q = \{f_0, f_1, \ldots, f_{m-1}\}$, and if $n = m$, take Q to be $\{f_1, \ldots, f_{m-1}\}$. In either case then, we have $|Q| < n$, and since Zer(Q) is nonempty (because it contains $\mathbf{0}$), Theorem 30.18 applies. It follows that Zer(Q) is infinite and hence certainly contains some point $\mathbf{x} \ne \mathbf{0}$. We have $f(\mathbf{x}) = 0$ for $f \in Q$, and what remains is to show that $f_0(\mathbf{x}) = 0$ in the case $n = m$, where we do not have $f_0 \in Q$. In this case, however, the hypothesis tells us that $M(\mathbf{x})$ is singular, and so $\det(M(\mathbf{x})) = 0$. Since $f_0(\mathbf{x}) = \pm \det(M(\mathbf{x}))$, the proof is complete.   ■

To prove Theorem 30.18, we appeal to the theory of noetherian rings, as presented in Chapter 27. Recall that if R is noetherian and $P \subseteq R$ is a prime ideal, then the height $\text{ht}(P)$ is the largest integer m such that there exist prime ideals P_i such that

$$P = P_0 > P_1 > \cdots > P_m .$$

The height of every prime is finite, and, in fact, by Krull's theorem (27.19) we know that if P is a minimal prime over an m-generator ideal, then $\text{ht}(P) \leq m$. We need a lemma.

(30.20) LEMMA.   *Let* $R = F[X_1, \ldots, X_n]$, *where* F *is a field. If* M *is any maximal ideal of* R, *then* $\text{ht}(M) \geq n$.

By Corollary 27.26 the maximum possible height for any prime of the ring R is n, and so we can strengthen Lemma 30.20 and assert that all maximal ideals have height exactly equal to n. For our present needs, the stated inequality is sufficient.

Proof of Lemma 30.20.   We work by induction on n, noting that when $n = 0$, we have $R = F$, and so the only maximal ideal is 0 with height 0. Suppose $n > 0$ and let $S = F[X_1, \ldots, X_{n-1}] \subseteq R$. Write $M_0 = M \cap S$ and note that M_0 is an ideal of S. Our first goal is to show that M_0 is a maximal ideal.

Use overbars to denote the canonical homomorphism $R \to R/M = \overline{R}$, and note that $\overline{R}$ is a field. Since $F \cap M = 0$, we have $\overline{F} \cong F$ and we identify $\overline{F}$ with F. Thus $\overline{R} = F[\alpha_1, \ldots, \alpha_n]$, where we have written α_i to denote $\overline{X_i}$. By Zariski's theorem (30.8) we deduce that the elements α_i are algebraic over F, and hence $\overline{S} = F[\alpha_1, \ldots, \alpha_{n-1}]$ is a field. Since $\overline{S} \cong S/M_0$, we conclude that M_0 is a maximal ideal of S, as desired.

By the inductive hypothesis, the height of M_0 in S is at least $n - 1$, and so we can find prime ideals M_i of S such that $M_0 > M_1 > \cdots > M_{n-1}$. We write $P_i = M_i[X_n] \subseteq R$ to denote the set of those polynomials in $R = S[X_n]$ that have all coefficients in the ideal M_i of S. The P_i are clearly ideals of R that satisfy

$$M \supseteq P_0 > P_1 > \cdots > P_{n-1} ,$$

and the proof will be complete if we can show that the P_i are prime ideals of R and that $M > P_0$.

Since $R = S[X_n]$ and $P_i = M_i[X_n]$, it is clear that R/P_i is isomorphic to the polynomial ring $(S/M_i)[X_n]$ over the domain S/M_i. It follows that R/P_i is a domain and hence P_i is a prime ideal. Furthermore, R/P_0 is not a field and hence P_0 is not maximal. Thus $P_0 < M$, as required.   ∎

Proof of Theorem 30.18.   Let I be the ideal of R generated by Q, and note that $I < R$ since $\text{Zer}(I) = \text{Zer}(Q)$ is nonempty. Suppose that P is a prime ideal minimal over I. By Krull's theorem, we have $\text{ht}(P) \leq |Q| < n$, and so Lemma 30.20 guarantees that P is not a maximal ideal of R.

Now consider the algebraic set $A = \text{Zer}(P)$. Since P is not a maximal ideal, it follows by Corollary 30.10 that A is not a minimal algebraic set, and so it contains more than one point. Clearly, A is the union of the singleton subsets $\{\mathbf{a}\}$ for $\mathbf{a} \in A$, and each of these is an algebraic set by Theorem 30.15. Now $\mathrm{Pol}(A) = P$ by Corollary 30.10, and since P is prime, A is irreducible by Theorem 30.11. It follows that A is not a finite union of proper algebraic sets. We conclude that A contains infinitely many points, and since $\mathrm{Zer}(Q) \supseteq \mathrm{Zer}(P) = A$, the result follows.   ∎

Problems

30.1 Let E be algebraically closed and give E^n the Zariski topology (working over some field $F \subseteq E$). Prove that E^n is compact.

HINT: Show that a collection of closed sets with all finite intersections nonempty has a nonempty intersection.

30.2 Suppose $F \subseteq E$, where E is a finite field. Show that the only irreducible algebraic sets over F in E^n are the minimal nonempty algebraic sets.

30.3 Let R be a commutative ring that is finitely generated as an algebra over a field. Show that the Jacobson radical of R is a nil ideal.

30.4 Let $R \subseteq S$ be commutative rings such that $S = R[s_1, s_2, \ldots, s_n]$. Let M be a maximal ideal of S, and assume that $M \cap R$ is maximal in R. Show that $M \cap T$ is maximal in T for every subring T of S such that $T \supseteq R$.

30.5 Let $E \supseteq F$ be fields and assume that E is algebraically closed. Let $A \subseteq E^n$ be maximal among proper irreducible algebraic subsets of E^n over F. Show that $A = \mathrm{Zer}(f)$, where f is some prime element in $F[X_1, \ldots, X_n]$.

30.6 Suppose $f(X, Y)$ and $g(X, Y)$ are polynomials with coefficients in some algebraically closed field F, and suppose they have infinitely many common zeros in F^2. Show that there is some nonconstant polynomial in $F[X, Y]$ that divides both f and g.

HINT: If $\{f, g\}$ is contained in infinitely many maximal ideals, show that this set is contained in some nonmaximal prime ideal. Show that a nonmaximal, nonzero prime in $F[X, Y]$ is necessarily principal.

30.7 Let $R = F[X_1, \ldots, X_n]$, where F is a field. Let M be a maximal ideal of R, and suppose that I is any ideal. Show that I/IM is finite dimensional as an F-vector space.

30.8 Let $R = F[X_1, \ldots, X_n]$, where F is a field, and suppose that $P \subseteq R$ is a prime ideal.

    a. Show that there exists a field extension $L \supseteq F$ and a point $\mathbf{x} \in L^n$ such that the transcendence degree $\mathrm{td}_F(L) \leq n$ and $P = \mathrm{Pol}(\mathbf{x})$.

    b. Suppose $E \supseteq F$, where E is algebraically closed and $\mathrm{td}_F(E) \geq n$. Show that every prime ideal of R has the form $\mathrm{Pol}(\mathbf{x})$ for some point $\mathbf{x} \in E^n$.

HINT: For part (a), observe that R/P would have to be embedded in L.

30.9 Let $A \subseteq E^n$ be an algebraic set over F. A point $\mathbf{x} \in E^n$ is called a *generic* point for A if $A = \mathrm{Zer}(\mathrm{Pol}(\mathbf{x}))$. (Note that necessarily, $\mathbf{x} \in A$ in this case.)

a. Show that only irreducible algebraic sets can have generic points.

b. If E is algebraic over F, show that the algebraic sets with generic points are precisely the minimal nonempty algebraic sets.

c. If every irreducible algebraic set in E^n has a generic point, show that either E is a finite field or the transcendence degree $\mathrm{td}_F(E) \geq n$.

d. If E is algebraically closed and $\mathrm{td}_F(E) \geq n$, show that every irreducible algebraic set has a generic point.

HINT: For part (c), show by induction on n that if E is infinite, then $\mathrm{Pol}(E^n) = 0$. Deduce that E^n is irreducible. For part (d), use Problem 30.8.

30.10 Let $R = F[X_1, \ldots, X_n]$ and suppose $E \supseteq F$. If every prime ideal of R has the form $\mathrm{Pol}(A)$ for some subset $A \subseteq E^n$, show that $\mathrm{Pol}(\mathrm{Zer}(I)) = \sqrt{I}$.

HINT: This problem can be done without appeal to the Nullstellensatz.

NOTE: Observe that if E is algebraically closed and $\mathrm{td}_F(E) \geq n$, then by Problem 30.8 the hypothesis is automatically satisfied. In particular, this provides a comparatively easy proof of the Nullstellensatz in the case where $F = \mathbb{Q}$ and $E = \mathbb{C}$.

30.11 Let F be algebraically closed and suppose that A is a not necessarily associative algebra over F with $\dim_F(A) = n < \infty$. Assume that for every element $x \in A$, there exists a nonzero element $y \in A$ such that $yx = 0$. Show that there exists a nonzero element $a \in A$ such that $(\cdots (((x)a)a) \cdots)a = 0$ for all $x \in A$, where there are n occurrences of a in this formula.

HINT: Use Theorem 30.19.

Index

Abel, N. 274
abelian group 11, 19, 37, 90–93, 139, 314
  characters of 229
abelian X-group 142–153, 156, 157, 167
ACC (ascending chain condition) 143–145,
  433
action
  of group 42–43
  on cosets 44, 45, 115
  via automorphisms 95, 112, 114
adjectives 163
adjoint matrix 439, 456
affine group of line 8, 29
afforded character 216
algebra (over a field) 166, 171, 172, 173, 213,
  295
algebraic closure 267–268, 271
  of finite degree 401–407, 415
algebraic element 254–259, 386, 398
algebraic field extension, definition of 256
algebraic geometry 493
algebraic integer 454, 463, 469, 471–472
algebraic number 256
algebraic number field 256
algebraic numbers, field of 256, 259, 268,
  410
algebraic point 494
algebraic set 493–505
algebraically closed field 267–269, 303, 355
algebraically independent elements 379–380,
  383, 398
alternating group 75
  simplicity of 77
annihilator (*see also* prime annihilator) 163,
  176, 177, 210–211, 440, 442

antichain 157
archimedian ordered field 412–413
Artin, E. 145–146
Artin-Rees theorem 461
Artin-Schreier theorem 401
Artin's theorem 260, 262, 389
artinian module 170, 444
artinian ring (*see* right artinian ring *or*
  left artinian ring)
artinian commutative ring 252, 428
artinian X-group 145–146, 148–149, 150,
  157
ascending central series (*see* upper central
  series)
ascending chain condition (*see* ACC)
associated factors 338
associated prime ideal 440–442, 450
associative property 9
augmentation ideal 173
augmentation map 173
automorphism
  of group 19–22
  of symmetric group 79–80
automorphism group
  of algebraically closed field 405
  of group 20, 28, 41, 310
axiom of choice 5, 144–145, 153, 269

basis for free module 202, 434
Berlekamp algorithm 333–339, 464
bijection 4
binary operation 9